ESSAYS IN MATHEMATICAL ECONOMICS

In Honor of Oskar Morgenstern

ESSAYS IN MATHEMATICAL ECONOMICS

In Honor of Oskar Morgenstern

EDITED BY MARTIN SHUBIK

S. AFRIAT
R. AUMANN
W. BAUMOL
K. BORCH
M. DAVIS
M. D. GODFREY
C. W. J. GRANGER
M. HATANAKA
H. KARREMAN
A. Y. C. KOO
H. W. KUHN
E. MARCUS
M. MASCHLER
J. MAYBERRY
K. MENGER
H. MILLS
K. MIZUTANI
K. MIYASAWA
S. NOBLE
D. ORR
B. PELEG
M. H. PESTON
J. PFANZAGL
L. S. SHAPLEY
M. SHUBIK
D. STERN
M. SUZUKI
G. THOMPSON
L. TÖRNQUIST
T. WHITIN

PRINCETON, NEW JERSEY
PRINCETON UNIVERSITY PRESS
1967

L.C. Card 65-17159

Printed in the United States of America

A book of this variety is the product of a highly cooperative effort. There are many colleagues, one-time colleagues and friends of Oskar Morgenstern, who although their articles do not appear, have participated in making the volume possible and, or have sent their well wishes. In particular Giles Mellon and Herman Karreman were of considerable assistance in planning and preparation.

A special debt of gratitude is owed to the many referees who were kind enough to give of their time so that the papers of all contributors could benefit from critical review.

Martin Shubik
New Haven
April, 1966

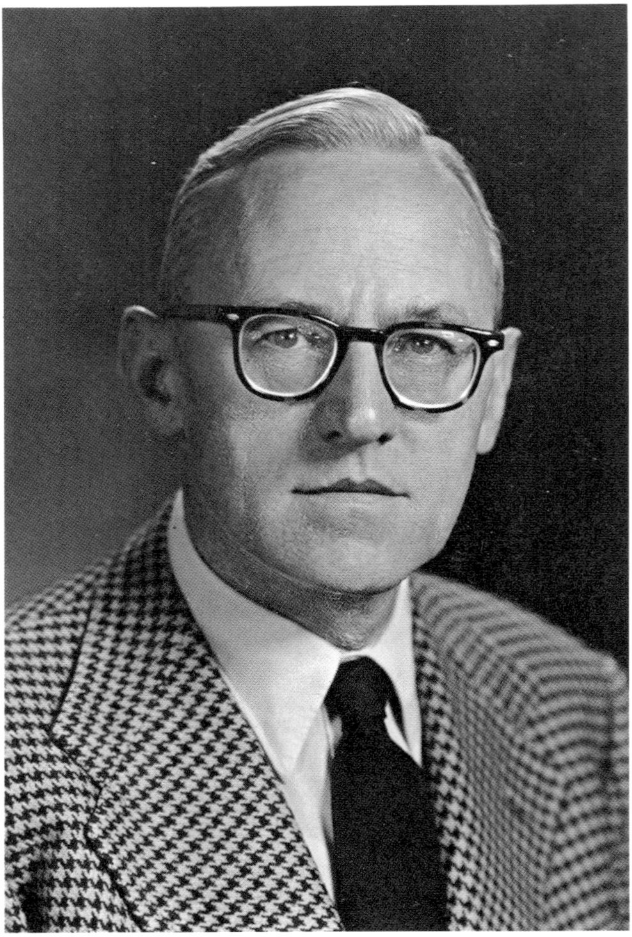

The Contribution of Oskar Morgenstern

Oskar Morgenstern was born in Goerlitz, Silesia, Germany, on January 24, 1902. After attending the gymnasium in Vienna, he received the degree of Doctor Rer. Pol. from the University of Vienna in 1925. In the same year he was awarded the Laura Spelman Rockefeller Fellowship and in the period 1925–1928 studied at the Universities of London, Paris, and Rome and at Harvard and Columbia Universities. In 1929 he returned to the University of Vienna as a privatdozent, and in 1935, at the age of thirty-three, was appointed Professor of Economics, a position he was to hold until moving to the United States in 1938.

In addition to his teaching duties, Professor Morgenstern during his years in Vienna served as Managing Editor of the *Zeitschrift für Nationalökonomie*, and from 1931 to 1938 was Director of the Austrian Institute for Business Cycle Research. He also served as an advisor to the Austrian National Bank from 1932 to 1938 and as an advisor to the Ministry of Commerce from 1936 to 1938. He was a member of the Committee of Danubian Experts of the International Institute of Intellectual Cooperation in Paris, and in 1936 became a member of the Committee of Statistical Experts of the League of Nations, a position he was to hold until the final dissolution of the League in 1945.

Although he had published several reviews in earlier years, Professor Morgenstern's first major publications, two papers on business cycles and obituaries of Edgeworth and von Wieser, appeared in 1927 when he was twenty-five. In the next eleven years, a steady stream of papers was published in various journals in Europe and the United States on business cycle theory, monetary policy, economic theory, and international trade. Professor Morgenstern's first major book, *Wirtschaftsprognose*, a study of the theory and applications of economic forecasting, was published in 1928. In it the author first considered a number of topics to which he was to return in later years. Never translated into English, the book received a thoughtful review by the late Arthur W. Marget in the *Journal of Political Economy* in June 1929, an article which even at this late date is worth rereading. A second major book, *Die Grenzen der Wirtschaftspolitik*, was published in 1934 and was translated into English in 1937 under the title, *The Limits of Economics*.

In 1938, like so many members of the faculties of the great Austrian and German universities, Oskar Morgenstern left Europe to begin the second phase of his career, as a member of the economics faculty of Princeton University. Appointed lecturer in 1938, he became an associate professor in 1941 and a full professor in 1944. He is currently Professor of Political Economy on the Class of 1913 Foundation. Dr. Morgenstern has been a visiting professor at the University of Muenster and the University of Basel. He received an Honorary Doctorate from the University of Mannheim in 1957 and also from the University of Basel in 1960.

In addition to his professorial duties, Oskar Morgenstern has been active in a large number of widely diverse undertakings. Since 1948 he has been Director of the Econometric Research Program at Princeton University. He was instrumental in the founding of, and since 1954 has been a co-editor of, the *Naval Research Logistics Quarterly;* since 1955 he has served as co-editor of the *Zeitschrift für Nationalökonomie*. He has also been a consultant to the Rand Corporation, the Atomic Energy Commission, the Sandia Corporation, and the White House, and has been active in business as a director of Mathematica and the Market Research Corporation of America. He is a co-editor of the Princeton Series in Mathematical Economics, published by Princeton University Press.

The substantial contributions of Oskar Morgenstern during the last twenty-five years have been many. The most significant grew out of his collaboration with the late John von Neumann which led to the publication of the celebrated *Theory of Games and Economic Behavior* in 1944, a book which must be counted among the truly great contributions to economics. A revised edition appeared in 1947, a third edition in 1953, and a German edition in 1961. Other major books by Professor Morgenstern include *On the Accuracy of Economic Observations*, first edition, 1950, completely revised edition, 1963; *International Financial Transactions and Business Cycles*, 1959; and *The Question of National Defense*, first edition, 1959, second edition, 1961, German edition, 1962. In addition, he was the editor of *Economic Activity Analysis*, 1954, and was one of the Committee of Experts responsible for the League of Nations' study, *Economic Stability in the Postwar World.* These books have been supplemented by over one hundred published papers on economic theory, business cycles, game theory, problems of defense, and statistical theory.

The contributions of Oskar Morgenstern to the science of economics, however, have not been limited to his own research and publications. The advice, guidance, and encouragement which he has given freely to colleagues and younger economists for nearly forty years are reflected in the hundreds of acknowledgements which appear in the literature of economics, and have been instrumental in the development of major contributions to economics, as in the case of the late Abraham Wald. Finally, as a testimonial to Professor Morgenstern's contribution to the training of younger economists, he was in 1963 chosen by the students of Princeton University to deliver a special series of lectures on the theory of games.

A Bibliography of the Work of Oskar Morgenstern

1927

1. "International vergleichende Konjunkturforschung," *Zeitschrift für die gesamte Staatswissenschaft*, Vol. 83, No. 2, June, pp. 261–290.
2. "Francis Y. Edgeworth," *Zeitschrift für Volkswirtschaft und Sozialpolitik*, Vol. 5, No. 10–12, October–December, pp. 646–652.
3. "Die andere Seite der Konjunkturforschung," *Oesterreichischer Volkswirt*, Vol. 19, No. 15, pp. 393–395.
4. "Friedrich von Wieser, 1851–1926," *American Economic Review*, Vol. 17, No. 4, December, pp. 669–674.

1928

5. "Qualitative und quantitative Konjunkturforschung," *Zeitschrift für die gesamte Staatswissenschaft*, Vol. 85, No. 1, pp. 54–88.
6. "Aufgaben und Grenzen der Institute für Konjunktforschung," in *Beitraäge zur Wirtschaftstheorie*. Part 2, *Konjunkturforschung und Konjunkturtheories*, ed. by Karl Diehl, Schriften des Vereins für Sozialpolitik, Vol. 172, Munich and Leipzig, Duncker und Humbolt, pp. 339–353.
7. *Wirtschaftsprognose: Eine Untersuchung ihrer Voraussetzungen und Möglichkeiten*, Vienna, Julius Springer, iv + 129 pp.
8. "Wirtschaftsprognose und Stabilisierung," *Wirtschaftsdienst*, Vol. 13, No. 47, November 23, pp. 1927–1930.

1929

9. "Allyn Abbott Young," *Zeitschrift für Nationalökonomie*, Vol. 1, May, pp. 143–145.

1930

10. "Ertekelmeletek," (Werttheorien), *Koezgazdasagi Enciklopedia*, Budapest, pp. 3–10.
11. "Nachwort zur Wirtschaftsprognose," *Allgemeines Statistisches Archiv*, Vol. 20, No. 2, pp. 273–277.
12. "Developments in the Federal Reserve System," *Harvard Business Review*, Vol. 9, Oct., pp. 1–7 (Translation in *Deutscher Volkswirt*, Berlin, 1930).

1931

13. "Offene Probleme der Kosten- und Ertragstheorie," *Zeitschrift für Nationalökonomie*, Vol. 2, No. 4, March, pp. 481–522.
14. "Bemerkungen über die Problematik der amerikanischen Institutionalisten," in: *Saggi di storia e di teoria economia, in onore e ricordo di Guiseppe Prato*, Turin, pp. 330–350. "Observations on the Problem of the American Institutionalists," transl. by George Stigler, mimeographed paper, circulated by University of Chicago.

15. "Die drei Grundtypen der Theorie des subjektiven Wertes," *Schriften des Vereins für Sozialpolitik*, Vol. 183, No. 1, pp. 1–43.
16. "Die Preise im Konjunkturzyklus," *Oesterreichischer Volkswirt*, Vol. 23, No. 52, pp. 1358–1361 (1931), Vol. 24, No. 1, pp. 18–20 (1932).
17. "Free and Fixed Prices in the Depression," *Harvard Business Review*, Vol. 10, No. 1, October, pp. 62–68.
18. "Mathematical Economics," in: *Encyclopedia of the Social Sciences*, New York, Macmillan, Vol. 5, pp. 364–368.
19. "Replik zur Antikritik der Ertragstheorie von W. Weddingen," *Zeitschrift für Nationalökonomie*, Vol. 3, No. 2, December, pp. 264–265.

1932

20. Edit. and Preface to: A. de Viti de Marco, *Grundlehren der Finanzwirtschaft*, Tübingen.
21. "Zolle," *Lehrbuch des Internationalen Handels*, by Sir William Beveridge, Vienna, pp. VI–VIII.
22. "Kapital- und Kurswertänderungen der an der Wiener Börse notierten österreichischen Aktiengesellschaften, 1913 bis 1930," *Zeitschrift für Nationalokönomie*, Vol. 3, No. 2, January, pp. 251–255.
23. Ed.: *Beiträge zur Konjunkturforschung*, Vol. 4 and subsequent, Vienna, Julius Springer.

1933

24. "Diskussionsbemerkungen," in: "Probleme der Wertlehre," *Schriften des Vereins für Sozialpolitik*, Vol. 183, No. 2, pp. 91–93.
25. "Der Stand und die nächste Zukunft der Konjunkturforschung," in: *Festschrift für Arthur Spiethoff*, Munich, Duncker und Humblot, pp. 193–198.
26. "Eine Bibliographie der allgemeinen Lehrgeschichten der Nationalökonomie," (with E. Schams), *Zeitschrift für Nationalökonomie*, Vol. 4, No. 3, March, pp. 389–397.

1934

27. "I tre tipi fondamentali della teoria del valore soggetivo," *Annali di scienze politiche*, Vol. 7, No. 3, pp. 205–256. (Translation of No. 15).
28. *Die Grenzen der Wirtschaftspolitik*, Vienna, Julius Springer, 136 pp.
29. "Das Zeitmoment in der Wertlehre," *Zeitschrift für Nationalökonomie*, Vol. 5, No. 4, September, pp. 433–458.

1935

30. "L'étude des conjonctures en Autriche," *Revue des sciences économiques*, April, pp. 2–8.
31. "Zur Theorie der Produktionsperiode," *Zeitschrift für Nationalökonomie*, Vol. 6, No. 2, June, pp. 196–208.
32. "Vollkommene Voraussicht und wirtschaftliches Gleichgewicht," *Zeitschrift für Nationalökonomie*, Vol. 6, No. 3, August, pp. 337–357. Engl. transl. by Frank H. Knight, mimeographed, distributed by University of Chicago.
33. "Organisation, Leistungen und weitere Aufgaben der Konjunkturforschung," *Public Statistical Institute for Economic Research*, Sofia, Vol. 1, No. 1, pp. 14–28.

1936

34. "Die Einordnung der Verkehrspolitik in die allgemeine Wirtschaftspolitik," *Oesterreichische Zeitschrift für Bankwesen*, Vol. 1, No. 1/2, February, pp. 11–25.
35. "Logistik und Sozialwissenschaften," *Zeitschrift für Nationalökonomie*, Vol. 7, No. 1, March, pp. 1–24.
36. "Il fattore tempo nella dottrina del valore," *Annali di statistica e di economia*, Vol. 4, No. 3, pp. 1–35. (Translation of No. 29.)
37. "Währung und Preise," *Oesterreichische Zeitschrift für Bankwesen*, Vol. 1, No. 5/6, October, pp. 166–171.
38. "Probleme der Devisenbewirtschaftung" *Public Statistical Institute for Economic Research*, Sofia, Vol. 2, pp. 21–36.

1937

39. "Entstehung und Abbau der österreichischen Devisenbewirtschaftung," *Nationalökonomisk Tidsskrift*, February, pp. 34–56.
40. *The Limits of Economics*, transl. by Vera Smith, London, W. Hodge and Co., Ltd., v + 151 pp. (Revised edition of No. 28)
41. "Goldpreisherabsetzung und Zinsfusspolitik," *Oesterreichische Zeitschrift für Bankwesen*, Vol. 2, No. 6/7 June, pp. 154–174.
42. "La introduzione e la abolizione dei controllo dei cambi esteri in Austria (1931–1934)" (introd. by Luigi Einaudi), *Rivista di storia economica*, Vol. 2, No. 4, June, pp. 1–21. (Translation of No. 39.)
43. "Free Exchange, the Experience of Austria," *London Times*, No. 47, 741, p. 17; No. 47, 742, p. 15, July 20/21.
44. "The removal of Exchange Control," in: *Gold: a World Economic Problem*, ed. by R. H. Brand, New York, Carnegie Endowment for International Peace, International Conciliation, pp. 678–679. (Reprint of No. 43.)

1938

45. "Aspects of Managed Currency in Europe," *Proceedings of the Association of Reserve City Bankers*, Twenty-seventh Annual Convention, March 28, 29, 30, 1938, pp. 53–61.

1939

46. "The Experience with Public Regulation and Public Monopoly Abroad," *Proceedings of the American Academy of Political and Social Sciences*, Vol. 202, No. 2, March, pp. 34–39.
47. "Scholarship and Value-Judgements," *Princeton Alumni Weekly*, Vol. 39, No. 28, May 5, pp. 647–648.

1941

48. "The Nature and Significance of Business Fluctuations," *Proceedings of the New Jersey Bankers' Association, Twelfth Mid-Year Conference, Princeton,* January 29–30, pp. 31–39.
49. "Unemployment: Analysis of Factors," *American Economic Review, Papers and Proceedings*, Vol. 30, No. 5, February, pp. 273–293.

50. "Professor Hicks on Value and Capital," *Journal of Political Economy*, Vol. 49, No. 3, June, pp. 361–393.

1943

51. "On the International Spread of Business Cycles," *Journal of Political Economy*, Vol. 51, No. 4, August, pp. 287–309.

1944

52. *Theory of Games and Economic Behavior* (with John von Neumann), Princeton, Princeton University Press, xviii + 625 pp.

1947

53. *Theory of Games and Economic Behavior* (with John von Neumann), 2nd edition, revised and enlarged, Princeton, Princeton University Press, xviii + 641 pp.
54. "German Economy," *New York Times*, Letters to the Times, Vol. 97, No. 32, 775, October 19, p. E10.

1948

55. "Demand Theory Reconsidered," *Quarterly Journal of Economics*, Vol. 62, February, pp. 165–201.
56. "Oligopoly, Monopolistic Competition and the Theory of Games," *American Economic Review, Papers and Proceedings*, Vol. 38, No. 2, May, pp. 10–18.
57. "Das Dollardefizit Europas," *Die Industrie*, Vol. 48, No. 33, August, pp. 5–6.

1949

58. "La reforma monetaria austriaca de 1947–1948," *Boletin del Banco Central de Venezuela*, Vol. 9, No. 47–48, January-February, pp. 19–23.
59. "Economics and the Theory of Games," *Kyklos*, Vol. 3, No. 4, pp. 294–308.
60. "Input-Output Analysis: Discussion," *American Economic Review, Papers and Proceedings*, Vol. 39, No. 3, May, pp. 238–240.
61. "The Theory of Games: Tool for Analysis of Social and Economic Behavior," *Scientific American*, Vol. 180, No. 5, May, pp. 22–25.
62. "The Accuracy of Economic Observations," *Linear Programming Conference*, Chicago, May 31, Document No. 704.
63. "Theorie des Spiels," *Die amerikanische Rundschau*, Vol. 5, No. 26, August-September, pp. 76–87. (Translation of No. 63.)
64. "La propagation internationale des cycles economiques," *Economie appliquée*, Vol. 2, No. 3–4, July–December, pp. 593–611.

1950

65. *On the Accuracy of Economic Observations*, Princeton, Princeton University Press, ix + 101 pp.
66. "The Stability of Inverses of Input-Output Matrices" (with Max A. Woodbury) (Abstract), *Econometrica*, Vol. 18, No. 2, April, pp. 190–192.

67. "Die Theorie der Spiele und des wirtschaftlichen Verhaltens," *Jahrbuch für Sozialwissenschaft*, Vol. 1, No. 2, pp. 113–139.
68. "The Computation of Economic Programs," Office of Naval Research, *Research Reviews*, June, pp. 21–27.
69. "Complementarity and Substitution in the Theory of Games" (abstract), *Econometrica*, Vol. 18, No. 3, July, pp. 279–280.

1951

70. "Logistica e scienze sociali," in: "Che puo la logistica per le scienze sociali? "(with Karl Menger), *L'industria*, No. 4, pp. 4–11.
71. "Obituary: Joseph A. Schumpeter," *Economic Journal*, Vol. 61, No. 241, March, pp. 197–202.
72. "Los calculos par los programas economicos," *Boletin del Banco Central de Venezuela*, Vol. 11, No. 73–74, March–April, pp. 13–17. (Translation of No. 70.)
73. "Notes on a Theory of Organization," *Logistics Papers*, George Washington University, Issue No. 5, 16 February–15 May, 24 pp.
74. "Note on the Formulation of the Study of Logistics," *RAND Corporation Report RM-614*, May 28, 12 pp.
75. "Limiti e condizioni dei programmi economici," *Studi economici*, No. 4–5, July-October, pp. 3–11.
76. "Abraham Wald, 1902–1950," *Econometrica*, Vol. 19, No. 4, October, pp. 361–367.
77. "La teoria dei giochi e del comportamento economico," *L'industria*, No. 3, pp. 315–346. (Translation of No. 67.)
78. "Prolegomena to a Theory of Organization," *RAND Corporation Report RM-734*, December 10, ii + 122 pp.

1952

79. "Ueber die Genauigkeit wirtschaftlicher Beobachtungen," transl. by V. Trapp, with preface by K. Wagner, *Einzelschriften der Deutschen Statistischen Gesellschaft*, No. 4, Munich, 129 pp. (Revised Translation of No. 65)
80. "An Economist in Europe," *Princeton Alumni Weekly*, Vol. 52, No. 21, March 21, pp. 12–13.
81. "Nuove considerazioni sulla teoria della domanda," *Studi economici*, May–August, pp. 1–36. (Translation of No. 55.)
82. "Note on the Role of Follow-ups in the Naval Supply System," *Progress Report*, Logistics Research Division, U.S. Naval Supply Research and Development Facility, Bayonne, New Jersey, 1 December, 7 pp.
83. "Oekonometrische Berechnungen im Grossen," Anderson Festschrift, *Mitteilungsblatt für mathematische Statistik*, Vol. 4, pp. 139–146.

1953

84. "Computaciones econometricas en gran escala," *Boletin del Banco Central de Venezuela*, Vol. 13, No. 95–97, January–March, pp. 27–31. (Translation of No. 83.)
85. "Aggregation and Errors in Input-Output Models" (with T. M. Whitin), *Logistics Papers*, George Washington University, Issue No. 9, 16 February–15 May, 8 pp.

86. *Theory of Games and Economic Behavior* (with John von Neumann), 3d edition, revised, Princeton, Princeton University Press, xx + 641 pp.
87. "When is a Problem of Economic Policy Solvable?" Amonn Festschrift, *Wirtschaftstheorie und Wirtschaftspolitik*', ed. by Wagner and Marbach, Berne, pp. 241–249.
88. "Remarks on Input-Output Relations," *Proceedings of a Conference on Inter-Industrial Relations Held at Driebergen, Holland*, Leyden, H. E. Stenfert Kroese N. V., pp. 23–32, 96–98, 100.

1954

89. "Econometric Computations in the Large," *Bulletin de l'Institut International de Statistique*, Vol. 34 No. 2, pp. 398–404.
90. "Squeeze on Japan," *The Wall Street Journal*, Vol. 144, No. 87, November 2 p. 8.
91. "Compressibility of Organizations and Economic Systems," *RAND Corporation Report*, *RM*-1325, 17 August, 19 pp.
92. Ed. and contributor, *Economic Activity Analysis*, New York, John Wiley & Sons, Inc., xviii and 554 pp.
93. "Experiment and Large Scale Computation in Economics," in: *Economic Activity Analysis* (see no. 92), pp. 484–549.
94. "Sperimentazione e calcolo su vasta scala in economica," *L'industria*, Part I in No. 3, pp. 289–312; Part II in No. 4, pp. 471–504. Also in: *Studi di Metodologia Economica* (see no. 99), pp. 15–74. (Translation of No. 95.)
95. "Keizai seisaku mondai no tokiuru hi wa itsuka?" *Kin-yu keizai* (Financial Report), No. 29, December 25, pp. 53–65. (Translation of No. 89.)
96. "Consistency Problems in Military Supply Systems," *RAND Corporation Report, RM-1296*, July 14, 31 pp. Revised and enlarged version in: *Naval Research Logistics Quarterly*, Vol. 1, No. 4, December, pp. 265–281.
97. "Capitalist Oasis," *The Wall Street Journal*, Vol. 144, No. 106, Nov. 30, p. 12.

1955

98. "The Economics of Input-Output Relations" (with T. M. Whitin), in: *Input-Output Analysis: An Appraisal*, Studies in Income and Wealth, No. 18, National Bureau of Economic Research, Princeton, Princeton University Press, pp. 128–135.
99. *Studi di Metodologie Economica*, ed. by F. di Fenizio, transl. by M. Talamona, Milan, 152 pp.
100. "To Abolish Military Scrip," *New York Times*, Letters to the Times, Section 4, Vol. 104, No. 25, 414, Jan. 9, p. 8E.
101. "La teoria de los juegos y del comportamiento economico," *Económica* (Buenos Aires), Vol. 1, No. 3–4, January–June, pp. 344–375. (Translation of No. 67.)
102. "Aggregation of Input-Output Tables," *Summaries of Lectures delivered at the Statistical Seminar held at Rome, September* 1953, The Hague, International Statistical Institute, 1955, pp. 77—80.
103. "Quando un problema di politica economica e risolubile?" *L'industria*, No. 1, pp. 1–10. Also in *Studi di Metodologie Economica* (see No. 99), pp. 143–152. (Translation of No. 87.)
104. "Sull'accuratezza delle osservasioni economiche," in: *Studi di Metodologia Economica*, (see No. 99), pp. 75–126. (Translation of No. 65.)

105. "Note on the Formulation of the Theory of Logistics," *Naval Research Logistics Quarterly*, Vol. 2, No. 3, pp. 129–36. (Revised version of No. 74.)
106. "The Validity of International Gold Movement Statistics," *Special Papers in International Finance*, No. 2, International Finance Section, Princeton University, 42 pp.

1956

107. "On the Equilibrium of a Linear Economic System with Non-dominant Outputs" (with Y. K. Wong) (abstract), *Econometrica*, Vol. 24, No. 2, April, pp. 200–201.
108. "Der theoretische Unterbau der Wirtschaftspolitik," *Arbeitsgemeinschaft für Forschung des Landes Nordrhein-Westfalen*, No. 63, July, 32 pp.
109. "Experiment und Berechnung grossen Umfangs in der Wirtschaftswissenschaft," *Weltwirtschaftliches Archiv*, Vol. 76, No. 2, pp. 179–239. (Translation of No. 93.)
110. "A Generalization of the von Neumann Model of an Expanding Economy" (with J. G. Kemeny and G. L. Thompson), *Econometrica*, Vol. 24, No. 2, April, pp. 115–135.
111. "Spieltheorie," in: *Handwörterbuch der Sozialwissenschaften*, Göttingen, Vol. 9, pp. 706–713.
112. "Methoden und Grenzen der Konjunkturpolitik," *Mitteilungen des Rheinisch-Westfälischen Instituts für Wirtschaftsforschung*, Vol. 7, No. 10, October, pp. 225–237.

1957

113. "A Study of Linear Economic Systems" (With Y. K. Wong), *Weltwirtschaftliches Archiv*. Vol. 79, no. 2, pp. 222–241.

1958

114. "John von Neumann, 1903–1957," *Economic Journal*, Vol. 68, March, pp. 170–174.
115. "Some Thoughts Bearing on National Defense Policy," *Sandia Corporation Research Colloquium* SCR-33, 18 pp.

1959

116. *International Financial Transactions and Business Cycles*, National Bureau of Economic Research, Princeton, Princeton University Press, xxvi + 591 pp.
117. Foreword to: Martin Shubik, *Market Strategy and Structure: Monopolistic Competition, Oligopoly and the Theory of Games*, New York, J. Wiley and Sons.
118. "The Game Theory in U. S. Strategy," *Fortune*, September 1959, pp. 126–7, 230–240.
119. *The Question of National Defense*, New York, Random House, November, xii + 306 pp.

1960

120. "The Theory of Games," *Challenge*, New York, Vol. 8, No. 6, March, pp. 35–43.
121. "Goal: An Armed, Inspected, Open World," *Fortune*, July 1960.

122. "Brass Hats and Striped Pants," *Foreign Service Journal*, Vol. 37, No. 7, July, pp. 21–23. (Extract from No. 119.)
123. "Goal: An Armed, Inspected, Open World," *The Executive*, Harvard University, Vol. 4, No. 2, July, pp. 14–15. (Reprint of No. 121.)
124. "How We Can Avert War: A Three-Step Proposal," *Life International*, August 29, 1960, pp. 70–77. (Adapted from No. 121.)
125. "The Theory of Games," reprinted from *The Scientific American*, May 1949 issue, in: *Personality, Dynamics and Effective Behavior*, Chicago, Scott, Foresman and Company, September, pp. 493–498.
126. "The Theory of Games," in: *Some Theories of Organization*, ed. by A. H. Rubenstein and C. J. Haberstroh, Homewood, Ill., The Dorsey Press, pp. 437–447. (Reprinted from No. 61.)
127. "Effective and Secure Deterrence: The Oceanic System," *Royal Canadian Air Force Staff College Journal*, Toronto, 1960, pp. 34–40.
128. "Decision Theory and the Department," *Foreign Service Journal*, December, Vol. 37, No. 12, pp. 19–22.
129. Foreword to: Michio Hatanaka, *The Workability of Input-Output Analysis*, Ludwigschafen am Rhein, Germany, Fachverlag für Wirtschaftstheorie und Oekonometrie, pp. i–iv.

1961

130. "A New Look at Economic Time Series Analysis," Princeton University, Econometric Research Program, *Research Memorandum No. 19*, January.
131. "The Cold War is Cold Poker," *New York Times Magazine*, February 5, 1961, pp. 20–22.
132. "A New Look at Economic Times Series Analysis," in: *Money, Growth and Methodology*, in honor of Johan Akerman, ed. by H. Hegeland, Lund, pp. 261–272.
133. *The Question of National Defense*, 2nd revised edition, New York, Vintage Books V-192, xiv + 328 pp.
134. "Almost Symmetric Solutions of Some Symmetric *n*-Person Games" (abstract), *Am. Math. Soc. Notices*, Vol. 8, No. 1, February, p. 69.
135. "Symmetric Solutions of Some General *n*-Person Games" (with J. von Neumann), *RAND Corporation Report P*-2169, March 2, 13 pp.
136. "The *n*-Country Problem," *Fortune*, March, pp. 136–137, 205–208.
137. "Nuclear Weapons Among the *n*-Powers," *Current*, April, pp. 35–36. (Extract from No. 136.)
138. "Peking as a Nuclear Power," *Current*, April p. 23. (Extract from No. 136.)
139. *Spieltheorie und wirtschaftliches Verhalten*, Würzburg, xxiv–668 pp. (German translation of No. 86, with new preface.)
140. "Nuclear Stalemate?" *Encounter*, July, pp. 70–71.
141. "Where are the Minds?" *The Virginia Quarterly Review*, Vol. 37, No. 3, July, pp. 450–453.
142. "The Navy/Industry Look at the Future," *Aerospace Engineering*, Vol. 20, No. 12, December, pp. 22–23, 47–50.
143. "A New Look at Economic Time Series Analysis," *L'industria*, No. 3, 12 pp.

1962

144. "Anschauliche und axiomatische Thoerie," in *Antidoron*, Edgar Salin zum 70. Geburtstag, Tübingen, pp. 80–90.

145. *Strategie-Heute*, Frankfurt, 323 pp. (German translation of No. 119, enlarged and new preface.)
146. "On the Application of Game Theory to Economics," The Princeton University Conference on Recent Advances in Game Theory, reprinted in *Giornale degli economisti e annali di economia*, Vol. 21 (new series), No. 1–2, January–February, pp. 47–60.
147. "Testimony of our Troubled Times," *The Virginia Quarterly Review*, Vol. 38, No. 3, pp. 510–513.
148. "Political Effects," in: *Space Flight Report to the Nation*, ed. by J. and V. Grey, New York, Basic Books, Inc., pp. 132–136.
149. "On the Application of Game Theory to Economics," *Giornale degli economisti e annali de economia*, Vol. 21 (new series), No. 1–2, January–February, pp. 47–60.
150. *Beikoku Kobubo no Shomondai*, Kagima Institute of Research, Tokyo, December, 383 pp. (Japanese translation of No. 119.)
151. "La scienza economica e la teoria dei giuochi," in: *Economisti Moderni*, ed. by F. Caffé, Milan, Garzanti, March, pp. 179–201.
152. "The Command and Control Structure," *The Proceedings of the Military Operations Research Symposia* (MORS), Vol. 2, No. 2, Part 1, Fall 1962.
153. "On the Accuracy of National Income and Growth Statistics," Princeton University Econometric Research Program, *Research Memorandum No. 43*, August.
154. "Spectral Analysis of New York Stock Market Prices" (with C. W. J. Granger), Princeton University, Econometric Research Program, *Research Memorandum No. 45*, September, 1962.

1963

156. "How to Plan to Beat Hell," *Fortune*, January, pp. 103, 200–208.
157. "Spectral Analysis of New York Stock Market Prices," (with C. W. J. Granger). *Kyklos* Vol. 16, no. 1, January, pp. 1–27.
158. "Limits to the Uses of Mathematics in Economics," Princeton University, Econometric Research Program, *Research Memorandum* No. 49, January.
159. "Sull'accuratezza delle statische del reddito nazionale," *L'industria*, No. 1, Jan.–March pp. 3–40.
160. "The Element of Time in Value Theory," and "Perfect Foresight and Economic Equilibrium," Princeton University, Econometric Research Program. *Research Memorandum* No. 55, April. (Reprint of No. 32.)
161. "Limits to the Uses of Mathematics in Economics," in: *Mathematics and the Social Sciences*, A symposium, sponsored by The American Academy of Political and Social Sciences, ed. by James C. Charlesworth, Philadelphia, June, pp. 12–29.
162. "Un progetto contro l'inferno," *Panorama*, No. 10, July, pp. 51–55. (Italian translation of No. 156.)
163. "Military Alliances and Mutual Security," in: *National Security: Political, Military, and Economic Strategies in the Decade Ahead*, ed. by David M. Abshire and Richard V. Allen, New York, Praeger, pp. 671–686.
164. "La validita delle statische riguardanti il progresso economice ed i tassi di progresso," *L'industria*, No. 2, April–June, pp. 178–195.
165. "Qui Numerare Incipit Errare Incipit," *Fortune*, October, pp. 142–144, 173–174, 178–180.

166. *On the Accuracy of Economic Observations*, 2nd completely revised edition, Princeton, Princeton University Press, v–viii + 322 pp.
167. *Spieltheorie und Wirtschaftswissenschaft*, Vienna, R. Oldenbourg, 200 pp.
168. "Limites à l'emploi des mathematiques en science économique," *Bulletin Sedeis*, No. 872, Supplement I, December 20, pp. 3–16. (Translation of No. 161.)
169. "Die Macht im Handel der Staaten: Ein Problem der Theorie des internationalen Handels," *Jahrbuch für Sozialwissenschaft*, Vol. 14, no. 3, pp. 48–55.

1964

170. "Qui Numerare Incipit Errare Incipit," *The Executive*, Vol. 7, No. 8, January, pp. 28–31. (Reprint of No. 165.)
171. "Pareto Optimum and Economic Organization," Princeton University, Econometric Research Program, *Research Memorandum* No. 63, January, 1964.
172. "Planung, Simulierung und Wirtschaftstheorie," in: *Planung ohne Planwirtschaft*, ed. by A. Plitzko, Basel, Kyklos Verlag, pp. 29–39, 80–81, 223–224, 257–258.
173. "The Random-Walk Hypothesis of Stock Market Behavior" (with Michael Godfrey and Clive W. J. Granger), *Kyklos*, Vol. 17, no. 1, January, pp. 1–30.
174. "Qui Numerare Incipit Errare Incipit," *President*, Vol. 2, No. 1, January, pp. 112–123. (Reprint of No. 165.)
175. Foreword to: C. W. J. Granger and M. Hatanaka, *Spectral Analysis of Economic Time Series*, Princeton, Princeton University Press, xviii + 299 pp.

Contents

PART I

Game Theory

CHAPTER 1

A Survey of Cooperative Games Without Side Payments*

By ROBERT J. AUMANN†

1. SCOPE OF THE PAPER

We begin by giving intuitive definitions of our terms. Start out with a game, either in extensive or in normal form. The game becomes *cooperative* if we allow the players to communicate before each play and to make binding agreements about the strategies they will use (either mixed or pure). We say that *side payments* are allowed if there is a common medium of exchange, such as money, which can be transferred between the players before or after the play. We say that *utility is transferable* if the increment to the payoff of a player caused by a transfer of money is proportional to the amount of money transferred [33]. The classical theory of n-person games as first conceived by von Neumann and Morgenstern [45], and later elaborated upon by many other writers, is concerned exclusively with cooperative games in which side payments are allowed and utility is transferable. It is commonly assumed that this involves an interpersonal comparison of utility, but this is false; it is only necessary that each individual's utility be an increasing linear function of money, and nothing need be said about the constant of proportionality (indeed any statement about the constant of proportionality is meaningless within the framework of N-M[1] utility theory). However, it *is* true that mathematically, N-M games can be treated as if the payoff were in money rather than in utility.

It is also often assumed that the N-M theory and its subsequent elaborations depend in an essential way on side payments and transferable utility; this is also false, as is shown by the small but growing body of recent work which parallels the N-M theory but deals with cooperative games in which side payments are either altogether forbidden, or are allowed but utility is not transferable. It is this body of work that I wish to survey here. Incidentally, recall that noncooperative games include cooperative games as a special case, cooperative games without side payments include cooperative games with side payments, and the case of transferable utility is the most special of all. Cooperative games without side payments and cooperative games with side payments but without transferable utility present many of the same problems, and since the former are more general we restrict much of our discussion to them.

* Revised version of a lecture delivered at the Princeton Conference on Recent Advances in Game Theory, October 1961.

† The Hebrew University of Jerusalem, Jerusalem, Israel.

[1] von Neumann—Morgenstern.

2. MOTIVATION

Cooperative games without side payments are of considerable importance in the applications. In some situations side payments are impossible because there is no common medium of exchange, or such a medium, if it exists, is irrelevant; think of the international situation. In other cases side payments are called bribes and are ruled out for ethical or legal reasons, while cooperation is considered perfectly all right. Finally, even when side payments are legal, utility usually is nonlinear in money, and this may result in a situation which is not covered by the N-M theory. It is this last fact that caused Luce and Raiffa to state that the N-M theory is "for many purposes next to useless" [19, p. 233]. We do not share this view, because if money is substituted for utility the N-M theory still applies to any situation in which probabilistic considerations are considered irrelevant;[2] but we do feel that an extension of the N-M theory to the no-side-payment case is useful.

3. THE CHARACTERISTIC FUNCTION

Let us now begin with a description of some of the work that has been done in this field. There are three widely used models for studying n-person games: the extensive form, the normal form, and the characteristic function.

The *extensive* form is essentially a mathematical representation of the rules of the game. The *normal* form is the "payoff matrix"—a list of strategies for each player, together with a payoff vector for each n-tuple of strategies. The *characteristic function* gives for each coalition the set of payoff vectors that that coalition can "assure" its players. There are, of course, connections between the various forms; the normal form can be calculated from the extensive form, and the characteristic function from the normal form. However, each form is suited for different kinds of investigations. We will begin our study of cooperative games without side payments with the characteristic function.

Let us represent the payoff to each player by a coordinate of Euclidean space; thus we will be working in Euclidean space of dimension equal to the number of players, and in its subspaces. We denote by N the set of players, by E^N the Euclidean space in which we are working, and by E^S the subspace of E^N spanned by the axes belonging to the players in a subset S of N. Points of E^N are called *payoff vectors*, of E^S payoff *S-vectors*. The *characteristic function* associates with each $S \subset N$ a subset $v(S)$ of E^S. Intuitively, $v(S)$ represents the set of payoffs that S can assure itself. When side payments are allowed and utility is transferable (this will henceforth be called

[2] Even when they are relevant the N-M theory applies in a much wider range of situations than has often been supposed; see §8.

the N-M case), $v(S)$ is the closed half-space

$$\left\{x \in E^S : \sum_{i \in S} x^i \leq f(S)\right\},$$

where $f(S)$ is the N-M characteristic function (i.e., the total amount of money that S can assure itself). A typical such half-space is illustrated in Figure 1 for the 2-player case; $v(S)$ is the whole area to the "southwest" of the line $x^1 + x^2 = f(S)$.

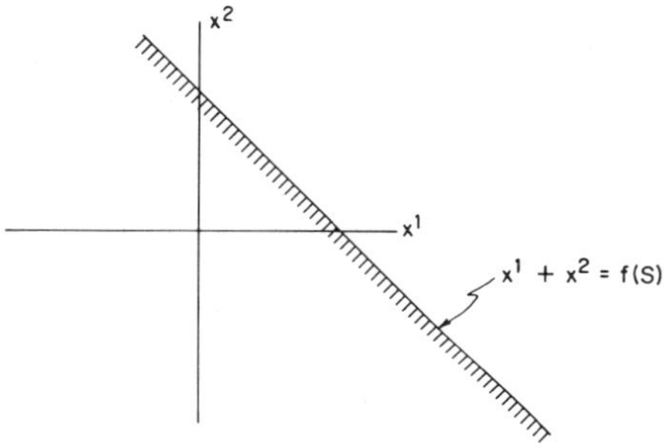

FIGURE 1

Returning to the no-side-payment case, we assume the following conditions for our characteristic function:

$$v(S) \text{ is convex, closed, and non-empty.} \tag{1}$$

$$x \in v(S),\ y \in E^S,\ x \geq y \Rightarrow y \in v(S). \tag{2}$$

$$v(S \cup T) \supset v(S) \times v(T) \text{ for } S \text{ and } T \text{ disjoint.} \tag{3}$$

The vector inequality in (2), like all subsequent vector inequalities, is meant to hold for each coordinate.

Intuitively, convexity follows from the fact that players can mix and correlate their strategies. Closedness is mainly a question of mathematical convenience and is satisfied in all applications that I can think of. Condition (2) says that if a coalition can assure itself of a payoff vector x, it can also assure itself of anything coordinate-wise less. The last condition is superadditivity; any vector whose components can be obtained by each of two disjoint coalitions acting separately can also be obtained by them when acting together. In Figure 2 we show a typical set of the form $v(S)$ in two dimensions.

We have defined a characteristic function; in order to define a *game in characteristic function form*, we need an additional concept that is not needed in the N-M theory. This is the set H of outcomes that "can actually occur." H has a very close connection with $v(N)$: its "top" coincides with

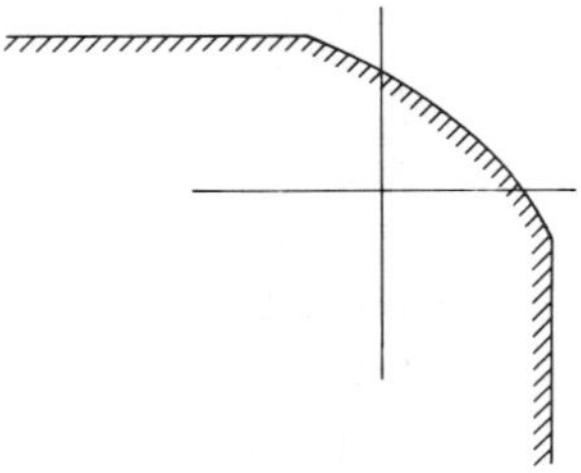

FIGURE 2

the "top" of $v(N)$ (see Figure 3), or in precise terms

$$v(N) = \{x \in E^N \colon \exists\ y \in H \text{ such that } y \geq x\}. \tag{4}$$

The difference between $v(N)$ and H is that $v(N)$ consists of those vectors x that N can "guarantee," in the sense that it can get *at least* x; whereas H is the set of vectors such that N can get *exactly* x.

Summing up, a game in *characteristic function form* is a pair (v, H), where v is a characteristic function obeying (1), (2), and (3), and H is a convex proper subset of E^N satisfying (4).

Sometimes it will be assumed that H is a convex compact polyhedron; this is justified if one thinks of the game in characteristic function form as being generated from a finite game in normal form. On other occasions, it is more convenient to assume that $H = v(N)$; this is justified, for example, if one makes an assumption of "free disposal." The latter assumption is the one more suited to the N-M case.

The set of conditions ((1) through (4)) that we have assumed for v and H is by no means the only possible one, and in fact almost every paper on the subject uses a different variant of the set of assumptions. In particular, for many purposes super-additivity (3) is unnecessary, sometimes convexity is not needed either, and for other purposes condition (4) is unnecessary. The conditions given here have been chosen for convenience in exposition,

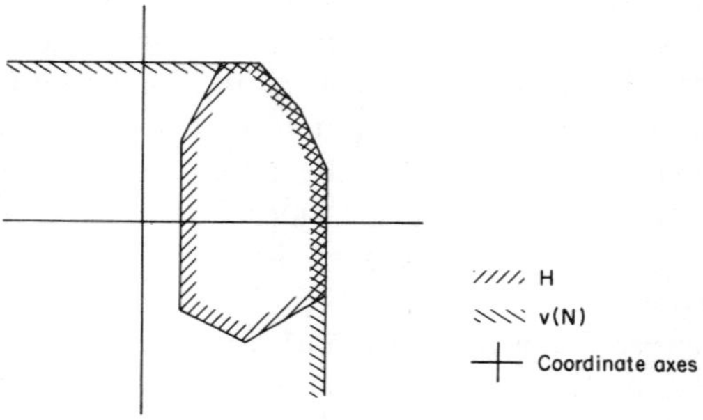

FIGURE 3

and so for many of the theorems they are stronger than necessary. The reader is referred to the original papers for statements of alternative conditions under which the various theorems hold.

In the N-M theory, various kinds of payoff vectors are distinguished according to their "rationality" attributes. A payoff vector is said to be *individually rational* if each player gets at least what he can guarantee himself, and *group rational* if the whole group cannot play in such a way that each player gets more. The same notions can be defined in our context; precisely, x is individually rational if for each player i, we have $x^i \geq \max v(\{i\})$; and x is group rational if there is no payoff vector $y \in v(N)$ such that $y > x$. Let us denote by H_{ig}, H_i, and H_g the subsets of H obtained by imposing individual and group rationality in various combinations; these sets, together with H, correspond to the sets of payoff vectors that have been studied in the N-M theory [39, 50]. In particular H_{ig} corresponds to what is usually called the set of "imputations."

Note that the characteristic function $v(S)$ does not necessarily have to be interpreted as the set of payoff vectors that S can *assure* itself; if preferred, it may be interpreted in any other way, such as what a coalition "thinks it can get." It is also possible that a game is given *a priori* in characteristic function form, like the following voting game:

Let the number of players be odd, and let C be a convex compact subset of E^N. The game consists of the players "voting" for a member of C by majority rule. If a majority agrees on a point x of C, then x is the payoff vector (to all players); otherwise each player i gets only his personal minimum m^i in C, i.e. $\min\{x^i: x \in C\}$. It is easy to see how this can be generalized to weighted majority games and to simple games in general.

To describe the characteristic function, let C^S denote the projection of C on E^S, and let m^S denote the S-vector $\{m^i\}_{i \in S}$. Then

$$v(S) = \begin{cases} \text{(if } S \text{ is winning) the set of all } S\text{-vectors that} \\ \quad \text{are} \leq \text{a member of } C^S; \\ \text{(if } S \text{ is losing) the set of all } S\text{-vectors that} \\ \quad \text{are} \leq m^S. \end{cases}$$

4. THE VON NEUMANN-MORGENSTERN SOLUTION

We can now develop a theory of games parallel to the N-M theory. The two most important elements of that theory are the solution and the core. First, we define domination:

Let $x, y \in E^N$, and let x^S denote the projection of x on E^S. Then

$$x \succ_S y \Leftrightarrow_{df} x^S \in v(S),\ x^S > y^S$$
$$x \succ y \Leftrightarrow_{df} x \succ_S y \textit{ for some } S.$$

Let $K \subset E^N$. Just as in the N-M theory, a *solution* of K is a subset V of K such that no two members of K dominate each other, and every member of K not in V is dominated by some member of V. The *core* of K (denoted by $C(K)$) is the set of members of K not dominated by other members of K.

THEOREM 1. *A solution of H_i is a solution of H_{ig}, and conversely.*

This corresponds to a theorem in the N-M theory first proved by Shapley [50]. The proof, which is not difficult, is given in [23]. Henceforth a *solution of a game* is a solution of H_{ig} for that game.

THEOREM 2. *Every 2-person game has a unique solution, namely all of H_{ig}. This is also the core of H_{ig}.*

This too is easy to prove. The first difficult theorem is:

THEOREM 3. *Every 3-person 0-sum game has a solution.*

A 3-person 0-sum game is one in which H is contained in the hyperplane $\sum_{i=1}^{3} x^i = 0$. Theorem 3 is proved in Peleg [23]. The 0-sum restriction may seem somewhat strange in a non-side-payment context; however, it makes sense if one assumes that the payoff to a game is in money, that no money enters or leaves the game from outside, that chance and mixed strategies are irrelevant, and that side payments, though obviously possible, are illegal. In addition, the proof was a considerable technical achievement, and pointed the way to the subsequent:

THEOREM 4. *Every 3-person game for which H is a polyhedron has a solution.*

This is proved in Stearns [30]. In the same place Stearns classifies all solutions to 3-person games.

The biggest problem left open by von Neumann and Morgenstern in their book [45] was that of the existence of a solution for an arbitrary *n*-person cooperative game with side payments and transferable utility. The problem remains unsolved to this day. One of the methods they used to attack this problem [45, pp. 266-271 and pp. 587-603] was to define the notion of solution for an abstract relation defined on an abstract set (abstracting from the game situation, where it is defined for the domination relation on the set of imputations). They then studied the solution notion in this abstract framework, seeking conditions of a general nature that would ensure the existence of a solution and that would be satisfied in the game context. This work was carried on by Richardson and others (see for example [40, 47]), but though many interesting sufficient conditions for existence were found, none could be proved to apply to the game context. In 1959, Kalisch and Nering [41] constructed a game with a countable infinity of players and showed that it has no solution, thus showing that the completely "abstract" approach to proving the existence theorem could not work. However, the imputation space in the Kalisch-Nering example is not compact. Thus there remained the hope that a "modified abstract" approach could be made to work, in which account would be taken of topological properties of the imputation space and the domination relation. This hope has recently been shattered by Stearns (unpublished[3]), who proved:

THEOREM 5. *There is a 7-person game with no solution.*

The original problem proposed by von Neumann and Morgenstern—for games with side payments and transferable utility—remains open.

We mention that it is possible to construct a theory of composition of games that parallels the N-M theory, but that yields simpler and more intuitive results [3, 5, 7].

Isbell [16] has constructed a theory of cooperative games without side payments in which he makes use of the notion of N-M solution. However, his work is not based on the characteristic function model presented in §3.

5. THE CORE

Let $K \subset E^N$. The *Core* of K (denoted by $C(K)$) is the set of members of K not dominated by other members of K.

THEOREM 6. *Assume either that H is a convex compact polyhedron, or that $H = v(N)$. Then $C(H) = C(H_i) = C(H_g) = C(H_{ig})$.*

In other words, all the "interesting" cores are equal, so we are justified in referring to the "core of a game." This is trivial in the N-M theory, but no longer so in the current theory. Under the first of the two assumptions, the proof was first published in [3]; subsequently it was considerably simplified by Stearns (unpublished). We sketch the simplified proof here.

The difficult part is to prove that imposing group rationality, either on H or on H_i, does not change the core. For example, take H; we must prove that $C(H) = C(H_g)$. $C(H) \subset C(H_g)$ is easily established. The crux of the proof is the opposite inclusion. For $x \in E^N$, denote $\max_i |x^i|$ by $\|x\|$. We need the following

LEMMA. *There is a positive number M (depending on H only) such that for all $z \in H - H_g$, there is a $\hat{z} \in H_g$ such that $\hat{z} > z$ and for each $i \in N$, $\hat{z}^i - z^i > \|\hat{z} - z\|/M$.*

In words, the lemma states that for each z in H that is not already in the top of H, we can find a ray that leads to the top of H, and that is increasing in all coordinates at a rate that is uniformly (i.e., independently of z) bounded away from 0. The lemma is true only because H is a polyhedron; for example, in Figure 4, as the points z approach the x^1-axis, the rate of increase of x^1 along the dotted lines tends to 0. Indeed, there are counter-examples to Theorem 5 if it is not assumed that H is a

[3] A previous published version [31] has the disadvantage that some of the $v(S)$ are empty.

polyhedron [3]. The proof of our lemma is given in [3], and will not be repeated here.

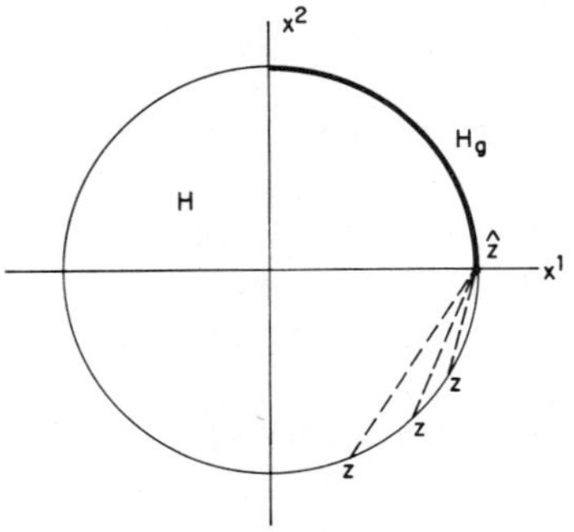

FIGURE 4

Let $x \in C(H_g)$. We will suppose that $x \notin C(H)$, i.e., x is dominated via some S by an element y of H, and will then construct an element $\hat{z}$ of H_g which also dominates x; this will be a contradiction. Roughly, this is done by taking an element z very close to x on the line segment $\overline{xy}$ which joins x and y, and constructing the corresponding $\hat{z}$. Now either

i) $\hat{z}$ is far from z, or

ii) $\hat{z}$ is close to z.

In the first case, it follows from the lemma that all the coordinates of $\hat{z}$ must be considerably greater than those of z; since z is close to x, it follows that $\hat{z} > x$, contradicting $x \in H_g$. In the second case, it follows that $\hat{z}$ is close to x. Hence from $y \succ_S x$ we deduce that $y \succ_S \hat{z}$, and hence it follows that $\hat{z}^S \in v(S)$ (from property (2) of the characteristic function). But $\hat{z}^S > z^S > x^S$, and therefore $\hat{z} \succ_S x$, which gives us the desired contradiction.

More precisely, we suppose without loss of generality that $x = 0$. Let $\sigma = \min_{i \in S} y^i$, where $y \succ_S x$. Let $z \in \overline{yx}$ be such that $|z| < \delta/(M+1)$. Then $\hat{z}^i - z^i > \|\hat{z} - z\|/M$ for all i. Hence if $\|\hat{z} - z\| \geq \delta M/(M+1)$, then $\hat{z}^i - z^i > \delta/(M+1)$. Then

$$\hat{z}^i = \hat{z}^i - z^i + z^i \geq \hat{z}^i - z^i - \|z\| > \delta/(M+1) - \frac{\delta}{(M+1)} = 0,$$

contradicting $0 \in H_g$. Hence $\|\hat{z} - z\| < \delta M/(M+1)$. Hence for all i, $\hat{z}^i = \hat{z}^i - z^i + z^i \leq \|\hat{z} - z\| + \|z\| < \delta M/(M+1) + \delta/(M+1) = \delta$. Hence for $i \in S$, $\hat{z}^i - y^i \leq \max_i \hat{z}^i - \min_i y^i < \delta - \delta = 0$. Hence $\hat{z}^S < y^S$. Hence $\hat{z}^S \in v(S)$, and $\hat{z} \succ_S x$, contradicting $x \in C(H_g)$.

Under the second of the two assumptions, Theorem 6 was proved by Burger [7]; the proof is simpler than under the first assumption. Burger's paper is the first to make systematic use of the assumption $H = v(N)$; this makes for a considerably simpler theory.

When is the core of a game non-empty? In the N-M case, a necessary and sufficient condition for the non-emptiness of the core has been given by Shapley [51], in terms of "balanced" collections of coalitions. A similar condition has been given (independently) by Bondareva [35]. Using this notion of balanced collections, Scarf [27] recently obtained a sufficient condition for the non-emptiness of the core in the no-side-payment case as well.

For each $S \subset N$, define a vector e_S in E^N by

$$e_S{}^i = \begin{cases} 1 & \text{if} \quad i \in S; \\ 0 & \text{if} \quad i \notin S. \end{cases}$$

A collection $\mathscr{S}$ of subsets S of N is called *balanced* if it is possible to assign to each S in $\mathscr{S}$ a non-negative number δ_S, such that

$$\sum_{S \in \mathscr{S}} \delta_S e_S = e_N.$$

For example, if $N = \{1, 2, 3\}$, then $\mathscr{S} = \{\{1, 2\}, \{2, 3\}, \{1, 3\}\}$ is balanced, where the δ_S are given by $\delta_{\{1,2\}} = \delta_{\{2,3\}} = \delta_{\{1,3\}} = \frac{1}{2}$.

Scarf's theorem may now be stated as follows:

THEOREM 7. *Let $H = v(N)$. Assume that for every balanced collection $\mathscr{S}$ of subsets of N, we have*

$$\bigcap_{S \in \mathscr{S}} (v(S) \times E^{N-S}) \subset v(N).$$

Then the core is non-empty.

The importance of this theorem may be illustrated by the fact that it implies the existence of a competitive equilibrium in a market (cf. §8); since the proof of Theorem 7 is "elementary" in the sense that it does not involve fixed point theorems, it follows that the existence of competitive equilibria may also be given an "elementary" proof.

6. VALUE

To motivate the notion of "value" as used in game theory, we can do no better than quote Shapley [49]: "At the foundation of the theory of games is the assumption that the players of a game can evaluate, in their utility scales, every 'prospect' that might arise as a result of a play. . . . One would normally expect to be permitted to include, in the class of 'prospects,' the prospect of having to play a game. The possibility of evaluating games is therefore of critical importance."

The value problem has been treated for games with side payments and transferable utility (the N-M case) by Shapley [49] and Selten [48], and in the no-side-payment case by Nash [21, 22], Harsanyi [9, 11, 12], Isbell [17], Miyasawa [20], and Shapley [28]. In all treatments the value assigns to each game one (or sometimes more than one) payoff vector. Often the treatment proceeds from the normal rather than the characteristic function form, and in at least one case [48] it proceeds direct from the extensive form. Here we will confine ourselves to discussing the Shapley value, for games in characteristic function form.

The Shapley value was defined for the N-M case in [49], as an imputation satisfying a certain set of axioms. It was proved to be unique, and was shown to have the following probabilistic interpretation: The Shapley value of player i is the expected value of the random variable $f(S \cup \{i\}) - f(S)$, where S is the set of players "before" i in a random ordering of all the players, and f is the N-M characteristic function. This definition clearly depends on a numerical value for f, and it is not at all clear how it can be generalized to a no-side-payment characteristic function as defined in §3.

Very recently Shapley [28] succeeded in giving an elegant definition of his values in the no-side-payment case by means of a reduction to the N-M case. His procedure is as follows: Let us be given a no-side-payment game with characteristic function v, and let us *imagine* what would happen if we were to allow side payments. We would then obtain an N-M game, and this would have a Shapley value, say $w = (w^1, \ldots, w^n)$. If it happens that $w \in v(N)$, i.e., that the players can attain w without actually making side payments, then w would be an excellent candidate for the Shapley value of the original game. Suppose that we now re-scale the original game, i.e., multiply the payoff of each player i by some non-negative constant p^i, and *then* allow side payments. We would then obtain another N-M game (generally different from the one discussed above), and this too would have a Shapley value. Shapley proved that for an appropriate choice of the scaling factors p^i, the Shapley value of the resulting N-M game is attainable by the players in the original (but re-scaled) no-side-payment game without actually making side payments. The scaling factors can then be eliminated and a Shapley value for the original (unscaled and no-side-payment) game results. We use the indefinite article advisedly; the value is no longer unique, because a number of different sets of scaling factors may yield attainable outcomes.

To simplify the formal description, we adopt the convention that if x and y are vectors, then xy denotes the vector whose ith coordinate is $x^i y^i$. For each vector $p \geq 0$, define a characteristic function v_p from the given one v by

$$v_p(S) = \{y \in E^S\colon \text{there is an } x \in v(S) \text{ such that } y \leq p^S x\}.$$

It may be verified that v_p satisfies the axioms for a characteristic function.

Now define an N-M characteristic function f_p by

$$f_p(S) = \max\left\{\sum_{i\in S} y^i : y \in v_p(S)\right\}; \tag{5}$$

it is not difficult to show, by using (4), that the maximum is attained. Let $w(p)$ be the Shapley value for f_p. Then a pair (p, w) is called a *valuation* of the original characteristic function v if $p \neq 0$, $w(p) = pw$, and $w \in v(N)$. Shapley's theorem is:

THEOREM 8. *Every game has a valuation.*

The first investigation of what amounts to a cooperative game without side payments is due to Nash [21]; Nash's "bargaining problem" is the same thing as a 2-person game in characteristic function form, in the sense of §3. Each such game has a unique valuation (p, w), in which w is the Nash solution. Like the Nash solution in the 2-person case, the valuation in the general case is derivable from a small number of abstract axioms [28].

7. THE BARGAINING SET M_1^i

In the context of the bargaining set, the object of interest is not a payoff vector, but a *payoff configuration*, i.e., a pair consisting of a payoff vector and a coalition structure (partition of the players into disjoint coalitions). Furthermore the possibility that certain coalitions are "forbidden" (for example because of legal restrictions or communication difficulties) is admitted. For the N-M case, M_1^i is defined elsewhere in this volume [37] as a set of payoff configurations enjoying certain stability properties. Peleg [46] has proved that in the N-M case it is non-empty for each choice of a coalition structure, i.e., for each coalition structure there is a payoff vector such that the resulting pair is stable in the required sense. Unlike the situation for Shapley values, it is here quite easy to generalize the definition of M_1^i to the no-side-payment case, and in fact this can be done in two ways; the more appropriate of the two generalizations is denoted $\tilde{M}_1^i$. However, the existence theorem does not generalize.

THEOREM 9. *There is a 4-person game for which $\tilde{M}_1^i$ is empty (for appropriate choice of coalition structure).*

The example is due to Peleg [24]. A positive result obtained by Peleg in the same paper is:

THEOREM 10. *In a game in which only 2-player coalitions are permitted, $\tilde{M}_1^i$ is non-empty for each coalition structure.*

The proof makes use of the Eilenberg-Montgomery fixed-point theorem [38].

8. GAMES WITH SIDE PAYMENTS BUT WITHOUT TRANSFERABLE UTILITIES

Consider a game given by an N-M characteristic function f, i.e., a numerical function defined on the set of all subsets of N satisfying the super-additivity condition

$$f(S \cup T) \geq f(S) + f(T) \quad \text{for} \quad S \cap T = \phi.$$

Give this game the following interpretation: Each coalition S may go to a "referee" and receive exactly $f(S)$ dollars, on condition that it has agreed beforehand on how this money should be divided.

For each player i, let $u_i(b)$ be[4] the utility of player i for b dollars, in the sense of N-M ([45, pp. 617–632]; also [19, pp. 12–38]). We will assume that u_i is bounded,[5] continuous, and strictly increasing in b. Define a function u from E^N to itself by

$$u^i(x) = u_i(x^i) \tag{6}$$

for all $x \in E^N$ and $i \in N$. Let v' be the function defined on the subsets S of N by

$$v'(S) = \Big\{ y \in E^S \text{: There is an } x \text{ in } E^N \text{ such that } \sum_{i \in S} x^i = f(S) \text{ and } y \leq u^S(x) \Big\}. \tag{7}$$

Intuitively, $v'(S)$ is the set of payoff vectors, *expressed in terms of utilities*, that are attainable by the coalition S. However, v' is not a characteristic function in the sense of §3, because $v'(S)$ may fail to be convex. We therefore replace $v'(S)$ by its convex hull; intuitively, this means that the coalition S will in general agree on a *lottery* that will determine the division of the $f(S)$ dollars, rather than agreeing on a specific division. We thus define a function v by

$$v(S) = \text{convex hull } v'(S). \tag{8}$$

Then v satisfies conditions (1)–(4) (where for convenience we take $H = v(N)$; it is of course neither compact nor polyhedral).

Suppose now that in the original game f, we exclude the possibility that the players will use lotteries to divide the payoffs. In that case the utility functions of the players become irrelevant, because their purpose is to represent preferences between *lotteries*. To represent preferences between actual sums of money (as distinguished from lotteries over such sums), utilities are not needed, as the dollar amount is a perfectly good measure for this purpose. And in fact, the reasoning leading to the N-M solution is then valid when the payoffs are expressed in money. Therefore we may calculate the N-M solutions (or the core, bargaining set, ψ-stable payoff configurations, and so on) of the characteristic function f as given, expressing the result in dollar terms, and the intuitive validity of the result is not

[4] u_i is determined only up to an additive and a positive multiplicative constant. These constants may be chosen independently for the various players (indeed there is no meaningful way of correlating them).

[5] The boundedness assumption is not strictly necessary but simplifies the discussion considerably; moreover it is intuitively very acceptable (cf. Isbell [16], p. 360).

based on any consideration of "linear utility," "transferable utility," "comparable utility," or indeed any utility whatsoever.

All this is based on the assumption that probabilistic considerations are for some reason excluded. If they are admitted, then utilities become relevant and indeed crucial; we must therefore replace the function f by the function v defined in (7) and (8), and use the corresponding definition of solution (§4). The question then arises: What is the relation, if any, between the solutions to f and the solutions to v?

THEOREM 11. *If the utility functions u_i are concave, then a subset F of E^N is a solution to f if and only if $u(F)$ is a solution to v.*

Theorem 11 asserts that if the utility functions are concave, then the same utility distributions—and so also the same money distributions—result when the characteristic function of §3 is used rather than the original N-M characteristic function. It follows that for the validity of solution theory as described in [45] it is not necessary to assume that utilities are *linear* in money, as is usually supposed,[6] but only that they are concave. The concavity assumption is an eminently reasonable one, and is often made in the literature.

The theorem is intuitively not surprising, because concave utilities mean that the players never prefer a gamble to its expectation, and hence the function v does not offer them different possibilities than the function f. The proof is very simple. From the concavity of the u_i it follows that $v' = v$, and hence u is a *domination-preserving* 1-1 correspondence from the space I_{ig} of imputations in the game f onto the corresponding space H_{ig} for v. Theorem 11 follows from this property of u.

It is rather curious that for simple games f (i.e., f taking the values 0 and 1 only), a result superficially similar to Theorem 11 holds in the diametrically opposed case, when the utility functions are all convex (i.e., the players always like a gamble at least as well as its expectation). Assume the utility functions are normalized so that $u_i(0) = 0$ and $u_i(1) = 1$. Then (in general) $u(I_{ig}) \neq H_{ig}$, and hence u does not provide a correspondence between I_{ig} and H_{ig}; but I_{ig} and H_{ig} are formally equal, and the *identity* is a domination preserving 1-1 correspondence between them. Here again, the result is easy to understand intuitively: a coalition of these gamblers will never split the money, always preferring a lottery in which one member gets all with a certain probability; the probabilities in the solutions to v then correspond to sums of money in the solutions to f.

Is there *always* a domination-preserving 1-1 correspondence between I_{ig} and H_{ig}? The answer is no. Consider the 3-person simple majority game ($f(S) = 0$ or 1 according as S has one or more members). Let the utility functions be the piecewise linear functions graphed in Figure 5.

[6] cf. the quotation from Luce and Raiffa in §2.

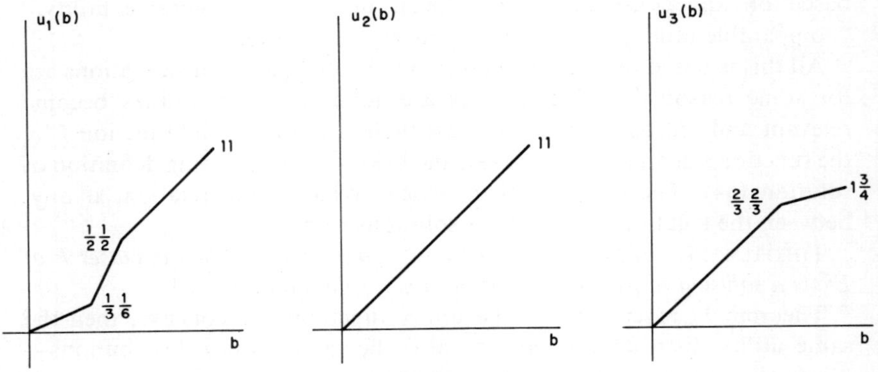

FIGURE 5

H_{ig} is pictured in Figure 6. A 1-1 domination-preserving mapping from H_{ig} onto I_{ig} would have to take both the points $(\frac{1}{6}, \frac{1}{3}, \frac{1}{2})$ and $(\frac{1}{4}, \frac{1}{4}, \frac{1}{2})$ of H_{ig} onto the point $(\frac{1}{4}, \frac{1}{4}, \frac{1}{2})$ of I_{ig}, an absurdity (parentheses and commas are omitted in the figures and henceforth in the text).

We close this section with an example of what happens when a pessimist (concave utility) plays a simple majority game with two optimists (convex

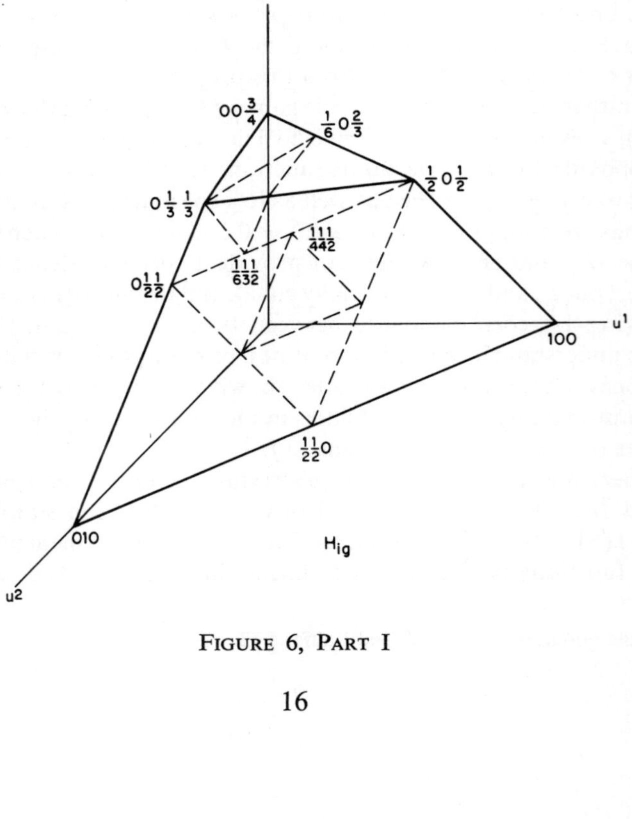

FIGURE 6, PART I

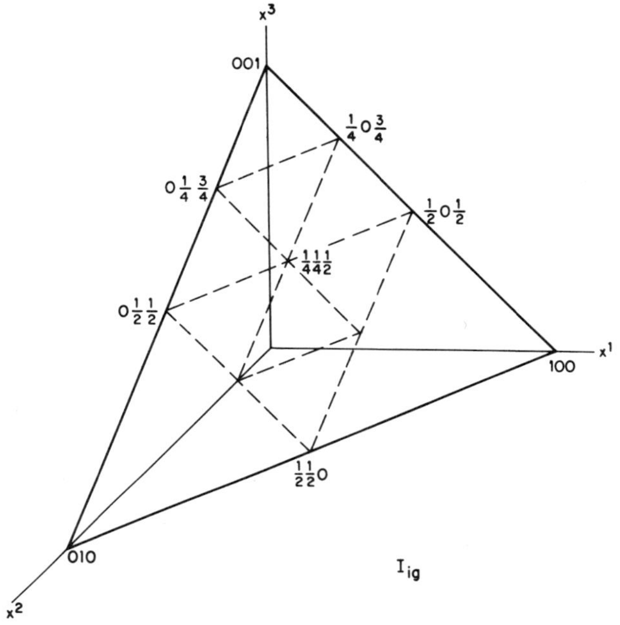

FIGURE 6, PART II

utility). Let f be as in the previous paragraph, and let the utility functions be as graphed in Figure 7. The 3-point solution of f is $\frac{1}{2}\frac{1}{2}0$, $\frac{1}{2}0\frac{1}{2}$, $0\frac{1}{2}\frac{1}{2}$, it being understood that the coordinates of the vectors in the solution are expressed in dollars. This solution applies when lotteries are excluded. When lotteries are admitted, we must pass from f to v. $u(I_{ig})$ is pictured in Figure 8; it may be seen that H_{ig} is formally equal to I_{ig}. Hence v also has the 3-point solution $\frac{1}{2}\frac{1}{2}0$, $\frac{1}{2}0\frac{1}{2}$, $0\frac{1}{2}\frac{1}{2}$, but this time the coordinates of the

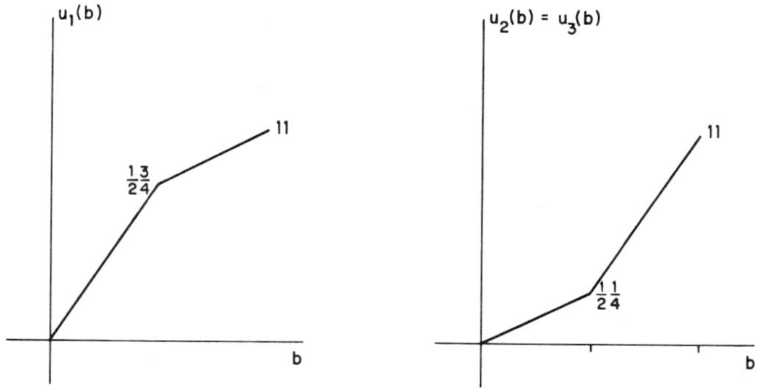

FIGURE 7

vectors are expressed in utility units rather than in dollars. If we translate back into dollars, we find that the solution is $\frac{1}{4}\frac{3}{4}0$, $\frac{1}{4}0\frac{3}{4}$, $0\frac{3}{4}\frac{3}{4}$. The point $0\frac{3}{4}\frac{3}{4}$ cannot be attained by a distribution of money, but is attained by a $\frac{1}{2}$-$\frac{1}{2}$ lottery for $f(23)$ between players 2 and 3. However, $\frac{1}{4}\frac{3}{4}0$ *is* attainable by a distribution of the sum $f(12)$, without recourse to lotteries. It follows that if the coalition 12—consisting of the pessimist and an optimist—forms,

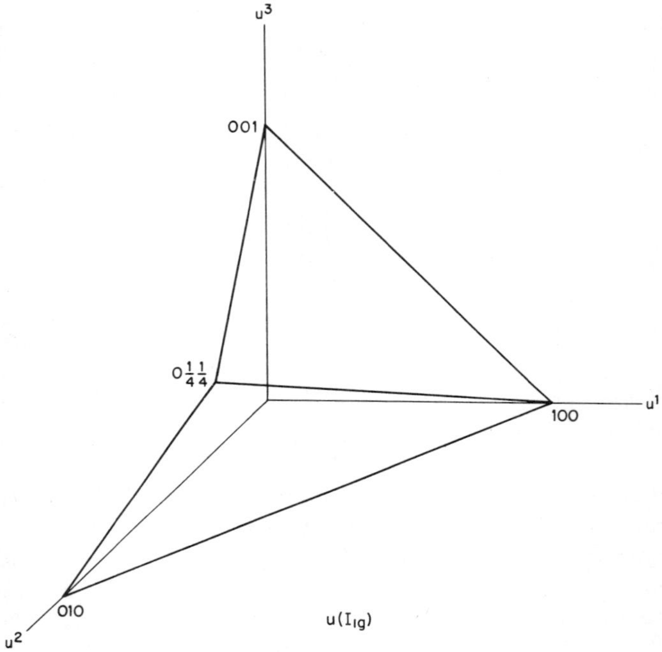

FIGURE 8

then the willingness of the optimist to take risks puts the pessimist at a very considerable material disadvantage, even though in the end no risks are taken by either player.

If *all* the points of the solution of v had been attainable without recourse to lotteries, then this phenomenon would not have occurred. This follows from the fact that in an arbitrary n-person game, *if a solution V of v is included in $u(I_{ig})$, then $u^{-1}(V)$ solves f.* The proof is an easy consequence of Theorem 1.

The results of this section, hitherto unpublished, are the outcome of conversations between M. Maschler and the author. Though they are not deep, they shed light on the relation between utilities and n-person games.

9. MARKET GAMES

Recently a good deal of attention has been paid to market games, which are in fact cooperative games without side payments, essentially in characteristic function form. The game-theoretic tool most significant in this connection is the core; it has been shown that for markets with "many" traders (this notion has been formalized in a number of different ways) the notion of core is essentially equivalent to that of Walrasian "competitive equilibrium." Under certain conditions it appears that Shapley's valuation (see §6) is closely connected with the competitive equilibrium as well. Finally, Scarf's core-theorem (Theorem 7) yields the non-emptiness of the core of a market game under wide conditions, and by using this an "elementary" existence proof for the competitive equilibrium can be obtained. For details, the reader is referred to the original papers (Debreu and Scarf [8], Aumann [4, 34], Vind [32], Shapley [28], Scarf [27]).

10. THE NORMAL FORM

The passage from the normal to the characteristic function form is not without its pitfalls. Even in the case of games with side payments and transferable utility (the N-M case), it is not generally agreed that the characteristic function as derived from the normal form by von Neumann and Morgenstern adequately represents the game; this is chiefly because for games that are not constant-sum, it does not always take adequate account of threats. Nevertheless the N-M definition is useful for some purposes, and we now examine how it can be generalized to the no-side-payment game.

In the N-M case, if f is the characteristic function and S is a coalition, then $f(S)$ is the maximum amount that S can guarantee itself; by the minimax theorem $N - S$ can prevent S from getting more. Here these two approaches—what S can guarantee itself, and what $N - S$ can prevent—are no longer equivalent. We write down definitions corresponding to both approaches.[7]

$v_\alpha(S)$ = the set of all payoff S-vectors x such that S can guarantee that it will get at least x.

$v_\beta(S)$ = the set of all payoff S-vectors x such that $N - S$ cannot prevent S from getting at least x.

By "getting at least x" we mean getting an amount that is at least x^i for each player i; and by "can guarantee" or "can prevent" we mean the existence of a single (correlated) strategy that guarantees or prevents, independent of the actions of the other players.

[7] These definitions were first explicitly given in [5].

When side payments are allowed,

$$v_\alpha(S) = v_\beta(S) = \left\{x \in E^S : \sum_i x^i \leq f(S)\right\},$$

where $f(S)$ is the N-M value of S. That v_α and v_β are in general different is seen by means of the 3-person game,[8]

1, −1	0, 0
0, 0	−1, 1

where the rows represent strategies of the coalition {1, 2}, the columns represent strategies of player 3, and the entries are payoff {1, 2}-vectors (the payoffs to player 3 are irrelevant). A picture of $v_\alpha(\{1, 2\})$ and $v_\beta(\{1, 2\})$ is given in Figure 9.

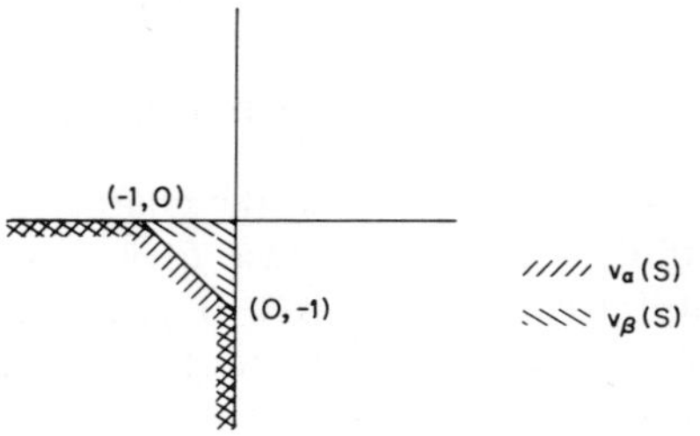

FIGURE 9

Whether the α-notion or the β-notion is preferable is a matter of taste. Both satisfy the axioms for characteristic functions [3]; that is an advantage of the axiomatic treatment. The α-notion seems to be intuitively more appealing, but as we shall see the β-notion has a certain technical advantage.

Some authors have considered the discrepancy between $v_\alpha(S)$ and $v_\beta(S)$ to be a disturbing phenomenon. We noted above that $v_\alpha(S) = v_\beta(S)$ for games with transferable utility side payments. Jentzsch [18] has investigated the possibility of obtaining a wider class of games with the same property. The general tenor of his result is negative, i.e., v_α and v_β cannot be expected to coincide unless one is talking about games that to start with are very similar to games with transferable utility side payments. To give a more precise description of his result, let us for the moment fix attention on a single coalition S. If we are interested in this coalition only, then in the (transferable utility) side-payment case we can substitute the following "adjusted normal form" for the usual normal form: The "adjusted normal form" is a matrix whose rows are pure strategies of S,

[8] See Jentzsch [18] and Aumann [3].

whose columns are pure strategies of $N - S$, and whose entries are the total payoff to S for the appropriate strategy n-tuples. What this really means is that after S and $N - S$ have chosen strategies, S can pick any vector whose sum does not exceed the entry (the redistribution is made possible by the side payments). Use of mixed (i.e., correlated) strategies on the parts of S and $N - S$ will yield a numerical payoff for S which is the appropriate mixture of the pure payoffs, and which S can also allocate between its members as it sees fit. It is exactly the "value" of this adjusted normal form, when considered as a 2-person 0-sum game, that gives the N-M characteristic function $f(S)$.

The adjusted normal form can be generalized to cover games with non-transferable utility side payments. Suppose that a pair (p, q) of strategies (of S and $N - S$) yields a payoff S-vector x. By the use of side payments, S can redistribute the income from x among its members, but because utility is not assumed to be transferable, total utility will not be conserved in the redistribution. The set of all payoff S-vectors that can be obtained from x by means of redistributions of this kind will be a subset of E^S that satisfies conditions (1) and (2) for characteristic functions (closedness, convexity, unboundedness towards the southwest); such subsets of E^S will be called S-catalogues, or simply *catalogues*.[9] The adjusted normal form for games with nontransferable utility side payments is thus a matrix whose entries are catalogues rather than numbers. As in the previous case, it is possible to consider mixed outcomes, corresponding to mixed strategies: a mixture of two catalogues (with specified probabilities) is simply the set of mixtures of its members (with the same probabilities), and is itself a catalogue.

If we now restrict the catalogues in the entries to be half-spaces of the form $\Sigma x^i \leq k$, then we are back in the transferable utility case, and it follows that $v_\alpha(S) = v_\beta(S)$. Jentzsch asked whether we could not still assure $v_\alpha(S) = v_\beta(S)$ by imposing a weaker restriction on the catalogues that appear in the adjusted normal form. More precisely, consider a family $\mathscr{F}$ of S-catalogues; let us call a game *adjusted to* $\mathscr{F}$ if in its adjusted normal form, all payoffs—including the mixed ones—are in $\mathscr{F}$. Then what conditions must be placed on $\mathscr{F}$ in order to ensure that for every game that is adjusted to $\mathscr{F}$, we have $v_\alpha(S) = v_\beta(S)$?

Let us call an $\mathscr{F}$ for which this holds *determinate*. Jentzsch showed that the condition of being determinate is a very strong one:

THEOREM 12. *Let us call a catalogue F regular if it has a supporting hyperplane in each positive direction,*[10] *and assume that every member of*

[9] Presumably because you can pick from them whatever you want.

[10] I.e., for each vector x with positive coordinates, there is a hyperplane that supports $\mathscr{F}$ and is perpendicular to x. Figure 2 illustrates a regular catalogue, and Figure 1 illustrates one that is not regular.

$\mathscr{F}$ is regular. If $\mathscr{F}$ is determinate, then of every three members of $\mathscr{F}$, one is a probability mixture of the other two (and in particular lies between the other two).

The theorem does not apply directly to the N-M case, because half-spaces are not regular. But in all practical cases side payments are limited, and this makes the catalogues regular and the theorem applicable.

Since the motivation for this work derives from an analysis of games with side payments but without transferable utility, it is natural to ask for conditions on the utility functions u_i of the players i in S that will lead to games for which[11] $v_\alpha(S) = v_\beta(S)$. Suppose, for example, that the utility function of each player for a (positive) amount of money b is that suggested by Bernoulli, namely $\log b$. Then if S has a total amount of money d to divide between its members (d positive), the resulting catalogue is

$$\left\{y \in E^S : \sum_{i \in S} \exp\,(y^i) \leq d\right\},$$

where exp is the exponential function ($\exp\,(b) = e^b$). Jentzsch remarked that it can be shown from his results that the family of all such catalogues, as d varies, is determinate. This means that in side-payment games played by players all of whom have the utility function $\log b$, we have $v_\alpha(S) = v_\beta(S)$.

What other S-tuples of utility functions have this property? This question was answered by B. Peleg in [25]. It turns out that there are very few of them. His chief result is as follows: Suppose that the utility functions of the players i in S are concave and have the property that $v_\alpha(S) = v_\beta(S)$ in any game in which these players participate, providing that the players all have personal minima of 0 (i.e., $v_\alpha(\{i\}) = v_\beta(\{i\}) = (-\infty, 0]$). Then *either* $u_i(b) = \log b$ for all i in S, *or* there is a λ obeying $0 < \lambda \leq 1$ such that for all i in S, $u_i(b) = b^\lambda$, *or* there is a λ obeying $\lambda < 0$ such that for all i in S, $u_i(b) = -b^\lambda$. This underscores the fact that equivalence between the α- and the β-notions is a very rare event.

The concepts we have described may be applied to supergames, i.e., long sequences of plays of a cooperative game without side payments. We look for stable behavior in such games. There are two ways of approaching this problem. One is to treat the entire supergame as a single game, and use stability criteria appropriate for a single game. The other way is to speculate as to what kind of behavior in the individual plays constituting the supergame would lead to stability in the long run. Now if we are going to follow the first method, then one of the concepts we could use would be Nash's equilibrium point [44]. Recall that this is a strategy n-tuple, or

[11] This question is not connected with that investigated in §6, where the characteristic function was given and there was no question of mixed strategies and of the difference between v_α and v_β.

"point," such that no individual can gain by deviating from it, while the others retain the same strategies they were previously using. Since we are discussing cooperative games, it would be more appropriate to consider a point such that no *coalition* can gain by deviating from it while the others retain the same strategies. Let us call such a point a *strong* equilibrium point.

Let us now try the other approach. If a coalition is expecting more plays, then the question of whether it can improve its lot for fixed strategies of the other players becomes irrelevant, because the other players will not keep their strategies fixed, but will take counteraction on subsequent plays. The question is: when can a coalition be *sure* of a higher payoff? Obviously some kind of core notion is involved here; the surprising fact is that it is not the core according to the α-notion but rather according to the β-notion.

THEOREM[12] 13. *The β-core of a game coincides with the set of payoff vectors to strong equilibrium points in its supergame.*

This is the "technical advantage" of the β-notion to which we previously referred.

We outline the proof of this theorem. For this purpose we should first define the payoff in the supergame. But the precise definition is complicated and not important at the moment; the general idea is that the payoff to a superplay is some kind of average of the payoffs to the individual plays, and this is all that we shall need.

First suppose that x is in the β-core of the game (because of Theorem 5 we do not have to specify which one of the β-cores). We will build a strong equilibrium point whose payoff is x. Now it is possible to prove[13] from the definition of v_β and with a little fussing that to say $x \in \beta$-core is equivalent to saying that *each coalition can prevent its complement from getting more than it (the complement) does at x.*

This being the case, let us construct a strong equilibrium point as follows: First find a correlated strategy n-tuple whose payoff is x; call this c^N. Next, for each coalition S, *let c^S be a correlated strategy for S that prevents the complement of S from obtaining more than it does at x.* Now each player adopts the following strategy in the supergame: He starts out by proposing c^N for the first play, and continues to propose this, play after play, as long as the other players agree. If, however, there is a set of players that disagree—let us call them "disloyal" players—then our player will propose c^S, where S is the set of loyal players. Once a player has become disloyal, he will no longer be accepted in the set of loyal players. The result is that if everybody plays along with this equilibrium point, then x is the result, but if some set of players does not, then eventually

[12] First proved in [1]. For a description that is more precise than the present one and more readable than that of [1], see the end of [3]; but no proof is sketched there.

[13] Cf. Lemma 9.1, p. 304 of [1].

each of its members will be found in the disloyal set; from then on all the loyal players S will be playing c^S against the disloyal players, and by the definition of c^S at least one disloyal player i will get no more than x^i. This shows (modulo some glossed-over difficulties) that the β-core is a subset of the strong equilibrium payoffs.

Conversely, suppose that x is *not* in the β-core; then it is β-dominated. This means that there is a coalition S and a $y \in E^S$ such that $N - S$ cannot prevent S from getting y, and $y > x^S$. Suppose there were a strong equilibrium point f in the supergame whose payoff is x. Let f^{N-S} denote the part that $N - S$ has in f, i.e., an $(N - S)$-tuple of strategies in the supergame, one for each member of $N - S$. On the first play f^{N-S} dictates a certain set of strategies in $N - S$. For this set of strategies, there exists a strategy for S which yields at least y (this is what is meant by saying that $N - S$ cannot prevent S from getting at least y). On the second play, f^{N-S} again dictates a strategy set for $N - S$, based on the history of the previous play. For this strategy, there again exists a strategy for S that yields at least y; it may be different from the strategy of S on the first play. We can continue in this way; no matter what f^{N-S} dictates, there exists a strategy for S that yields at least y on every play. Since $y > x^S$, this shows that by deviating from f, S can gain, so f cannot be a strong equilibrium point.

This last part of the proof has a curious flavor, because of course S cannot know which supergame strategy $N - S$ is using. However, it does definitely prove that f cannot be in equilibrium, which really involves nobody wanting to deviate even if he knows what the others are playing. Theorem 13 is related to Blackwell's work on games with vector payoffs [6].

The Zermelo-von Neumann-Kuhn theorem about pure-strategy equilibrium points in games of perfect information has the following analogue for supergames of cooperative games with side payments.

THEOREM 14. *If the supergame of a game of perfect information has any strong equilibrium points at all, then it also has strong equilibrium points which involve only pure strategies.*[14]

In §6, we discussed the close connection between no-side-payment characteristic functions and Nash's bargaining problem [21], and pointed out that Shapley's valuation generalizes Nash's solution to the bargaining problem. Nash followed up the work in [21] by a paper on 2-person games in normal form [22], for which he proposed a "value" taking threat possibilities into account. This work was generalized by Raiffa in [26], but he too treated only 2-person games. The problem of defining a value for n-person games (both with and without side payments) that will take threats into account has been treated by several authors (cf. §6). Isbell [16] has constructed a theory of games without side payments that parallels the N-M solution theory but takes threat possibilities into account.

[14] For the precise statement and proof see [2].

9. HISTORICAL REMARKS

Shapley and Shubik [29] were the first to suggest that N-M solutions could be defined even in the absence of a transferable utility. Their definition of dominance is very similar to ours; but rather than explicitly using a characteristic function, it depends directly on the α-notion. Shapley and Shubik imply that they must have (nontransferable utility) side payments to make their definition work, but actually it is perfectly general. They proved no theorems, confining themselves to general definitions.

Luce and Raiffa [19, p. 234] also gave a definition of dominance and solution for cooperative games without side payments. Their definition has some restrictive and complicating features, which in the light of later work turn out to have been unnecessary. They, too, proved no theorems, mentioning that "next to nothing is known about these definitions."

Functions that are very similar in form to the characteristic functions of §3 were first described by Isbell [16, 17]. He called them *end-games*, and used them to characterize, for each given payoff vector, the redistributions of utility that are made possible by means of nontransferable utility side payments. This use is related to Jentzsch's catalogues rather than to the development of §3; the latter is due to Aumann and Peleg [5]. The form of v in this survey differs slightly from that in [5]; the $v(S)$ of [5] would be $v(S) \times E^{N-S}$ in the notation of this paper.

Other historical references may be found in the body of the paper.

10. A Bibliography of Cooperative Games Without Side Payments

[1] R. J. Aumann, "Acceptable points in general cooperative n-person games," in [43], pp. 287–324.

[2] ——, "Acceptable points in games of perfect information," *Pac. J. Math.* 10 (1960), pp. 381–417.

[3] ——, "The core of a cooperative game without side payments," *Trans. Amer. Math. Soc.* 98 (1961), pp. 539–552.

[4] ——, "Markets with a continuum of traders," *Econometrica* 32 (1964), pp. 39–50.

[5] R. J. Aumann and B. Peleg, "Von Neumann-Morgenstern solutions to cooperative games without side payments," *Bull. Amer. Math. Soc.* 66 (1960), pp. 173–179.

[6] D. Blackwell, "An analog of the minimax theorem for vector payoffs," *Pac. J. Math.* 6 (1956), pp. 1–8.

[7] E. Burger, "Bemerkungen zum Aumannschen Core-Theorem," *Zeitschrift für Wahrscheinlichkeitstheorie* 3 (1964), pp. 148–153.

[8] G. Debreu and H. Scarf, "A limit theorem on the core of an economy," *Int. Econ. Rev.* 4 (1963), pp. 235–246.

[9] J. C. Harsanyi, "A bargaining model for the cooperative n-person game," in [43], pp. 325–356.

[10] ——, "Measurement of social power in *n*-person reciprocal power situations," *Behavioral Science* 7 (1962), pp. 81–91.
[11] ——, "A simplified bargaining model for the *n*-person cooperative game," *Int. Econ. Rev.* 4 (1963), pp. 194–220.
[12] ——, "Approaches to the bargaining problem before and after the theory of games: a critical discussion of Zeuthen's, Hicks' and Nash's theories," *Econometrica* 24 (1956), pp. 144–157.
[13] ——, "On the rationality postulates underlying the theory of cooperative games," *J. Conflict Resolution* 5 (1961), pp. 179–196.
[14] ——, "Bargaining in ignorance of the opponent's utility function," *J. Conflict Resolution* 6 (1962), pp. 29–38.
[15] ——, "A bargaining model for social status in informal groups and formal organizations," to appear in *Behav. Sci.*
[16] J. R. Isbell, "Absolute games," in [43], p. 357–396.
[17] ——, "A modification of Harsanyi's bargaining model," *Bull. Amer. Math. Soc.* 66 (1960), pp. 70–73.
[18] G. Jentzsch, "Some thoughts on the theory of cooperative games," in [36], pp. 407–442.
[19] R. D. Luce and H. Raiffa, *Games and Decisions*, New York, John Wiley, 1957.
[20] K. Miyasawa, "The *n*-person bargaining game," in [36], pp. 547–575.
[21] J. F. Nash, "The bargaining problem," *Econometrica* 18 (1950), pp. 155–162.
[22] ——, "Two-person cooperative games," *Econometrica* 21 (1953), pp. 128–140.
[23] B. Peleg, "Solutions to cooperative games without side payments," *Trans. Amer. Math. Soc.* 106 (1963), pp. 280–292.
[24] ——, "Bargaining sets of cooperative games without side payments," *Israel J. Math.* 1 (1963), pp. 197–200.
[25] ——, "Utility functions of money for clear games," *Nav. Res. Log. Quart.* 12 (1965), pp. 57–63.
[26] H. Raiffa, "Arbitration schemes for generalized two-person games," in [42], pp. 361–387.
[27] H. Scarf, "The core of an *n*-person game," Cowles Foundation Discussion Paper No. 182—Revised, September 7, 1965.
[28] L. S. Shapley, "Values of large market games: Status of the problem," The RAND Corporation RM3957PR, February 1964.
[29] L. S. Shapley and M. Shubik, "Solutions of *n*-person games with ordinal utilities" (abstract), *Econometrica* 21 (1953), p. 348.
[30] R. Stearns, "Three-person cooperative games without side payments," in [36], pp. 377–406.
[31] ——, "On the axioms for a cooperative game without side payments," *Proc. Amer. Math. Soc.* 15 (1964), pp. 82–86.
[32] K. Vind, "Edgeworth-allocations in an exchange economy with many traders," *Int. Econ. Rev.* 5 (1964), pp. 165–177.

Other References

[33] R. J. Aumann, "Linearity of unrestrictedly transferable utilities," *Nav. Res. Log. Quart.* 7 (1960), pp. 281–284.
[34] ——, "Existence of competitive equilibria in markets with a continuum of traders," *Econometrica* 34 (1966), pp. 1–17.

[35] O. Bondareva, "The core of an n-person game," *Vestnik Leningrad Univ.* 17 (1962), pp. 141–142.

[36] M. Dresher, L. S. Shapley, and A. W. Tucker, editors, *Advances in Game Theory*, Ann. of Math. Study 52, Princeton, New Jersey, Princeton University Press, 1964.

[37] M. Davis and M. Maschler, "Existence of stable payoff configurations for cooperative games", this volume.

[38] S. Eilenberg and D. Montgomery, "Fixed point theorems for multi-valued transformations," *Amer. J. Math.* 68 (1946), pp. 214–222.

[39] D. B. Gillies, "Solutions to general non-zero-sum games," in [43], pp. 47–85.

[40] F. Harary and M. Richardson, "A matrix algorithm for solutions and r-bases of a finite irreflexive relation," *Nav. Res. Log. Quart.* 6 (1959), pp. 307–314.

[41] G. K. Kalisch and E. D. Nering, "Countably infinitely many person games," in [43], pp. 43–45.

[42] H. W. Kuhn and A. W. Tucker, editors, *Contributions to the Theory of Games II*, Ann. of Math. Study 28, Princeton, New Jersey, Princeton University Press, 1953.

[43] R. D. Luce and A. W. Tucker, editors, *Contributions to the Theory of Games IV*, Ann. of Math. Study 40, Princeton, New Jersey, Princeton University Press, 1959.

[44] J. F. Nash, "Non-cooperative games," *Ann. of Math.* 54 (1951), pp. 286–295.

[45] J. von Neumann and O. Morgenstern, *Theory of Games and Economic Behavior*, Princeton, New Jersey, Princeton University Press, 1944, Third Edition rev., 1953.

[46] B. Peleg, "Existence theorem for the bargaining set $M_1^{(i)}$," this volume.

[47] M. Richardson, "Solutions of irreflexive relations," *Ann. of Math.* 58 (1953), pp. 573–590.

[48] R. Selten, "Valuation of n-person games," in [36], pp. 577–626.

[49] L. S. Shapley, "A value for n-person games," in [42], pp. 307–317.

[50] ——, "Notes on the n-person game III: Some variants of the von-Neumann Morgenstern definition of solution," The RAND Corporation, RM817, April 1952.

[51] ——, "On Balanced Sets and Cores," The RAND Corporation, RM-4601-PR, June, 1965.

CHAPTER 2

On Games of Fair Division

By HAROLD W. KUHN*

The subject of this note is unusual in two respects. On the one hand, the question of which political and social systems ensure the equitable distribution of economic goods underlies a large segment of welfare theory and hence is of the utmost importance. On the other hand, the theoretical literature on such systems is almost non-existent, consisting in the main of isolated paradigms for special and simple situations, often unaccompanied by any rigorous proof. We shall not remedy this situation but rather propose to reopen some old questions, giving them new formulations, and adding a minor bit of precision here and there.

Any rigorous discussion of the problem must begin with some definitions. Those presented below may seem overly elaborate for the immediate application given to them; however, they have been designed to provide a framework for a quite general development of the subject.

DEFINITION 1. A *fair division problem* is defined by $(S, N, \mathscr{P}, \mathscr{F})$ where S is a set to be divided among the players $N = \{1, \ldots, n\}$ and $\mathscr{P}$ is a family of *partitions* $P = \{S_1, \ldots, S_n\}$ into n-subsets. The family $\mathscr{F}$ consists of classes $\mathscr{F}_i$ of subsets T of S for $i \in N$. Player i considers any set $T \in \mathscr{F}_i$ to be *acceptable* or *fair* for his share.

DEFINITION 2. A *legal division* is an assignment of the sets S_j of a partition $P \in \mathscr{P}$ to the players $i \in N$. If S_{j_i} is assigned to player i and $S_{j_i} \in \mathscr{F}_i$ for $i = 1, \ldots, n$, then the legal division is called a *fair division*. Such an assignment may be denoted by $X = (x_{ij})$ where $x_{ij} = 1$ if S_j is given to player i and $x_{ij} = 0$ otherwise.

DEFINITION 3. A *fair division scheme* is the extensive form of a game in which all outcomes are legal divisions and in which each player can assure himself a fair share through the use of an appropriate strategy.

All discussion of "fair-division" begins with the classical method for dividing a finely divisible object (such as a cake) into shares for two individuals. In terms of the definitions given above, S is the cake, $N = \{1, 2\}$, $\mathscr{P} = \{\{S_1, S_2\}\}$ is the set of physically realizable partitions of the cake into two parts, and each player has a rule for deciding which of the physically realizable pieces T are acceptable or fair for him. The extensive form of a game to resolve this problem is as follows:

Move 1. The players are assigned the roles of Divider and Chooser through the toss of a fair coin.

* Princeton University.
This research was supported by the National Science Foundation.

Move 2. The Divider selects $P = \{S_1, S_2\} \in \mathscr{P}$.

Move 3. The Chooser selects $X = (x_{ij})$ where $i, j = 0$ or 1 and $\sum_i x_{ij} = \sum_j x_{ij} = 1$ for $i, j = 1, 2$.

Payoff: Player i receives share S_j if and only if $x_{ij} = 1$.

In order to insure that this scheme is fair, we need only assume that:

(1) no matter how the cake is divided into two pieces, each player will find at least one of the pieces to be acceptable;

(2) either player is able to divide the cake so that both pieces are acceptable to him.

(Note that we have introduced an additional feature of an initial chance move to insure that the outcome does not depend on the labeling of the players; this is not required to verify that the division scheme is fair according to Definition 3.)

Steinhaus [8] has proposed a generalization of this scheme for three players. We shall present his method as the extensive form of a game.

Move 1. The players are assigned the roles of one Divider and two Choosers by means of a chance device yielding equal probabilities to the three possible assignments.

Move 2. The Divider selects $P = \{S_1, S_2, S_3\} \in \mathscr{P}$.

Move 3. Each Chooser i announces which of the shares S_j are acceptable to him, i.e., which $S_j \in \mathscr{F}_i$.

Move 4. If a share can be assigned to each Chooser that is acceptable to him, then this is done and the remaining share is given to the Divider. Otherwise, some share is unacceptable to both Choosers. This share is assigned to the Divider and the Choosers divide the remaining shares according to the two-player fair division scheme.

In order to insure that this scheme is fair, we need only assume that:

(1) no matter how the cake is divided into three pieces, each player will find at least one of the pieces acceptable to him;

(2) each of the players is able to divide the cake so that each of the three pieces is acceptable to him;

(3) if a piece found unacceptable by both of the Choosers is assigned to the Divider then the remaining parts are considered by the Choosers as a fair amount to divide among themselves (i.e., (1) and (2) hold for the resulting two-player fair division scheme).

Steinhaus [3] reports a "divide-and-choose" scheme for n players due to B. Knaster and S. Banach. Their solution is:

> "The partners being ranged 1, 2, 3, . . . , n, 1 cuts from the cake an arbitrary part. 2 has now the right, but is not obliged, to diminish the

slice cut off. Whatever he does, 3 has the right (without obligation) to diminish still the already diminished (or not diminished) slice, and so on up to n. The rule obliges the "last diminisher" to take as his part the slice he was the last to touch. This partner thus disposed of, the remaining $n - 1$ persons start the same game with the remainder of the cake. After the number of participants has been reduced to two, they apply the classical rule [one divides while the other chooses] for halving the remainder."

This scheme, which seems to be equivalent to the method in which an Umpire moves a knife across the cake until some player calls "stop" and receives the resulting piece, is very ingenious. However, it is not a direct generalization of the schemes discussed above. Our first result is a scheme which extends the two- and three-player schemes in a direct manner. We begin with some preparatory definitions.

Let $T \subset S$ and let $M \subset N$ contain m players. The restriction of $\mathscr{P}$ to T and M will be denoted by $\mathscr{P}(T, M)$ and consists of all partitions $\{S_1, \ldots, S_m\}$ of T such that there exists $\{S_1, \ldots, S_m, S_{m+1}, \ldots, S_n\} \in \mathscr{P}$. (Note that $\mathscr{P}(S, N) = \mathscr{P}$.) Suppose $P = \{S_1, \ldots, S_m, S_{m+1}, \ldots, S_n\} \in \mathscr{P}$ is such that $\bigcup_{j=1}^{m} S_j = T$ and $S_j \notin \mathscr{F}_i$ for $i \in M = \{i_1, \ldots, i_m\}$ and $j = m + 1, \ldots, n$. Then $\mathscr{P}(T, M)$ is called a *fair restriction* of $\mathscr{P}$ to T and M. (Note that $\mathscr{P}(S, N)$ is a fair restriction.)

We shall assume, for all fair restrictions $\mathscr{P}(T, M)$ of $\mathscr{P}$ to T and M, that:

(1) for all $i \in M$ and all $P \in \mathscr{P}(T, M)$, $P \cap \mathscr{F}_i \neq \phi$;
(2) for all $i \in M$, there exists a $P_i \in \mathscr{P}(T, M)$ such that $P_i \subset \mathscr{F}_i$.

To see the relation of these assumptions to those given previously, note that a "fair restriction" is a reduction of the division problem to a (possibly smaller) set of m players by removing $n - m$ parts of S, none of which is acceptable to the m players remaining. Assumptions (1) and (2) say that, in a (possibly fairly restricted) problem with m players, if T is divided in any legal manner into m parts, then each player will find at least one of the parts acceptable and, furthermore, each player is able to divide T legally into m parts, all of which are acceptable to him.

Before we can state the generalization of the two- and three-player schemes to n players we need a combinatorial lemma.

Definition. Given an n by n matrix $A = (a_{ij})$ with entries 0 or 1, an *assignment* is a set $\{(i_1, j_1), \ldots, (i_r, j_r)\}$, where $I = \{i_1, \ldots, i_r\}$ and $J = \{j_1, \ldots, j_r\}$ each contain $r \geq 1$ distinct indices, and $a_{ij} = 1$ for $i \in I$ and $j \in J$. An assignment is *complete* if $r = n$.

Lemma. *Let $A = (a_{ij})$ be an n by n matrix with entries 0 or 1 such that*

$a_{1j} = 1$ for $j = 1, \ldots, n$ and $\sum_j a_{ij} \geq 1$ for $i = 2, \ldots, n$. Then there exists an assignment such that $a_{ij} = 0$ for $i \notin I$ and $j \in J$.

PROOF. Either there is a complete assignment or, by the theorem of Frobenius-König, there exists an s by t rectangle of zeros with $s + t = n + 1$. Choose s to be a maximum. If $s = n - 1$ then

$$A = \left[\begin{array}{c|ccc} 1 & 1 & \ldots & 1 \\ \hline 0 & a_{22} & \ldots & a_{2n} \\ \cdot & \cdot & & \cdot \\ \cdot & \cdot & & \cdot \\ \cdot & \cdot & & \cdot \\ 0 & a_{n2} & \ldots & a_{nn} \end{array}\right]$$

by reordering the rows and columns and $\{(1, 1)\}$ is an assignment. Otherwise, we may reorder the rows and columns of A so that:

$$\begin{array}{r|c|c|} & \overbrace{\qquad\qquad}^{t} & \overbrace{\qquad\qquad}^{s-1} \\ \hline 1\{ & 1 \ \ldots \ 1 & 1 \ \ldots \ 1 \\ \hline s\left\{\begin{array}{c} \\ \\ \end{array}\right. & 0 & \\ \hline t-2\left\{\begin{array}{c} \\ \end{array}\right. & M & \\ \hline \end{array}$$

(Note that M has at least one row and three columns since $s + t = n + 1$ and $s \leq n - 2$ imply $t - 2 \geq 1$.) Since s is a maximum, there is no k by l rectangle of zeros in M with $(s + k) + l = n + 1$. That is, there is no k by l rectangle of zeros in M with $k + l = (n + 1) - s = t$. Now consider the $(t - 1)$ by t matrix

$$\begin{array}{|c|} \hline 1 \ \ldots \ 1 \\ \hline \\ M \\ \\ \hline \end{array}$$

and delete the first column. In this $(t-1)$ by $(t-1)$ matrix, there is no k by l rectangle of zeros with $k + l = (t-1) + 1$. Hence there exists a complete assignment for this matrix which is an assignment for A satisfying the condition of the Lemma that $a_{ij} = 0$ for $i \notin I$ and $j \in J$.

Q.E.D.

FAIR DIVISION SCHEME. Let a Player, say 1, be designated Divider by an equiprobable chance device. Then player 1 selects $P = \{S_1, \ldots, S_n\} \in \mathscr{P}$ such that $S_j \in \mathscr{F}_1$ for all j. (This is possible by (2).) Define $A = (a_{ij})$ by $a_{ij} = 1$ if $S_j \in \mathscr{F}_i$ and $a_{ij} = 0$ otherwise. The matrix A satisfies the hypotheses of the Lemma. (Note that $\sum_j a_{ij} \geq 1$ for $i = 2, \ldots, n$ by (1).) Hence there exists an assignment $\{(i_1, j_1), \ldots, (i_r, j_r)\}$ with $I = (i_1, \ldots, i_r\}$, $J = \{j_1, \ldots, j_r\}$ with $S_j \in \mathscr{F}_i$ for $i \in I$ and $j \in J$. Assign S_j to Player i for $i \in I$ and $j \in J$. Then, if $T = \bigcup_{j \notin J} S_j$ and $M = \{i \mid i \notin I\}$, the restriction of $\mathscr{P}$ to T and M is a fair restriction and the process may be repeated. Since $r \geq 1$, at least one player i receives a part $S_j \in \mathscr{F}_i$ at each iteration, and hence the scheme terminates. Whenever S_j is assigned to Player i we have $S_j \in \mathscr{F}_i$ and hence the scheme is fair.

Knaster has also suggested (cf. Steinhaus [3]) a method of division applicable to situations in which S is indivisible, i.e., assumption (2) does not hold. To illustrate his scheme, suppose that three heirs must share four objects among themselves. They are first asked to reveal the monetary value of each object to an Umpire. Denote the value of object i to heir j by v_{ij}. An example is tabulated below:

		Heir 1	Heir 2	Heir 3
Object	1	\$1,000	5,000	1,000
	2	3,000	1,000	4,000
	3	3,000	2,000	1,000
	4	5,000	4,000	6,000

If we assume that the monetary value of sets of objects is additive, then heir j believes his fair share is $\frac{1}{3} \sum_i v_{ij}$ which is \$4,000 for each j in our example. Knaster's rule assigns each object to an heir who values it most highly and then defines sidepayments (summing to zero) among the heirs so as to insure an equal surplus over the fair share for each. The amount of this surplus is exactly

$$\frac{1}{n}\left(\sum_i \left(\max_j v_{ij}\right) - \sum_j \frac{1}{n} \sum_i v_{ij}\right).$$

Applying this rule to the example:

$$\text{Heir}\begin{cases}1\\2\\3\end{cases}\text{receives object(s)}\begin{cases}3\\1\\2,4\end{cases}\text{valued at}\begin{cases}\$3{,}000\\5{,}000\\10{,}000\end{cases}\text{and a side payment of}\begin{cases}\$3{,}000\\1{,}000\\-4{,}000.\end{cases}$$

Note that since the fair share of each is \$4,000, the outcome gives each heir the same surplus of \$2,000. It is a simple and straightforward matter to verify that this rule provides a non-negative surplus no matter what the valuations are. However, we propose to show how Knaster's rule can be "discovered" through linear programming and simultaneously demonstrate the properties claimed for it.

To establish some notation, let v_{ij} be the *valuation* of object i by heir j, where $i = 1, \ldots, m$ and $j = 1, \ldots, n$. Then $v_j = \frac{1}{n}\sum_i v_{ij}$ is the *fair share* of heir j. As before $X = (x_{ij})$ will be an *assignment* matrix defined by $x_{ij} = 1$ if object i is assigned to heir j and $x_{ij} = 0$ otherwise. Finally, let y_j denote the *sidepayment* to heir j.

PRIMAL PROGRAM. Find (x_{ij}), (y_j), and z so as to maximize z subject to $x_{ij} \geq 0$ and

$$\sum_i v_{ij}x_{ij} + y_j - v_j \geq z \qquad (j = 1, \ldots, n)$$

$$\sum_j x_{ij} = 1 \qquad (i = 1, \ldots, m)$$

$$\sum_j y_j = 0.$$

The interpretation of the program should be clear. We seek to maximize the surplus z over the fair share which can be given to all players, using side payments y_j which sum to zero. A straightforward application of the theory of duality yields:

DUAL PROGRAM. Find (s_j), (r_i) and t so as to minimize $-\sum_j v_js_j + \sum_i r_i$ subject to $s_j \geq 0$ and

$$\sum_j s_j = 1$$

$$-v_{ij}s_j + r_i \geq 0 \qquad (i = 1, \ldots, m \text{ and } j = 1, \ldots, n)$$

$$-s_j + t = 0 \qquad (j = 1, \ldots, n).$$

Since all $s_j = t$, $nt = 1$ and $s_j = t = \frac{1}{n}$ for all j and we may rewrite the dual:

Find (r_i) so as to minimize $\sum_i r_i - \frac{1}{n}\sum_j v_j$ subject to

$$r_i \geq \frac{1}{n} v_{ij} \qquad (i = 1, \ldots, m \text{ and } j = 1, \ldots, n).$$

The solution of this program is immediate. Set

$$n\bar{r}_i = \max_j v_{ij}.$$

Then

$$n^2\bar{r}_i \geq \sum_j v_{ij}$$

$$n^2 \sum_i \bar{r}_i \geq \sum_{i,j} v_{ij} = n \sum_j v_j$$

$$\sum_i \bar{r}_i \geq \frac{1}{n} \sum_j v_j$$

which implies that the optimal value of the dual program is nonnegative. This completes an *existence* proof for the primal, ignoring for the moment the question of whether the x_{ij} take on integral values. However, we may solve the primal explicitly.

For each i, set $\bar{x}_{ij} = 1$ for one j for which $v_{ij} = \max_j v_{ij}$. The primal program then becomes: Find (y_j) and z so as to maximize z subject to $\sum_j v_{ij}\bar{x}_{ij} + y_j - v_j \geq z \; (j = 1, \ldots, n)$

$$\sum_j y_j = 0.$$

Summing the first set of constraints,

$$\sum_{i,j} v_{ij}\bar{x}_{ij} - \sum_j v_j \geq nz.$$

If we set

$$\bar{z} = \frac{1}{n}\left(\sum_i \left(\max_j v_{ij}\right) - \sum_j v_j\right)$$

this gives equality in the sum and hence equality in all of the inequality summands. Note that $\bar{z} \geq 0$, by our result for the dual program. Moreover,

$$\bar{y}_j = v_j - \sum_i v_{ij}\bar{x}_{ij} + \frac{1}{n}\left(\sum_i\left(\max_j v_{ij}\right) - \sum_j v_j\right)$$

clearly sum to zero. The three terms in $\bar{y}_j$ are, respectively, the fair share of j, the value of the objects assigned to j, and the surplus. Thus the Knaster procedure is completely verified.

The feeling of euphoria which may have been created by the two results proved above may be dispelled by the following two examples.

EXAMPLE 1. Suppose S consists of six pieces of cake, one with a cherry and the others plain, to be divided by P_1 and P_2. No piece of cake can be divided further. The acceptable sets are defined by giving additive utility functions. Namely, both P_1 and P_2 value a plain piece of cake at 1 utile, P_1 places a value of -1 utile on the piece with the cherry, and P_2

values the piece with the cherry at 3 utiles. If we denote the resulting utility functions by u_1 and u_2, respectively,

$$\mathscr{F}_1 = \{T \mid u_1(T) \geq 2\}$$
$$\mathscr{F}_2 = \{T \mid u_2(T) \geq 4\}.$$

Clearly, if A divides and B chooses, A will select $S_1 = \{$2 pieces of plain cake$\}$ and $S_2 = \{$3 pieces of plain cake, 1 piece with the cherry$\}$, and the utility outcome will be $u_1 = 2$, $u_2 = 6$. On the other hand, if B divides and A chooses, the obvious division yields the utility outcome $u_1 = 4$, $u_2 = 4$. If *either* divider has knowledge of the chooser's utility function, it is easy to verify that the utility outcome is $u_1 = 3$, $u_2 = 5$. Simple as our example is, it illustrates phenomena which are not covered by our previous theory namely the "chooser's advantage" and the moderating effect of the knowledge of the chooser's preferences.

EXAMPLE 2. Let us modify our example of the three heirs and the four objects by assuming that Heir 2 "cheats" by announcing false values to the Umpire. The following table gives these false values:

		Heir 1	Heir 2	Heir 3
Object	1	$1,000	1,001	1,000
	2	3,000	3,999	4,000
	3	3,000	2,999	1,000
	4	5,000	6,001	6,000

Application of Knaster's rule yields:

Heir	receives object(s)	valued at	and a side payment of
1	3	$3,000	$1,445
2	1, 4	7,002	−1,890
3	2	4,000	445

Note that the true value to Heir 2 is $9,000 − $1,890 = $7,110 compared to the $6,000 which he received originally. The surplus value to the other players has been reduced to $445.

The numbers in this example have been chosen only to exhibit the advantages that can accrue to a player who falsely portrays his own valuations with a knowledge of the other player's true valuations. It points up a clear need for an analysis of the strategic opportunities of this situation.

This note was begun with a disclaimer that no general theory of fair division would be presented. The truth of this should be obvious to the reader who has penetrated this far. However, some of the needs are clear and the field seems open to the diligent researcher. To aid anyone who would heed this call, a bibliography of sources known to the author is appended.

BIBLIOGRAPHY

[1] Knaster, B. and Steinhaus, H., *Ann. de la Soc. Polonaise de Math.*, *19*, 228–31, 1946.
[2] Steinhaus, H., "The Problem of Fair Division," *Econometrica*, *16*, 101–104, 1948.
[3] ——, "Sur la division pragmatique," *Econometrica*, *17* (supplement), 315–319, 1949.
[4] G. Th. Guilbaud, "Les problèmes de partage," *Économie Appliqué*, *I*, 93–137, 1952.
[5] Luce, R. D. and Raiffa, H., *Games and Decisions*, New York, 1957.
[6] Flood, M., "Some Experimental Games," *Management Science*, *5*, 5–26, 1958.
[7] Shubik, M., *Strategy and Market Structure; Competition, Oligopoly, and the Theory of Games*, New York, 1959.
[8] Steinhaus, H., *Mathematical Snapshots*, 2nd edition, New York, 1960.
[9] Dubins, L. E. and Spanier, E. H., "How to Cut a Cake Fairly," *American Math Monthly*, *68*, 1–17, 1961.
[10] Singer, E., "Extension of the Classical Rule of 'Divide and Choose,'" *Southern Economic Journal*, *38*, 1962.

CHAPTER 3

Existence of Stable Payoff Configurations for Cooperative Games

By MORTON DAVIS and MICHAEL MASCHLER*

1. INTRODUCTION

In R. J. Aumann and M. Maschler [1], a theory was developed to attack the following general question: If the players in a cooperative n-person game have decided upon a specific coalition-structure, how then will they distribute among themselves the values of the various coalitions in such a way that some stability requirements will be satisfied. Several criteria for the "stable" splits were given, centering upon the idea that a "stable" payoff should offer the players some security in the sense that each "objection" could be met by a "counter objection." A variety of concepts of objections and counter objections were suggested, and one of them was studied in more detail. This one, and some of the others, had the feature that for some coalition-structures there were no stable payoffs, and therefore these coalition-structures could not be used by those players who wished stability in this sense. (See also [6].) Moreover, cases were established in which even a coalition-structure that yields the maximum total amount to all the players had no stable outcome. In particular, an example was given in [1] of a game with a superadditive, non-negative, non-identically zero characteristic function, in which no outcome was stable unless each player received a zero amount.

It is conceivable that many would reject such an outcome on the ground that "rational" players in a superadditive game would agree only on outcomes which are Pareto optimal.

We do not share this opinion, for we feel that often a desire for security is stronger than a wish to make some extra profit. In fact, many profitable coalitions in everyday life are never realized because the "players" do not consider them safe. Nevertheless, we do believe that in some cases, especially if large profits are at stake, people may be willing to *relax* their safety requirements in order to make more out of a game.

It is therefore of interest to develop a theory in which safety requirements are so relaxed that there always exist stable outcomes in a superadditive game, which are Pareto optimal. We shall prove that this is indeed the case for one of the variants proposed in [1]. Moreover, we conjecture

* Rutgers University and The Hebrew University of Jerusalem.
The research described in this paper was supported partially by the Office of Naval Research and partially by the Carnegie Corporation of New York. This paper was written while both authors were associated with the Econometric Research Program, Princeton University.

that this variant always provides stable outcomes for each choice of a coalition-structure. We are able to prove this conjecture for those coalition-structures in which each coalition does not contain more than three players.

The key theorem, very interesting in itself, states that each outcome induces a partial "order" relation among the players which, although it is asymmetric and acyclic (i.e., never intransitive), is not necessarily transitive. This phenomenon, which, e.g., does not occur in the von Neumann-Morgenstern concept of domination, is "just enough" for proving various existence theorems. The necessary definitions are stated to make the paper self-contained.

We are grateful to Dr. Martin Arkowitz for helpful discussions during our research.

LATER ADDITIONAL REMARKS

Shortly after this paper was written, B. Peleg succeeded in an ingenious way to advance the proof of the conjecture stated above. We now know three proofs of the conjecture, the most elegant of which is presented in Peleg's paper in this volume.

We have decided, however, to retain Sections 5 and 6 of this paper, which provide only a partial success in this direction. The reason is that we hope that eventually it will be proved that Corollaries 5.1 and 5.2 are true also for coalitions of more than 3 players. Such a result will not only provide an essentially different proof of the conjecture but will also reveal very interesting properties of the bargaining set.

2. BASIC DEFINITIONS

We consider an n-person cooperative game Γ, described by a set $N \equiv \{1, 2, \ldots, n\}$ of n *players* and a real function $v(B)$ defined for each non-empty subset B of N. B is called a *coalition* and $v(B)$ is its value. The function $v(B)$ is known as the *characteristic function* of the game.[1] It is not necessarily superadditive.

There will be no essential loss of generality if we assume that[2]

$$v(B) \geq 0 \quad \text{and} \quad v(i) = 0 \quad \text{for each } i, \quad i = 1, 2, \ldots, n. \tag{2.1}$$

An *outcome* of a game Γ is represented by a *payoff configuration* (p.c.), which is an expression of the form

$$(\mathbf{x}; \mathscr{B}) \equiv (x_1, x_2, \ldots, x_n; B_1, B_2, \ldots, B_m), \tag{2.2}$$

[1] The theory allows also for the possibility that some non-1-person coalitions are not permissible. If B is such a coalition, we simply agree that $v(B) = 0$, and modify slightly the permissible outcomes.

[2] By strategic equivalence we can assume $v(i) = 0$. If, then, $v(B) < 0$ for a coalition B, we replace $v(B)$ by zero. To get the bargaining set (see Definition 2.1) of the original game, all one has to do is to remove the payoff configurations of the bargaining set of the modified game in which B appears in the coalition structure.

where $\mathscr{B} \equiv \{B_1, B_2, \ldots, B_m\}$ is the *coalition-structure*, and hence satisfies

$$B_j \cap B_k = \emptyset \text{ for all } j, k, j \neq k, \text{ and } \bigcup_{j=1}^{m} B_j = N, \tag{2.3}$$

and $\mathbf{x} \equiv (x_1, x_2, \ldots, x_n)$ represents the *payoff vector* according to which player i receives in the outcome the amount x_i, $i = 1, 2, \ldots, n$; and it is also assumed that each coalition makes full use of its value, and therefore $\mathbf{x}$ is required to satisfy

$$\sum_{i \in B_j} x_i = v(B_j), \qquad j = 1, 2, \ldots, m, \tag{2.4}$$

If one requires, in addition, that each outcome is *individually rational*, i.e., that

$$x_i \geq 0 \quad \text{for each } i, \quad i = 1, 2, \ldots, n, \tag{2.5}$$

then $(\mathbf{x}; \mathscr{B})$ is called an *individually rational payoff configuration* (i.r.p.c.). Thus, for each fixed coalition-structure $\mathscr{B} \equiv B_1, B_2, \ldots, B_m$, the set of all possible payoff vectors in an i.r.p.c. consists of a cartesian product of m simplices

$$S = S_1 \times S_2 \times \ldots \times S_m, \tag{2.6}$$

where, in view of (2.4) and (2.5),

$$S_j \equiv \left\{\{x_i\}_{i \in B_j} \,\middle|\, \sum_{i \in B_j} x_i = v(B_j),\ x_i \geq 0\right\}, \qquad j = 1, 2, \ldots, m. \tag{2.7}$$

Let $(\mathbf{x}; \mathscr{B})$ be an individually rational payoff configuration (i.r.p.c.), (2.2)-(2.5), in a game Γ, and let k and l be two distinct members of a coalition[3] B_j of $\mathscr{B}$.

For a coalition C and a distribution $\{y_i\}$, $i \in C$, of its value among its members, the pair $(\{y_i\}; C)$ is called an *objection* of k against l in $(\mathbf{x}; \mathscr{B})$, if

$$k \in C, \quad l \notin C, \qquad k, l \in B_j, \tag{2.8}$$

$$\sum_{i \in C} y_i = v(C), \tag{2.9}$$

$$y_k > x_k, y_i \geq x_i \quad \text{for all } i, \quad i \in C. \tag{2.10}$$[4]

Let $(\mathbf{x}; \mathscr{B})$ be an i.r.p.c. (2.2)—(2.5), in a game Γ and let $(\{y_i\}; C)$ be an objection of a player k against a player l in $(\mathbf{x}; \mathscr{B})$, (2.8), (2.9), and (2.10). For a coalition D and a distribution $\{z_i\}$, $i \in D$, of its value among its members, the pair $(\{z_i\}; D)$ is called a *counter objection* to the above

[3] This requires, of course, that B_j contains more than one player.
[4] No loss of generality will be caused if we assume that all the inequalities are strict.

objection, if

$$l \in D, k \notin D, \tag{2.11}$$

$$\sum_{i \in D} z_i = v(D), \tag{2.12}$$

$$z_i \geq x_i \quad \text{for all } i, \quad i \in D, \tag{2.13}$$

$$z_i \geq y_i \quad \text{for all } i, \quad i \in D \cap C. \tag{2.14}$$

DEFINITION 2.1. An i.r.p.c. $(\mathbf{x}; B)$ in a game Γ is called *stable* ($\mathscr{M}_1^{(i)}$ — stable), if for each objection there exists a counter objection.

The set of all the stable p.c.'s is called the *bargaining set*[5] $\mathscr{M}_1^{(i)}$.

It will be of advantage to introduce a "strength" relation among the players, which corresponds to each i.r.p.c.

DEFINITION 2.2. Let $(\mathbf{x}; B)$ be an i.r.p.c. (2.2)—(2.5), for a game Γ. Let k and l be two players in a coalition B_j of $\mathscr{B}$. We say that player k is *stronger* than player l in $(\mathbf{x}; \mathscr{B})$, and we denote this by $k \succ l$, if player k has an objection against player l, which cannot be countered.

We say that a player k is equipollent to player l in $(\mathbf{x}; \mathscr{B})$, and denote this by $k \sim l$, if $k \not\succ l$ and $l \not\succ k$. ($\not\succ$ means "not stronger than").

Obviously, *an i.r.p.c.* $(\mathbf{x}; \mathscr{B})$ *is stable in a game* Γ *if and only if in each coalition of* $\mathscr{B}$, *each player is equipollent to each other player who belongs to the same coalition.*

In the next section we shall study some properties of the relation $\succ$.

3. ACYCLICITY

DEFINITION 3.1. A binary relation $\mathscr{R}$ (not necessarily complete) is called *acyclic* if it is never intransitive, i.e., if

$$A_1 \mathscr{R} A_2, A_2 \mathscr{R} A_3, \ldots, A_{\alpha-1} \mathscr{R} A_\alpha \Rightarrow \sim A_\alpha \mathscr{R} A_1. \tag{3.1}$$

It will be shown subsequently that $\succ$ is such a relation, hence this relation may enter everyday situations in a natural way.

Certainly $\mathscr{R}$ can be imbedded in a partial order relation $\mathscr{R}^*$ by defining $A_1 \mathscr{R}^* A_\alpha$ whenever $A_1 \mathscr{R} A_\alpha$ or a sequence $A_1, A_2, \ldots, A_\alpha$ exists, which satisfies the left-hand side of (3.1). However, it is not always advisable to replace $\mathscr{R}$ by $\mathscr{R}^*$, if one wishes to derive theorems concerning $\mathscr{R}$ itself.

It follows from (3.1) that acyclicity is an asymmetric and an irreflexive relation.

Let $\mathscr{S}$ be a binary relation defined by:

$$A_\nu \mathscr{S} A_\mu \text{ if and only if } \sim A_\nu \mathscr{R} A_\mu \text{ and } \sim A_\mu \mathscr{R} A_\nu, \tag{3.2}$$

[5] This is one of several variants mentioned in R. J. Aumann and M. Maschler [1]. Although formulated differently, it is actually the same as $\mathscr{M}_1$ of [1], with the coalitional rationality requirement being replaced by individual rationality. Tihe definition n [1], however, "sounds" more general. (See [6].)

then $\mathscr{S}$ is a reflexive and symmetric relation (but not necessarily transitive). Certainly, the relation [$\mathscr{R}$ or $\mathscr{S}$] is complete.

Let $(\mathbf{x}; \mathscr{B})$ be an i.r.p.c., (2.2)—(2.5), for a game Γ, and let C be a coalition. Then the expression

$$e(C) \equiv v(C) - \sum_{i \in C} x_i \tag{3.3}$$

will be called the *excess of the coalition* C in $(\mathbf{x}; \mathscr{B})$. Clearly, this excess, if it is positive, is the supremum of the amounts with which a player in C can "manoeuvre," if he claims an objection by forming the coalition C.

LEMMA 3.1. *Let* $(\mathbf{x}; \mathscr{B})$ *be an* i.r.p.c., (2.2)—(2.5), *for a game* Γ, *and let k and l be two distinct players in a coalition B_j of $\mathscr{B}$. Suppose that player k has an objection $(\{y_i\}; C)$ against player l, and that this objection cannot be countered. Under these conditions, any coalition D for which*

$$l \in D, \quad e(D) \geq e(C) \tag{3.4}$$

must contain player k.

PROOF. Certainly, by (2.9) and (2.10), $e(C) > 0$ and therefore $e(D) > 0$. If $k \notin D$, then (2.11) is satisfied. Player l can then counter-object by $(\{z_i\}; D)$, where

$$z_i = \begin{cases} x_i & \text{for} \quad i \in D - C,\ i \neq l, \\ y_i & \text{for} \quad i \in D \cap C \\ v(D) - \sum_{i \in D - \{l\}} z_i & \text{for} \quad i = l. \end{cases} \tag{3.5}$$

Indeed, it remains to show that (2.13) is satisfied for $i = l$. Actually

$$\begin{aligned} z_l - x_l &= v(D) - \sum_{i \in D - \{l\}} z_i - x_l = v(D) - \sum_{i \in D \cap C} y_i - \sum_{i \in D - C} x_i \\ &= v(D) - v(C) + \sum_{i \in C - D} y_i - \sum_{i \in D - C} x_i \\ &\geq v(D) - v(C) + \sum_{i \in C - D} x_i - \sum_{i \in D - C} x_i = e(D) - e(C) \geq 0. \end{aligned}$$

This contradicts the assumption that the objection cannot be countered.

THEOREM 3.1. *Let* $(\mathbf{x}; \mathscr{B})$ *be an* i.r.p.c., (2.2)—(2.5), *for a game* Γ; *then the relation* $\succ$ *in* $(\mathbf{x}; \mathscr{B})$ (see Definition 2.2) *induces an acyclic binary relation* (see Definition 3.1) *among the members of each coalition in* $\mathscr{B}$.

PROOF. Let B_j be a coalition in $\mathscr{B}$, and suppose that the relation $\succ$ is not acyclic among the players in a coalition B_j of $\mathscr{B}$. Without loss of generality we can assume that B_j contains the players $1, 2, \ldots, t$, and that in $(\mathbf{x}; \mathscr{B})$,

$$1 \succ 2, \quad 2 \succ 3, \quad \ldots, \quad t - 1 \succ t, \quad t \succ 1. \tag{3.6}$$

We know, therefore, that an objection $(\{y_i^\nu\}; C^\nu)$, of player ν against player $(\nu + 1) \pmod t$, exists, which cannot be countered, $\nu = 1, 2, \ldots, t$. Let C^{ν_0} be a coalition among the C^ν's, which has the maximum excess (see (3.3)). We shall show that C^{ν_0} contains all the players $1, 2, \ldots, t$, and this will furnish the contradiction, because, by (2.8), C^{ν_0} cannot contain player $(\nu_0 + 1) \pmod t$. We proceed by induction: By (2.8), $\nu_0 \in C^{\nu_0}$. Suppose that a player ν belongs to the coalition C^{ν_0}; then, by Lemma 3.1, replacing k, l, C, D by $(\nu - 1) \pmod t$, ν, $C^{(\nu-1)(\mathrm{mod}\, t)}$, C^{ν_0}, respectively, we find that player $(\nu - 1) \pmod t$ also belongs to C^{ν_0}. This completes the proof.

EXAMPLE 3.1. Let Γ be a 5-person game with the characteristic function $v(123) = 30$, $v(14) = 40$, $v(35) = 20$, $v(245) = 30$, $v(B) = 0$ otherwise; and consider the p.c. (10, 10, 10, 0, 0; 123, 4, 5). In this p.c., $1 \succ 2$, because player 1 can object against player 2 by ((11, 29); 14) and this objection cannot be countered. Similarly, $2 \succ 3$, the objection being ((11, 1, 18); 245). On the other hand $1 \sim 3$. This example shows that *the relation $\succ$ is not necessarily transitive.*

EXAMPLE 3.2. Let Γ be a 5-person game with the characteristic function: $v(123) = 30$, $v(14) = 30$, $v(34) = 20$, $v(25) = 30$, $v(B) = 0$ otherwise. Clearly, $1 \sim 2$, $2 \sim 3$, but $1 \succ 3$ in the p.c. (10, 10, 10, 0, 0; 123, 4, 5). This shows that the relation $\sim$ is not necessarily transitive.

4. MAKING A COALITION STABLE

DEFINITION 4.1. Let $(\mathbf{x}; \mathscr{B})$ be an i.r.p.c., (2.2)—(2.5), for a game Γ, and let B_j be a coalition in $\mathscr{B}$. We shall say that the coalition B_j is *stable* with respect to $(\mathbf{x}; \mathscr{B})$, if each player in B_j is equipollent to each other player in B_j.

Clearly, *an i.r.p.c. $(\mathbf{x}; \mathscr{B})$ is stable if and only if all the coalitions in $\mathscr{B}$ are stable.*

THEOREM 4.1. *Let $(\mathbf{x}; \mathscr{B})$ be an* i.r.p.c., (2.2)—(2.5), *for a game Γ, and let B_j be a fixed coalition in $\mathscr{B}$. It is possible to modify the payoffs to the players in B_j, without changing the other payoffs and the coalition-structure, in such a way that B_j will be stable with respect to the modified* i.r.p.c.

PROOF. There is no loss of generality in assuming that the coalition B_j consists of the players $1, 2, \ldots, t$. We know that all the possible payoffs to the members of B_j constitute the simplex S_j defined by (2.7), (j being fixed). To each point $\mathbf{x}^* = (x_1^*, x_2^*, \ldots, x_t^*)$ in S_j there corresponds an i.r.p.c. $(\hat{\mathbf{x}}; \mathscr{B})$, where

$$\hat{x}_i = \begin{cases} x_i^* & \text{for all } i, \quad i \in B_j \\ x_i & \text{for all } i, \quad i \notin B_j. \end{cases} \tag{4.1}$$

Let $E_\nu \equiv E_\nu(\{x_i\}_{i \notin B_j}; \mathscr{B})$, $\nu = 1, 2, \ldots, t$, be the set of points $\mathbf{x}^*$, $\mathbf{x}^* \in S_j$, for which player ν is stronger than or equipollent to ($\succsim$) all the players i, $i \in B_j$, in the p.c. ($\hat{\mathbf{x}}$; $\mathscr{B}$). The theorem will be proved if we show that

$$M_j \equiv M_j\left(\{x_i\}_{i \notin B_j};\ \mathscr{B}\right) \equiv \bigcap_{\nu=1}^{t} E_\nu \neq \phi. \tag{4.2}$$

In order to show this, note first that the face $x_\nu = 0$ of the simplex S_j is contained in E_ν, $\nu = 1, 2, \ldots, t$. Indeed, if $x_\nu = 0$ in ($\hat{\mathbf{x}}$; $\mathscr{B}$), then, by (2.1), player ν can counter object to each objection raised against him (if such exists) by ($\{0\}$; ν).

We shall now show that

$$\bigcup_{\nu=1}^{t} E_\nu = S_j. \tag{4.3}$$

Indeed, suppose that there exists a point $\mathbf{x}^*$ in S_j which is not in this union, then there exist players $i_1, i_2, \ldots, i_t$ in B_j such that in ($\hat{\mathbf{x}}$; $\mathscr{B}$)

$$1 \prec i_1, 2 \prec i_2, \ldots, t \prec i_t.^6 \tag{4.4}$$

This violates the acyclicity property of the relation $\succ$. (See Theorem 3.1.) Thus, (4.3) holds. In Theorem 4.2, it will be shown that the sets E_i, $i \in B_j$, are closed point sets in S_j. Assuming this and applying the lemma of B. Knaster, C. Kuratowski, and S. Mazurkiewicz [4], usually used to prove in a direct way the Brouwer fixed-point theorem (see also Kuratowski [5]), (4.2) follows immediately. This completes the proof of the theorem.

An important consequence of this lemma is that there always exists a payoff vector $\mathbf{x}$, such that ($\mathbf{x}$; N) is stable. If the characteristic function is superadditive, then $\mathbf{x}$ lies on the Pareto optimum.

We conjecture that to each coalition-structure $\mathscr{B}$, there exists a payoff vector $\mathbf{x}$ such that ($\mathbf{x}$; $\mathscr{B}$) is stable. It seems that in order to prove this, one has to know more properties of M_j. We shall state some of the properties we have in mind in the next section, and verify them in some cases.

THEOREM 4.2. *The set M_j, defined by* (4.2), *is a union of a finite number of closed convex polyhedra.*

PROOF. Let $F_{\mu\nu} \equiv F_{\mu\nu}(x_{t+1}, x_{t+2}, \ldots, x_n; \mathscr{B})$ be the set of points $\mathbf{x}^*$, $\mathbf{x}^* \in S_j$, for which player μ is stronger than or equipollent to player ν, in the p.c. ($\hat{\mathbf{x}}$: $\mathscr{B}$) (see (4.1)). $\mu, \nu \in B_j$. If we prove that $F_{\mu\nu}$ is a union of a finite number of closed convex polyhedra, then so also will M_j be, because

$$M_j = \bigcap_{i=1}^{t} E_i = \bigcap_{i=1}^{t} \bigcap_{\nu=1}^{t} F_{i\nu}. \tag{4.5}$$

By a well-known theorem in logic it follows (see [1], Theorem 2.1) that $F_{\mu\nu}$ is a union of a finite number of convex polyhedra. We shall prove that it is closed by showing that its complement is open. Indeed, if $\mathbf{x}^*$ belongs to the complement of $F_{\mu\nu}$, with respect to S_j, then player ν has an objection

[6] We define $\prec$ in the obvious way.

$(\{y_i\}: C)$ against player μ, which cannot be countered. Without loss of generality, we can assume that $y_i > x_i$ for all i, $i \in C$. (See footnote to (2.10).) Let z_μ be the maximum amount that player μ can assure himself by paying each other member of a coalition D, $\mu \in D$, $\nu \notin D$, the amount x_i if this member is in $D - C$ and y_i if he is in $D \cap C$. $0 \leq z_\mu < x_\mu$, because the objection cannot be countered. Let $\delta = \text{Min}\{x_\mu - z_\mu, y_i - x_i; i \in C\}$, then $\delta > 0$. Any point of S_j which is in a δ/n-neighborhood of $\mathbf{x}^*$ also belongs to the complement of $F_{\mu\nu}$, because at such a point $(\{y_i\}: C)$ is still an objection which cannot be countered. This completes the proof.

COROLLARY 4.1. *The set* $G_{\mu\nu} \equiv G_{\mu\nu}(x_{t+1}, \ldots, x_n; \mathscr{B})$ *of points* $\mathbf{x}^*$ *in* S_j, *for which player* μ *is stronger than player* ν *in* $(\hat{\mathbf{x}}; \mathscr{B})$ *is open in* S_j $\nu, \mu \in B_j$.

5. THE EXISTENCE PROBLEM

We shall now generalize somewhat a theorem due to von Neumann [7]. We shall employ Kakutani's method of proof [3], but we shall make use of the sharper S. Eilenberg and D. Montgomery fixed-point theorem [2]:

LEMMA 5.1. (Von Neumann theorem for $m = 2$.)

Let $S_1, S_2, \ldots, S_m$ *be* m *bounded closed acyclic polyhedra*[7, 8] *in the euclidean spaces* $R^{n_1}, R^{n_2}, \ldots, R^{n_m}$, *respectively. Let us consider their cartesian product* $T \equiv S_1 \times S_2 \times \ldots \times S_m$ *in* $R^{n_1+n_2+\cdots+n_m}$, *and let* $T_i \equiv S_1 \times S_2 \times \ldots \times S_{i-1} \times S_{i+1} \times \ldots \times S_m$ *be the respective cartesian product in* $R^{n_1+n_2+\cdots+n_{i-1}+n_{i+1}+\cdots+n_m}$, $i = 1, 2, \ldots, m$. *Let* $U_1, U_2, \ldots, U_m$ *be* m *closed subsets of* T *such that for each point* $\mathbf{x}^{(i)} \equiv (\mathbf{x}_1, \mathbf{x}_2, \ldots, \mathbf{x}_{i-1}, \mathbf{x}_{i+1}, \ldots, \mathbf{x}_n\}$ *in* T_i, *the set* $V^{(i)}(\mathbf{x}^{(i)})$ *of all the points* $\mathbf{x}_i$, $\mathbf{x}_i \in S_i$, *such that* $\{\mathbf{x}_1, \mathbf{x}_2, \ldots, \mathbf{x}_m\} \in U_i$ *is a non-empty closed acyclic polyhedron,* $i = 1, 2, \ldots, m$. *Under these assumptions, the sets* $U_1, U_2, \ldots, U_m$ *have a non-empty intersection.*

PROOF. We define a point-to-set mapping $\mathbf{x} \to \phi(\mathbf{x})$, of T into itself, as follows:

$$\phi(\mathbf{x}) \equiv \phi(\mathbf{x}_1, \mathbf{x}_2, \ldots, \mathbf{x}_m) = V^{(1)}(\mathbf{x}^{(1)}) \times V^{(2)}(\mathbf{x}^{(2)}) \times \ldots \times V^{(m)}(\mathbf{x}^{(m)}). \tag{5.1}$$

This mapping is upper-semi-continuous because the sets $U_1, U_2, \ldots, U_m$ are closed. The image of each point is a cartesian product of acyclic closed polyhedra; hence it is an acylic polyhedron, and so is T itself. Therefore, by the Eilenberg and Montgomery fixed-point theorem [2], there exists a point $\xi \equiv \{\xi_1, \xi_2, \ldots, \xi_m\}$ in T, such that $\xi \in \phi(\xi)$. In other words, the components $\xi_1, \xi_2, \ldots, \xi_m$ satisfy $\xi_i \in S_i$, $\xi_i \in V^{(i)}(\xi^{(i)})$, $i = 1, 2, \ldots, m$; therefore, $\xi \in U_1 \cap U_2 \cap \ldots \cap U_m$. This completes the proof.

[7] I.e., connected polyhedra whose homology groups of order ≥ 1 vanish. This should not be confused with acyclicity defined in §3 for binary relations.

[8] This lemma can further be applied for absolute neighborhood retracts.

THEOREM 5.1. *Let* $\mathscr{B} \equiv \{B_1, B_2, \ldots, B_m\}$ *be a fixed coalition-structure* (2.3) *for a game* Γ, *and let* $(\mathbf{x}; \mathscr{B})$ *be an arbitrary* i.r.p.c., (2.2)—(2.5). *Let* $M_j \equiv M_j(\{x_i\}_{i \notin B_j}; \mathscr{B})$ *be the set of points* $\mathbf{x}^*$, $\mathbf{x}^* \in S_j$, *defined by* (2.7), *for which* B_j *is stable with respect to* $(\hat{\mathbf{x}}; \mathscr{B})$, *defined by* (4.1). *If, for each choice of* $j, j = 1, 2, \ldots, m$, *and for each choice of* $(\mathbf{x}; \mathscr{B})$, *the set* M_j *is acyclic,*[9] *then there exists a stable* p.c. $(\xi_1, \xi_2, \ldots, \xi_n; \mathscr{B})$ *having* $\mathscr{B}$ *as a coalition-structure.*

PROOF. Let U_t, $j = 1, 2, \ldots, m$ be the set of points $\mathbf{x}$ in $S = S_1 \times S_2 \times \ldots \times S_m$ for which B_j is stable. Clearly, the sets $V^{(j)}$ defined in Lemma 5.1 are now the sets M_j, $j = 1, 2, \ldots, m$. Thus, the sets U_j, $j = 1, 2, \ldots, m$, have a non-empty intersection. This intersection is precisely the set of points $\mathbf{x}$ in S such that $(\mathbf{x}; \mathscr{B})$ is stable.

In some cases we are able to show that M_j is indeed acyclic. The following lemma will be of much use.

LEMMA 5.2. *Let* $\mathscr{B} \equiv \{B_1, B_2, \ldots, B_m\}$ *be a fixed coalition-structure for a game* Γ, *and suppose that* B_1 *consists of the players* $1, 2, \ldots, t$. *Let* $(\mathbf{x}\colon \mathscr{B}) \equiv (x_1, x_2, \ldots, x_t, \; x_{t+1}, \ldots, x_n; B_1, B_2, \ldots, B_m)$ *and* $(\xi; \mathscr{B}) \equiv (\xi_1, \xi_2, \ldots, \xi_t, x_{t+1}, \ldots, x_n; B_1, B_2, \ldots, B_m)$ *be two* i.r.p.c.'s. *Denote by* P *the set of players* i, *different from player* 2, *for which* $\xi_i > x_i$. *If*

$$\xi_1 \leq x_1, \quad \xi_2 \geq x_2, \tag{5.2}$$

$$x_1 - \xi_1 \geq \sum_{i \in P} (\xi_i - x_i), \tag{5.3}$$

$$1 \succ 2 \quad \text{in the p.c.} \quad (\mathbf{x}; \mathscr{B}), \tag{5.4}$$

then

$$1 \succ 2 \quad \text{also in the p.c.} \quad (\xi; \mathscr{B}). \tag{5.5}$$

PROOF. Since $\xi_1 + \xi_2 + \ldots + \xi_t = x_1 + x_2 + \ldots + x_t = v(B_1)$ (see (2.4)), (5.3) is equivalent to

$$\xi_2 - x_2 \geq \sum_{i \in Q} (x_i - \xi_i), \tag{5.6}$$

where Q is the set of players i, different from player 1, for which $x_i > \xi_i$. Intuitively, (5.3) and (5.6) will make it "easier" for player 1 to object against player 2, and "more difficult" to counter object.

Let $(\{y_i\}; C)$ be an objection of player 1 against player 2 in $(\mathbf{x}; \mathscr{B})$, which cannot be countered. Thus, (2.8), (2.9), and (2.10) hold for $k = 1$, $l = 2$. We shall form an objection $(\{\eta_i\}; C)$ of player 1 against player 2 in $(\xi; \mathscr{B})$ as follows:

$$\eta_i = \begin{cases} \text{Max}\,(y_i, \xi_i) & \text{for all } i, \quad i \neq 1, i \in C \\ v(C) - \sum_{i \in C - \{1\}} \eta_i & \text{for} \quad i = 1. \end{cases} \tag{5.7}$$

[9] By Theorem 4.2, we know that it is a closed polyhedron.

Clearly, (2.8) and (2.9) are satisfied for $k = 1$, $l = 2$, and so is (2.10) for $i \neq 1$. Checking the case $i = 1$, we find, by (5.7), that

$$\eta_1 = v(C) - \sum_{i\in C-\{1\}} \eta_i = v(C) - \sum_{i\in E} y_i - \sum_{i\in F} \xi_i,$$

where E (F) is the set of players i, $i \neq 1$, $i \in C$, for which $y_i \geq \xi_i$ ($\xi_i > y_i$). Certainly $F \subset P$, hence, by (2.9), (2.10), and (5.3),

$$\eta_1 - \xi_1 = y_1 + \sum_{i\in F} y_i - \sum_{i\in F} \xi_i - \xi_1 > x_1 - \xi_1 - \sum_{i\in F} (\xi_i - x_i) \geq 0.$$

This objection cannot be countered. Indeed, if $(\{\zeta_i\}; D)$ is a counter objection, then (2.11)—(2.14) are satisfied for $k = 1$, $l = 2$, x_i, y_i, z_i being respectively replaced by ξ_i, η_i, ζ_i, $i \in D$. Consider the payoff $\{z_i\}$, $i \in D$, defined by

$$z_i = \begin{cases} x_i & \text{for all } i, \quad i \neq 2,\ i \in D - C \\ y_i & \text{for all } i, \quad i \in D \cap C \\ v(D) - \sum_{i\in D-\{2\}} z_i, & \text{for } i = 2. \end{cases} \tag{5.8}$$

We shall arrive at a contradiction by showing that $(\{z_i\}; D)$ is a counter objection to the objection $(\{y_i\}; C)$ in $(\mathbf{x}; \mathscr{B})$. Indeed, (2.11), (2.12), and (2.14) are satisfied and so is (2.13) for $i \neq 2$. Checking for $i = 2$, we find that, by (5.8), (2.11)—(2.14) applied to $(\{\zeta_i\}; D)$, and by (5.7) and (5.6),

$$\begin{aligned} z_2 - x_2 &= v(D) - \sum_{i\in D-C} x_i - \sum_{i\in D\cap C} y_i = \sum_{i\in D} \zeta_i - \sum_{i\in D-C} x_i - \sum_{i\in D\cap C} y_i \\ &\geq \sum_{i\in D-C} \xi_i + \sum_{i\in D\cap C} \eta_i - \sum_{i\in D-C} x_i - \sum_{i\in D\cap C} y_i \\ &= \sum_{i\in D-C} (\xi_i - x_i) + \sum_{i\in D\cap C\cap F} (\xi_i - y_i) \\ &\geq \sum_{i\in D-C} (\xi_i - x_i) \geq (\xi_2 - x_2) - \sum_{i\in Q} (x_i - \xi_i) \geq 0. \end{aligned}$$

This completes the proof.

COROLLARY 5.1. *Let* $(\mathbf{x}; \mathscr{B})$ *be an arbitrary* i.r.p.c. *for a game* Γ, *and let* B_j *be a coalition in* $\mathscr{B}$ *which contains* 2 *players. Then the set* $M_j = M_j(\{x_i\}_{i\notin B_j}; \mathscr{B})$ *of the points* $\mathbf{x}^*$, $\mathbf{x}^* \in S_j$, *which make* B_j *stable in* $(\hat{\mathbf{x}}; \mathscr{B})$, *defined by* (4.1), *is a closed interval.*[10]

PROOF. We may assume that B_j consists of the players 1, 2. Let **ab** be the simplex S_j, where $x_1 = 0$ at **a** and $x_2 = 0$ at **b**. If $\mathbf{c}^* \equiv (c_1^*, c_2^*)$ is a point in **ab** having the property that $1 \succ 2$ in $(\hat{\mathbf{c}}; \mathscr{B})$, then, by Lemma 5.2, all the points $\mathbf{x}^* \equiv (x_1^*, x_2^*)$ of the closed interval **ac** have the same property. Thus, the set of points $\mathbf{x}^*$ with this property is either empty or

[10] Possibly a point. See (2.7), (4.2), and Definition 4.1.

consists of an interval with **a** being one of its end points. By Corollary 4.1, this interval is open with respect to **ab**. Similarly, the set of points $\mathbf{x}^*$, having the property that $2 \succ 1$ in $(\hat{\mathbf{x}}; \mathscr{B})$ is either empty or consists of an open interval with **b** as an end point. (See Figure 1.) Since the relation $\succ$ is asymmetric, this implies that M_j is a non-empty closed interval; therefore it is acyclic.

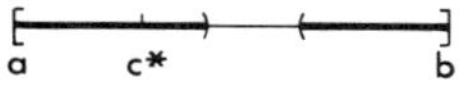

FIGURE 1

If B_j contains more than 2 players, M_j is not necessarily a convex set.

EXAMPLE 5.1. Let Γ be a 5-person game with the characteristic function $v(123) = 10$, $v(15) = 100$, $v(24) = 100$, $v(34) = 98$, $v(B) = 0$ otherwise. It is easy to verify that the coalition 123 is stable both in (10, 0, 0, 0, 0; 123, 4, 5) and in (0, 6, 4, 0, 0; 123, 4, 5), but not in (5, 3, 2, 0, 0; 123, 4, 5), where $2 \succ 3$. Thus, (10, 0, 0) and (0, 6, 4) belong to $M_1 \equiv M_1(0, 0; 123, 4, 5)$ but (5, 3, 2) does not. (See Figure 2, where the points of M_1 are marked.) Here $\{B_1, B_2, B_3\} = \{123, 4, 5\}$.

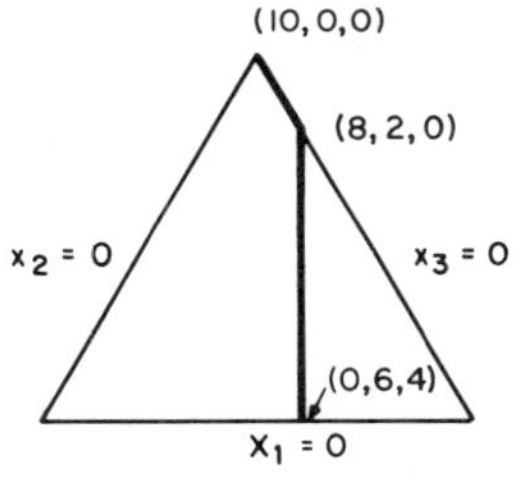

FIGURE 2

COROLLARY 5.2. *Let* $(\mathbf{x}; \mathscr{B})$ *be an arbitrary* i.r.p.c. *for a game* Γ, *and let* B_j *be a coalition in* $\mathscr{B}$ *which contains* 3 *players. Then the set* $M_j \equiv M_j(\{x_i\}_{i \notin B_j}; \mathscr{B})$ *of the points* $\mathbf{x}^*$, $\mathbf{x}^* \in S_j$, *which make* B_j *stable in* $(\hat{\mathbf{x}}; \mathscr{B})$, *defined by* (4.1), *is an acyclic closed polygon.*

PROOF. We know by Theorem 4.2 that M_j is a closed polygon.

(i) Let $B_j = (1, 2, 3)$. Let **abc** be the simplex S_j, where $x_1 = 0$ on the face **bc**, $x_2 = 0$ on **ac** and $x_3 = 0$ on **ab**. If $\mathbf{d}^* = (d_1^*, d_2^*, d_3^*)$ is a point in S_j having the property that $1 \succ 2$ in $(\hat{\mathbf{d}}; \mathscr{B})$, draw parallels through $\mathbf{d}^*$ to the faces **ac** and **bc**. By Lemma 5.2, all the points $\mathbf{x}^* = (x_1^*, x_2^*, x_3^*)$ in the shaded region[11] of Figure 3 have the same property. (Actually, by Corollary 4.1, there exists a neighborhood of this region whose points have the same property.)

[11] Characterized by $x_1^* \leq d_1^*$, $x_2^* \geq d_2^*$.

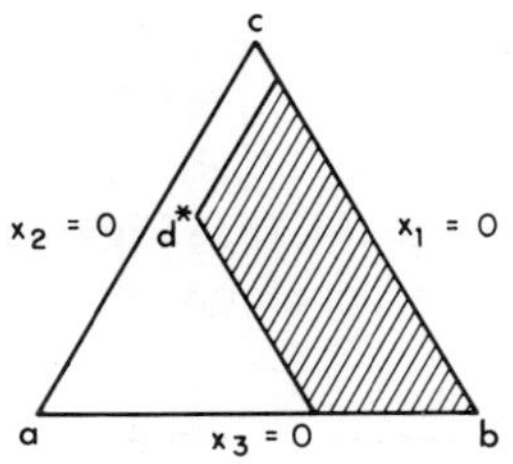

FIGURE 3

(ii) We shall first show that M_j is always a connected set. Indeed, if this is not the case, let $\mathbf{e}^*$ and $\mathbf{f}^*$ be the two nearest points in two nearest distinct components of M_j. By definition, $1 \sim 2$, $1 \sim 3$, $2 \sim 3$ hold both in $(\hat{\mathbf{e}}; \mathscr{B})$ and in $(\hat{\mathbf{f}}; \mathscr{B})$.

CASE A. Suppose that $\mathbf{e}^*$ and $\mathbf{f}^*$ lie on a line parallel to a 1-face, say **bc**, and let $\mathbf{x}^*$ be any point of the segment $\mathbf{e}^*\mathbf{f}^*$. (Figure 4.) If $1 \succ 2$ in $(\hat{\mathbf{x}}; \mathscr{B})$, then in view of (i), $1 \succ 2$ also in $(\hat{\mathbf{f}}; \mathscr{B})$, contrary to our assumption. If $2 \succ 1$ in $(\hat{\mathbf{x}}; \mathscr{B})$ then $2 \succ 1$ also in $(\hat{\mathbf{e}}; \mathscr{B})$, contrary to our assumption. In a similar fashion one proves that no strong relation holds between any other pair among the players 1, 2, and 3. Thus the segment $\mathbf{e}^*\mathbf{f}^*$ belongs to M_j, contrary to the assumption that $\mathbf{e}^*$ and $\mathbf{f}^*$ belong to distinct components of M_j.

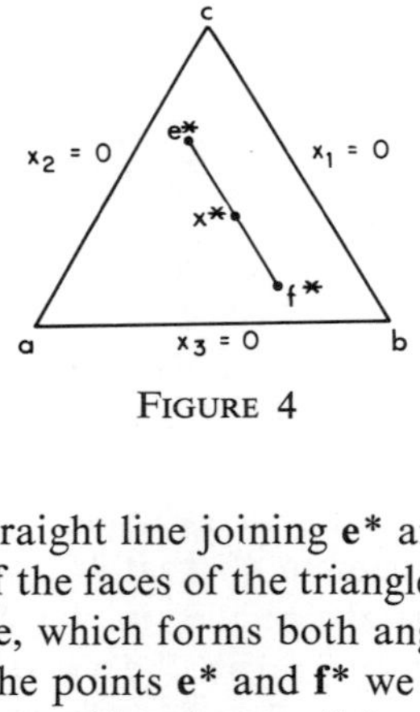

FIGURE 4

CASE B. Draw the straight line joining $\mathbf{e}^*$ and $\mathbf{f}^*$, now assuming that it is not parallel to any of the faces of the triangle. There exists exactly one side, say l, of the triangle, which forms both angles greater than 60° with this line. From each of the points $\mathbf{e}^*$ and $\mathbf{f}^*$ we draw lines parallel to the 2 sides other than l, and consider the parallelogram formed by them. We may assume that the situation is as shown in Figure 5. Obviously, the parallelogram $\mathbf{e}^*\mathbf{g}^*\mathbf{f}^*\mathbf{h}^*$ belongs to S_j. By applying the results stated in (i), one observes immediately that $1 \sim 2$ and $1 \sim 3$ in $(\hat{\mathbf{x}}; \mathscr{B})$, if $\mathbf{x}^*$ lies in this closed parallelogram. Moreover, $2 \succsim 3$ in $(\hat{\mathbf{g}}; \mathscr{B})$ and $3 \succsim 2$ in $(\hat{\mathbf{h}}; \mathscr{B})$. Take any closed path which lies in the parallelogram and joins the points

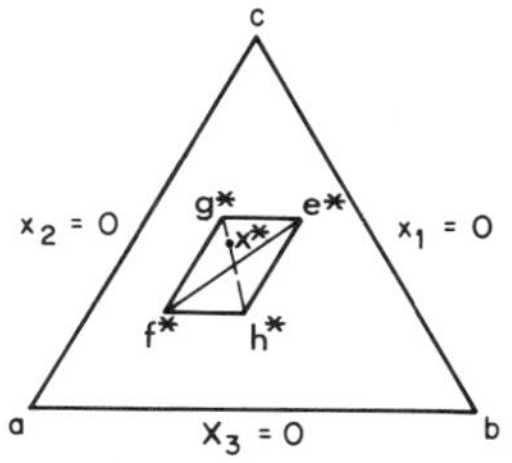

FIGURE 5

$\mathbf{g}^*$ and $\mathbf{h}^*$. Then, in view of Corollary 4.1, and the fact that $\succ$ is an asymmetric relation, it follows that there exists a point $\mathbf{x}^*$ on the path with the property that $2 \sim 3$ in $(\hat{\mathbf{x}}; \mathscr{B})$. Therefore, $\mathbf{x}^* \in M_j$. Obviously, $\mathbf{x}^*$ is closer to $\mathbf{e}^*$ than $\mathbf{f}^*$, and this contradicts our assumption. Therefore, M_j is connected.

(iii) We shall now show that any 1-cycle in M_j bounds.[12] Indeed, if α is a 1-cycle of M_j which does not bound, then there exists a point $\mathbf{d}^*$ in S_j, which is surrounded by the carrier α^* of α, and $\mathbf{d}^* \notin M_j$. If, say, $1 \succ 2$ in $(\hat{\mathbf{d}}; \mathscr{B})$, then, by the result stated in (i), there is a region of points $\mathbf{x}^*$ having the property that $1 \succ 2$ in $(\hat{\mathbf{x}}; \mathscr{B})$. This region connects $\mathbf{d}^*$ to the face $x_1 = 0$ and hence it intersects α^*. This is impossible, since $\alpha^* \in M_j$, and we have arrived at a contradiction. This completes the proof of the Corollary.

From Theorem 5.1, Corollaries 5.1 and 5.2, we deduce:

THEOREM 5.2. *Let* $\mathscr{B} \equiv \{B_1, B_2, \ldots, B_m\}$ *be a coalition structure* (2.3) *for a game* Γ, *such that each* B_j, $j = 1, 2, \ldots, m$, *does not contain more than 3 players. Then, there exists a payoff* $\mathbf{x} = (x_1, x_2, \ldots, x_n)$ *such that* $(\mathbf{x}; \mathscr{B})$ *is* $\mathscr{M}_1^{(i)}$*-stable.*

6. MISCELLANEOUS

Let $\mathscr{B}$ be a fixed coalition-structure for a game Γ, and let $(\mathbf{x}; \mathscr{B})$ be an i.r.p.c. We shall show that any intersection of the form $H \equiv \bigcap_{s=1}^{r} F_{\mu\nu_s}$, μ fixed, $\mu\nu_s \in B_j$, $B_j \in \mathscr{B}$, $r \geq 1$ (see Theorem 4.2), is acyclic. In particular, E_μ is acyclic. Indeed, if a point $\mathbf{x}^*$ belongs to H, then, by decreasing x_μ and increasing the other components of $\mathbf{x}^*$ in any arbitrary way, we always get points of H, because, by Lemma 5.2, μ will remain stronger than or equipollent to each ν_s, $s = 1, 2, \ldots, r$. Moreover, the face $x_\mu = 0$ obviously belongs to H. Hence H is contractible over itself to a point, and therefore it is acyclic. By similar considerations, one can prove that the set I of points $\mathbf{x}^*$ having the property that a player μ is weaker than or equipollent to the players $\nu_1, \nu_2, \ldots, \nu_r$ is acyclic.

[12] Assuming that M_j is now triangulated.

REFERENCES

[1] R. J. Aumann and M. Maschler, "The Bargaining set for cooperative games," *Advances in Game Theory*, M. Dresher, L. S. Shapley and A. W. Tucker eds., Annals of Mathematics Studies, No. 52, Princeton, Princeton University Press (1964), pp. 443–476.

[2] S. Eilenberg and D. Montgomery, "Fixed point theorems for multi-valued transformations." *Amer. J. Math.* Vol. 68 (1946), pp. 214–222.

[3] S. Kakutani. "A generalization of Brouwer's fixed point theorem." *Duke Math. J.* Vol. 8 (1941), pp. 457–459.

[4] B. Knaster, C. Kuratowski, S. Mazurkiewicz. "Ein Beweis des Fixpunktsatzes für *n*-dimensionale Simplexe." *Fund. Math.* Vol. 14 (1929), pp. 132–137.

[5] K. Kuratowski. *Introduction to Set Theory and Topology*. New York, Pergamon Press (1961).

[6] M. Maschler. "*n*-Person games with only 1, *n*-1, and *n*-person permissible coalitions." *J. Math. Analysis and Appl. Vol.* 6 (1963), pp. 230–256.

[7] J. von Neumann. "Über ein ökonomisches Gleichungssystem und eine Verallgemeinerung des Brouwerschen Fixpunktsatzes." *Erg. Math. Kolloqu.* Vol. 8 (1937), pp. 73–83.

CHAPTER 4

Existence Theorem for the Bargaining Set $M_1^{(i)}$

By BEZALEL PELEG*

In [2] M. Davis and M. Maschler conjecture that for each coalition structure b in an n-person cooperative game there is a payoff vector $\mathbf{x}$ such that the payoff configuration $(x, b) \in M_1^{(i)}$. In this paper we prove their conjecture.[1] The proof that we give is somewhat shorter than the proof that was outlined in [4].

1. DEFINITIONS

In this section we give only the definitions that we need in this paper. Further introductory material on bargaining sets is to be found in [1].

DEFINITION. An *n-person game* is a pair (N, v), where $N = \{1, \ldots, n\}$ is a set with n members, and v is a real valued function defined on the power set of N.

There will be no loss of generality if we assume that $v(\{i\}) = 0$, $i = 1, \ldots, n$, and $v(B) \geq 0$ for all $B \subset N$.

Let (N, v) be an n-person game.

DEFINITION. A *coalition structure* (c.s.) is a disjointed set of subsets of N whose union is N.

Let $B \subset N$. A *B-vector* x^B is a real function defined on B whose value at $i \in B$ is x^i. The superscript N will be omitted.

DEFINITION. An *individually rational payoff configuration* (i.r.p.c.) is a pair (x, b), where b is a c.s. and x is an N-vector that satisfies $\sum_B x^i = v(B)$ for all $B \in b$, and $x^i \geq 0$, $i = 1, \ldots, n$.

Let b be a c.s. The set of all N-vectors x such that (x, b) is an i.r.p.c. is denoted by $X(b)$. $X(b)$ is a non-void compact and convex subset of some euclidean space.

DEFINITION. Let (x, b) be an i.r.p.c. and $i, j \in B \in b$, $i \neq j$. A Q-vector y^Q is an *objection* of i against j in (x, b), if: $i \in Q$, $j \notin Q$, $y^i > x^i$, $y^k \geq x^k$ for $k \in Q$, and $\sum_Q y^k = v(Q)$.

DEFINITION. Let (x, b) be an i.r.p.c., $i, j \in B \in b$, $i \neq j$, and y^Q an objection of i against j in (x, b). An R-vector z^R is a *counter objection* to y^Q, if: $j \in R$, $i \notin R$, $z^k \geq x^k$ for $k \in R$, $z^k \geq y^k$ for $k \in R \cap Q$, and $\sum_R z^k = v(R)$.

An i.r.p.c. (x, b) is *stable* if for each objection in (x, b) there is a counter objection. The set of all stable i.r.p.c.'s is the *bargaining set* $M_1^{(i)}$.

* This work was done at the Hebrew University in Jerusalem under the supervision of Dr. R. J. Aumann.

[1] Another proof was given by M. Davis, M. Maschler, and the author.

An objection in an i.r.p.c. is *justified* if it cannot be countered. Let (x, b) be an i.r.p.c. and $i, j \in B \in b$, $i \neq j$. We write $i \succ j$ in (x, b), if i has a justified objection against j in (x, b); we write $i \succsim j$ in (x, b), if j has no justified objection against i in (x, b).

If $B \subset N$ we denote by $|B|$ the number of members of B.

2. AUXILIARY LEMMAS

Let (N, v) be an n-person game.

LEMMA 2.1. *Let b be a* c.s. *and $i \in B \in b$. If $E_i = \{x: x \in X(b), i \succsim j$ in (x, b) for all $j \in B - \{i\}\}$, then E_i is compact and $E_i \supset \{x: x \in X(b), x^i = 0\}$.*

PROOF. To show that E_i is compact we prove that $X(b) - E_i$ is open. Let $x \in X(b) - E_i$. There is a $j \in B - \{i\}$ that has a justified objection y^Q against i in (x, b). We may assume that $\min_{k \in Q} (y^k - x^k) > 0$. Denote $d_{ij} = \{D: D \subset N, i \in D \text{ and } j \notin D\}$. We define, for $z \in X(b)$, $f(z) = z^i + \max_{D \in d_{ij}} \left(v(D) - \sum_{D \cap Q} y^k - \sum_{D - Q} z^k \right)$. $f(x)$ is the maximum that i can get when he tries to counter object y^Q, and therefore $f(x) < x^i$. So there is an $\epsilon > 0$ such that if $z \in X(b)$ and $|z^k - x^k| < \epsilon$, $k \in N$, then $f(z) < z^i$ and $y^k > z^k$, $k \in Q$; it follows that y^Q is a justified objection of j against i in (z, b) and therefore $z \notin E_i$.

To see that $E_i \supset \{x: x \in X(b), x^i = 0\}$, observe that if $x \in X(b)$ and $x^i = 0$, then i can counter object any objection against him in (x, b) by forming a single person coalition.

LEMMA 2.2. *Let (x, b) be an* i.r.p.c.; *then the partial relation $\succ$ in (x, b) is acyclic.*

For proof see [2] Theorem 3.1 or [3] Theorem 1.

COROLLARY 2.3. *Let b be a* c.s. *For each $x \in X(b)$ and $B \in b$ there is an $i \in B$ such that $x \in E_i$.*

PROOF. If the above assertion is false then there are $x_0 \in X(b)$ and $B_0 \in b$ such that for each $i \in B_0$ there is a $j \succ i$ in (x_0, b), contradicting Lemma 2.2.

LEMMA 2.4. *Let b be a* c.s. *and let $c^1(x), \ldots, c^n(x)$ be n non-negative continuous real functions defined on $X(b)$. If for each $x \in X(b)$ and $B \in b$ there is an $i \in B$ such that $c^i(x) \geq x^i$, then there is an $x_0 \in X(b)$ such that $c^j(x_0) \geq x_0^j$, for $j = 1, \ldots, n$.*

PROOF. Suppose, per absurdum, that no $x \in X(b)$ satisfies $x^i \leq c^i(x)$ for $i = 1, \ldots, n$; this is equivalent to the assumption that

$$\delta = \min_{x \in X(b)} \max_{i \in N} (x^i - c^i(x)) > 0.$$

We now define a continuous function f: $X(b) \to X(b)$, by setting for

$x \in X(b)$ and $i \in B \in b$

$$(f(x))^i = v(B) \frac{c^i(x) + \dfrac{\delta}{2} + \left| x^i - c^i(x) - \dfrac{\delta}{2} \right|}{\sum_B c^k(x) + \dfrac{|B|\,\delta}{2} + \sum_B \left| x^k - c^k(x) - \dfrac{\delta}{2} \right|}.$$

By Brouwer's fixed point theorem there is an $x_0 \in X(b)$ such that $f(x_0) = x_0$. Recalling the definition of δ we see that there is an $m \in N$ such that $x_0{}^m \geq c^m(x_0) + \delta$. $m \in B_0 \in b$. We know that there is a $j \in B_0$ such that $c^j(x_0) \geq x_0{}^j$, which implies that $c^j(x_0) + \dfrac{\delta}{2} > x_0{}^j$. Now

$$\begin{aligned} x_0{}^m &= v(B_0) \frac{c^m(x_0) + \dfrac{\delta}{2} + \left| x_0{}^m - c^m(x_0) - \dfrac{\delta}{2} \right|}{\sum_{B_0} c^k(x_0) + \dfrac{|B_0|\,\delta}{2} + \sum_{B_0} \left| x_0{}^k - c^k(x_0) - \dfrac{\delta}{2} \right|} \\ &= v(B_0) \frac{x_0{}^m}{\sum_{B_0} c^k(x_0) + \dfrac{|B_0|\,\delta}{2} + \sum_{B_0} \left| x_0{}^k - c^k(x_0) - \dfrac{\delta}{2} \right|} \\ &< v(B_0) \frac{x_0{}^m}{\sum_{B_0} c^k(x_0) + \dfrac{|B_0|\,\delta}{2} + \sum_{B_0} \left(x_0{}^k - c^k(x_0) - \dfrac{\delta}{2} \right)} = x_0{}^m, \end{aligned}$$

a contradiction which proves that the assumption $\delta > 0$ is false.

COROLLARY 2.5. *Let b be a* c.s. *and let $A_1, \ldots, A_n$ be compact subsets of $X(b)$. If $A_i \supset \{x\colon x \in X(b),\ x^i = 0\}$, $i = 1, \ldots, n$, and for each $B \in b$ $\bigcup_{i \in B} A_i = X(b)$, then $\bigcap_{i \in N} A_i \neq \varnothing$.*

PROOF. For $x, y \in X(b)$ we set $\rho(x, y) = \max_{1 \leq i \leq n} |x^i - y^i|$. Now define, for $i \in N$ and $x \in X(b)$, $c^i(x) = x^i - \rho(x, A_i) = x^i - \min_{y \in A_i} \rho(x, y)$. Since $A_i \supset \{x\colon x \in X(b),\ x^i = 0\}$ $c^i(x)$ is non-negative. $c^i(x)$ is continuous. If $x \in X(b)$ and $B \in b$ then there is a $j \in B$ such that $x \in A_j$; so $\rho(x, A_j) = 0$ and $c^j(x) = x^j$. So we can apply Lemma 2.4 to obtain an $x_0 \in X(b)$ such that $c^j(x_0) \geq x_0{}^j$, $j = 1, \ldots, n$. Now, $\rho(x_0, A_j) = 0$, $j = 1, \ldots, n$, and therefore $x_0 \in \bigcap_{j \in N} A_j$.

3. THE EXISTENCE THEOREM

Let (N, v) be an n-person game.

THEOREM. *For each* c.s. *b there is an N-vector $x \in X(b)$ such that the* i.r.p.c. *$(x, b) \in M_1^{(i)}$.*

PROOF. Let b be a c.s. $E_1, \ldots, E_n$, as defined in Lemma 2.1, are compact subsets of $X(b)$; also $E_i \supset \{x\colon x \in X(b),\ x^i = 0\}$, $i = 1, \ldots, n$. By

Corollary 2.3, for each $B \in b$, $\bigcup_{i \in B} E_i = X(b)$. So we can apply Corollary 2.5 to show that $\bigcap_{i \in N} E_i \neq \emptyset$. Let $x_0 \in \bigcap_{i \in N} E_i$; it is clear from the definition of the sets $E_1, \ldots, E_n$ that $(x_0, b) \in M_1^{(i)}$.

REFERENCES

[1] R. J. Aumann and M. Maschler. *The Bargaining Set for Cooperative Games.* Annals of Mathematical Studies No. 52, Princeton University Press, Princeton, New Jersey, 1964 pp. 443–476.

[2] M. Davis and M. Maschler. *Existence of Stable Payoff Configurations for Cooperative Games*. This book.

[3] M. Davis and M. Maschler. *Existence of Stable Payoff Configurations for Cooperative Games*, Bull. Amer. Math. Soc. Vol. 69 (1963) pp. 106–108.

[4] B. Peleg. *Existence Theorem for the Bargaining Set* $M_1^{(i)}$, Bull. Amer. Math. Soc. Vol. 69 (1963) pp. 109–110.

CHAPTER 5

On Solutions That Exclude One or More Players

By L. S. SHAPLEY*

1. INTRODUCTION

In this note we examine the possibility of solving n-person games by using the solutions of certain reduced games, obtained by eliminating one or more players from the characteristic function. The central result is a simple-sounding, neccssary and sufficient condition for the existence of completely discriminatory, or "exclusive" solutions. Stated informally, the condition is that the reduced game must be solvable, and every "exclusive" imputation in the original game must be enforceable by at least one effective coalition consisting entirely of nonexcluded players (Theorem 3, below). This result is perhaps familiar to others[1] who have worked with the von Neumann-Morgenstern theory, but we know of no full account in the literature.

From the general theorem, a number of known results concerning solutions and solvability can be easily deduced; these are discussed in the final section.

2. TERMINOLOGY

Let v be the characteristic function of an n-person game with player set $N = \{1, \ldots, n\}$. A solution[2] X of v will be said to *exclude* player i if $x_i = v(\{i\})$ for all imputations x in X. This is an extreme form of what von Neumann and Morgenstern call "discrimination,"[3] which requires only that x_i be constant but not necessarily minimal. Solutions that exclude one or more players occur quite frequently; they are found, e.g., in almost all simple games, and they probably exist in all constant-sum games.

For convenience, let us assume that v is in 0, 1-normalized form, thus: $v(\{1\}) = \ldots = v(\{n\}) = 0$ and $v(N) = 1$. Let R be a subset of N having at least two members. Define v_R on the subsets of R as follows:

$$v_R(T) = \begin{cases} v(T) & \text{for } T \subset R, \\ 1 & \text{for } T = R.^4 \end{cases}$$

* The RAND Corporation.

[1] For example, Peleg makes an observation in [5] that is essentially equivalent to our Theorem 2. (See p. 382, Eq. (4).)

[2] The elementary definitions of solution theory will not be repeated here. See [2], [3], [4], or [6].

[3] See [6], p. 288 ff.

[4] Without normalization, $v_R(R)$ would be $v(N) - \sum_{N-R} v(\{i\})$.

This function is superadditive, and hence it is the characteristic function of a game with player set R. Call that game the *R-fraction* of v. It is usually non-constant-sum, the only exception being when all the players in $N - R$ are dummies. Note that "fraction of" is a transitive relation, i.e., we have $(v_R)_S = v_S$ if $S \subseteq R \subseteq N$.

There is a natural embedding of the imputation space A_R of v_R in the imputation space A of v, according to which the former appears as an $(|R| - 1)$-dimensional simplicial face of the latter. We shall speak as though this embedding had been carried out, once and for all, and make no distinction in notation between the elements of A_R and the corresponding elements of A (with zeros filled in for the members of $N - R$).

THEOREM 1. *A solution of v is a solution of v_R if and only if it excludes the players of $N - R$.*

The proof is immediate from the definition of solution.

A *winning coalition* is a set of players S with $v(S) = 1$.[5] Any superset of a winning coalition is also winning. If R is winning, then (and only then) the fraction V_R is simply the restriction of the function v to the subsets of R; such a fraction will be called a *winning fraction.*

One more group of definitions is needed: A game is said to be *weak* if it possesses an undominated imputation, i.e., if its "core" (see [2], p. 50) is not empty. If the core has an interior point, we shall call the game *strictly weak*; this means there is an imputation for which no proper subset of the players is effective. A game that is not strictly weak will be called *effective.* For example, constant-sum games are effective, quota games are effective, essential 2-person games are strictly weak, etc. It is easily verified that if a game is weak then all its fractions are weak, and if a game is strictly weak then all its fractions are strictly weak. However, an effective game need not have any (proper) fractions that are effective.[6]

3. CHARACTERIZATION OF EXCLUSIVE SOLUTIONS

THEOREM 2. *A solution of v_R is a solution of v if and only if v_R is winning or effective.*

[5] Without normalization, the defining condition would be $v(S) = v(N) - \sum_{N-S} v(\{i\})$.

[6] Adding a dummy player to a strictly weak game transforms it into an effective game, since it raises the dimension of the imputation space without enlarging the core. This infelicitous consequence of our definitions can be avoided if the definition of "strictly weak" is relativized to the smallest face of the imputation simplex that contains the core. For example, the 3-person game with $v(1) = v(2) = v(3) = 0$, $v(12) = v(13) = 1/3$, $v(23) = v(123) = 1$ would then qualify as strictly weak, the core being a one-dimensional set in the boundary of the imputation triangle. It can be shown that a game that is strictly weak only in the new, relative sense must possess a proper winning coalition. (In the example above, 23 is winning.) It follows that any "relatively effective" fraction is either effective (in the old sense) or winning.

REMARK. To say that v_R is either winning or effective is the same as saying that for each imputation in A_R at least one subset of R is effective. In other words, all exclusions are enforceable.

PROOF. Let $X \subseteq A_R$ be a solution of v_R. Consider the domination relation as applied to the full game v. It is clear that X dominates $A_R - X$ and does not dominate any element of itself. Therefore the only remaining question is whether or not X dominates $A - A_R$.

Case I: v_R is winning. Take $y \in A - A_R$ and let z be its perpendicular projection on A_R, so that every player in R prefers z to y. Since R is winning, we have $z \underset{R}{\succ} y$. If $z \in X$, then $X \succ y$. If, on the other hand, $z \notin X$, then there is an $x \in X$ and an $S \subset R$ such that $x \underset{S}{\succ} z$. This implies $x \underset{S}{\succ} y$; hence again $X \succ y$. Therefore X is a solution of v.

Case II: v_R is effective. Take $y \in A - A_R$ and let z be its perpendicular projection on A_R as above. If $z \in X$, let $T \subset R$ be an effective coalition for z. Then $z \underset{T}{\succ} y$, and hence $X \succ y$. If, on the other hand, $z \notin X$, then proceed as in Case I.

Case III: v_R is neither winning nor effective. Let z be an interior point of the core of v_R. Neither R nor any proper subset of R is effective for z. The same statement holds for all points y in $A - A_R$ that are sufficiently close to z. We assert that such points are undominated by X, and that X is therefore not a solution. In fact, $x \underset{T}{\succ} y$ is impossible for $T \subseteq R$, for any x, by the noneffectiveness of T. Furthermore, $x \underset{T}{\succ} y$ is impossible for $T \nsubseteq R$ if $x \in X$, since $x_i = 0$ for $i \notin R$. This completes the proof.

THEOREM 3. *A necessary and sufficient condition that the game v have a solution that excludes the set of players S is that the fraction v_{N-S} be solvable and either winning or effective.*

PROOF. If v has an S-exclusive solution, then v_{N-S} is solvable by Theorem 1; and by Theorem 2 it is winning or effective. Conversely, if v_{N-S} is solvable, and winning or effective, then by Theorem 2 its solutions are S-exclusive solutions of v.

4. APPLICATIONS

A number of known results come out as special cases, or as simple consequences, of the foregoing theorems.

COROLLARY 1 (von Neumann). *The solutions of a general-sum game are precisely the solutions of its zero-sum extension that exclude the added player.*[7]

[7] See [6], p. 526.

PROOF. In a zero-sum n-person game the $(n-1)$-person coalitions are winning.

COROLLARY 2 (Peleg). *If there are three players i, j, k in a game v in* 0, 1-*normalized form such that*

$$v(\{i,j\}) + v(\{i,k\}) + v(\{j,k\}) \geq 2,$$

then v has a solution.[8]

PROOF. The $\{i, j, k\}$-fraction, like all 3-person games, is solvable; the given inequality guarantees that it is effective.

COROLLARY 3. *Any game with a* 2- *or* 3-*person winning coalition is solvable.*

COROLLARY 4. (Gillies and others.) *Every simple game is solvable.*

PROOF. The R-fraction v_R, where R is a minimal winning coalition, is trivially solved by its core, A_R.[9]

COROLLARY 5. (von Neumann). *A constant-sum game Γ that decomposes into two essential components Δ and H has a solution that excludes the players of Δ if and only if* H *is solvable and* $|\mathrm{H}|_2 \geq |\Delta|_1$.[10]

PROOF. Let v be the characteristic function of Γ, and let J and K denote the player sets of Δ and H, respectively. Now $|\Delta|_1$ is defined as $v(J) - \sum_J v(\{j\})$, while $|\mathrm{H}|_2$ is defined as the largest number that can be added to $v(K)$ without making the game v, restricted to subsets of K, a strictly weak game. Thus, the K-fraction v_K will be strictly weak if and only if

$$\begin{aligned} |\mathrm{H}|_2 &< v_K(K) - v(K) \\ &= v(N) - \sum_{N-K} v(\{j\}) - v(K) \\ &= v(J) - \sum_J v(\{j\}) \\ &= |\Delta|_1. \end{aligned}$$

The proof is completed by noting that K cannot be winning in Γ without making Δ inessential.

We have included this last corollary in order to highlight the resemblance between the notions of "weak" and "strictly weak" games, as used here,

[8] See [5]. In the -1, 0 normalization of [6], the right-hand side of the inequality would read $2(n-3)$. Without any normalization, it would read

$$2\left[v(N) - \sum_{N-\{ijk\}} v(\{l\})\right].$$

[9] See [1]. The more general Theorem 20 of [2], p. 73, concerning "semi-simple" games, can be derived similarly.

[10] The terminology is that of [6], Chapter IX, "Composition and Decomposition of Games." This corollary is a special case of the comprehensive theorem on solutions of decomposable games, given on pp. 393–394, *loc. cit.*

and the "detached" and "fully detached" imputations of [6], *loc. cit.* The heuristic meaning of the "$|H|_2$" limit established in [6] is this:

> The players comprising one component of a *decomposable* game can maintain a discriminatory excess over their opponents, provided that it is not so great as to create fully detached imputations in their component game.

In much the same way, our Theorem 2 states:

> The players comprising a nonwinning coalition in an arbitrary game can maintain *complete* discrimination over their opponents, provided that their fraction of the game is not strictly weak.

Corollary 5 is, so to speak, the overlap of these two theorems. One wonders whether there might not be a common generalization as well, covering incomplete discrimination in indecomposable games.

Our theorems provide a method that reduces the solving of large games, in certain cases, to the solving of smaller games. One might ask what progress this represents on the major, unresolved problem of *n*-person game theory: to prove that every game has a solution. The answer is disappointing. Not all games are so reducible; in fact, "most" games are not.

Going in the other direction, we may take a game that is known to be solvable and readily construct the class of larger games of which it is a winning or effective fraction. (This is the idea behind Corollaries 2 and 3, above.) However, iteration of this process yields nothing more, because the solvable extensions of the larger games will already be included in the class, by virtue of being solvable extensions of the original game.

REFERENCES

[1] Gillies, D. B., "Some Theorems on *n*-Person Games," Thesis, Princeton University, 1953.

[2] Gillies, D. B., "Solutions to General Non-Zero-Sum Games," pp. 47–85 *in* A. W. Tucker and R. D. Luce (eds.), *Contributions to the Theory of Games*, Annals of Mathematics Study No. 40, Princeton University Press, Princeton, N.J., 1959.

[3] Luce, R. D., and Howard Raiffa, *Games and Decisions*, John Wiley and Sons, Inc., New York. 1957.

[4] McKinsey, J. C. C., *Introduction to the Theory of Games*, McGraw-Hill Book Company, Inc., New York, 1952.

[5] Peleg, B., "On the Set of Solvable *n*-Person Games," *Bulletin of the American Mathematical Society*, Vol. 65, 1959, pp. 380–383.

[6] von Neumann, J., and O. Morgenstern, *Theory of Games and Economic Behavior*, Princeton University Press, Princeton, N.J., 1944, 1947, 1953.

CHAPTER 6

Concepts and Theories of Pure Competition

By L. S. SHAPLEY and MARTIN SHUBIK*

1. INTRODUCTION

Underlying much of basic economic theory concerning the formation of prices and the operation of markets is the idea that if the number of individuals trading in all markets is sufficiently large, and if there are no institutional bounds or opportunities for cooperative arrangements between groups, then a competititve price system will emerge. Each individual may proceed to maximize his own welfare, utilizing only his knowledge of the price levels and of his own desires and assets, with the overall result that all markets will be cleared and the resulting imputation of goods and services will meet certain broad requirements of optimality.

At the other end of the spectrum, Edgeworth [1], and many others since, have observed that in bargaining among a *few* individuals (of more or less comparable strength), there will be a whole range of outcomes that are optimal, in the appropriate sense, and economic conditions alone will not determine a specific outcome.

A third possibility is a situation in which individuals of one type are few in number, or are organized in some manner, while the others are not. In this case it has been observed that the few can employ "monopolistic practices" against the others, and that the resultant outcome will be determinate but not optimal. Institutionally, such a situation may arise when a few firms confront many customers in a market; the natural setting for this type of analysis is within the framework of the theory of oligopoly.

An interpretation and unification of these three different viewpoints is presented here in terms of the theory of games. The results described are based primarily upon the joint findings of Shapley and Shubik [2], but also upon recent papers by Shubik [3], Shapley [4], Scarf [5], Aumann [6], and Debreu [7].

2. BILATERAL MONOPOLY

2.1. THE EDGEWORTH MODEL. For simplicity we consider a market with two traders, trading in two commodities. The initial holding of the first trader is $(a, 0)$ and that of the second is $(0, b)$. Let their preferences be represented by two families of continuously differentiable, convex indifference curves, denoted by $\psi(x, y)$ and $\varphi(x, y)$, respectively, where x and y are the amounts held of the first and second commodity. By superimposing the two families of curves, with coordinates oppositely oriented, we obtain

* The RAND Corporation and Yale University.

the familiar "Edgeworth box" [1], as illustrated in Figure 1. Any point in the box represents a jointly attainable trade and will have coordinates $(a - x, b - y)$ for the first trader and (x, y) for the second. The point R represents the initial position of both players prior to trading. The first trader's zero point is at O', and his initial holdings are represented by the vector $O'R$, which has length a. His goal is to carry out trades that move the outcome in a "south-west" direction as far as possible, i.e., away from

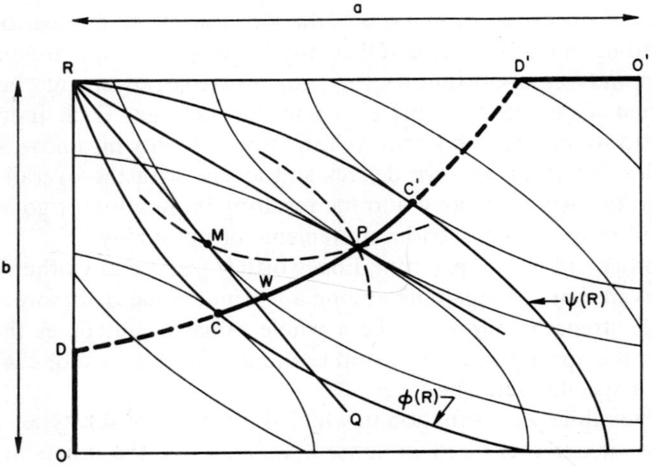

FIGURE 1

O'. The zero point for the second trader is at O, and OR represents his initial stock of b units of the second commodity. He wishes to trade in a manner that moves the outcome in a "north-east" direction as far as he can.

There are three basic models, or "games," which can be formulated in bilateral monopoly to illustrate three strategically very different trading procedures; we shall denote them by the symbols $(1, 1)_0$, $(1, 1)_1$, and $(1, 1)_2$. In Figure 1 they will lead to the outcomes indicated by the point P, the point M, and the curve CC', respectively.

2.2. THE GAME $(1, 1)_0$: THE COMPETITIVE EQUILIBRIUM. The essence of the economic model of pure competition is that all participants act as *price-takers*. One assumes that in some manner or other a schedule of prices has been established in the market, and that each individual takes note of these prices and buys or sells accordingly. He does not actively control or influence the prices. When the number of participants in the market is large, it is usually assumed either implicitly or explicitly that a dynamic market mechanism produces the prices, but in our "game," which at present has just two players, we shall unrealistically assume that the omniscient referee performs his calculations and names an appropriate set of prices, then acts as a clearing house for all transactions. By the rules of the game, the players are strategically constrained to act as price-takers. The subscript on the symbol $(1, 1)_0$ is meant to convey that neither of the two traders has sufficient freedom of strategy to manipulate price.

We know from basic economic theory [8] that when consumer's tastes are independent and can be represented by convex indifference sets (as is the case here), then a *competitive equilibrium* will always exist. This is illustrated in Figure 1 by the equilibrium price line RP and the competitive allocation point P. In this two-commodity example, the slope of the line RP gives the relative price of one commodity in terms of the other. If prices in this ratio are announced by the referee, then trading will continue until the point P is reached. No further gain can be made by either side because of the tangency between the price line and each traders' indifference curve. This so-called "competitive" solution exists under quite general conditions, but it is not necessarily unique.[1]

The point P is *Pareto optimal*, which means that, given the final distribution of resources, it is not possible to improve the welfare of any one individual without decreasing the welfare of another. The extended curve $ODD'O'$ describes the full Pareto-optimal set in Figure 1. (The interior portion DD' is the locus of the points of tangency between the two sets of indifference curves.) Any point *off* this curve can be dominated by a point *on* it. For instance, M and W in Figure 1 are indifferent to the second trader but W is preferred to M by the first trader. Hence M is not Pareto optimal since the welfare of at least one trader can be improved without damage to the other.

2.3. THE GAME $(1, 1)_1$: THE MONOPOLISTIC SOLUTION. The second case we consider is $(1, 1)_1$, where one of the traders, say the first, is strategically empowered to name the price, while the other is restricted to maximizing his own welfare, taking price as given. Any ray through the initial point R represents a set of (relative) prices. The object of the first trader will be to select a ray such that the final trade is as favorable as possible to himself, knowing that the second player will trade up to the point of tangency between the price ray and his family of indifference curves. This is indicated in Figure 1 for the ray RQ at the point M. A monopolistic trading curve may be drawn—the curve MP in the figure—which is the locus of points of tangency between the price rays and the indifference curves of the second

[1] Somewhat unfortunately, the term "competitive" has been attached by established usage to just that solution concept, out of the three or four that we shall be considering, that has the least to do with game theory. Paradoxically, the competitive equilibrium involves no interplay between the "competitors," and might better be described as an "administered-price" solution.

trader. (There will be a similar "response curve" for the first player, intersecting the other at *P*.) The optimum for the first player is in fact the point *M*, where his indifference curve is tangent to the second player's response curve.

We see that this "monopolistic" solution is not Pareto optimal, and that it favors the price-naming monopolist as compared to the solution *P* of the previous case.

2.4. THE GAME $(1, 1)_2$: THE CONTRACT CURVE. In the third case, denoted by $(1, 1)_2$, we assume that both players are strategically capable of naming price and may negotiate in any manner they choose. There is no specification of the dynamic process—the particular "rules of the game"—beyond noting that any bargaining or haggling is permitted. The solution propounded by Edgeworth to this model of bilateral monopoly was the *contract curve* CC' in Figure 1. This is not the entire Pareto-optimal set, but just that part of it where neither side can force a more favorable distribution by a refusal to trade. The two indifference curves through *R* provide the bounds *C* and C'.

This Edgeworth solution does not yield a unique prediction of the imputation of resources; it merely indicates a range. The price-taking model $(1, 1)_0$, on the other hand, did produce a determinate, Pareto-optimal outcome; the latter was located on the contract curve but it was obtained by imposing somewhat unrealistic restrictions on the trading possibilities, considering the small number of traders.

3. MARKETS WITH MANY TRADERS. THE GAME $(n, n)_0$

Implicitly or explicitly in all models of trading, assumptions have to be made concerning the institutional nature of the market. In the broad sweep of economics when many individuals are involved, we expect markets to be more or less insensitive to minor institutional differences, but when two individuals engage in face to face bargaining we suspect that personality, cultural factors, psychological details, the fine structure of moves and timing, etc., all play a major role.

This being the case, it is entirely appropriate to use the "game" $(1, 1)_2$ when we deal with bilateral monopoly. When we discuss markets with many traders on each side, however, we often tend to use something like $(m, n)_0$ as the model under investigation. In other words, we effectively assume that in markets with many participants on all sides, the individual is constrained to act as a price-taker. There are many good reasons for this, such as the cost of communication, the lack of time to talk to everyone, and other organizational factors that drive toward impersonal mechanisms of trade.

On the other hand, a sociologist or an anthropologist might point out that in spite of numerical size and communication problems, patterns can exist and persist in a society that rely on the overt or covert coordination of many individuals. There may be stable configurations involving the compliance of large groupings or coalitions within the society. Furthermore, both long-run socioeconomic considerations and a study of information processes indicate that although many costs and much expenditure of time and energy may be incurred in creating a set of intricate coalitions *ab initio*, yet given their existence, little effort need be spent in maintaining them. In view of these considerations, it will be worth investigating the characteristic properties of games of the form $(m, n)_1$ and $(m, n)_2$, as well as $(m, n)_0$.

For simplicity in the sequel, we shall not only restrict our attention to bilateral markets, but we shall assume $m = n$. Traders of the same type are assumed to have identical preferences and initial holdings. Under these assumptions, the discussion of $(n, n)_0$ becomes especially easy.

Indeed, in going from $(1, 1)_0$ to $(n, n)_0$, the form of the solution remains the same, since the competitive equilibrium point (illustrated for the two-person case as the point P in Figure 1) has a direct $2n$-dimensional analogue. In fact, the outcome will be the same (under our highly symmetric assumptions) as though n games of the form $(1, 1)_0$ were being conducted independently and simultaneously. As n becomes large, the competitive equilibrium point, from the viewpoint of the economist, will be much more "reasonable" as a solution than it was in the case of bilateral monopoly, for the reasons already given.

4. COOPERATIVE GAMES. THE GAME $(n, n)_2$

4.1. CORES AND SOLUTIONS. Before we proceed to $(n, n)_1$ or $(n, n)_2$, a brief digression into the subject of cooperative n-person game theory will be necessary. For ease and clarity the exposition will be mostly in terms of games where the players have transferable, measurable, and comparable utilities. (It is as though the players all attach the same worth to money and have a constant marginal utility for it.) It must be emphasized that the results we shall give concerning $(n, n)_2$ are independent of these utility restrictions.

In order to explain the concepts of *cooperative solution* and *core*, we must also define what is meant by the following terms: the *characteristic function* of a game, an *imputation*, an *effective set* of players, and *domination* of one imputation by another.

The *characteristic function* specifies the worth that a coalition can achieve if they limit their trades strictly to themselves. Mathematically it is a function $v(S)$ defined on sets of players S, with the properties

$$v(\theta) = 0,$$

$$v(S \cup T) \geq v(S) + v(T), \quad \text{whenever } S \cap T = \theta.$$

The first condition merely states that the amount achievable by the null set is nothing. The second condition is the fundamental economic property of superadditivity: if two separate groups having commerce only amongst themselves are joined together, the resultant group is at least as effective as were the two independent groups. Beyond these two conditions there is nothing more than can be said *a priori* about a characteristic function.

If we denote the set of all players in a game by N, then $v(N)$ specifies the total amount that the whole group can obtain by cooperation.[2] A reasonable form of "cooperative" behavior would be for the players to agree to maximize jointly, and then to decide how the proceeds are to be apportioned, or "imputed." We define an *imputation* α to be a division of the proceeds from the jointly optimal play of the game among all the n players:

$$\alpha = (\alpha_1, \alpha_2, \alpha_3, \ldots, \alpha_n),$$

where

$$\alpha \geq v(\bar{i}) \quad \text{and} \quad \sum_{i=1}^{n} \alpha_i = v(N).$$

The condition $\alpha_i \geq v(\bar{i})$ embodies the principle that no individual will ever consent to a division that yields him less than he could obtain by acting by himself. It is often convenient to normalize the individual scales so that $v(\bar{i}) = 0$.

A set of players is said to be *effective* for an imputation if by themselves they can obtain at least as much as they are assigned in that imputation. Symbolically, S is effective for α if and only if

$$v(S) \geq \sum_{i \in S} \alpha_i.$$

If ">" rather than "=" holds, we shall say that S is *strictly effective.*

An imputation α *dominates* an imputation β if there exists an effective set S for α such that for all members of S, $\alpha_i > \beta_i$. Following the notation of von Neumann and Morgenstern [9], we write

$$\alpha \succ \beta.$$

In other words, if a set S of players is in a position to obtain by independent action the amounts that they are offered in the imputation α, and if, when they compare the amounts offered in α to the amounts offered in β, all of them prefer the former, then α dominates β. There is a potential coalition that prefers α to β and is in a position to do something about it. Note that S is necessarily *strictly* effective for β, the dominated imputation.

[2] Their utilities being transferable, this is properly represented by a single number, which denotes maximum obtainable welfare. If utilities were not transferable, $v(N)$ would instead have to represent the Pareto-optimal surface, and similarly for smaller coalitions. (See [6], [4].)

Finally, we may define two "solution" concepts. The *core* of an n-person game is the set of undominated imputations, if any. A von Neumann-Morgenstern *solution*, on the other hand, consists of a set of imputations which do not dominate each other but which collectively dominate all alternative imputations. There is at most one core, but there may be many solutions. All solutions contain the core, if it exists.

4.2. SOME EXAMPLES. A series of simple, three-person games will illustrate these concepts. Consider first the game in which any player acting by himself obtains nothing, but any pair of players acting in concert can demand three units to share between them, while all three players in coalition are also awarded three. The characteristic function of this game is

$$v(\theta) = 0,$$
$$v(\bar{1}) = v(\bar{2}) = v(\bar{3}) = 0,$$
$$v(\bar{1}\bar{2}) = v(\bar{1}\bar{3}) = v(\bar{2}\bar{3}) = 3,$$
$$v(\bar{1}\bar{2}\bar{3}) = 3,$$

where $\bar{1}\bar{2}$ means "the set consisting of players 1 and 2," etc.

We may represent the imputations in this game by triangular coordinates, as shown in Figure 2. The vertices P_1, P_2, P_3 represent the imputations (3, 0, 0), (0, 3, 0), and (0, 0, 3), respectively. The point $\omega = (1, 1, 1)$ is the center of the triangle. Consider the two imputations $\alpha = (1.9, 0, 1.1)$ and $\beta = (0, 1.5, 1.5)$. The set is $\bar{2}\bar{3}$ effective for β, and furthermore both 2 and 3 are better off in β than in α. Hence $\beta \succ \alpha$.

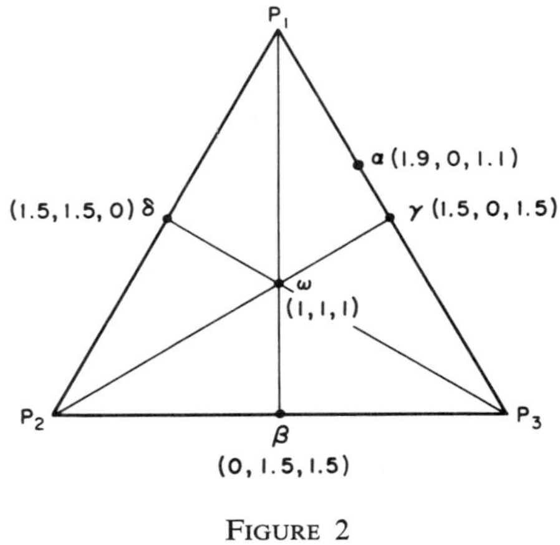

FIGURE 2

The trio of imputations β, γ, and δ forms a solution set to this particular game. Any other imputation gives two of the players less than 1.5 apiece and thus is dominated by one of these three imputations, but the three do not dominate each other. (There are other solution sets which we need not discuss.) *This game has no core*, since the imputations β, γ and δ, dominating all the rest, are themselves dominated by others. For example, the imputation α, which was dominated by β, in turn dominates δ via the effective set $\overline{13}$. Note that domination is not a transitive relation: $\beta \succ \alpha$ and $\alpha \succ \delta$ do not entail $\beta \succ \delta$.

We now consider three closely related games differing from the previous one only in what the two-person coalitions obtain. In the first variant we have

$$v(\overline{12}) = v(\overline{13}) = v(\overline{23}) = 0.^{3}$$

In this case, *all* imputations are in the core. The only set of players that is effective, for most imputations, is the three-person set; however, this is useless for domination since, on examining the distribution of welfare from the viewpoint of all three players, we see that if one player prefers one of two imputations, then at least one of the other players will prefer the other, the sum of the allotments being constant. In fact, it suffices to point out that no set of players is *strictly* effective for any imputation—hence there is no domination. The core is therefore as large as possible, and is also the unique von Neumann-Morgenstern solution.

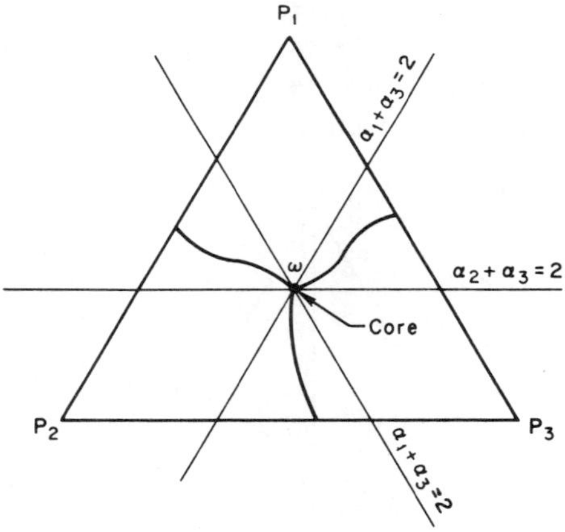

FIGURE 3

[3] This all-or-nothing type of characteristic function, like the previous one, is associated more with political than economic processes [10]. The previous game was a majority-take-all situation; the present one is a veto situation, since if one member wishes to be the "dog in the manger," he can prevent the others from obtaining any payoff. In economics such extremes—called *simple games*—are not typical. We shall presently consider variants in which the two-person coalitions obtain intermediate amounts, reflecting the more usual situation in which any new adherent to a coalition means added possibilities for profit.

In our third example we assume

$$v(\overline{1}\overline{2}) = v(\overline{1}\overline{3}) = v(\overline{2}\overline{3}) = 2.$$

As shown in Figure 3, the lines which describe the amount obtainable by each coalition of two players intersect in a single point, the imputation ω with coordinates (1, 1, 1). This is the only undominated imputation of the game, and thus constitutes a single-point core. Since ω fails to dominate the three small triangles adjoining it in the diagram, however, it is not a von Neumann-Morgenstern solution by itself. To get a solution we must add some more or less arbitrary curves, as shown, traversing the three triangular regions (see [9], pp. 550-554).

In the final variant, we assume that the two-person coalitions are only half as profitable as in the preceding example. That is, we have

$$v(\overline{1}\overline{2}) = v(\overline{1}\overline{3}) = v(\overline{2}\overline{3}) = 1.$$

The lines indicating the ranges of effectiveness of these coalitions are spread apart, as shown in Figure 4, revealing a large, hexagonal core. All imputations in that area are undominated. As in the second example, this core is the unique solution.

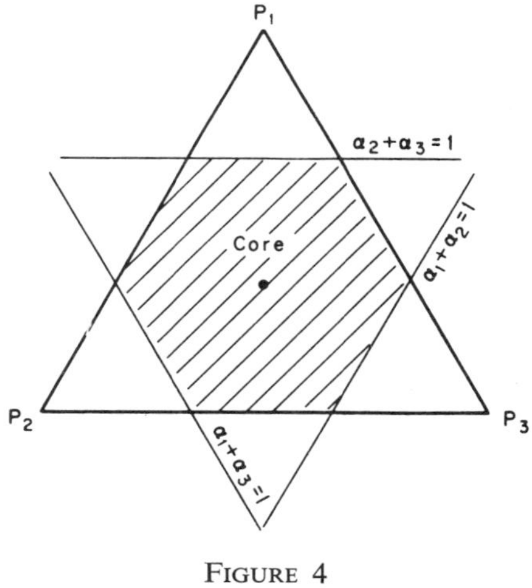

FIGURE 4

A superficial examination of these four examples suggests a relationship between the size of the core and the "fatness" of the coalitions in a game, i.e., how much they can promise their members per capita as compared to the per capita amount available in the whole game. In all four instances, the latter amount was $v(\overline{1}\overline{2}\overline{3})/3 = 1$. Denote $v(\overline{i}\overline{j})/2$ by f_2. In the first game, f_2 was 1.5, which is greater than 1, and there was no core. In the third game, f_2 was exactly 1, and the core was a single point. In the fourth game,

f_2 was 1/2, and there was a large core, while in the second game, f_2 was 0, and every imputation was in the core.

Of course, in a less symmetric situation, this principle would not reveal itself in such a clean-cut manner. However, a general rule of thumb seems to persist: the more power there is in the hands of the middle-sized groups, the more narrowly circumscribed is the range of outcomes of the cooperative game. This rule, vague as it is, applies to solutions as well as cores.

When we do not permit the transfer of utility, we can no longer talk about the amount attainable by a group of players as a single number; nevertheless, cores [6] and solutions [11] can still be defined, and the idea of the per-capita gain in one coalition being larger than in another can still be utilized in a vectorial sense.

4.3. THE EDGEWORTH MARKET GAME, $(n, n)_2$. Let us return to the Edgeworth bilateral-monopoly game $(1, 1)_2$. Even without the simplifying assumption of transferability of utility, it is not hard to see that the Edgeworth contract curve is the "cooperative solution," in the spirit of the von Neumann-Morgenstern definition, but with weaker utility assumptions [11]. It is also the core. The reason is simply that with just two players, there are no coalitions of intermediate size, between the individual and the whole group, and hence no domination occurs between outcomes that are both Pareto optimal and individually rational.

It can be shown that the whole contract curve (considered in the higher dimensions) will remain as a cooperative solution to the game $(n, n)_2$ for any number of traders. However, it will no longer be the core if there are more than two players, and there will be many other von Neumann-Morgenstern-type solutions, which may in general be quite difficult to compute (see [4]).

It has been shown that as the number of players increases, the core shrinks down upon the competitive equilibrium point (or set of points) [3, 5, 12]. Hence, if we regard the core as our concept of cooperative solution, the game $(1, 1)_2$ has the contract curve as its solution, while the game $(n, n)_2$, as n grows large, has an increasingly determinate solution consisting of a small neighborhood of the competitive equilibrium.

Figure 5 illustrates the shrinking of the core as the number of players is increased. It is similar to Figure 1 of Edgeworth ([1], pp. 20–25). The line RP is the competitive price ray on the exchange diagram for $2n$ traders, consisting of n of each type. The arc CC' is the range of the two-person contract curve, and I is a typical point on that curve. The shaded area between I and I_1 indicates a domain that is preferred to I by a player of the first type; similarly the shaded area between I and I_2 is preferred to I by a player of the second type. These areas are bounded by the two indifference curves which are tangent at I; each area includes a portion of the price ray RQ only if I is *not* the competitive allocation P. We shall describe the conditions under which I (or more precisely, the imputation associated with I) is dominated by the outcome of some trade among a subset of the players.

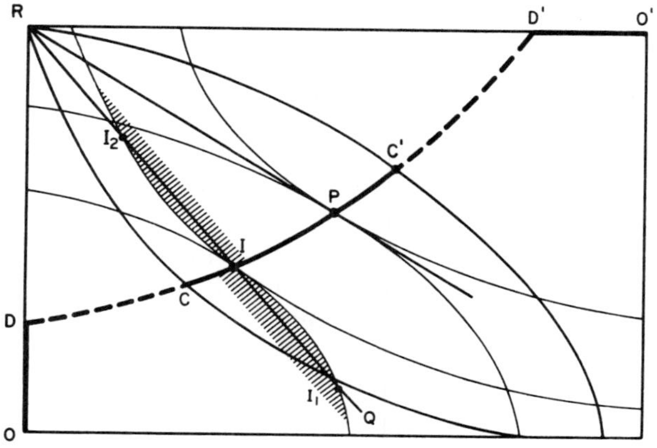

FIGURE 5

When only two traders enter into contract, exchanging x units of the first good for y of the second, the final distribution of assets between them is given by $(a - x, y)$ and $(x, b - y)$. This can be represented by a single point on the Edgeworth diagram. Suppose that a coalition of k traders of the first type and m traders of the second type forms, and suppose that traders of the same type decide to divide their gains equally. Together the coalition controls ka units of the first commodity and mb of the second. After trade, a player of the first type will have $\left(a - \frac{x}{k}, \frac{y}{k}\right)$, and a player of the second type will have $\left(\frac{x}{m}, b - \frac{y}{m}\right)$, for some x, y. This outcome can be represented on the Edgeworth diagram by two points, corresponding to the two types of players, which fall on the line RQ, one on either side of the contract curve. If $k = m$, the two points coincide at I. If the ratio of k to m is only slightly different from unity, it will be possible for the point corresponding to each type of player to lie in the shaded region in which that type of player gains, in comparison to the distribution offered at I. The imputation corresponding to I will therefore be dominated. As n grows larger, the available ratios k/m become denser around 1, and the portion of the contract curve that escapes domination shrinks down to an arbitrarily small neighborhood of the competitive equilibrium point. (If the latter is not unique, the convergence will be to the set of competitive equilibria.)

It is remarkable that such widely different sets of modeling assumptions as were used in $(n, n)_0$ and $(n, n)_2$ should lead, in the limit, to the same solution. In the one case, the participants operate on a minimum of information, they have essentially no strategic initiative, and they are prevented from cooperating or even interacting with their fellows. In the other case, information is freely available, exchanges of goods may be made in all possible ways, without regard for prices, and collusion not only is permitted but is in fact essential to the maintenance of stability.

Viewed as a limit of cores, the competitive equilibrium for large n is seen to be "sociologically neutral." No coalition is effective against it. It is not a von Neumann-Morgenstern-type solution by itself, even in the limit, since it does not dominate the other points on the contract curve; nevertheless, every such solution will include the competitive imputation. Thus, without any dynamic assumptions regarding prices or other mechanisms of the market place, and without special hypotheses concerning costs of communication, information, and so forth, the competitive equilibrium plays an important and distinctive role purely on the ground of social stability.

5. NONCOOPERATIVE GAMES. THE COURNOT OLIGOPOLY GAME

5.1. NONCOOPERATIVE EQUILIBRIUM POINTS. An approach highly different from that of cooperative game theory, and connected less directly to welfare considerations than to problems of oligopoly and control of industry, is the idea of the *noncooperative* solution to a game. The spirit of much of the classical discussion of oligopoly behavior and equilibrium in monopolistic competition has been to regard the firms as powerful and their customers as weak. This conforms to our commonsense notions of the oligopolistic market place. The automobile or tobacco companies, for example, have some form of control over output, prices, brands, and so forth, but their customers are, for the most part, price-takers.

Both the classical and many of the more recent works in oligopoly theory have made use of "open" models, in the sense that they investigate the behavior of a group of firms or an industry, taking the behavior of customers and suppliers as given. Underlying the writings of Cournot [13], Bertrand [14], Edgeworth [15], Hotelling [16], Stackelberg [17], Chamberlin [18], and others, is the concept of the attainment of a noncooperative equilibrium by the firms in competition. On the mathematical side, Nash [19] was first to develop the basic idea of noncooperative solution in the abstract framework of the theory of games. The crucial concept in the noncooperative theory is that of "equilibrium point," which may be defined as follows:

Consider a game with n players. Let player i have a class of possible "strategies" S_i, and let s_i denote a particular strategy belonging to the class. The payoff to the ith player is denoted by $P_i(s_1, s_2, s_3, \ldots, s_n)$, a function of the strategies of all the players. A strategy vector $(\bar{s}_1, \bar{s}_2, \bar{s}_3, \ldots, \bar{s}_n)$ is said to constitute an *equilibrium point* if, for all i, the function

$$P_i(\bar{s}_1, \bar{s}_2, \ldots, \bar{s}_{i-1}, s_i, \bar{s}_{i+1}, \ldots, \bar{s}_n)$$

is maximized by setting $s_i = \bar{s}_i$. In other words, a set of strategies, one for each player, forms an equilibrium point if each player, knowing the strategies of all others, will not be motivated to change. If the equilibrium point is unique, as it proves to be in many of the classical economic models, it is termed the "noncooperative solution" to the game.

5.2 THE COURNOT MODEL. An early economic example of noncooperative equilibrium was presented by Cournot [13]. Suppose that two firms are constrained to name amounts q_1, q_2 offered for sale. These quantities are their strategic variables. Furthermore, suppose that there is a market mechanism of some sort which selects a price that exactly clears the market, say $p = D(q_1 + q_2)$. Let the cost functions of the two firms be $C_1(q_1)$ and $C_2(q_2)$, respectively. The payoffs are then

$$P_1 = q_1 D(q_1 + q_2) - C_1(q_1),$$

$$P_2 = q_2 D(q_1 + q_2) - C_2(q_2)$$

respectively. An equilibrium point will be a pair of strategies $(\bar{q}_1, \bar{q}_2)$ such that

$$P_1(q_1, \bar{q}_2) \text{ is maximized at } q_1 = \bar{q}_1,$$

$$P_2(\bar{q}_1, q_2) \text{ is maximized at } q_2 = \bar{q}_2.$$

This is the "Cournot" solution [20].

In the game described above, the welfare of the customers was included only implicitly, through the workings of the demand relation; it is thus an open model. Considering only the welfare of the two duopolists, it is easy to see that the noncooperative equilibrium is not Pareto optimal in the open model. They could both improve matters by joint action, restricting the amounts offered to be sold.

It is nevertheless conceivable that the noncooperative equilibrium might be Pareto optimal in the closed model, in which the welfares of the customers are also taken into account; however, even this is not the case in general. We already have a simple example of this in the game $(1, 1)_1$, which can be viewed as a monopolist dealing with a compliant, price-taking customer. In this case, quantity naming is usually equivalent to price naming. Referring to Figure 6, we see that if the monopolist (origin at 0′) offers the amount e to be sold, for example, the price mechanism

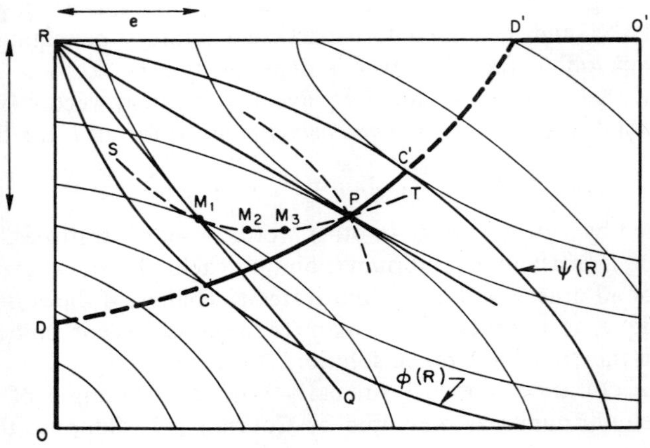

FIGURE 6

will determine the price ray *RQ* that just clears the market when the other player optimizes, selecting the point *M*, where the ray intersects his response curve *ST*. Under suitable assumptions, this derived price ray will be unique, for any value of *e*, giving the monopolist effective control over prices.

Returning to our point about optimality, we note that the best outcome from the viewpoint of the monopolist is the point M_1, which is clearly not Pareto optimal in the closed model since it lies off the contract curve. In this instance of monopolistic exploitation, all could have their welfare improved by negotiation.

5.3. THE GAME $(n, n)_1$. In order to define the game $(n, n)_1$, we must specify the market mechanism more fully, making explicit the scope of strategic choice. There are many ways in which this can be done. We might assume that the firms playing the role of oligopolists are in a position to name price. This is the manner in which Bertrand and Edgeworth handled oligopoly. Though complicated, it is possible to specify a mechanism whereby each oligopolist names both a price and a limit to the amount he is willing to trade. In order to complete the specification, the nature of the market-clearing process, in the presence of possibly different prices, would also have to be made explicit.

An easier model to define, involving fewer arbitrary institutional assumptions, is the analogue to Cournot's approach, in which the oligopolists are free to specify the *amounts* to be offered for trade, but not the prices. A mechanism then sums together all the offers and calculates the prices that will exactly clear the market. The traders of the second type are constrained to be price-takers, as before. We shall adopt this quantity-naming model as the most natural generalization of the game $(1, 1)_1$, in which, as already noted, price naming and quantity naming are essentially equivalent.

Let us assume that the "response curve" (ST in Figure 6) of the second player in the two-person game $(1, 1)_1$ can be expressed as a single-valued function $y = r(x)$, where (x, y) is that player's final holding, after trade. In addition, we shall assume that $r(x)$ has a bounded derivative. (The purpose of these assumptions is to exclude certain difficulties related to the possibility of different prices yielding the same demand.) Then, in the game $(n, n)_1$, the price that clears the market is given by

$$p = \frac{b - r(\bar{q})}{\bar{q}},$$

where $\bar{q} = \sum_i q_i/n$ is the average quantity offered of the first good. The final holdings will be

$$(a - q_i, pq_i),\ i = 1, 2, \ldots, n,$$

for the players of the first type, and

$$(\bar{q}, b - p\bar{q})$$

for each player of the second type. The noncooperative equilibrium is found by requiring the function $\psi(a - q_i, pq_i)$ to be a maximum with respect to q_i, for each i.

This maximum will generally not be found at the point M_1 in Figure 6, where $r(x)$ is tangent to the family ψ. The reason is that when a player varies his offer q_i, the *average* offer $\bar{q}$, which is what determines the price, varies by only $1/n$ as much. In fact, the oligopolists will tend to offer larger quantities than they would if they were in monopolistic collusion, and the noncooperative equilibrium outcome M_n will be displaced along the response curve in the direction of the competitive equilibrium outcome P. (See Fig. 6.)

To see why M_n converges to P in the limit, we merely set $\partial\psi(a - q_i, pq_i)/\partial q_i = 0$, and obtain

$$p = \frac{\psi_x}{\psi_y} - q_i\frac{\partial p}{\partial q_i} = \frac{\psi_x}{\psi_y} + \frac{q_i}{n\bar{q}}(r' + p),$$

a relation which must hold at M_n. But r' is bounded by assumption, as is q_i, and it can be shown that $\bar{q}$ does not go to zero; therefore

$$p \to \frac{\psi_x}{\psi_y} \quad \text{as} \quad n \to \infty.$$

This implies that M_∞ lies on the response curve of the players of the first type as well as of those of the second type. Hence $M_\infty = P$.[4]

The intuitive idea behind this argument is that when n becomes large, the effect of one individual on the price structure becomes negligible; i.e., $\partial p/\partial q_i \rightarrow 0$.

6. CONCLUSION

We began with three different models of bilateral monopoly, based on the same data. Three different solution concepts were advanced, each appropriate to one of the models, and three qualitatively different outcomes were predicted. As the number of traders on both sides of the market was increased, the three solutions merged into one as regards predicted outcome, but their *rationales* remained quite distinct. A relationship was thereby demonstrated among (1) administered price stability (the *competitive equilibrium*), (2) noncollusive oligopolistic exploitation (the *equilibrium point* of a noncooperative game), and (3) unrestricted bargaining between coalitions (the *core* of a cooperative game).

Each model embodied radically different assumptions concerning the strategies and information available to the participants. A scrutiny of the differences would reveal many places where sociological and institutional assumptions might be slipped in, perhaps inadvertently, in the construction of models of markets.

REFERENCES

[1] Edgeworth, F. Y., *Mathematical Psychics*, Kegan Paul, London, 1881.

[2] Shapley, L. S., and M. Shubik, manuscript in progress on applications of *n*-person game theory.

[3] Shubik, Martin, "Edgeworth Market Games," in: *Contributions to the Theory of Games*, Vol. 4, Annals of Mathematics Study No. 40, Princeton, Princeton University Press, 1959, pp. 267–278.

[4] Shapley, L. S., "The Solutions of a Symmetric Market Game," in: *Contributions to the Theory of Games*, Vol. 4, Annals of Mathematics Study No. 40, Princeton, Princeton University Press, 1959, pp. 145–162.

[5] Scarf, Herbert, "An Analysis of Markets with a Large Number of Participants," *Recent Advances in Game Theory* (proceedings of a conference held at Princeton University, October 4–6, 1961), Princeton University, 1962, pp. 127–155.

[6] Aumann, R. J., "The Core of a Cooperative Game without Side Payments," *Trans. Amer. Math. Soc.*, Vol. 98, 1961, pp. 539–552.

[7] Debreu, Gerard, *On a Theorem of Scarf*, Cowles Foundation Discussion Paper No. 130, November, 1961.

[4] More precisely, since uniqueness is not assured, any limit point of any sequence $\{M_n\}$ of noncooperative equilibria is a competitive allocation. Our assumptions regarding $r(x)$ are essential to this result.

[8] Debreu, Gerard, *Theory of Value, An Axiomatic Analysis of Economic Equilibrium*, Cowles Foundation Monograph 17, New York, John Wiley & Sons, Inc. 1959.

[9] von Neumann, J. and O. Morgenstern, *Theory of Games and Economic Behavior*, Princeton, Princeton University Press, 3rd Ed., 1953.

[10] Shapley, L. S., "Simple Games: An Outline of the Descriptive Theory," *Behavioral Science*, Vol. 7, 1962, pp. 59–66.

[11] Shapley, L. S., and M. Shubik, "Solutions of *n*-person Games with Ordinal Utilities" (abstract) *Econometrica*, Vol. 21, 1953, pp. 348–349.

[12] Shapley. L. S., and M. Shubik, *The Core of an Economy with Nonconvex Preferences*, The RAND Corporation, RM–3518, February, 1963.

[13] Cournot, Augustin A., *Researches into the Mathematical Principles of the Theory of Wealth*, New York, Macmillan and Co., 1897, pp. 79–80, 84.

[14] Bertrand, J., "Théorie Mathématique de la Richesse Sociale" (review), *Journal des Savants*, Paris, September 1883, pp. 499–509.

[15] Edgeworth, F. Y., *Papers Relating to Political Economy*, I. Macmillan and Co., London, 1925, pp. 111–142, 118, 120.

[16] Hotelling, H., "Stability in Competition," *The Economic Journal*, March 1929, p. 41.

[17] Stackelberg, H. von, *Marktform und Gleichgewicht*, Berlin, Julius Springer, 1934.

[18] Chamberlain, Edward H., *The Theory of Monopolistic Competition*, Cambridge, Harvard University Press, 6th Ed., 1950.

[19] Nash, J. F., Jr., "Non-Cooperative Games," *Annals of Mathematics*, Vol. 54, September 1951, pp. 286–295.

[20] Mayberry, J. P., J. F. Nash, and M. Shubik, "A Comparison of Treatments of a Duopoly Situation," *Econometrica*, Vol. 21, 1953, pp. 141–154.

PART II

Mathematical Programming

CHAPTER 7

A Property and a Use of Output Coefficients of a Leontief Model

By STEDMAN B. NOBLE*

This paper introduces output coefficients of a Leontief model, demonstrates a formal relationship between these coefficients and the standard input coefficients, and cites a method of model construction where output coefficients could be used. The functional dependence that is assumed in the Leontief model is represented by input coefficients; hence, a main purpose of this paper is to prove a proposition on the relationship between an inverse matrix based upon output coefficients and one based upon input coefficients. This proposition shows that it is meaningful to invert a matrix of output coefficients, while holding to the causal interpretation of input coefficients.

Output coefficients can be used to suggest portions of a Leontief matrix that need to be disaggregated. A major limitation of the Leontief model is the existence of errors due to aggregation. Rather strict criteria must be satisfied for there to be no error due to aggregation, and empirically it is essentially inevitable that there will be some aggregation error somewhere in a Leontief model. However, different aggregations will place the error in different places, so the problem is to locate the error where it does as little harm as possible.

One way to focus the problem is to use a Leontief model in order to solve for only one industry, adjusting the aggregation error so as to have little effect upon this industry. The amount of error that is possible for a given industry depends upon the connections of this industry to other industries and upon the size of these connections. Output coefficients represent the proportion of output of an industry that goes to each user, and computation based upon output coefficients represents the size of the connections. Using output coefficients, an analyst can reconstruct (or construct) a model for a given industry by determining its connections with other industries, seeing the size of the connections, and reclassifying the industries that appear to introduce the major aggregation errors.

* Research Analysis Corporation.

This work was done while the author was at The George Washington University Logistics Research Project and it was supported by the Office of Naval Research, Contract Nonr 761(05). Reproduction in whole or in part is permitted for any purpose of the United States Government. The author is indebted to the support and comment of numerous people, including Jerome Bracken, W. Duane Evans, Edward B. Hincks, W. H. Marlow and David Rosenblatt, without extending to them responsibility for limitations or errors.

1. THE STATIC LEONTIEF MODEL

The static Leontief matrix [1, 3, 4, 12] is derived from a table of balanced accounts which can be represented by the following equations:

$$p_i x_i = \sum_j p_i t_{ij} + p_i f_i \quad i = 1, \ldots, n, \tag{1}$$

$$p_j x_j = \sum_i p_i t_{ij} + v_j x_j \quad j = 1, \ldots, n. \tag{2}$$

p_i is the price of commodities sold by industry i;[1]
x_i is the output level of industry i;
t_{ij} is the purchases by industry j from industry i and is identically equal to the sales from i to j.
f_i is the exogenous demand (called "final demand") of industry i. This includes items that "balance" the matrix [13] such as net changes in inventories.
v_j is the "value added" of industry j. This includes "balancing" items such as profit margins.

By convention $x_i > 0$ and $p_i > 0$, and all elements of the accounts except the balancing items are nonnegative, hence $t_{ij} \geq 0$. On empirical grounds, let us further assume $v_j \geq 0$.[2]

An element $t_{ij} \neq 0$ represents a flow of commodities that has passed from industry i to industry j; hence there is a relationship between i and j. To obtain a model, one desires a relationship that is stable enough so that predictions can be made. Such a relationship may occur for purchases since some inputs are necessary in order to produce, that is[3]

$$t_{ij} = f(x_j). \tag{3}$$

More specifically, the assumption that has been used is proportionality

$$t_{ij} = a_{ij} x_j \tag{3'}$$

[1] These prices are price indices for the composite commodities sold by the account. Furthermore, the indices can represent either of two prices: purchaser's price or producer's price, which differs by the "spread items" such as transportation costs [4]. In a Leontief table, these costs are assigned to the purchaser or the producer, resulting in one price index for each industry.

[2] Wage payments are typically exogenous to the model, and empirically they would outweigh any negative profits industry by industry, hence $v_j \geq 0$. The elements f_i are unrestricted in sign since an industry may have inventory decreases that are larger than other portions of its final demand.

[3] Two aspects of this assumption can be distinguished: an industry is assumed to have (1) homogenous output for which there is (2) a single method of production. When the second aspect is improper, a linear programming model can be used [3]. The first aspect implies the need for detailed industries. In practice, industries of a Leontief table are composed of aggregations of many outputs, introducing errors into the resulting model. For an important test of this model, see [7].

where a_{ij} is a non-negative input coefficient that is stable through time.

The assumption (3) cannot be made for some sectors, hence these sectors are treated as exogenous (f_i) which means that the demand of these sectors must be specified apart from the model. For the sectors where the assumption is used, assumption (3′) can be inserted into the equations of balanced accounts (1) and (2) to obtain the two[4] static Leontief models

$$x_i - \sum_j a_{ij}x_j = f_i \quad i = 1, \ldots, n, \tag{4}$$

$$p_j - \sum_i p_i a_{ij} = v_j \quad j = 1, \ldots, n. \tag{5}$$

In matrix notation, these equations are $(I - A)x = f$ and $p(I - A) = v$.

The inverse matrix is given by the Neumann series which is the sum of the powers of the matrix [16]:

$$B \equiv I + \sum_{n=1}^{\infty} A^n \tag{6}$$

where a_{ij}^n (the ij term of the matrix A^n) is the sum of the products of all sets of n coefficients that "connect" i to j [5, 12, 13]. A set of coefficients is connected if the adjacent indices are the same. An example: $a_{ij}^3 \equiv \sum_g \sum_h a_{ih}a_{hg}a_{gj}$. To every term $a_{ij}^n \neq 0$ there can be associated a directed graph that connects the indirect supplier i to the producer j,[5] and i is attainable from j in n steps.

Clearly the terms of A^n go to zero as n goes to infinity if and only if the inverse matrix exists. When the inverse matrix exists, then the equations (4) and (5) can be transformed into $x = (I - A)^{-1}f$ and $p = v(I - A)^{-1}$. It has long been known that an inverse matrix will exist if $v_jx_j > 0$ for every j; or if the matrix is indecomposable and $\sum_j v_jx_j > 0$ [14].[6] A sharper theorem is that, if $\sum_j v_jx_j > 0$, then there will exist an inverse matrix of the principal submatrix (possibly the entire matrix) composed of the set of industries that is connected to any positive element of p_if_i [12]. Such a positive element will exist, since from (1) and (2)

$$\sum_i p_if_i = \sum_i p_ix_i - \sum_i \sum_j p_it_{ij} = \sum_j p_jx_j - \sum_j \sum_i p_it_{ij} = \sum_j v_jx_j.$$

[4] Since the balanced condition depends upon prices as well as industry levels, a solution of each model is required to predict a balanced table [1, p. 352, 13].

[5] The direction of flow in [12] represents purchases, whereas in the present paper, the direction of flow represents sales.

[6] A basic paper by Wong [17] gives ten equivalent conditions, one of which is the convergence of a power series. [18] is a review of the numerous contributions of Wong to the mathematics of the Leontief model.

2. OUTPUT COEFFICIENTS[7]

An output coefficient is defined by

$$c_{ij} = t_{ij}/x_i, \tag{7}$$

that is, it is the sales of industry i to industry j divided by the output of industry i. The relationship between output coefficients (7) and input coefficients (3′) is $a_{ij}x_j = t_{ij} = x_i c_{ij}$. From this follows the *Main Proposition*:

$$d_{ij}(x_i/x_j) = b_{ij}, \tag{8}$$

where

- a_{ij} is an input coefficient of the matrix A,
- b_{ij} is an element of the matrix $B = (I - A)^{-1}$,
- c_{ij} is an output coefficient of the matrix C, and
- d_{ij} is an element of the matrix $D = (I - C)^{-1}$.

PROOF.[8] Let X be a diagonal matrix with the positive elements x_i on the diagonal. By definition, $AX = T = XC$. From this it follows that

$$A^n X = (TX^{-1})^{n-1}T = T(X^{-1}T)^{n-1} = XC^n, \quad n = 1, \ldots, \infty.$$

But by (6),

$$B \equiv (I - A)^{-1} = I + \sum_{n=1}^{\infty} A^n.$$

Hence

$$BX = X + \sum_{n=1}^{\infty} A^n X = X + \sum_{n=1}^{\infty} XC^n = XD,$$

and the proposition follows directly.

The proposition states that the inverse of a matrix of output coefficients can be transformed into the inverse of the matrix of input coefficients (of the same table of accounts) by multiplying by the ratio x_i/x_j.

The related matrix $d_{ij}(f_j/x_j)$ is easier to form than an inverse matrix and has some desirable properties. The relevance of this form is seen by substituting (7) into (1)[9] to obtain

$$x_i = \sum_j x_i c_{ij} + f_i, \qquad \text{or} \qquad 1 - \sum_j c_{ij} = f_i/x_i,$$

[7] Output coefficients were used by Wood and Horton [19] in defining an almost-triangular arrangement of a Leontief matrix.

[8] The proof was simplified by discussions with W. D. Evans and David Rosenblatt. The proposition is an example of normalization, other examples for the Leontief model being the conversion from a net to a gross matrix [2] and the adjustment for prices so that a solution is a set of balanced accounts [1, p. 352, 13].

[9] Substitution of (7) into (2) complicates the price equations.

or from the definition of an inverse matrix and the definition (6),

$$1 = \sum_j \sum_{n=0}^{\infty} c_{ij}{}^n f_j / x_j = \sum_j d_{ij} f_j / x_j. \tag{9}$$

Since the sum of the products in the power series equals unity, any partial set of elements of a power series can be summed, and one can determine the proportion of sales represented by the partial set. This is a useful property of this form of matrix.

The *Main Proposition* proved above implies that

$$d_{ij} f_j / x_j = b_{ij} / x_i f_j. \tag{10}$$

The elements f_j and x_j are for the base year: the year for which the table was constructed. To solve for a future year, future elements of exogenous demand $\hat{f}_j$ are used and these predict future output levels $\hat{x}_i$. Then

$$\sum_j \frac{b_{ij} f_j}{x_i} \frac{\hat{f}_j}{f_j} = \frac{\hat{x}_i}{x_i} \tag{11}$$

since $\sum_j b_{ij} \hat{f}_j = \hat{x}_i$.

To restate this analysis verbally, an important way to use output coefficients is to form an inverse matrix based upon them and weight the terms to accord with normalized exogenous demand. The result is a row for each industry showing the proportion of output accounted for by each exogenous demand. This row is identically equal to a corresponding expression based upon input coefficients. These terms can be multiplied by an index of the proportionate change in the elements of final demand $\hat{f}_j / f_j$ in order to obtain an index of the proportionate change in output levels of the industry, $\hat{x}_i / x_i$.

3. A USE OF OUTPUT COEFFICIENTS

The main proposition establishes a formal equivalence between an inverse matrix based upon input coefficients and one based upon output coefficients. Hence inversions and solutions can be obtained using either kind of coefficient, and the result can be transformed if necessary, by multiplying by the ratio of the output levels. This ratio can be far from unity. For example, for cadmium going into automobiles, the ratio is approximately one to one thousand: very little cadmium is put into automobiles, but a considerable proportion of the purchases of cadmium is made by the automobile industry. This difference in size is important in the computation of approximate solutions, where it may be efficient to ignore small coefficients [8]. A Leontief model predicts the output of an industry, so the approximation should reflect significance in terms of the proportion of output accounted for by any flow. For example, the automobile industry is a significant purchaser of cadmium so a model for cadmium should include an explicit sector for automobiles (and perhaps should disaggregate it into more detailed sectors).

One reason for obtaining an approximate solution might be the cost of computation for a large matrix. More important, however, is that any solution of a Leontief matrix is approximate because of errors due to aggregation. Empirically, it has been impossible to quantify or to solve a matrix in sufficient detail to prevent errors due to aggregation.[10] While any single error can be corrected by disaggregation, it is not possible to correct all such errors. What is needed is a criterion as to where disaggregation should occur. An important aspect of this criterion is the significance that is measured by output coefficients.

Operationally, the correction of aggregation errors can involve the following four-part program.

1. Select a definite industry i whose solution is desired.[11]
2. Compute the terms of row i of the inverse matrix. These terms can be multiplied by base year exogenous demand, shown in equation (9), to see which terms have accounted for significantly large proportions of sales in the past. This measurement of significance needs to be adjusted, as suggested by equation (11), for proportionate changes in exogenous demand for the predicted year. In practice, the base year weights can be treated as surrogates of future weights except where a large proportionate change in weights is anticipated.
3. The previous analysis can be thought of as having computed a directed graph of all flows that originate in industry i [8]. For the flows that are significantly large, this graph can be analyzed to identify connections that are incorrect.
4. Industries can be redefined and quantities can be reallocated that seem to introduce incorrect connections and noticeable quantitative errors.[12]

This procedure requires the use of judgment at a number of points, and the resulting models include aggregate industries. However, the use of output coefficients delineates the significantly large sales for an industry, permitting the elaboration of the model where it may have a noticeable effect upon accuracy. Hence, the use of output coefficients focuses the judgment and may permit models to be constructed that can, with reasonable accuracy, predict for a given industry of the economy.

[10] There is an extensive literature on criteria for there to be no errors due to aggregation. Several writers [6, 15] admit these will not be satisfied in practice.

[11] Of course, the analysis can be repeated for as many industries as is desired.

[12] Another procedure is to obtain the directed graph first, and then associate quantities with this graph. This is the method of "material flow graphs" [9, 10].

REFERENCES

[1] R. G. D. Allen, *Mathematical Economics*, London, Macmillan, 1956, Chapter 11.
[2] Judith B. Balderston and T. M. Whitin, "Aggregation in the Input-Output Model," in *Economic Activity Analysis*, editor O. Morgenstern, New York, Wiley, 1954, pp. 89–90.
[3] Hollis B. Chenery and Paul G. Clark, *Interindustry Economics*, New York, Wiley, 1959.
[4] W. Duane Evans and Marvin Hoffenberg, "The Interindustry Relations Study for 1947," *Review of Economic Statistics*, Vol. 34, 1952, pp. 102–104.
[5] Richard M. Goodwin, "Dynamical Coupling with Especial Reference to Markets Having Production Lags," *Econometrica*, Vol. 15, 1947, pp. 181–204.
[6] Michio Hatanaka, "Note on Consolidation Within a Leontief System," *Econometrica*, Vol. 20, 1952, pp. 301–303.
[7] ——, *The Workability of Input-Output Analysis*, Ludwigshafen, 1960.
[8] Martin Hershkowitz and Stedman B. Noble, "Finding the Inverse and Connections of a Type of Large Sparse Matrix," *Naval Research Logistics Quarterly*, Vol. 12, March 1965.
[9] Stedman B. Noble, "Material Flow Graphs," abstract in *Econometrica*, Vol. 30, 1962, pp. 580–581.
[10] ——, "Some Flow Models of Production Constraints," *Naval Research Logistics Quarterly*, Vol. 7, 1960, pp. 401–419.
[11] ——, "Structure and Classification in Resource Flow Models," Serial T-100/59, Logistics Research Project, The George Washington University, 24 April 1959.
[12] David Rosenblatt, "On Linear Models and the Graphs of Minkowski-Leontief Matrices," *Econometrica*, Vol. 25, 1957, pp. 325–338.
[13] ——, "On Some Aspects of Models of Complex Behavioral Systems," in *Information and Decision Processes*, editor R. E. Machol, New York, McGraw-Hill, 1960, pp. 62–86.
[14] Olga Taussky, "A Recurring Theorem on Determinants," *American Mathematical Monthly*, Vol. 56, 1949, pp. 672–676.
[15] H. Theil, "Linear Aggregation in Input-Output Analysis," *Econometrica*, Vol. 25, 1957, pp. 111–122.
[16] Y. K. Wong, "Some Mathematical Concepts for Linear Economic Models," in *Economic Activity Analysis*, editor O. Morgenstern, New York, Wiley, 1954, pp. 291–293.
[17] ——, "Some Properties of the Proper Values of a Matrix," *Proceedings of the American Mathematical Society*, Vol. 6, 1955, pp. 891–899.
[18] Y. K. Wong and Oskar Morgenstern, "A Study of Linear Economic Systems," *Weltwirtschaftliches Archiv*, Band 79, 1957, pp. 222–239.
[19] Marshall K. Wood and H. Burke Horton, "An Experimental Dynamic Model of the U.S. Economy," presented at the Annual Meeting of the American Statistical Association, 29 December 1950.

CHAPTER 8

Some Approaches to the Solution of Large-Scale Combinatorial Problems

By GERALD L. THOMPSON*

1. INTRODUCTION

In a paper on demand theory [16], Oskar Morgenstern criticized classical demand theory, pointing out some of its imperfections. One of the criticisms he made was that the classical assumption of infinite divisibility of goods was impossible for such goods as newspapers, where demand must necessarily be for 0 or 1 unit. The same criticism goes for other consumer goods (e.g., automobiles, swimming pools, yachts, etc.) for which the demand must necessarily be an integral number (typically small). To change these into infinitely divisible goods would alter essentially their character.

Another of Morgenstern's criticisms was that a given individual's demand curves for several different goods are not independent since they compete with each other for the individual's total spendable income in a given period. This is true regardless of whether or not all or some of the goods are infinitely divisible. A price change for one good will cause a complicated sequence of time-dependent changes in demand functions for each of the other goods. Thus the classical economic-budget problem of maximizing an individual's total utility subject to a budget constraint (which might be an inequality constraint instead of an equality constraint) becomes, in part, a combinatorial problem. And the more indivisible goods for which the individual has positive demand, the bigger the combinatorial problem it is.

Morgenstern's further criticisms involve problems of aggregation and game-theoretic questions that will not concern us here. Suffice to say that the paper was of considerable importance and stimulated many researches such as those of Martin Shubik [22], and others.

For our purposes it is enough that Morgenstern isolated in economics a *combinatorial decision problem*, that is, the problem of choosing from a very large set of objects (e.g., the set of all possible commodity bundles that an individual can afford) one that is best when measured relative to a given criterion function (such as total utility).

Management science (which overlaps economics) abounds in combinatorial decision problems, which are also called discrete optimization problems. The set of objects from which a choice is to be made is typically very large, for instance, on the order of the number of elementary particles

* Carnegie Institute of Technology.

in the universe. Hence the mathematically acceptable solution of enumerating all possibilities and selecting an optimal one is computationally infeasible even for the fastest electronic computers. Nevertheless, these problems are presently being faced and "solved" in actual practice. Returns that result from even a small improvement over present practices are considerable.

The purpose of this article is to discuss some of the recent attempts to obtain, within present computational abilities, near-optimum or good solutions to such problems. We discuss the construction of algorithms and heuristic programs for their solutions and give some actual applications in which successful programs have been written. Some experimental results are also discussed.

2. CODES, ALGORITHMS, HEURISTICS

Before going further into the description of combinatorial problems, it is necessary to define terms to be used. All sets to be considered are finite sets, and the only computer codes that are of interest are ones that obtain answers in a reasonable length of time on presently available computers.

A *recursive function* is a function with the following property: Given an argument in the domain of the function, the corresponding value of the function can be calculated by carrying out a finite number of machine steps (e.g., arithmetic, logical, syntactic, etc.). A *code* (sometimes called a program) is a description of a recursive function in some convenient language (e.g., FORTRAN, ALGOL). A *digital computer* is a recursive function evaluator.

In mathematics there are many instances in which it is possible to give a constructive computational procedure for solving a class of problems. Such procedures can be coded for computers and are frequently called algorithms. We suggest a definition of the word algorithm which will suffice for our purposes: An *algorithm* is a code that, when applied to any one of a class (set) of problems, will terminate in a finite number of steps with an (optimum) answer. The word optimum, which could also be replaced by the word correct, is put in parentheses because it frequently happens that in actual practice only approximately optimal answers are obtained with the use of an algorithm. Difficulties arise because of the finite word length of digital machines, round-off errors, data imperfections, etc. The important thing about an algorithm is that there exists a mathematical proof that the computational procedure will, in principle, find the correct answer.

Thus an algorithm is a rule, or procedure, for selecting from a set of possible answers, a correct answer. Typically, the amount of effort involved in finding a correct answer may be very large, and may increase exponentially, or factorially, or as the cube of the size of the problem.

In contrast, a *heuristic* is any rule for selecting an element from a set.

The element so selected may, or may not, have desirable mathematical properties. Thus an algorithm is a heuristic, but most heuristics are not algorithms. A synonym for heuristic is *decision rule*, which is particularly applicable when a set of decisions is under consideration. Examples of heuristics are: the "first-in first-out" rule in inventory control, the "shortest immanent operation" rule in job shop scheduling, etc.

Heuristics simplify the environment of a decision-maker by permitting him to make decisions quickly without considering the astronomical number of ways in which each decision could be made. Various authors (including the present one) have suggested that the way in which business organizations work in highly complex environments is by the use of heuristics. See [2, 14, 21, 23].

The use of heuristics typically requires relatively little effort even for large-sized problems, compared with that involved in the use of algorithms. Codes can be written that employ heuristics to solve classes of problems by selecting elements from sets when necessary. This gives rise to the following definition: A *heuristic program* is a code that, when applied to any one of class of problems, will terminate in a finite number of steps with a (good) answer. Note that the only difference between an algorithm, as we have defined it, and a heuristic program is that the word (optimal) has been replaced by the word (good). Here the evaluation of an answer as being "good" is used in a rather loose sense. Essentially, an answer is good if it is better than current practice.

The principal difference between the two concepts is that a heuristic program can be constructed rather freely, in a commonsensical, intuitive manner; but an algorithm must be based on a mathematical model, and a proof that it yields an optimal answer must be supplied.

Algorithms are always to be preferred when they exist and when they yield answers in a reasonable amount of time. However, there are many important problems—and the list is growing longer daily—for which no workable algorithm exists. The purpose of this paper is to discuss some recent heuristic programs developed for the solution of large-scale combinatorial problems that do not have algorithmic solutions.

3. THE TRAVELING SALESMAN PROBLEM AND VARIANTS

A salesman must visit each of n cities. It is desired to plan his route so that he will go through each city exactly once and so that the total distance he travels is minimized.

This is the so-called *traveling salesman problem*. In the given form it sounds rather frivolous, but there are practical applications of it and problems similar to it. For instance, the *wiring problem* is the following: There are a set of terminals on various panel boards of, say, an electronic computer; it is desired to connect them all by wire, with not more than two wires meeting at any terminal, in such a way that the total length of wiring is minimized. This problem is like the traveling salesman problem except that it is not necessary to run a wire back to the starting terminal, whereas it is necessary for the salesman to return home.

The set of objects in the traveling salesman problem is the set of all acyclic permutations of the cities, i.e., the set of feasible tours. The number of these turns out to be bounded by $(n-1)!/2$, which is an extremely large number for moderate n. For instance, for $n = 10$ the number is about 180,000 and for $n = 11$ it is nearly 2 million. Moreover, of all these feasible tours, exactly one is optimal, barring some rather unlikely arithmetical coincidences. Several exact mathematical solutions of this problem have been proposed, but they amount to "sensible" complete enumeration of the alternatives, that is, enumeration of the more likely tours, omitting those obviously incorrect. Such methods seem to work up to about $n = 20$, and then break down because of excessive demands upon computer time.

Karg and Thompson have found [9] that two rather simple heuristics will suffice to select very good answers to large problems of the traveling salesman type in a relatively short computation time.

The first heuristic constructs a possible tour of the cities by starting with an initial tour on three randomly chosen cities, and adding the remaining cities one at a time to form a sequence of partial tours, the last one having all n cities in it. More specifically, it can be described in the following series of steps:

1. Choose any three cities to form an initial tour of length 3. They can be listed in any order, the tour is the same in each case.
2. Choose any city from the list of remaining cities and try inserting it between each pair of the cities on the tour of length 3 in order to compute the length of the resulting tour. Select that tour on 4 cities that has shortest length.
3. Select any city from the list of remaining cities and insert it between each pair of cities on the tour of length 4 in order to compute the length of the resulting tour. Select that tour on 5 cities that has minimum length.
4. Repeat the above steps until a complete tour on n cities has been constructed.

This heuristic rule will construct a variety of tours depending upon the order in which new cities are introduced in the process. To sample from these, we choose a random order in which to introduce the cities, and repeat the above construction process a number of times, say 100, saving the shortest one found as the answer. By increasing the number of times the heuristic is tried, better and better answers are found. But the rate at which better answers turn up decreases. Moreover, although the heuristic as just described is sufficient to solve relatively small problems, it does not do very well on larger problems. We have therefore been forced to add another heuristic rule to the program.

When we examined the population of tours produced by the first heuristic we found that the better ones tended to agree rather well at the extreme "ends" of the tour. For instance, in Figure 1 we have given the optimal

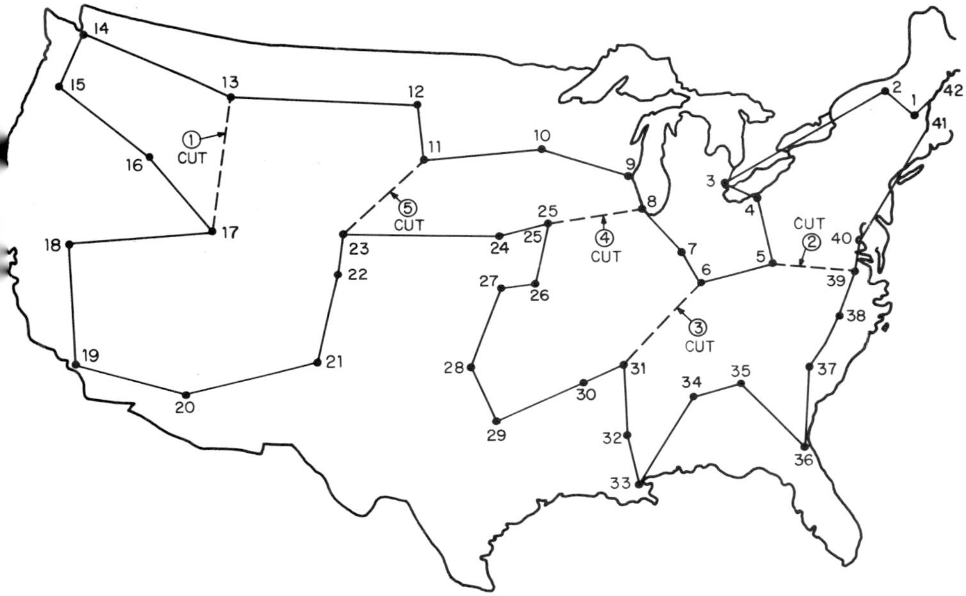

FIGURE 1

tour of a salesman who must visit each of the capitals of the states in continental United States (only 42 of the 48 capitals are shown on the map because 6 of the capitals on the east coast lie on the straight line between cities 40 and 41 and are not specifically shown). The extreme ends of the tour are the northeast, the southeast, the northwest, and the southwest. Moreover, we noted that most of the better tours were, in fact, correct at these extreme ends, even though they were badly wrong in the "center" of the tour. We therefore devised a second heuristic to cut off extreme ends of the tour and then to solve the smaller problem defined with the remaining cities.

The second heuristic can be described as follows:

1. Run the first heuristic 100 times, saving the best tour found.
2. In the best tour locate one of the extreme ends, and determine a cut

that will eliminate some of the cities. (In Figure 1, the first cut eliminated cities 14, 15, and 16).

3. Run the first heuristic 100 times more on the reduced problem, saving the best tour found.

4. In that best tour locate one of the extreme ends, and make another cut. (In Figure 1 the second cut eliminated cities 1, 2, 3, 4, 40, 41, and 42.)

5. Continue this process until no further cutting is possible.

Figure 1 shows the results of an actual run of the program. When the first cut was made, the correct tour was not known. In fact, the tour from which the first cut was made has a number of mistakes in the middle west, but it was correct in the region where the first cut was made. By eliminating cities, subproblems of smaller size (and hence easier to solve) are created, and heuristic 1 has a better chance of obtaining the correct answer. In fact, on this particular run after the first cut, the correct answer was found in the next 100 tries of heuristic 1, so that from there on the program never changed its answer. A total of five cuts was needed, and the program ended with the answer shown in Figure 1, which has already been proved to be correct by other means. This computation can be carried out in about a minute on the current fastest computers.

It should be noted that the program itself made all the cuts and defined all the subproblems that it worked on. In a sense, it learned from its own computational experience. It can learn well or badly, depending on how much time it is given to apply heuristic 1 after a subproblem has been defined. The program has worked successfully on a number of problems, and always gives a very good answer. In a 57-city problem its best answer so far is 30 miles longer than the shortest known answer of 12,955 miles.

4. THE CRITICAL PATH METHOD WITH RESOURCE LIMITATIONS

A project is a set of jobs that must be completed for a given purpose. For instance, consider the jobs involved in building a house. Some of these jobs can be worked on simultaneously (in parallel), but others must be done consecutively (in series). For each job we know the time it will take and all its predecessors, that is, all the jobs which must be finished before the given job can be started. A project graph can be drawn of the jobs in the project (see the example in Figure 2) which shows the technological order in which the jobs must be completed. The question asked first is: What is the earliest time that the project can be completed? Second: What is a feasible schedule that will complete all the jobs by the given time? The graph of a project involving 11 jobs is shown in Figure 2. There, the jobs are represented as arrows and the numbers beside each arrow are the

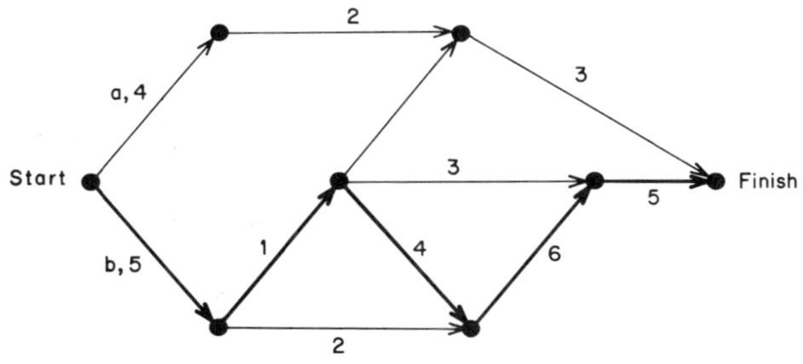

PROJECT GRAPH
FIGURE 2

lengths of time the jobs will take. The dots in the diagram are called *nodes*. Notice the node marked Start, which has no arrows pointing into it, and the node marked Finish, which has no arrows pointing out of it. These indicate the beginning and end of the project. Every other node has some arrows pointing into it and some pointing out of it. The jobs whose arrows point into a node must be *finished* before the jobs whose arrows point out of that node can *begin*.

The mathematical problem to be solved is to find the longest path from Start to Finish, where the length of such a path is defined as the sum of the job times on that path. In Figure 2, the longest path (shown by a heavier line) is 21 time units. Clearly, the length of the longest path gives the shortest time in which the project can be finished. The longest path (or paths) is called the *critical path* (*paths*) and jobs on the critical path (paths) are called *critical jobs*. If critical jobs are delayed, then necessarily the entire project will be delayed since the length of the longest path will thereby be increased.

More detailed, but still elementary expositions of the critical-path method are given in references [11] and [12]. In those references a simple algorithm is given for finding the critical paths of a project graph.

Such elementary analyses do not hold when jobs compete for resources. For instance, in Figure 2, if jobs *a* and *b* each require 5 carpenters, and only 7 carpenters are available, then both cannot be worked on simultaneously, even though there is no technological restriction preventing it. When such resource constraints exist, the simple algorithm previously mentioned no longer solves the problem, and it becomes instead a combinatorial decision problem. Now it is necessary to make a schedule of the order in which jobs are to be worked on in such a way that (i) the technological restrictions are observed, and (ii) available resources are not exceeded

Several different approaches to heuristic programs for solving this problem have been worked out by Levy, Thompson, Wiest, and others

[13, 26]; see the history in [26]. We describe next some of the heuristics used.

First, the project is analyzed as if there were no resource limitations, and all jobs are assigned to start as soon as it is technologically possible for them to do so. Then the resulting resource requirements for each day and each kind of resource are calculated. The solid graph in Figure 3 illustrates

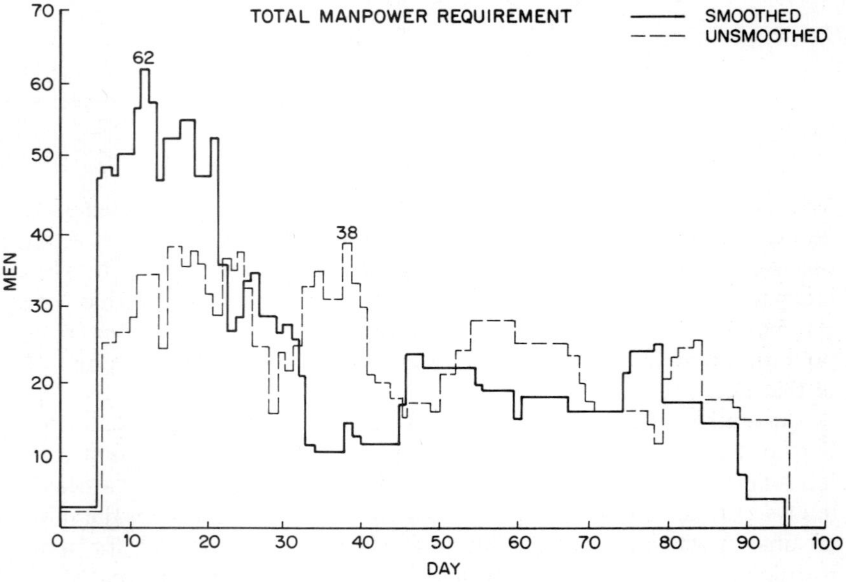

FIGURE 3

the total manpower requirements on an actual construction project involving 75 jobs. Notice that the peak requirement for men occurs early in the project, and then tapers off as the project nears completion.

After the initial manpower loading is calculated, the heuristic program tries to decrease the peak requirements by using the following heuristics:

1. Delay peak jobs so that the peaks will be decreased and valleys will be filled in.

2. When several different choices can be made, choose at random among them.

3. Employ overtime on jobs that are predecessors of peak jobs in order to move those jobs ahead of the peak.

4. After a tentative schedule is partially made and peaks are still observed, back up and reschedule to see if the peaks can be reduced.

5. Smooth manpower requirements (i) resource by resource, or (ii) in all resources at once.

6. Move the due date ahead in order to have more flexibility in shifting jobs back and forth.

The dotted lines in Figure 3 show the smoothed resource requirements for the total manpower requirements for the project. Note that peaks have been considerably reduced and the total manpower requirements come much closer to the ideal of a rectangle than the original plot. Note also that no increase in due date was needed to accomplish the smoothing.

The above program has been employed on a number of actual construction jobs, resulting in cost savings. Many computer manufacturers and consulting firms now have programs of this kind to solve smoothing problems faced by their customers.

5. THE JOB SHOP SCHEDULING PROBLEM

A plant consists of several different work centers, each with special capabilities. Suppose there is a given list of products to be completed. Each of the products to be made has a given technological routing through the work centers which is a simple order, i.e., the first operation on a product must be completed before the second can begin, and so on down the list. A specific machine is designated to carry out each of these operations, with dates for the completion of each of the products possibly having been assigned. How shall the products be scheduled on the machines in such a way that (i) due dates are met whenever possible, or (ii) the total time to complete all jobs is minimized, or (iii) some other criterion function is optimized?

An example of a job shop scheduling problem having 6 goods and 6 machines is given in Figure 4. The goods numbers are listed in the left-hand

Good						
1	3(1)	1(3)	2(6)	4(7)	6(3)	5(6)
2	2(8)	3(5)	5(10)	6(10)	1(10)	4(4)
3	3(5)	4(4)	6(8)	1(9)	2(1)	5(7)
4	2(5)	1(5)	3(5)	4(3)	5(8)	6(9)
5	3(9)	2(3)	5(5)	6(4)	1(3)	4(1)
6	2(3)	4(3)	6(9)	1(10)	5(4)	3(1)

FIGURE 4

column, and the numbers to the right indicate the sequence of machines to which the good must go in its manufacturing process. The numbers in parentheses indicate the times that the jobs will require. Thus, Good 1 must go to Machine 3 first where it requires 1 hour processing time; then it goes to Machine 1 with 3 hours processing time; then to machine 2 with 6 hours processing time, etc.

Giffler, Van Ness, and Thompson have completely enumerated (see Chapter 3 of [17]) all the active schedules there are for the problem consisting of the first 5 columns of the table in Figure 4 and found that there were 84,802. Hence the complete problem of Figure 4 has several million possible schedules and is thus, much too big for complete enumeration. The problem has been solved by hand, and one of the optimal (e.g., shortest) schedules appears in Figure 5. There the number at the left

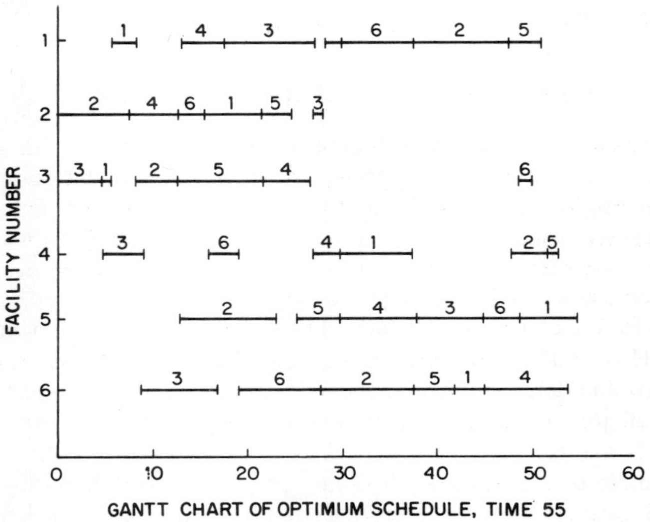

FIGURE 5

indicates a machine, and the line segments opposite indicate hours during which the machine is working. The number of the good being worked on during a given time is marked above the line segment. Such a chart is called a Gantt chart, and it is the most common way of indicating to a shop foreman how his resources are to be scheduled during a given period of time.

Story and Wagner [17, Ch. 14] report on a rather discouraging experience with problems simpler than the one in Figure 4 for which they tried integer linear programming as a solution technique.

Over the past few years, Crowston, Glover, Fischer, Thompson, and Trawick have been engaged in experimenting with heuristic programs for solving such problems, [1] and [3]. In order to describe these programs it is necessary to distinguish between local and global scheduling rules.

Local rules. These are rules that can be applied by an operator of a machine simply by looking at the queue of jobs that he has yet to complete. Some of these are:

(a) SIO or "shortest immanent operation" rule. This rule says, Work on that job next that you can finish soonest.

(b) LRT or "longest remaining time" rule, which says, Work on that job that has the most hours of work remaining to be done on it.

(c) FIFO or "first in first out rule," which says, Work on that job that has been waiting longest.

(d) LIFO rule, which says, Work on that good that arrived in the queue most recently.

These rules are quite commonly used in industry, the SIO rule being perhaps the most common.

Global rules. In order to apply a global rule, it is necessary to observe not only the queue in front of a given machine, but also the entire configuration of jobs in the shop. Thus global rules would have to be applied by a scheduler or shop foreman, rather than by a machine operator. Examples are:

(a) Machine slack, which says, Work on that good that will go next to a machine that has the most slack, that is, the largest number of idle hours.

(b) Look-ahead rule, which says, Work out all possible schedules a few moves ahead, and select the move that leads to the best situation.

None of these rules is invariably good. Sometimes a given rule gives good results on a problem, and sometimes not. The situation is something like that of a game; given a rule, it is not hard to devise both problems for which the rule will do well and problems for which the rule will do badly.

Fischer and Thompson [3] showed that a *probabilistic combination* of local rules was better than either taken singly on the above example. This result was confirmed by later work of Crowston, Glover, Thompson, and Trawick, [1]. A probabilistic combination of two rules is made as follows: When a decision is necessary as to which of the two rules is to be used, the choice between the two is made at random with prescribed probability weights. For example, in the problem of Figure 4 neither the SIO nor the LRT rule applied alone will give the optimum schedule shown in Figure 5. Instead, they produce feasible schedules that are 15 to 75 per cent longer than the optimum. However, we have found that when a 50-50 combination of the two rules is taken and applied 100 times to the problem, an optimum schedule is observed, on the average, 2 or 3 times out of the 100 (see [3]).

A second alternative is to write a program to *learn* what is a good probabilistic combination of the rules. In such a program, a 50-50 combination of the rules is used initially and tried a number of times. The program then analyzes the actual choices made in the construction of the best schedule found and counts how many of them were SIO choices and how many were LRT choices. Depending on the result, it then changes the probability weights to favor whichever rule gave better results. In this way the program continuously modifies the probability weights until it locates the best combination. For the problem of Figure 1, the best combination is about 30 percent SIO choices and 70 percent LRT choices. A typical learning curve is shown in Figure 6.

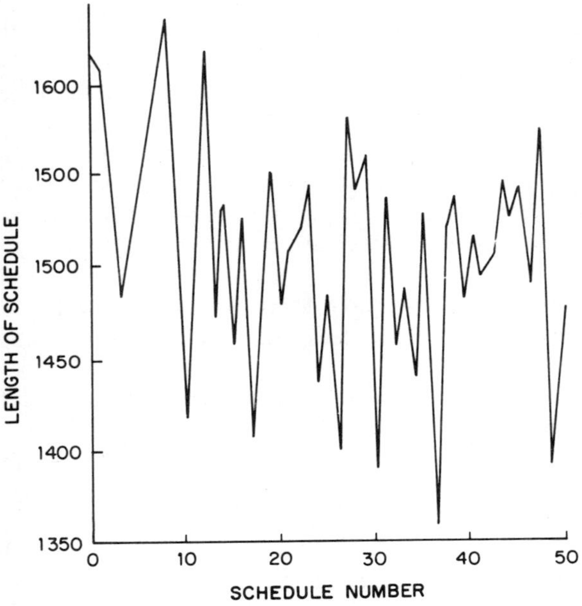

FIGURE 6

Notice that in Figure 6 the learning is not progressively better and better but instead there is an average tendency for shorter schedules to be produced as time goes on. As might be expected, the learning procedure takes a considerable amount of computational time, and there is a serious question as to whether or not learning programs are practical. References [1] and [3] give results of various experiments with such programs.

7. CONCLUSIONS

The history of the uses of the electronic computer is extremely interesting. The computer was originally developed to solve problems of numerical analysis such as differential equations, partial differential equations, integrals, etc. Hence it was first thought of as a "super slide rule." Then about 1955 it was realized that the computer is a general symbol manipulator, that is, a recursive function evaluator. Since that time it has been used for such widely diverse problems as translating from one language to another, playing chess, proving theorems in elementary logic and mathematics, choosing investment portfolios, simulating business operations for business games, composing music. The list (see [15] for a fuller discussion) is now so long that a lengthy book would be needed to dicuss them all.

The application of heuristic programming techniques to the solution of large-scale combinatorial problems in management science is yet another application. It is now a vigorous and fast moving field and promises to become an important area of operations research in the future.

At the beginning of the paper, the classic economic problem of maximizing an individual's satisfaction, subject to his budget constraints, was considered. The author knows of no existing heuristic solution to this problem, but such a solution could be constructed on the basis of some of the ideas developed in this paper.

REFERENCES

[1] Crowston, W., F. Glover, G. L. Thompson, J. D. Trawick, "Heuristic Programming Approaches to the Job Shop Scheduling Problem," ONR Research Memo. No. 117, Nov. 1963, Carnegie Institute of Technology.

[2] Cyert, R. M., and J. G. March, *A Behavioral Theory of the Firm*, Prentice-Hall, 1963.

[3] Fischer, H., and G. L. Thompson, "Probabilistic Learning Combinations of Local Job Shop Scheduling Rules," Chapter 15 of reference [17].

[4] Garvin, W. W., *Introduction to Linear Programming*, McGraw-Hill, 1960.

[5] Gere, W., "A Heuristic Approach to Job Shop Scheduling," Ph.D. Thesis, Carnegie Institute of Technology, 1962.

[6] Giffler, B., G. L. Thompson, and V. Van Ness, "Numerical Experience with the Linear and Monte Carlo Algorithms for Solving Production Scheduling Problems," Chapter 3 of Reference [17].

[7] Gomory, R., "An All-Integer Integer Programming Algorithm," Chapter 13 of Reference [17].

[8] Graves, R. L., and P. Wolfe (Editors), *Recent Advances in Mathematical Programming*, McGraw-Hill, 1963.

[9] Karg, R., and G. L. Thompson, "A Heuristic Approach to Solving Traveling Salesman Problems," *Management Science* (1964).

[10] Kuehn, A. A., and M. J. Hamburger, "A Heuristic Program for Locating Warehouses," *Management Science* (1963)

[11] Levy, F. K., G. L. Thompson, and J. D. Wiest, "The ABC's of the Critical Path Method," *Harvard Business Review*, Sept.–Oct. (1963) pp. 98–108.

[12] Levy, F. K., G. L. Thompson, and J. D. Wiest, "Introduction to the Critical-Path Method," Chapter 20 of Reference [17].

[13] Levy, F. K., G. L. Thompson, and J. D. Wiest, "Multi-Ship, Multi-Shop Workload Smoothing Program," *Naval Research Logistics Quarterly* 8 (1962) pp. 37–44.

[14] March, J. G., and H. A. Simon, *Organizations*, Wiley, 1958.

[15] Minsky, M., "Steps Toward Artificial Intelligence," *Proc. I.R.E.*, 49:1 (1961) pp. 8–30.

[16] Morgenstern, Oskar, "Demand Theory Reconsidered," *Quarterly Journal of Economics*, 17 (1948) pp. 165–201.

[17] Muth, J. F., and G. L. Thompson (editors), *Industrial Scheduling*, Prentice-Hall, 1963.

[18] Simon, H. A., and A. Newell, "Heuristic Problem Solving: The Next Advance in Operations Research," *Operations Research*, 6 (1958) pp. 1–10.

[19] Simon, H. A., A. Newell, and J. C. Shaw, "Chess-Playing Programs and the Problem of Complexity," *IBM Journal of Research and Development*, 2 (1958) pp. 320–335.

[20] Simon, H. A., A. Newell, and J. C. Shaw, "Report on a General Problem Solving Program," in *Proceedings of the International Conference on Information Processing*, London, Butterworth's, 1960, pp. 256–265.

[21] Simon, H. A., *Administrative Behavior*, Second Edition, The Macmillan Co., 1957.

[22] Shubik, Martin, *Strategy and Market Structure*, Wiley, 1959.

[23] Thompson, G. L., "Commentary on: The Budgetary Process and Management Control," in, *Management Controls: New Directions in Basic Research*, edited by C. P. Bonini, R. K. Jaedicke, and H. M. Wagner, McGraw-Hill, 1964.

[24] Tonge, F., "Summary of a Heuristic Line Balancing Procedure," *Management Science*, 7 (1960) pp. 21–42.

[25] Van Woermer, T., "The Trimmer: A Heuristic Solution to the Trim Problem in the Corrugated Container Industry," Ph.D. Thesis, Carnegie Institute of Technology, 1963.

[26] Wiest, J. D., "The Scheduling of Large Projects with Limited Resources," Ph.D. Thesis, Carnegie Institute of Technology, 1963.

CHAPTER 9

Minimaxing and Optimal Programming

By LEO TÖRNQUIST*

We shall study the following problems:

I. The minimaxing problem.

Find

$$\tilde{x} = (\hat{x}_J, \check{x}_I) = \{\hat{x}_1, \ldots, \hat{x}_n, \check{x}_{n+1}, \ldots, \check{x}_N\}$$

when

$$W(\tilde{x}) = \max_{x_J} (\min_{x_I} W(x_J, x_I)).$$

II. The optimal programming problem.

Find $\hat{x}_J$ when

$$U(\hat{x}_J) = \max_{x_J} U(x_J); \quad a_i^{(-)} \leqq v_i(x_J) \leqq a_i^{(+)} \qquad \text{for all } i \in I$$

where (;) denotes the expression "under the conditions . . .". The upper bounds $a_i^{(+)}$ and the lower bounds $a_i^{(-)}$ are independent of x_J. The functions v_i are continuous for all x_J.

III. The minimax programming problem.

Find $\tilde{x}$ when

$$W(x) = U(x_J) + \sum_{i \in I} x_i(a_i^{(\text{sign } x_i)} - v_i(x_J)).$$

If $\tilde{x} = (\hat{x}_J, \check{x}_I)$ is a solution to the minimax programming problem (III), $\hat{x}_J$ is a solution to the optimal programming problem (II). When the conditions of the optimal programming problem are fulfilled we have

$$x_i(a_i^{(\text{sign } x_i)} - v_i(x_J)) \geq 0$$

for all $i \in I$ and thus $\min_{x_I} W(x) = U(x_J)$, while $\min_{x_I} W(x) = -\infty$ if some of the conditions

$$x_i(a_i^{(\text{sign } x_i)} - v_i(x_J)) \geq 0$$

are not fulfilled. If $W(\tilde{x}) \neq -\infty$ we have $W(\tilde{x}) = U(\hat{x}_J)$, where

$$U(\hat{x}_J) = \max_{x_J} U(x_J); \quad a_i^{(-)} \leq v_i(x_J) \leq a_i^{(+)}$$

for all $i \in I$.

The minimax programming problem is a special case of the minimaxing problem (I). If we can find a method for solving the general minimaxing problem (I), this method will also give a solution to the optimal programming problem (II).

* University of Helsinki.

Definitions and assumptions.

DEFINITION 1. We call the point $\tilde{x} = (\hat{x}_J, \check{x}_I)$ a minimax point of $W(x)$. The value of $W(x)$ for $x = \tilde{x}$ we denote $\tilde{W}$.

DEFINITION 2. The function $J(h) = 1$ for $h \in J$, $J(h) = -1$ for $h \in I$. We define $I(h) = -J(h)$. If $h \notin I + J$ we put $J(h) = 0$, $I(h) = 0$. We call the functions $J(h)$ and $I(h)$ the partition functions of $I + J$.

DEFINITION 3. The function

$$\check{U}(x_J) = \min_{x_I} W(x_J, x_I) = W(x_J, \check{x}_I(x_J))$$

we call the preference function of the minimaxing problem (I).

The definition 1 can be stated more precisely as follows:

DEFINITION 1′. The point $x = \tilde{x} = (\hat{x}_J, \check{x}_I)$ is a minimax point of $W(x)$ if and only if

1) $x_I = \check{x}_I(x_J)$, that is if $\Delta_{x_I} W(x) \geq 0$ for every Δx_I where

$$\Delta_{x_I} W(x) = W(x_J, x_I + \Delta x_I) - W(x)$$

and if

2) $x_J = \hat{x}_J$, that is if $\Delta_{x_J} \check{U}(x_J) \leq 0$ for every Δx_J where

$$\begin{aligned}\Delta_{x_J} \check{U}(x_J) &= \check{U}(x_J + \Delta x_J) - \check{U}(x) \\ &= W(x + \Delta x) - W(x), \\ \Delta x &= (\Delta x_J, \Delta_{x_J} \check{x}_I(x_J)).\end{aligned}$$

We make the following regularity assumptions about the function $W(x)$:

ASSUMPTION 1. The function $W(x)$ is a continuous real function of x uniquely defined for all real x.

ASSUMPTION 2. The partial derivatives

$$\frac{\partial W(x)}{\partial x_h} = a_h^{(\text{sign } x_h)} - v_h(x), \quad h \in I + J$$

are uniquely determined and continuous for $x_h \neq 0$. The real numbers $a_i^{(+)}$ and $a_i^{(-)}$ can be equal and some of them may be $\pm\infty$.

We make the convention:

$$x_h \frac{\partial W(x)}{\partial x_h} = 0 \qquad \text{for } x_h = 0.$$

Under the regularity assumptions made we have

$$\Delta_x W(x) = \sum_h \frac{\partial W(\dot{x})}{\partial \dot{x}_h} \Delta x_h \qquad \text{for some } \dot{x}$$

fulfilling $0 < |\dot{x}_h - x_h| < |\Delta x_h|$ whenever $x_h(x_h + \Delta x_h) \geq 0$. This formula remains valid for all Δx_h if we in the case when $x_h(x_h + \Delta x_h) < 0$, let

$\dfrac{\partial W(\dot{x})}{\partial \dot{x}_h}$ for $\dot{x}_h = 0$ denote some suitably chosen mean value of the limes values of this partial derivative for $\dot{x}_h \to 0$, $\dot{x}_h > 0$ and for $\dot{x}_h \to 0$, $\dot{x}_h < 0$.

In the following we restrict if not otherwise stated our studies to functions $W(x)$ fulfilling the regularity assumptions made above.

Theorems.

THEOREM 1. *If the point x has the property $J(h)\dfrac{\partial W(\dot{x})}{\partial \dot{x}_h}\Delta x_h \leq 0$ for every h and Δx_h and some $\dot{x} = \dot{x}(x, \Delta x)$, $|\dot{x}_h - x_h| < |\Delta x_h|$ such that $\Delta_x W(x) = \sum_h \dfrac{\partial W(\dot{x})}{\partial \dot{x}_h}\Delta x_h$, then x is a minimax point of $W(x)$.*

PROOF. To prove this theorem it is sufficient to show that $x = \dot{x} + \Delta\dot{x}$, $|\Delta\dot{x}_h| < |\Delta x_h|$ fulfills the conditions of Definition 1′. We consider separately the cases $J(h) = -1$ and $J(h) = 1$. According to Definition 2 $J(h) = -1$ if $h \in I$ and $J(h) = 1$ if $h \in J$. We have according to our assumptions

$$\sum_{h\in I} J(h)\frac{\partial W(\dot{x})}{\partial \dot{x}_h}\Delta x_h = -\sum_{h\in I}\frac{\partial W(\dot{x})}{\partial \dot{x}_h}\Delta x_h = -\Delta_{x_I}W(x) \leq 0$$

for some $\dot{x}$ and all Δx_I and thus $\Delta_{x_I}W(x) \geq 0$, that is the condition 1) of Definition 1′ is fulfilled for x. For $h \in J$ we have $J(h) = 1$ and

$$\begin{aligned}\Delta_{x_J}\check{U}(x_J) &= W(x + \Delta x) - W(x)\\ &= \Delta_{x_J}W(x) + \min_{\overset{\shortparallel}{x}_I}\,(W(x_J + \Delta x_J, \overset{\shortparallel}{x}_I) - W(x_J + \Delta x_J, x_I)).\end{aligned}$$

The first term $\Delta_{x_J}W(x) = \sum_{h\in J} J(h)\dfrac{\partial W(\dot{x})}{\partial \dot{x}_h}\Delta x_h \leq 0$ according to the assumption of Theorem 1, while the second term $\leq W(x_J + \Delta x_J, x_I) - W(x_J + \Delta x_J, x_I) = 0$. Thus the condition (2) of Definition 1′ is fulfilled.

REMARK. A minimax point fulfilling the conditions of Theorem 1 we call a strong minimax point. A minimax point which is not strong we call a weak minimax point.

We can generalize the concept of a strong minimax point to all real functions $W_R(x)$ by defining a strong minimax point as a point $x = \tilde{x}$ for which $\Delta_{x_J}W_R(x) \leq 0$ and $\Delta_{x_I}W_R(x) \geq 0$ for all $\Delta x = (\Delta x_J, \Delta x_I)$.

If a strong minimax point exists, such a point is thus a common solution to the problem to find $\tilde{x}$ for which

$$W(\tilde{x}) = \max_{x_J}\min_{x_I} W(x_J, x_I)$$

and the problem to find $\bar{\bar{x}}$ for which

$$W(\bar{\bar{x}}) = \min_{x_I}\max_{x_J} W(x_J, x_I).$$

When $\tilde{x} = \bar{\bar{x}}$ and $W(x)$ is uniquely defined, we have $W(\tilde{x}) = W(\bar{\bar{x}})$.

THEOREM 2. *If $\tilde{x}$ is a strong minimax point, we have for $x = \tilde{x}$, either $x_h = 0$ or $\frac{\partial W(x)}{\partial x_h} = 0$, and thus $x_h \frac{\partial W(x)}{\partial x_h} = 0$, whenever Assumption 2 is valid.*

PROOF. According to the conditions of Theorem 1 $\frac{\partial W(\dot{x})}{\partial \dot{x}_h}$ and Δx_h change their signs simultaneously. If $\frac{\partial W(\dot{x})}{\partial \dot{x}_h}$ is continuous for $\dot{x} = \tilde{x}$, we must have $\frac{\partial W(\dot{x})}{\partial \dot{x}_h} = 0$ for $\dot{x} = \tilde{x}$, while a stepwise change in $\frac{\partial W(\dot{x})}{\partial \dot{x}_h}$ can occur only for $\dot{x}_h = \tilde{x}_h = 0$. Theorem 2 is thus proved.

THEOREM 3. *For a strong minimax point we have when Assumption 2 is valid*

$$a_h^{(J(h))} \leq v_h(\tilde{x}) \leq a_h^{(I(h))}$$

and

$$\tilde{x}_h(a_h^{(\operatorname{sign} \tilde{x}_h)} - v_h(\tilde{x})) = 0.$$

PROOF. Substituting in $J(h) \frac{\partial W(\dot{x})}{\partial \dot{x}_h} \Delta x_h \leq 0$ the expression of $\frac{\partial W(x)}{\partial x_h}$ given in Assumption 2 and using the assumptions about the regularity of $v_h(x)$ we obtain after some calculation the inequalities of Theorem 3.

THEOREM 4. *If the function $W(x)$ fulfills the assumptions required for the mean value theorem, that is if*

$$\Delta_{x_k} W(x) = \frac{\partial W(\dot{x})}{\partial \dot{x}_k} \Delta x_k \qquad \text{where } \dot{x} = x + \delta_{x_k} x,\ |\delta x_k| < |\Delta x_h|,$$

$\Delta x_k \delta x_k \geq 0$

for every Δx_k, $k \in J + I$, we have

$$J(k)\, \Delta_{x_k} W(x) = J(k) \frac{\partial W(\dot{x})}{\partial \dot{x}_k} \Delta x_k \leq 0$$

for every $k \in I$ if $x_I = \check{x}_I(x_J)$
and for every $k \in J$ if $x_J = \hat{x}_J$ and $x_I = \check{x}_I(\hat{x}_J + \Delta_{x_k} x_J)$.

REMARK. The minimax points

$$\tilde{x}^{(s_k,k)} = \{\hat{x}_J, \check{x}_I^{(s_k,k)}\}$$

where $s_k = \operatorname{sign} \Delta x_k$; $|\Delta x_k| \to 0$ and

$$\check{x}_I^{(s_k,k)} = \operatorname*{limes}_{|\Delta_k x| \to 0} (\check{x}_I(x_J + \Delta_{x_k} x_J)); \quad s_k = \operatorname{sign} \Delta x_k$$

will not necessarily be equal for $s_k = +$ and $s_k = -$. For all minimax points $\tilde{x}^{(s_k,k)}$ we have

$$W(\tilde{x}^{(s_k,k)}) = \tilde{W} = \max_{x_J} \min_{x_I} W(x_J, x_I) = U(\hat{x}_J)$$

and

$$J(k) s_k \left[\frac{\partial W(x)}{\partial x_k}\right] \leqq 0 \quad \text{for } x = \tilde{x}^{(s_k,k)}.$$

If $\tilde{x}^{(+,k)} = \tilde{x}^{(-,k)}$: we have

$$\tilde{x}_k^{(s_k,k)}\left[\frac{\partial W(x)}{\partial x_k}\right] = 0 \quad \text{for } x = \tilde{x}^{(s_k,k)}.$$

For $\tilde{x}^{(+,k)} \neq \tilde{x}^{(-,k)}$ this equation need not be fulfilled.

For every $h \in I$ and every minimax point $\tilde{x}$ we must as proved in connection with Theorem 2 have

$$\check{x}_h\left[\frac{\partial W(x)}{\partial x_h}\right]_{x=\tilde{x}} = 0.$$

Proof. For a minimax point (x_J, x_I) we have according to Definition 1′ and Definition 3 $\Delta x_k W(x) \geq 0$ for $k \in I$, $J(k) = -1$, $x_I = \check{x}_I(x_J)$ and

$$\Delta_{x_k}\check{U}(x_J) = \Delta_{x_k} \min_{x_I} W(x_J, x_I) \leq 0.$$

We always have for $x = \tilde{x}$, $\Delta_{x_k}x_I = \check{x}_I(x_J + \Delta_{x_k}x_J) - \check{x}_I(x_J)$, $J(k) \neq +1$

$$\begin{aligned}[\Delta_{x_k}\check{U}(x_J)]_{x_J=\hat{x}_J} &= [W(\hat{x}_J + \Delta_k x_J, \check{x}_I + \Delta_{x_k}x_I) \\ &\quad - W(\hat{x}_J, \check{x}_I + \Delta_{x_k}x_I)] \\ &\quad + [W(\hat{x}_J, \check{x}_I + \Delta_{x_k}x_I) - W(\hat{x}_J, \check{x}_I)] \leq 0,\end{aligned}$$

where

$$W(\hat{x}_J + \Delta_k x_J, \check{x}_I + \Delta_{x_k}x_I) = \check{U}(\hat{x}_J + \Delta_k x_J) = \min_{x_I} W(\hat{x}_J + \Delta_k x_J, x_I),$$

and $W(\hat{x}_J, \check{x}_I + \Delta_{x_k}x_I) - W(\hat{x}_J, \check{x}_I) \geq 0$ since $W(\hat{x}_J, \check{x}_I) = \check{U}(\hat{x}_J) = \min_{x_I} W(\hat{x}_J, x_I)$. We thus have

$$W(\hat{x}_J + \Delta_k x_J, \check{x}_I + \Delta_{x_k}x_I) - W(\hat{x}_J, \check{x}_I + \Delta_{x_k}x_I) \leq 0.$$

Applying the mean value theorem to this result we obtain Theorem 4.

Theorem 5. *The point $\tilde{x}$ is a minimax point of $W(x)$ if and only if*

$$\tilde{W} = W(\tilde{x}) \geq W(x)$$

whenever $W(x_J, x_I + \Delta x_I) \geq W(x)$ for every Δx_I.

This theorem is only a reformulation of Definition 1′.

Theorem 6. *If $\underline{W}(W(x_J(y_{\jmath}), x_I(y_{\jmath}, y_{\imath}))) = \overline{W}(y_{\jmath}, y_{\imath})$ where $\underline{W}(W)$ is a strictly increasing function of W and $y_{\jmath}, y_{\imath}$ are such new independent variables that to every x_J corresponds some $y_{\jmath}$ for which $x_J = x_J(y_{\jmath})$ and to every x_I and $y_{\jmath}$ corresponds some $y_{\imath}$ for which $x_I = x_I(y_{\jmath}, y_{\imath})$, and $\{\hat{y}_{\jmath}, \check{y}_{\imath}\}$ is a minimax point of $\overline{W}(y_{\jmath}, y_{\imath})$, then $\{\tilde{x}_J = x_J(\hat{y}_{\jmath}), \check{x}_I = x_I(\hat{y}_{\jmath}, \check{y}_I)$ is a minimax point of $W(x_J, x_I)$.*

Proof. Since $\underline{W}$ is a strictly increasing function of W, both $\underline{W}$ and W must have the same minimax points. Thus if $\tilde{x}$ is a minimax point of W it is also a minimax point of $\underline{W}$. Since the dependent variables $\{x_J, x_I\}$ take all values when the new independent variables vary, we have

$$
\begin{aligned}
\overline{W}(\hat{y}_{\acute{J}}, \check{y}_{\acute{I}}) &= \max_{y_{\acute{J}}} \min_{y_{\acute{I}}} \overline{W}(y_{\acute{J}}, y_{\acute{I}}) \\
&= \max_{y_{\acute{J}}} \min_{y_{\acute{I}}} \underline{W}(W(x_J(y_{\acute{J}}), x_I(y_{\acute{J}}, y_{\acute{I}}))) \\
&= \underline{W}(W(x_J(\hat{y}_{\acute{J}}), x_I(\hat{y}_{\acute{J}}, \check{y}_{\acute{I}}))) \\
&= \underline{W}(W(\hat{x}_J, \check{x}_I)) = \max_{x_J} \min_{x_I} \underline{W}(W(x_J, x_I)).
\end{aligned}
$$

REMARK. By a suitable transformation ($x_J = x_J(y_{\acute{I}})$, $x_I = x_I(y_{\acute{J}}, y_{\acute{I}})$) it is possible to simplify the problem of finding minimax points and avoid difficulties connected with minimax points for which some coordinates of x or $W(x)$ are infinite.

General discussion of possibilities of finding minimax points.

In principle we could try to find the solutions of the system of equations and inequalities mentioned in the Theorems 1–4. Every such solution fulfilling the conditions of Definition 1′ is then a solution to our minimax problem. This method for finding the solution of our minimaxing problem we call the direct method.

We could try to solve our problem also by making use of such transformations of the minimaxing problem to another minimaxing problem that the new minimaxing problem is easier to solve by means of the direct method. If the transformation used fulfills the conditions of Theorem 6, we can from the solution of the transformed problem obtain the solution of the original problem. Methods using a finite number of transformations we call transformation methods. If the method needs in principle an infinite number of transformations for obtaining the solution we call it a sequential approximation method. The rules required for choosing the successive transformations are called transition rules. A method with given transition rules usually leading to a minimax point we call a minimax algorithm. In the following we suggest some minimax algorithms which may be useful in practice.

Linear minimax algorithms.

Assuming that strong minimax points exist we can try to use linear minimax algorithms for finding these points.

In such an algorithm the transform rules are of the following form

$$x_h = b_{h0}^{(\nu)} + \sum_{p \in P_{h\nu}} b_{hp}^{(\nu)} y_p^{(\nu)}$$

where ν is the ordering number of the last transition made and $h \in J + I$.

The function to be minimaxed is denoted by $W(x_J, x_I)$ and the transformed minimax function by

$$W_\nu(y^{(\nu)}_{J_\nu}, y^{(\nu)}_{I_\nu}) = W(x_J, x_I).$$

The transition $\nu \to \nu + 1$ is defined by the formulas

$$y^{(0)} = x.$$
$$y^{(\nu)}_p = a^{(\nu)}_{p0} + \sum_{q \in Q_p} a^{(\nu)}_{pq} y^{(\nu+1)}_q$$

where $a^{(\nu)}_{pq} = 0$ if $q \in J_{\nu+1}$ and $p \in I_{\nu+1}$ and $a^{(\nu)}_{pp} \neq 0$, $p \in Q_{p\nu}$.

These substitutions fulfill the conditions of Theorem 6. We obtain the following transition rule for the coefficients $b^{(\nu)}_{ho}$ and $b^{(\nu)}_{hp}$

$$b^{(\nu+1)}_{h0} = b^{(\nu)}_{h0} + \sum_{q \in P_{h\nu}} b^{(\nu)}_{hq} a^{(\nu)}_{q0}$$
$$b^{(\nu+1)}_{hp} = \sum_{q \in P_{h\nu}} b^{(\nu)}_{hq} a^{(\nu)}_{qp}, \quad P_{h,\nu+1} = Q_{h\nu}$$

To obtain a completely specified minimax algorithm we have to define the numbers $a^{(\nu)}_{po}, a^{(\nu)}_{pq}$. The rules for determining these numbers must be somehow related to the partial derivatives of first and second order of $W_\nu(y_{J_\nu}, y_{I_\nu})$ for $y_{J_\nu} = y_{I_\nu} = 0$. By constructing these rules the following guiding principles may be useful.

PRINCIPLE 1. If we can prove that in a minimax point of $W_\nu(y_{J_\nu}, y_{I_\nu})$, $\tilde{y}_h = 0$ for some h, we put $y_h = 0$ in the following steps thus reducing the number of independent variables.

PRINCIPLE 2. If it seems to be likely that $\tilde{y}^{(\nu)}_h = 0$ for some h, we use in the following step the transform $y^{(\nu)}_h = y^{(\nu+1)}_h$ for these h.

PRINCIPLE 3. If it seems to be likely or if it can be proved that $\tilde{y}^{(\nu)}_h \neq 0$, $\tilde{x}_h \neq 0$ for $h \in H_\nu$ and consequently according to the remark of Theorem 4 we expect or otherwise can prove that $\dfrac{\partial W_\nu(y^{(\nu)})}{\partial y^{(\nu)}_h} = 0$ for $h \in H_\nu$, $y^{(\nu)} = \tilde{y}$, we have to use such a transform for $\nu \to \nu + 1$ that $\dfrac{\partial W_{\nu+1}(y^{(\nu+1)})}{\partial y^{(\nu+1)}_h} \approx 0$ for $h \in H_\nu$ when $y^{(\nu+1)}_h \approx 0$.

PRINCIPLE 4. Taking into account Principles 1–3 we try to minimize the work to be done in finding sufficiently good approximations $b^{(\nu)}_{ho}$ for the minimax values $\tilde{x}_h$ sought.

The crucial problem is to determine the sets H_ν mentioned in Principle 3. Algorithms for finding such sets for linear programming problems are presented in my paper "Some New Principles for Solving Linear Programming Problems" published in *Bulletin de L'Institut International de Statistique*, Tome 36-3. Livraison (pp. 335–356).

PART III

Decision Theory

CHAPTER 10

Alternate Prior Distributions in Statistical Decision Theory

By JOHN P. MAYBERRY*

1. INTRODUCTION

a. The most frequent criticism made of statistical decision theory is that the computations are based on an assumed prior distribution, which is at best rather subjective and is often rather arbitrary. A common way of overcoming this criticism is to perform a "sensitivity analysis" by assembling several reasonable prior distributions and computing for each of those prior distributions the optimal policy, i.e., what experiment to perform, and what action to take for each possible outcome.

The sensitivity analysis consists in investigating the effect of the various prior distributions on the optimum policy. If the optimal policies are not much different for the several prior distributions considered, our inability to choose the "correct" prior distribution is seen to be practically unimportant. (This is often the situation when the cost of an experiment is very small, relative to the payoff; for any reasonable prior distribution, we will perform many tests, and the action will then be largely determined by the test results.) On the other hand, the optimal policies, for the various prior distributions, may be very different; this will be more likely when testing is expensive relative to the payoff, because for reasonable prior distributions there will be little testing (or perhaps none) and the posterior distribution will inherit many of the characteristics of the prior distribution. In this case, the usual sensitivity analysis ends with the conclusion: "Yes, it *matters* which of these distributions you choose."

The present paper proposes a scheme for deriving a reasonable policy in the second case, where the prior distribution *matters*, but we are not willing to state that we *know* it.

b. It is unfortunate that statistical decision theory, which explicitly considers testing costs, is most necessary when testing costs are an important factor, so that the need to use a single prior distribution is often the weakest point in the whole procedure. When we have chosen a prior distribution, we can calculate the optimal policy (i.e. the best choice of experiment, and the best action to take for each possible result); but we may wonder uneasily what policy would have resulted if we had assumed a different prior distribution.

c. This paper suggests that we compute the optimal policy for each of several reasonable prior distributions, and then compute the expected

* Operations Analysis, Headquarters United States Air Force.

payoff for each of those policies with respect to each of the prior distributions. Now we confront a decision-process more abstract than the one we confronted before: This time we must choose a policy (rather than a final action) and we may imagine Nature choosing a prior distribution (rather than one of the alternatives which we previously called a "state of nature"). Several reasonable methods have been proposed for making decisions in case of a "game against nature." This paper considers the implications of two of them: Wald's "minimax" [11] and Savage's "minimax-regret" [9], and a variant on the second, which we call the "modified-regret." All three of these methods are first applied (in Section 3) directly to the particular missile-testing problem described in Section 2 (which was also used as an example in [3]). Then the application of those methods to the more abstract decision-process is described in subsection 3.c.

d. We conclude that this general approach is feasible and that the computations can be performed with reasonable ease if we do not use too many alternative prior distributions. (Even two or three distributions give us some protection against arbitrariness in the prior distributions.) Although "minimax-regret" approaches lead to mixed strategies, it appears that the two "minimax-regret" approaches are the most reasonable, and that the "modified-regret" is most appropriate for our particular example. (See Section 6 below for a suggestion on avoiding mixed strategies.)

2. BRIEF STATEMENT OF OUR MISSILE-TESTING PROBLEM

The viewpoint we take in our problem is that there is (or will be) a stockpile of N missiles of a certain type, and we wish to allocate them so that the number of targets for which there is 90% assurance of destruction is as large as possible.[1] As described in [3], we assume: that all targets are of equal value; that all missiles have the same probability R of functioning and reaching the target area; that a missile which does reach the target area will be practically certain to destroy the target; and that not more than six missiles will be allocated per target. Consequently, we must decide whether to allocate 1 missile to each target, 2 missiles to each target, or 3, 4, 5, or 6 missiles to each target; but we may procure some additional missiles and subject them to operational test firing, before making the decision.

If we knew the value of the "reliability" R, we should allocate k missiles to each target, where

$$k = \text{the least integer satisfying } 1 - (1 - R)^k \geq 0.90 \tag{2.1}$$

[1] Although these particular calculations have not been performed for values of the "assurance coefficient" different from 90%, there is both theoretical and computational evidence that very similar results would be obtained if different values, e.g. 75% or 95%, were used.

The first two rows of Table 1 show how k, the number of missiles per target required, depends on R.

Since the outcome will be uncertain, we must assign a numerical value, or "payoff," to each possible outcome. The payoffs can then be averaged with respect to the anticipated probabilities of their occurring. The payoff

TABLE 1

Payoff Matrix: Average Number of Targets, Correctly Identified as Killed with 90% Assurance, per Missile Allocated

	k, Number of missiles *required* per target						
	1	2	3	4	5	6	
Boundaries of corresponding ranges of R	1.000 --- .900	--- .684	--- .536	--- .438	--- .369	--- .319	
C, Number of missiles *allocated* per target							
1	1.000	0	0	0	0	0	
2	.500	.500	0	0	0	0	
3	.333	.333	.333	0	0	0	
4	.250	.250	.250	.250	0	0	
5	.200	.200	.200	.200	.200	0	
6	.167	.167	.167	.167	.167	.167	

must depend on the number C of missiles allocated per target, and on the true value of the reliability R; the payoff-function is

$$P(C, R) = \begin{cases} \dfrac{1}{C} & \text{if } C \geq k, \\ 0 & \text{if } C < k, \end{cases} \tag{2.2}$$

where k is given by (2.1) above. Table 1 also shows values of the payoff function. Since the number N of operational missiles is assumed given, the critical measure of effectiveness is the ratio of targets successfully attacked to missiles possessed; this is $1/C$ if C missiles was enough, and 0 if C missiles was not enough. The unit of this payoff $P(C, R)$ is therefore "targets per missile."

It is clear that because we will not know the true value of R, we must "hedge" by allocating "more missiles than we feel are necessary" to avoid the large losses associated with allocating too few. Consequently, it may be desirable to gain information about R by testing missiles; the value of the information obtained will depend on the number of missiles in the stockpile, and the cost will be simply the cost of testing. In order to calculate the optimal policy without the application of human judgment, it is necessary to express the cost of testing in units compatible with the payoff.[2] We employ a cost-function different from that of [3], paragraph 4.b.(6)(b), and based on the idea that fewer operational missiles may be procured if more money is spent on the testing. Then, if N is the number of missiles planned for our operational stockpile, and if S test missiles can be procured for the cost of one operational missile, testing E missiles would result in an operational force of $N - (E/S)$ missiles. If the resulting payoff, calculated on the assumption we test E missiles, is $R(E)$ targets per missile, the net outcome of testing E is $\left(N - \frac{E}{S}\right) \cdot R(E) = N \cdot R(E) - E \cdot \frac{R(E)}{N \cdot S}$, so that the cost $C(E)$ of testing E missiles, in units compatible with the $R(E)$ function, is

$$C(E) = \frac{E \cdot R(E)}{N \cdot S}$$

Without (for obvious reasons) making any pretense at realism, we might use $S = 3$ and $N = 100$ in our example, so that

$$C(E) = \frac{E}{300} \cdot R(E).$$

We must finally decide on a policy, which comprises a number E of experiments (missile tests) to be performed, and a number C of missiles to be assigned to each target. Since C will depend on the number NS of those tests which are successful, we describe a policy as a set (E; C_{NS} for $NS = 0, 1, \ldots, E$).[3] For example, the policy (3; 4, 2, 2, 1) would mean that 3 missiles would be tested; that 4 missiles would be assigned to each target if none of the three tests were successful; that 2 missiles would be allocated to each target if one or two of the three were successful; and that 1 missile would be allocated per target if all three tests were successful.

Several rational ways of making such a decision have been proposed; one of the most interesting is to behave as if the prior distribution were known, and to employ the policy which maximizes the expected outcome. This is the usual statistical decision theory approach.

Other rational approaches involve behaving as if Nature would "choose" the reliability in some way. For example, the Wald minimax proposal [11] is to assume that Nature's interests are opposite to ours; in that case, we

[2] Decision-theory may still be used, by applying judgment, without expressing the cost in such units; see paragraph 4.b.(6)(a) of [3].

[3] We are excluding possible sequential policies, assuming that production for all the tests must be planned before any tests are performed.

should act so as to protect ourselves (as much as possible) against the worst alternative. In contrast, Savage [9] has proposed a minimax-regret principle wherein we try both to exploit the possibilities of the favorable states of nature and to protect ourselves against the worst consequences of the unfavorable states of nature. We also describe a modification of the minimax-regret principle, which may be more appropriate for certain problems.

All four of those approaches will be applied, first to the case where we must make our allocation without any tests, and then to the more general situation where we must choose a policy for testing and allocation.

3. SUMMARY OF CALCULATIONS

The four approaches to this problem are presented in paragraph 3.a., as they would apply to the allocation portion of our problem if no testing were to be performed. Then the more general (and much more complicated) problem of including the testing option is briefly mentioned in paragraph 3.b. Finally, in paragraph 3.c., we calculate the new approach which this paper proposes, making use of the techniques of paragraph 3.a. on a more abstract level as explained in Section 1.

a. USING RAW PAYOFF FUNCTION (NO TESTING). In this case we must observe the payoff matrix and choose an allocation-strategy without testing any missiles. Recall that the experimenter is to choose C, the number of missiles allocated to each target, that the state of nature is determined by the reliability R, and that $P(C, R)$ is given by equation (2.2), Section 2.

(1) *Minimax.* Wald proposed in Ref. 11 that a minimax strategy be played in games against nature; in other words, the desirability of each policy should be assessed according to its consequence in the worst case. This may be regarded as a desire to be conservative, even if unlucky; or it may be interpreted as a game where Nature plays the part of a malevolent opponent. Formally, we write

$$P_{\text{minimax}} = \max_{s_i} \min_{r_j} \sum_{i,j} (s_i \cdot P(C_i, R_j) \cdot r_j), \quad \text{with } \sum_i s_i = \sum_j r_j = 1, \quad (3.1)$$

where the s_i and r_j allow for the fact that we may have to randomize our decision, and allocate C_i with probability s_i; we imagine Nature also randomizing and choosing R_j with probability r_j.

Referring to the payoff matrix in Table 1, we see that (whatever the experimenter does) the worst case for him is the case where 6 missiles were in fact required. Consequently, the best the experimenter can assure himself of is a payoff of 1/6 target per missile, by allocating 6 missiles to each target.

For comparative purposes, we present the set of payoffs which will be obtained (if the experimenter chooses to allocate six missiles) in the six possible states of nature (respectively: 1 missile required, 2 required, 3, 4, 5, and 6):

$$(.167, .167, .167, .167, .167, .167).$$

The minimax strategy lives up to its conservative reputation; we will never do worse than 1/6. However, it has two serious disadvantages:

First, it cannot take advantage of any possible gain (however large) which implies the possibility of a loss (however small).

Second, whenever there is a "uniformly worst" state of nature (as in this example), it is that worst state which *entirely* governs the minimax strategy. Since we can always imagine an alternative both less favorable and less likely than any we had considered before, we see that our whole strategy is determined by our (necessarily arbitrary) decision on just how pessimistic we should be.

We now examine the consequences of other principles.

(2) *Minimax-regret.* J. Savage [9] suggested using a minimax-regret principle in games against nature. He defines the "regret" as the difference between the payoff actually attained, and the payoff which would have been possible if we had known the state of nature. Thus

$$P_{\text{reg}}(C, R) = P(C, R) - \max_{C} P(C, R) \tag{3.2}$$

Table 2 shows the P_{reg} calculated from the payoff function of Table 1. One view of the minimax-regret principle has been expressed as: "Nature doesn't really want to hurt us; she just wants to make us feel bad."

The minimax-regret solution, of course, gives us

$$\max_{s_i} \min_{r_j} \sum_{i,j} s_i \cdot P_{\text{reg}}(C_i, R_j) r_j. \tag{3.3}$$

TABLE 2

Regret, $P_r(C, R)$

	R, Number required					
C, Number allocated	1	2	3	4	5	6
1	0	−.500	−.333	−.250	−.200	−.167
2	−.500	0	−.333	−.250	−.200	−.167
3	−.667	−.167	0	−.250	−.200	−.167
4	−.750	−.250	−.083	0	−.200	−.167
5	−.800	−.300	−.133	−.050	0	−.167
6	−.833	−.333	−.167	−.083	−.033	0

In the case of our example, we find that the minimax-regret policy for the experimenter is to allocate one missile per target with probability 0.500, two missiles per target with probability 0.333, and three missiles per target with probability 0.167. Under these conditions the set of payoffs which will be obtained in the six possible states of nature 1, 2, 3, 4, 5, or 6 missiles were in fact required) is:

$$(0.722, 0.222, 0.056, 0, 0, 0).$$

This set of numbers seems to have little reason behind it, but the corresponding set of "regret" values explains things:

$$(-.278, -.278, -.278, -.250, -.200, -.167).$$

Note that the "strategy" played by Nature uses only the first three assumed states of nature. We observe that our minimax-regret policy gives us less in the unfavorable cases where three or more missiles are actually required, and more in the favorable cases where only one or two are required. This is characteristic of minimax-regret, since it attempts to obtain some advantage from the presence of more favorable possibilities—and thereby necessarily incurs a loss in the less-favorable cases.

(3) *Modified regret.* One of the basic reasons for considering the minimax-regret case is that differences between results of various actions are usually important, but the absolute size of the numbers in a payoff matrix is usually not important. Consequently, there is justification in general for setting a "zero-level" in each state of nature independently of the "zero-level" in the other states of nature, but keeping the unit of payoff constant. Then, if we equate all the "best actions," we obtain the usual minimax-regret criterion. On the other hand, we may decide that the payoff *units* in different states of nature[4] are unrelated. In the case of our specific missile-testing problem there are two concepts definable in all states of nature; first, having no deterrent capability (recall that we are proceeding as if this missile were our only deterrent) and, second, getting the most possible deterrent out of our investment in these missiles. The first could reasonably be given a payoff of 0 in each state of nature, and the second a payoff of 1.00 in each state of nature; a suitable linear transformation, chosen separately for each column, can provide payoffs with those values.[5]

There emerges the suggestion, which seems to be a new one, that the payoffs be modified in a somewhat more complicated way to give what we

[4] (Instead of, or in addition to, the zero levels).

[5] Transformations more general than linear transformations can hardly be considered because the payoff is a measure of utility, which must be averaged; nonlinear transformations would not be preserved under an arbitrary averaging operation.

shall call the "modified-regret (mr) function,"[6]

$$P_{mr}(C, R) = \frac{P(C, R)}{\max\limits_{C} P(C, R)} \tag{3.4}$$

Table 3 shows the modified regret P_{mr} calculated from the payoff matrix of Table 1.

TABLE 3

MODREG (*C*,*R*)

C, Number allocated	*R*, Number required 1	2	3	4	5	6
1	1.000	0	0	0	0	0
2	.500	1.000	0	0	0	0
3	.333	.667	1.000	0	0	0
4	.250	.500	.750	1.000	0	0
5	.200	.400	.600	.800	1.000	0
6	.167	.333	.500	.667	.833	1.000

It is easy to see from the above formula that every column-maximum of $P_{mr}(C, R)$, will be equal to 1. It is also clear that any pair C, R for which there is no deterrence, i.e. where there are *no* targets killed with the desired 90% assurance, had $P(C, R) = 0$ and thus $P_{mr}(C, R) = 0$ also. Of course, P_{mr} was defined in order to have just those properties.

If now we wish to choose the policy which maximizes the guaranteed outcome, in the sense of the modified regret, we must compute

$$\max_{s_i} \min_{r_j} \sum_{i,j} s_i \cdot P_{mr}(C_i, R_j) \cdot r_j, \tag{3.5}$$

and the desired policy is determined by the numbers s_i which achieve this minimax value.

In the example, we find $s_1 = .204$, $s_2 = .136$, $s_3 = .102$, $s_4 = .082$, $s_5 = .068$, and $s_6 = .408$. The resulting payoffs, in the six possible states of nature, are respectively[7]

$$(.408, .204, .136, .102, .082, .068).$$

[6] Note that our $P(C, R)$ will be non-negative. This shows that the denominator will not be equal to 0 unless there is an R_0 with $P(C, R) = 0$ for all suboptimal C. We should simply ignore any such R_0, since our action (if chosen from among the suboptimal actions C) would have *no* influence on the payoff if R_0 was the true state of nature.

[7] The fact that the set of payoff values are the same numbers as the weights s_i is an accident caused by the particular payoff function used in this example.

This modified-regret policy also does better than the Wald minimax policy in the two most favorable states of nature, and less well in the four least favorable, but not to such an extent as the Savage minimax-regret policy. In addition, the modified-regret policy uses more of the possible pure strategies (in this example, it uses all of them). Both of these characteristics are likely to be generally true.

Another renormalization might also be reasonable in certain special cases; one could define a second modified-regret function as

$$P_{\mathrm{mr}_2}(C, R) = \left(\frac{P(C, R) - \min_C P(C, R)}{\max_C P(C, R) - \min_C P(C, R)} \right). \tag{3.4'}$$

This form of modified regret has the properties that

$$\min_C P_{\mathrm{mr}_2}(C, R) = 0 \quad \text{for all } R, \quad \text{and}$$
$$\max_C P_{\mathrm{mr}_2}(C, R) = 1 \quad \text{for all } R.$$

Note that in our example the form differs only slightly from the modified regret proposed in formula (3.4). Note that, again, the denominator of the fraction cannot equal zero because we ignore states of nature in which the payoff is unaffected by our choice of action.

(4) *Statistical decision theory.* The essence of statistical decision theory is that the experimenter assigns probabilities to the various states of nature before he begins to perform, or even to design, his experiments. In this way the several states of nature are replaced by a single state of nature, weighting the others as dictated by those *prior probabilities.* From this point on, there is comparatively little conceptual difficulty, although choosing a prior distribution can be a problem.

For our example, we suppose that the prior probability is defined as an incomplete beta-function distribution over the parameter R.[8] If we take $\rho = 1$ and $\nu = 2$, we get the so-called "uniform distribution on R," and the prior probabilities of the various values of k are as follows (recall that k is the number of missiles required):

k =	1	2	3	4	5	6	>6[9]
prior (k) =	.100	.216	.147	.098	.069	.049	.319

Against this prior distribution, the result is that the best policy is to allocate 2 missiles per target, which results in an expected outcome of .158. Against

[8] The incomplete beta-function with parameters ρ and ν is the distribution we would infer if the only evidence we possessed about the parameter R was that we had tested $(\nu - 2)$ missiles and obtained $(\rho - 1)$ successes.

[9] Since the incomplete beta-functions give non-zero probability of any R, we must, in this and the following sections, take account of the possibility that $k > 6$.

states of nature 1 through 6 respectively, allocation of 2 missiles gives the expected outcomes:

$$(.500, .500, 0, 0, 0, 0).$$

b. INCORPORATION OF THE TESTING PROBLEM. In subsection 3.a. above, the problem was to choose one of the six possible policies, each of which specified a number of missiles (from 1 through 6) which would be allocated against each target. When we begin to consider testing, and subsequent allocation, we are faced with a large number of possible policies, even for moderate numbers of missiles to be tested. For example, if we test three missiles, there are 120 "reasonable" policies (i.e. policies in which the number allocated actually depends on the results of the test but which never allocate more missiles when more successes are observed). Those 120 policies must be considered if we are to test 3 missiles; furthermore, we probably also want to consider the results of a test of 0, 1, or 2, implying a total of 191 policies.[10] It appears that the Wald minimax scheme will again say "look only at the most unfavorable state of nature." although a proof is less obvious. Application of any one of the other three (regret-type) decision-schemes described above, to those 191 policies vis-à-vis the 6 states of nature, is a complex combinational problem, which we pass over since it does not represent the main point of this paper.

(1) *Decision-theory.* In this case we have a prior distribution on the parameter R, and we can calculate the optimal policy without the complications implied in the other approaches.

We describe the process briefly. For any proposed number E of missiles to be tested, and any possible number NS of successful trials, we can use Bayes' theorem to infer a posterior distribution on the parameter R. From that posterior distribution we may calculate the value of that maximum expected payoff, and determine which choice of allocation would obtain it.

Now consider the same experiment, and its other possible results; for each of those we may also calculate the payoff obtainable by making the optimum allocation. We also can calculate (in view of the prior distribution) the probability of obtaining each of those possible results and thus the expected payoff if we perform that experiment and afterwards make the optimal allocation. Now, in view of the costs of the various experiments, we need only choose that experiment which maximizes the quantity "expected payoff minus cost"; the result of the optimal policy consists of the allocations recommended for each possible outcome. We shall refer to those policies, each of which is optimal against one of the prior distributions, as suboptimal policies.

[10] The number of reasonable policies in which E are tested, if the number allocated can be as great as k, is

$$k \text{ if } E = 0, \text{ and } \frac{(E + k)!}{(E + 1)!\,(k - 1)!} - k \text{ if } E > 0.$$

c. ALTERNATE PRIOR DISTRIBUTIONS. By using the method suggested in this paper, we may take account of several prior distributions. The optimal policy is calculated for each of those prior distributions. Next we calculate the expected payoffs, for each of those prior distributions, when each of the policies is employed. We obtain a matrix, whose columns correspond to the prior distributions and whose rows correspond to the policies. Now we have a decision-problem more abstract than the problem considered in subsection 3.a., for now the experimenter must choose a

TABLE 4

Augmented Payoff Matrix*

	k, Number of missiles required per target						
	1	2	3	4	5	6	>6
Boundaries of corresponding ranges of R	1.000 --- .900	--- .684	--- .536	--- .438	--- .369	--- .319	--- .000
C, Number allocated							
1	1.000	0	0	0	0	0	0
2	.500	.500	0	0	0	0	0
3	.333	.333	.333	0	0	0	0
4	.250	.250	.250	.250	0	0	0
5	.200	.200	.200	.200	.200	0	0
6	.167	.167	.167	.167	.167	.167	0

* Compare with Table 1

policy (rather than an *action*) and we imagine Nature choosing a *prior distribution* (rather than a *state of nature*).

The first three methods described in subsection 3.a. may now be applied to this new, more abstract problem.

We have used incomplete beta-functions for the prior distributions in our example. Rather than distort the prior distributions by assuming that $R < 0.319$ was impossible,[11] we instead assumed that a seventh state of nature existed, with 0 payoff regardless of allocation. Table 4 gives the augmented payoff matrix. Recall that the incomplete beta-function over R, with parameters (ρ, ν), is proportional to the likelihood, if $\nu - 2$ missiles

[11] $R < 0.319$ is just the condition that k be bigger than 6; see formula (2.1), and footnote 9. Of course, 0.319 is an approximation to

$$1 - (1 - .9)^{1/6}.$$

had been tested, that just $\rho - 1$ of them would have been successful if the true reliability were R. For each of the four cases considered in our hypothetical example, Table 5 presents the parameters (ρ, ν) of the prior distribution and summarizes the results of the conventional decision-theory calculations as follows: The parameters (ρ, ν) of the beta-distribution assumed as the prior distribution; the average value of R implied by those

TABLE 5

Alternative Prior Distributions Used and Results of Decision Theory Calculations

Prior distribution number	(ρ,ν)	Average value of R	Corresponding suboptimal policy	Net value, in targets per missile
1	(9,13)	0.69	(20: 6, 6, 6, 6, 6, 6, 5, 5, 4, 4, 4, 3, 3, 3, 3, 2, 2, 2, 2, 2, 2)	0.324
2	(3,4)	0.75	(26: 6, 6, 6, 6, 6, 6, 6, 6, 6, 6, 5, 5, 4, 4, 4, 3, 3, 3, 3, 2, 2, 2, 2, 2, 1, 1, 1)	0.427
3	(9,12)	0.75	(18: 6, 6, 6, 6, 5, 5, 4, 4, 4, 3, 3, 3, 3, 2, 2, 2, 2, 2, 1)	0.377
4	(9,11)	0.82	(22: 6, 6, 6, 6, 6, 6, 5, 5, 4, 4, 4, 3, 3, 3, 3, 2, 2, 2, 2, 2, 2, 1, 1)	0.463

parameters (which is ρ/ν); the net value (payoff minus cost) of that optimal policy.

Note that those policies vary from "test 20" to "test 26," although the variations in *average* value of R are rather small. Those six distributions were chosen as representatives of prior distributions for a missile with "about 70–80% chance of success."

Table 6 presents the 16 possible payoffs. To illustrate the meaning of these numbers, consider the item in row 4, column 1. Referring to Table 5, we see that distribution No. 1 is defined by $\rho = 9$, $\nu = 13$, and that policy No. 4 is:

> "Test 22 missiles: If none, one or up to five succeed, allocate 6 per target; if six or seven succeed, allocate 5 per target; if eight to ten succeed, allocate 4 per target; if eleven to fourteen succeed, allocate 3 per target; if fifteen to twenty succeed, allocate 2 per target; and if twenty-one or twenty-two succeed, allocate 1 per target."

TABLE 6

Payoffs for Various Sub-Optimal Policies vis-à-vis Various Distributions

Suboptimal policy number	True distribution number 1	2	3	4
1	.324	.355	.371	.415
2	.307	.427	.359	.456
3	.320	.406	.377	.453
4	.309	.421	.370	.463

The entry in Table 6 shows that using policy No. 4 will result in an expectation of 0.309 targets per missile if the true state of nature is as described by distribution No. 1.

Note that using the figures in Table 6 to attempt to *predict* what will happen is a very unprofitable pastime; the numbers range from 0.307 to 0.463, a spread of nearly 50%. Nevertheless, such numbers *can* be used to *decide* what to do.

(1) *Minimax.* Here again the minimax solution is quite uninteresting, since there is one choice for Nature (distribution No. 1) which is uniformly worst. Then the minimax solution says in effect: "Be conservative; assume distribution No. 1, therefore use policy No. 1." The payoffs for the four prior distributions are respectively

(.324, .355, .371, .415).

(2) *Minimax-regret.* It is also possible to employ the minimax-regret approach to the matrix of Table 6. The regret matrix is shown in Table 7. The resulting solution is that policies No. 2 and No. 3 should be used with

TABLE 7

Regret Matrix Derived from Table 6

Suboptimal policy number	Distribution number 1	2	3	4
1	0	−72	−6	−48
2	−17	0	−18	−7
3	−4	−21	0	−10
4	−15	−6	−7	0

Note: This matrix has been multiplied by 1000 for convenience in reading.

equal probability. The resulting regrets are respectively

$$(-.010, -.010, -.009, -.008),$$

and the payoffs are respectively

$$(.314, .416, .368, .454).$$

(3) *Modified regret.* Applying the modified-regret approach, we multiply each column of Table 6 by a constant so that the maximum becomes 1.000, and obtain the modified regret matrix shown in Table 8. The minimax solution to this matrix is that policies No. 2 and No. 3 should be

TABLE 8
Modified-Regret Matrix
Derived from Table 6

1.000	.831	.984	.896
.948	1.000	.952	.985
.988	.951	1.000	.978
.954	.986	.982	1.000

employed with probabilities 0.415 and 0.585 respectively. (Note that this is very close to the usual regret solution.) The resulting modified regrets for the various prior distributions are

$$(.971, .971, .980, .981),$$

and the resulting payoffs are

$$(.315, .415, .369, .454).$$

4. SUMMARY OF RESULTS

Table 9 gives the results of the various principles in tabular form. It appears that the results of the two minimax-regret approaches do not differ

TABLE 9

	True distribution number			
Payoff with	1	2	3	4
Policy No. 1	0.324	0.355	0.371	0.415
Policy No. 2	0.307	0.427	0.359	0.456
Policy No. 3	0.320	0.406	0.377	0.453
Policy No. 4	0.309	0.421	0.370	0.463
Best policy	0.324	0.427	0.377	0.463
Worst policy[a]	0.307	0.355	0.359	0.415
Minimax	0.324	0.355	0.371	0.415
Minimax regret	0.314	0.416	0.368	0.454
Minimax modified regret	0.315	0.415	0.369	0.454

[a] (i.e. worst of the four suboptimal policies; other policies, of course, will be worse still)

significantly, and that either will produce a much greater payoff in the favorable states of nature, with only a small decrease in the payoff in the unfavorable cases.

5. CONCLUSIONS

First, we conclude that the proposal leads to calculations now feasible on large-scale computers.

Second, we conclude from Tables 5 and 6 that different prior distributions, even though apparently very similar, may differ significantly in the absolute payoff attainable, and may have radical differences in their optimal policies.

Third, we suspect from Table 6 (and other experiments not described here) that the influence of the prior distribution on the payoff typically outweighs the influence of the specific test-policy used. In other words, it is usually better to have a better missile and a reasonable test policy than a poorer missile and an optimal test policy.

Fourth, in spite of the previous conclusion, it appears that the techniques proposed here can add measurably to the efficiency of decision-theory uses. There is no doubt that this technique makes decision-theory more palatable to those who object to the arbitrariness of the single prior distribution.

Fifth, there are certain decision-problems, of which our example is one, for which the Wald minimax approach seems unreasonably conservative. Either of the two minimax-regret approaches seems acceptable; at least they are free from the obvious flaws of the simpler minimax approach.

6. ADDENDUM

The referee has suggested using the optimal mixed strategy for nature as a preferred or compromise prior distribution, and optimizing (in the usual decision-theory sense) against *that* prior distribution. This idea frees us from the need to propose mixed strategies for the experimenter.

An analogous approach was used by the author in [5] but has not been employed in the example of this paper because, in general, a linear combination of incomplete beta-functions is not itself an incomplete beta-function.

REFERENCES

[1] David Blackwell and M. A. Girshick, *Theory of Games and Statistical Decisions*, Wiley, 1954.

[2] Herman Chernoff and Lincoln E. Moses, *Elementary Decision Theory*, Wiley, 1959.

[3] W. L. Deemer and J. P. Mayberry, *The Application of Statistical Decision Theory to a Missile Testing Problem*, Operations Analysis Paper No. 5, Headquarters, USAF (AFGOA), Dec. 1961.

[4] R. D. Luce and Howard Raiffa, *Games and Decisions*, Wiley, 1957.
[5] J. P. Mayberry, *Alternate Payoff-Functions in Statistical Decision Theory*, to appear in Proceedings of the Third International Conference on Operational Research, Oslo, 1963.
[6] J. W. Milnor, *Games Against Nature*, (Chapter IV in [9]). (An earlier version was published as Project RAND Research Memorandum RM-679, RAND Corp., September 1951.)
[7] Robert Schlaifer, *Probability and Statistics for Business Decisions*, McGraw-Hill, 1959.
[8] Howard Raiffa and Robert Schlaifer, *Applied Statistical Decision Theory*, Graduate School of Business Administration, Harvard University, 1961.
[9] R. M. Thrall, C. H. Coombs and R. L. Davis, *Decision Processes*, Wiley, 1954.
[10] P. J. M. van den Bogaard and J. Versluis, "Design of Optimal Committee Decisions," *Statistica Neerlandica*, v. 16 (1962) No. 3, p. 271.
[11] Abraham Wald, *Statistical Decision Functions*, Wiley, 1950.

CHAPTER 11

Smoothing in Inventory Processes

By HARLAN D. MILLS*

1. INTRODUCTION

A typical inventory or production operation, in a decentralized multi-echelon supply or manufacturing system, acquires, stores, and supplies inventory on the basis of local information and criteria. Consider the requisition problem at such local inventory decision points. We visualize an inventory operation which has a certain inventory available, makes certain requisitions at the beginning of each administrative time period (months, weeks, days, etc.), and then is confronted with certain demands during the period. The inventory of the following period is determined by the events of the last, requisitions are determined by choice in the operation, and the demands are governed by an external process.

If the demands on the inventory point have definite statistical properties, then any given requisition policy will induce statistical properties in requisitions and inventory levels as well. Variability in requisitions and inventory levels often reflects directly into costs or service. Increased requisition variability can mean increased costs through inefficient production or inventory operations at previous stages of the inventory. Increased inventory variability increases the chance of stock outs when the average inventory level is held constant, or requires an increased average inventory level to maintain a given level of servicing demands. Thus, it is often desired to formulate requisition policies which lead to high degrees of stability in the derived time series of requisitions and inventory levels. The physical nature of the operations involved will dictate more specific requirements; requisition stability may be more important to some operations than inventory stability, or conversely. In a multi-echelon supply system, the requisitions of one echelon literally become the demands of the next, so stability of the system as a whole depends in an obvious way on the requisition policies.

The ability of a requisition policy to induce stability in the derived time series of requisitions and inventory levels is called "smoothing." This is our main interest; given a time series of demands, with known statistical properties, we seek decision policies which smooth the derived time series of requisitions and inventory levels. Our approach parallels that of Simon

* International Business Machines Corporation.

This research was supported by the Bureau of Supplies and Accounts, United States Navy, under contract No. Nonr 2928(00) with Mathematica.

[4], Vassian [5], and Pinkham [3], in seeking requisition policies with desirable servo-statistical properties. Part of these results are formulated in a more general setting in [2].

Our main results deal with the case in which demands are independent, identically distributed random variables. In this case, under *mean square criteria*, we find (1) the ultimate ability (or capacity) of requisition policies simultaneously to smooth requisition and inventory levels; (2) a class of linear requisition policies which achieve this maximal smoothing; (3) the effect of information delays on smoothing capacities and performances of requisition policies.

These results are developed, first, for the relatively simple case in which inventories are allowed to be negative, as backorders, and extended later to the more difficult case in which inventories are required to be non-negative, and "sales" are lost when demands exceed inventories.

Section 2 formulates the problems of interest, while Section 3 analyzes examples of requisition policies. The main results are derived in Section 4 and extended to non-negative inventory formulations in Section 5.

2. CONSERVATIVE INVENTORY PROCESSES

We define a conservative inventory process to be a sequence of random variables [1]

$$P = (i_1, r_1, s_1, i_2, r_2, s_2, \ldots),$$

where we interpret

- i_t as *inventory* in an operation at the beginning of time period t,
- r_t as inventory added to the operation, or *requisitions*, immediately after the beginning of period t,
- s_t as inventory deleted from the operation, or *shipments*, during period t,

such that P, in conjunction with a *history* of numbers,

$$I = (\ldots, i_{-1}, r_{-1}, s_{-1}, i_0, r_0, s_0)$$

is characterized by the following properties, for $t = 0, 1, 2, \ldots$,

$$i_{t+1} = i_t + r_t - s_t, \tag{1}$$

(with Probability 1),

$$r_{t+1} = R(\ldots, s_{t-1}, i_t, r_t, s_t), \tag{2}$$

(with Probability 1),

$$\text{Prob}\,\{s_{t+1} \leq x\} = S(x; t), \tag{3}$$

where R is an arbitrary function, and S is an arbitrary distribution in x.

Equation (1) represents a material balance in the operation; (2) specifies a requisition policy, R; (3) describes the random nature of shipments.

Equations (1) and (3) together will be referred to as a "conservative inventory model." The term "conservative" refers to the conservation of shipping requirements, in the form of backorders, from period to period if necessary. Notice we require R to be independent of explicit reference to t, while S is not. This is a common problem in servo-mechanisms; even though the environment may be known to be changing, no distinguishable time reference exists, so that such changes must be detected and responded to by time-independent policies. More generally, the servo-mechanism itself may influence the environment, though we do not formulate that here.

The following notation indicates means, variances, auto- and cross-correlation functions in process P, namely

$$\bar{i}_t = E(i_t), \qquad \bar{r}_t = E(r_t), \qquad \bar{s}_t = E(s_t)$$

$$\sigma_i^2(t) = E[(i_t - \bar{i}_t)^2], \text{ etc.}$$

$$\nu_{ir}(t, u) = \frac{1}{\sigma_i(t)\sigma_r(u)} E[(i_t - \bar{i}_t)(r_u - \bar{r}_u)], \text{ etc.}$$

A process P is defined as *stable* if all first- and second-order moments approach definite limits, if

$$\lim_{t\to\infty} \bar{i}_t = \bar{i}, \qquad \lim_{t\to\infty} \bar{r}_t = \bar{r}, \qquad \lim_{t\to\infty} \bar{s}_t = \bar{s},$$

$$\lim_{t\to\infty} \sigma_i^2(t) = \sigma_i^2, \text{ etc.}$$

$$\lim_{t\to\infty} \nu_{ir}(t, t + u) = \nu_{ir}(u), \text{ etc.}$$

In a stable process, we will be interested in δ_i^2, σ_r^2 and σ_s^2, and specifically in (assuming $\sigma_s^2 > 0$)

$$k_i = \sigma_i/\sigma_s, \qquad k_r = \sigma_r/\sigma_s. \tag{4}$$

The ratios k_i and k_r measure the transformations of shipment variability into inventory and requisition variabilities. To denote their dependence on the requisition policy R, we may write, on occasion,

$$k_i = k_i(R), \qquad .k_r = k_r(R).$$

In most instances, we seek to make either or both k_i and k_r small. If, say for two policies R_1, R_2, we have

$$k_i(R_1) < k_i(R_2), \qquad k_r(R_1) < k_r(R_2),$$

we say R_1 is "better" than R_2; that is, R_1 does a better job of smoothing, or suppressing the variability which is imposed on the process.

In some instances, the variability between successive requisitions may be of more interest than their variability about their means, and we may be concerned with

$$\lim_{t\to\infty} E(r_{t+1} - r_t)^2 = 2(1 - \nu_{rr}(1))\sigma_r$$
$$= k_\Delta{}^2\sigma_s{}^2,$$

say, so that

$$k_{\Delta r}{}^2 = 2(1 - \nu_{rr}(1))\, k_r{}^2.$$

3. SOME ILLUSTRATIVE PROCESSES

In order to illustrate our interests concretely, we formulate and analyze some specific inventory processes. In each case we will suppose shipments s_t to be independently and identically distributed, with finite means and variances.

"k MONTHS OF SUPPLY" INVENTORY POLICIES. A standard and ordinary inventory policy attempts to maintain a given number of periods of supply on hand or on order, using a running average to define a period of supply. If we seek to maintain k periods of supply, using averages over n periods, (2) becomes

$$r_t = k\left(\frac{1}{n}\sum_{j=1}^{n} s_{t-j}\right) - i_t. \tag{5}$$

We shall allow, in illustration, negative requisitions (or returns of inventory), and negative inventories (or backlogs); non-negative restrictions in both instances will be formulated later.

Now (1) and (5) can be solved by straightforward techniques to obtain

$$i_t = -s_{t-1} + \frac{k}{n}\sum_{j=1}^{n} s_{t-1-j}$$

$$r_t = \left(1 + \frac{k}{n}\right)s_{t-1} - \frac{k}{n}\, s_{t-n-1}.$$

We note that $\{i_t, r_t, s_t\}$ constitutes a stable process. Taking expected values, we obtain

$$i = (k-1)\bar{s}, \qquad \bar{r} = \bar{s},$$

as is to be expected, and (for $n \geq 2$)

$$\sigma_i^2 = \sigma_s{}^2 + \frac{k^2}{n^2}\sum_{j=1}^{n}\sigma_s{}^2 = \left(1 + \frac{k^2}{n}\right)\sigma_s{}^2,$$

$$\sigma_r{}^2 = \left[\left(1 + \frac{k}{n}\right)^2 + \left(\frac{k}{n}\right)^2\right]\sigma_s{}^2,$$

$$E(r_{t+1} - r_t)^2 = 2\left[\left(1 + \frac{k}{n}\right)^2 + \left(\frac{k}{n}\right)^2\right]\sigma_s{}^2.$$

In our notation above, these become

$$k_i^2 = 1 + \frac{k^2}{n}, \quad k_r^2 = \left(1 + \frac{k}{n}\right)^2 + \left(\frac{k}{n}\right)^2, \quad k_{\Delta r}^2 = r\left[\left(1 + \frac{k}{n}\right)^2 + \left(\frac{k}{n}\right)^2\right]. \tag{6}$$

We shall see shortly that these values are rather distressing—that a great deal of unnecessary variability has been introduced into the requisitions and inventory levels of the process.

"LINEAR SERVO" INVENTORY POLICIES. An extremely simple requisition policy can be devised, adjusting r_t about a fixed level r^* linearly with respect to deviations of i_r from a fixed level i^*; (2) then becomes

$$r_t - r^* = \alpha(i_t - i^*). \tag{7}$$

Now (1) and (7) can be solved, to obtain

$$i_t = \left(i^* + \frac{r^*}{\alpha}\right)[1 - (1 - \alpha)^{t+1}]\sum_{j=0}^{t-1}(1 - \alpha)^j s_{t-1-j} + (1 - \alpha)^t i_0 \tag{8}$$

$$r_t = \alpha\sum_{j=0}^{t-1}(1 - \alpha)^j s_{t-1-j} - \alpha(1 - \alpha)^t i_0 + (r^* + \alpha i^*)(1 - \alpha)^t, \tag{9}$$

and a stable process $\{i_t, r_t, s_t\}$ results when $|1 - \alpha| < 1$. Then, we find, taking expectations,

$$\bar{i} = i^* - \frac{1}{\alpha}(\bar{s} - r^*), \quad \bar{r} = \bar{s},$$

and

$$\sigma_i^2 = \sum_{j=0}^{\infty}(1 - \alpha)^{2j}\sigma_s^2 = [1/1 - (1 - \alpha)^2]\sigma_s^2$$

$$\sigma_r^2 = \alpha^2\sum(1 - \alpha)^{2j}\sigma_s^2 = [\alpha^2/1 - (1 - \alpha)^2]\sigma_3^2,$$

$$\lim_{t\to\infty} E(r_{t+1} - r_t)^2 = \alpha^2\left[\sigma_s^2 + \alpha^2\sum_{j=0}^{\infty}(1 - \alpha)^{2j}\sigma_s^2\right] = \alpha^2\left[1 + \frac{\alpha^2}{1 - (1 - \alpha)^2}\right]\sigma_s^2,$$

the second moments, surprisingly enough, being independent of r^* and i^*. In our previous notation, these become

$$k_i^2 = \frac{1}{1 - (1 - \alpha)^2}, \quad k_r^2 = \frac{\alpha^2}{1 - (1 - \alpha)^2}, \quad k_{\Delta r}^2 = \alpha^2\left(1 + \frac{\alpha^2}{1 - (1 - \alpha)^2}\right). \tag{10}$$

We shall see shortly that these values are, in a certain sense, the best possible. This simple class, for $0 < \alpha < 1$ will be shown to be an optimal class under the criteria of minimizing k_i and k_r simultaneously.

"RANDOM WALK" INVENTORY POLICIES. A simple nonlinear policy can be devised to requisition a fixed amount r^* except when inventory departs by more than an amount a from a given level i^*, and then to requisition the fixed amount plus the amount required to bring the inventory level back to i^*. Then (2) becomes

$$r_t = \begin{cases} r^* & \text{if } |i_t - i^*| \leq a \\ r^* + i^* - i_t & \text{if } |i_t - i^*| > a. \end{cases} \tag{11}$$

Thus $e_t = i_t - i^*$, the inventory deviation from i^*, is a random walk, set back to r^* each time it proceeds past a barrier at a or $-a$.

We shall not be able to obtain an explicit characterization of this process. We do note, however, that it is stable, and that k_i will depend on a in an intuitively apparent way; the larger a is, the larger $\sigma_i^2 = \sigma_e^2$, for example. When $a = 0$, then $\sigma_i^2 = \sigma_s^2$, or $k_i = 1$.

To obtain some additional insight into this process, we note, if the $x_t = s_t - r^*$ are normally distributed with zero means ($r^* = \bar{s}$), that the expected duration d of a "run" between "special requisitions" has been approximated by Wald [6],

$$d = a^2/\sigma_s^2.$$

Hence, since $r^* = \bar{s} = \bar{r}$, we have,

$$(r_t - \bar{r})^2 = \begin{cases} (a + y)^2 & \text{with probability } \dfrac{1}{1 + d} = \dfrac{\sigma_s^2}{a^2 + \sigma_s^2} \\ 0 & \text{with probability } \dfrac{d}{1 + d} = \dfrac{a^2}{a^2 + \sigma_s^2} \end{cases}$$

where y is the random excess over a (or less than $-a$) when the random walk exceeds its barrier. We get an underestimate of $(a + y)^2$ by taking $y = 0$, and an overestimate by taking $y = |z|$ to be "half-normal;" z normally distributed. These provide expectations

$$\begin{aligned} a^2 \leq E(a + y)^2 &\leq E(a + z)^2 \\ &= a^2 + 2aE(z) + E(z)^2 \\ &= a^2 + 2a(2/\sqrt{2\pi})\,\sigma_s + \sigma_s^2 \\ &\leq (a + \sigma_s)^2. \end{aligned}$$

These bounds, with the estimated probabilities above, give approximate bounds

$$\frac{a^2\sigma_s^2}{a^2 + \sigma_s^2} \leq \sigma_r^2 \leq \frac{(a + \sigma_s)^2\sigma_s^2}{a^2 + \sigma_s^2}$$

or,

$$\frac{a^2}{a^2 + \sigma_s^2} \leq k_r^2 \leq \frac{(a + \sigma_s)^2}{a^2 + \sigma_s^2}, \tag{12}$$

which approach each other as a becomes large; when $a = 0$, k_r^2 equals the upper bound.

The ratio of these bounds is $1 + \sigma_s/a$, so they are quite sharp for typical values of a. Thus, surprisingly enough, k_r must be close to 1 for large values of a and equals 1 exactly for $a = 0$.

Similarly, we can also estimate $E(r_{t+1} - r_t)^2$, for $d \gg 1$, so that the probability of passing the barrier on successive occasions can be ignored. Then, we have

$$(r_{t+1} - r_t)^2 = \begin{cases} (a + y)^2 & \text{with probability } \dfrac{2}{1 + d^2} = \dfrac{2\sigma_s^2}{a^2 + \sigma_s^2} \\ 0 & \text{with probability } \dfrac{d - 1}{1 + d} = \dfrac{a^2 - \sigma_s^2}{a^2 + \sigma_s^2} \end{cases}$$

and, we have

$$\frac{2a^2\sigma_s^2}{a^2 + \sigma_s^2} \leq E(r_{t+1} - r_t)^2 \leq \frac{2(a + \sigma_s)^2\sigma_s^2}{a^2 + \sigma_s^2}$$

or

$$\frac{a^2}{a^2 + \sigma_s^2} \leq \frac{k_{\Delta r}^2}{2} \leq \frac{(a + \sigma_s)^2}{a^2 + \sigma_s^2} \tag{13}$$

Thus, $k_{\Delta r}/\sqrt{2}$ is subject to the same bounds as k_r.

Some numerical comparisons. In order to develop a better feel for the behavior of policies listed above, we compute or estimate some of their numerical characteristics, using (6), (10), (12), and (13).

k Months of Supply Inventory Policy			k_i	k_r	$k_{\Delta r}$
a.	$k = 5$	$n = 10$	1.8	1.6	2.2
b.	$k = 10$	$n = 5$	4.6	3.6	5.1
Linear Servo Inventory Policy					
c.	$\alpha = 0.2$		1.7	0.3	0.2
d.	$\alpha = 0.5$		1.2	0.6	0.6
e.	$\alpha = 0.8$		1.02	0.8	1.0
f.	$\alpha = 1.0$		1.0	1.0	1.4
Random Walk Inventory Policy					
g.	a arbitrary		≥ 1	~ 1	~ 1.4

It is quite apparent that the Linear Servo policies have a decided advantage in minimizing k_i, k_r, and $k_{\Delta r}$. This is especially apparent in plotting these values in the (k_i, k_r) plane (Figure 1). But the question remains as to whether more ingenuity might devise policies which out-perform any so far considered. This question, with regard to the criteria of k_i and k_r, is answered in the following section.

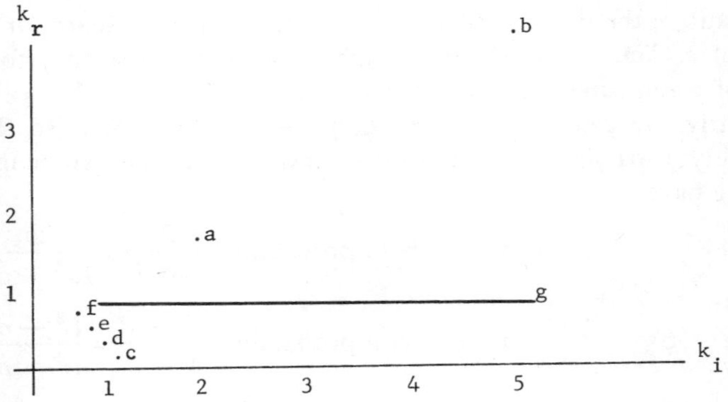

FIGURE 1

We shall not prove any general properties about policies which minimize $k_{\Delta r}$. But since we can write

$$k_{\Delta r}{}^2 = 2[1 - \nu_{rr}(1)]k_r{}^2,$$

policies which minimize k_r and correlate r_t and r_{t+1} positively would appear to do quite well. Notice that Linear Servo policies do just this, and with rather favorable results as illustrated in the numerical comparisons.

4. BASIC SMOOTHING THEOREMS

Three theorems follow which organize information about the possibilities and techniques of smoothing in conservative inventory models with independent, identically distributed shipments. These theorems, in turn, will serve, in the following section, as models for analogous theorems in more complex models, e.g. models with non-negative restrictions on inventories and requisitions.

Theorem 1 establishes the fact that a limit exists in the amount of smoothing that can be accomplished by any policy *R*. This limit can be described in terms of a boundary (curve) in the (k_i, k_r) plane, beyond which (at any point on the curve) k_i and k_r cannot both be decreased. This boundary represents the smoothing capacity of servo-mechanisms in the process. Incidentally, it supplies quantitative information on the rate at which smoothing in inventories and in requisitions can be traded off against each other. In general, inventory variation can be decreased at the expense of increased requisition variation, and conversely.

Theorem 2 discovers, happily enough, that an extremely simple class of linear decision policies achieves the maximal smoothing established in Theorem 1. This is the class of Linear Servo Inventory Policies already used in illustration. Several equivalent forms of this optimal policy class are presented.

Theorem 3 describes the effect of time delays of information (or of delays in filling requisitions) in such processes. Again, the results have quite a simple and useful form. An immediate general consequence is that time delays hurt "good" policies much more than "poor" ones, and within the optimal policy class, time delays hurt more those policies that are more concerned with inventory smoothing than requisition smoothing.

THEOREM 1. (Smoothing Capacity) *In a conservative inventory model with independent, identically distributed shipments, any requisition policy which generates a stable process must satisfy the relation*

$$k_i \geq \frac{1}{2}\left(k_r + \frac{1}{k_r}\right). \tag{14}$$

PROOF. By (1) we have

$$i_{t+1} = i_i + r_t - s_t$$

and s_t is independent of i_t and r_t. Since the process is stable, we have necessarily

$$\bar{i}_{t+1} = \bar{i}_t + \bar{r}_t - \bar{s}_t.$$

Then (1) becomes

$$(i_{t+1} - \bar{i}_{t+1}) = (i_t - \bar{i}_t) + (r_t - \bar{r}_t) - (s_t - \bar{s}_t).$$

Square both sides of this equation, take expectations and limits, to obtain

$$\sigma_i^2 = \sigma_i^2 + \sigma_r^2 + \sigma_s^2 + 2\sigma_i\sigma_r\nu_{ir}, \ \nu_{ir} = \nu_{ir}(0),$$

where $-1 < \nu_{ir} < 1$; using the definition of k_i and k_r, this becomes

$$0 = [k_r^2 + 1 + 2k_ik_r\nu_{ir}]\,\sigma_s^2$$

which can be rearranged (since $\sigma_s^2 > 0$), as

$$k_i \geq -k_i\nu_{ir} = \frac{1}{2}\left(k_r + \frac{1}{k_r}\right),$$

as was to be proved.

The boundary provided by Theorem 1 is diagrammed in Figure 2. Note that part of this boundary is drawn as a solid curve, part as a dotted curve. We are interested primarily in the solid curve, since this is the actual boundary to the joint minimization of k_i and k_r. Note that on this boundary, $k_r = \alpha k_i$, $0 < \alpha \leq 1$. Theorem 2 establishes a class of policies which generates this boundary.

THEOREM 2. (Optimal Policy Class) *In a conservative inventory model with independent, identically distributed shipments the requisition policy*

$$r_t - r^* = \alpha(i_t - i^*) \tag{7}$$

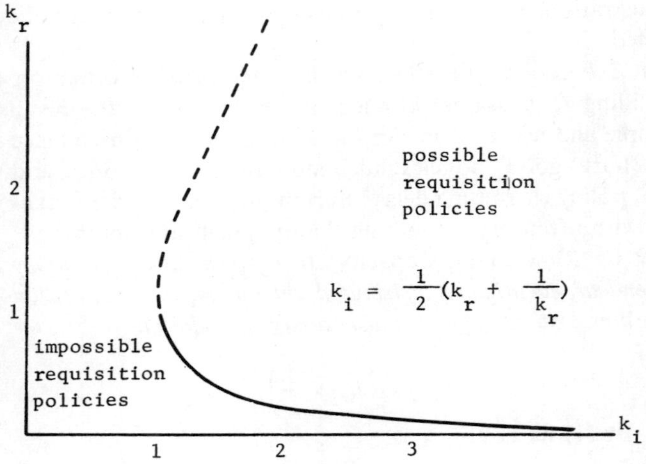

FIGURE 2

where r^* and i^* are arbitrary constants and $0 < \alpha \leq 1$, *determines a stable process such that*

$$k_i = \frac{1}{2}\left(k_r + \frac{1}{k_r}\right) \quad \text{and} \quad k_r = \alpha k_i.$$

PROOF. We have already seen, in (8) and (9), that the requisition policy of (7) determines a stable process when $|1 - \alpha| < 1$, or certainly when $0 < \alpha \leq 1$, with the values,

$$k^2 = \frac{1}{1 - (1 - \alpha)^2}, \quad k_r^{\,2} = \frac{\alpha^2}{1 - (1 - \alpha)^2} \tag{10}$$

Now both statements of the theorem can be verified immediately from (10). This completes the proof of the theorem.

The policy class of Theorem 2,

$$r_t - r^* = -\alpha(i_t - i^*), \tag{7}$$

has many equivalent forms, of which two are

$$r_t = \alpha s_{t-1} + (1 - \alpha)r_{t-1} \tag{15}$$

[eliminating i_t and i_{t+1} from (1) by means of (7)], and, as previously found

$$r_t = \alpha \sum_{j=0}^{t-1} (1 - \alpha)^j s_{t-1-j} - \alpha(1 - \alpha)^t(i_0 - i^*) + (r^* + \alpha i^*)(1 - \alpha)^t. \tag{9}$$

It is clear from the form of (15) that $r_t \geq 0$ when $r_0 \geq 0$, $s_t \geq 0$, $t \geq 0$.

Thus far, we have assumed the immediate filling of requisitions. Suppose now that requisitions are not filled immediately, but T periods after they are made. Then (1) should read

$$i_{t+1} = i_t + r_{t-T} - s_t. \tag{1'}$$

On the other hand, suppose every piece of information is delayed T periods (but requisitions are filled immediately). Then (2) should be revised as

$$r_{t+1} = R(\ldots, r_{t-T}, s_{t-T}, i_{t-T+1}). \tag{2'}$$

It is clear, on a moment's reflection, that the two sets of properties, (1′), (2), (3) and (1), (2′), (3), determine precisely the same recursive structure for determining an inventory process; it is a trivial matter to verify the matter formally. Thus, a delay in the filling of requisitions and an equal delay in information are equivalent, as is any combination of delays which total the original amount. For convenience, then, we can phrase delays in filling requisitions as equivalent delays in information. The effects of such delays are characterized in the following theorem.

THEOREM 3. (Information Delay) *Let R^T be identical with policy R except that information is delayed T periods, i.e.*

$$R^T(\ldots, r_t, s_t, i_{t+1}) = R(\ldots, r_{t-T}, s_{t-T}, i_{t-T+1});$$

then if R and R^T with independent identically distributed shipments determine stable conservative inventory processes P and P^T, they have the properties

$$k_i(R^T) = \sqrt{k_i^2(R) + T}, \qquad k_r(R^T) = k_r(R). \tag{16}$$

PROOF. First, we show that P^T, the process which satisfies (1) and (2′), can be obtained from P in a simple way. Let $P = \{i_t, r_t, s_t\}$ be the stable conservative inventory process determined by R. Then

$$i_{t-T+1} = i_{t-T} + r_{t-T} - s_{t-T}. \tag{1}$$

Add, now, $-s_{t-T+1} - \ldots - s_t$ (T terms) to both sides of this equation, and regroup as

$$(i_{t-T+1} - s_{t-T+1} - \ldots - s_t) = (i_{t-T} - s_{t-T} - \ldots - s_{t-1}) + r_{t-T} - s_t.$$

Thus, $P^T = \{i_{t-T} - s_{t-T} - \ldots - s_{t-1}, r_{t-T}, s_t\}$ satisfies (1) and

$$r_{t-T+1} = R^T(\ldots, r_t, s_t, i_{t+1}). \tag{2'}$$

Now, since shipments are uncorrelated, statement (16) of the Theorem can be computed directly from the form of P^T above. This completes the proof of the theorem.

5. EXTENDED SMOOTHING THEOREMS

We define a *semi-conservative inventory process* to be a sequence of random variables,

$$P = (i_1, r_1, d_1, s_1, i_2, r_2, d_2, s_2, \ldots),$$

where we interpret

- i_t as *inventory* in an operation at the beginning of time period t,
- r_t as inventory added to the operation, or *requisitions*, immediately after the beginning of period t,
- d_t as *demands* placed on inventory in the operation during period t,
- s_t as inventory deleted from the operation, or *shipments*, during period t,

such that P, in conjunction with a history of numbers,

$$I = (\ldots, i_{-1}, r_{-1}, d_{-1}, s_{-1}, i_0, r_0, d_0, s_0),$$

is characterized by the following properties, for $t = 0, 1, 2, \ldots,$

$$i_{t+1} = i_t + r_t - s_t \geq 0 \tag{1}$$

(with Probability 1)

$$r_{t+1} = R(\ldots, s_{t-1}, i_t, r_t, d_t, s_t) \geq 0 \tag{17}$$

(with Probability 1)

$$\text{Prob}\,\{d_{t+1} \leq x\} = D(x, t) \tag{18}$$

$$s_{t+1} = \min\,(i_{t+1}, d_{t+1}) \tag{19}$$

(with Probability 1)

where R is an otherwise arbitrary function, and D is an arbitrary distribution in x except that $D(0, t) = 0$ and $D(b, t) = 1$ for some finite b. This states that the demand is bounded by 0 and b.

Equation (1) represents the material balance in the operation; (17) specifies a requisition policy; (18) describes the random nature of demands; (19) determines shipments when no back-ordering is permissible. Equations (1), (18), and (19) together will be referred to as a "semi-conservative inventory model." The term semi-conservative refers to the failure of shipping requirements to be conserved in the operation, even though inventories are. This ban on backorders, as we shall see presently, complicates our analysis of the variability of the process. Results analogous to our previous ones will hold in asymptotic form. But now, for example, linear requisition policies which induce a stable process do not, in general, even exist.

Theorems 4 and 5 below are the counterparts of Theorems 1 and 2. We do not formulate the counterpart to Theorem 3 because it provides no essentially new information. The structure of the proof of Theorem 3 follows that of Theorem 1, except that an inequality in stochastic variables has to be dealt with rather than an equation, and this fact leads to some additional considerations. In keeping with our previous notation, but recognizing that demand variability is of fundamental importance here, we define $k_i' = \sigma_i/\sigma_d$, $k_r' = \sigma_r/\sigma_d$, $k_s' = \sigma_s/\sigma_d$ for a stable process P.

THEOREM 4. *In a semi-conservative inventory model with independent, identically distributed demands, any requisition policy which generates a stable process must satisfy the relation*

$$k_i' \geq \frac{1}{2}\left(k_r' + \frac{a}{k_r'}\right) \tag{20}$$

where

$$a = 1 - (\bar{d} - \bar{s})(2b + 2\bar{i} - \bar{d} + \bar{s})/\sigma_d^2 \tag{21}$$

PROOF. We note to begin with, from (1), that for a stable process,

$$\bar{i} = \bar{i} + \bar{r} - \bar{s}, \quad \text{or} \quad \bar{r} = \bar{s}.$$

Combining (1) and (19), we have

$$i_{t+1} \geq i_t + r_t - d_t,$$

and, with Probability 1,

$$i_{t+1} + b \geq i_t + r_t - d_t + b \geq 0,$$

or,

$$(i_{t+1} - \bar{i}_{t+1}) + (b + \bar{i}_{t+1}) \geq (i_t - \bar{i}_t) + (r_t - \bar{r}_t) - (d_t - \bar{d}_t) + (b + \bar{i}_t + r_t - \bar{d}_t) \geq 0$$

Square both expressions (preserving the inequality), take expectations and limits, to obtain

$$\sigma_i^2 + (b + \bar{i})^2 \geq \sigma_i^2 + \sigma_r^2 + \sigma_d^2 + 2\nu_{ir}\sigma_i\sigma_r + (b + \bar{i} + \bar{r} - \bar{d})^2,$$
$$\nu_{ir} = \nu_{ir}(0)$$

where $-1 \leq \nu_{ir} \leq 1$; using the definition of k_i' and k_r', we obtain

$$0 \geq \left(k_r'^2 + 1 + \frac{(b + \bar{i} + \bar{r} - \bar{d})^2 - (b + \bar{i})^2}{\sigma_d^2} + 2k_i' k_r' \nu_{ir}\right)\sigma_d^2,$$

which can be rearranged as

$$k_i \geq -k_i\nu_{ir} \geq \frac{1}{2}\left(k_r' + \frac{a}{k_r'}\right)$$

where

$$a = 1 + \frac{(b + \bar{i} + \bar{r} - \bar{d})^2 - (b + \bar{i})^2}{\sigma_d^2}$$
$$= 1 + \frac{(\bar{r} - \bar{d})(\frac{1}{2}b + 2\bar{i} + \bar{r} - \bar{d})}{\sigma_d^2}$$
$$= 1 - \frac{(\bar{d} - \bar{s})(\frac{1}{2}b + 2\bar{i} + \bar{s} - \bar{d})}{\sigma_d^2},$$

as was to be proved.

The boundary provided by Theorem 4 is, for $a > 0$, qualitatively like the boundary for the conservative case of Theorem 1, except that the units are the factor $\sqrt{a}$ in this case, as diagrammed in Figure 3. When $a < 0$, no

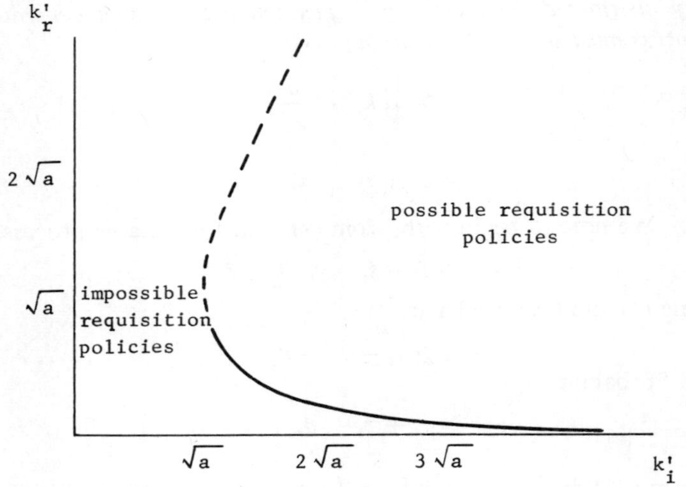

FIGURE 3

effective boundary exists. It may be possible to completely stabilize the operation.

This latter eventuality, as well as the structure of this derived parameter a, is instructive. Notice

$$a = 1 - \frac{(\bar{d} - \bar{s})(2b + 2\bar{i} - \bar{d} + \bar{s})}{\sigma_d^{\,2}}$$

has the properties

(i) $$a \leq 1,$$

since $\bar{d} - \bar{s} \geq 0$, from (19) and $2b + 2\bar{i} - \bar{d} + \bar{s} = (b - \bar{d}) + b + 2\bar{i} + \bar{s} \geq 0$ because all terms are non-negative

(ii) $$\lim_{\bar{s} \to \bar{d}} a = 1.$$

Thus, the more the demands are supplied in the process, the nearer the relation of Theorem 4 approaches the relation of Theorem 1. On the other hand, if some demands are left unsatisfied, the variability of the process can be decreased, as the following illustration shows. Suppose demands are never less than a constant c (where c may be > 0). Then the process determined by

$$i_0 = 0,$$
$$R(\ldots, s_t) = c$$

will provide the process with

$$i_t = r_t = s_t = c, \text{ all } t$$

and $k_i' = k_r' = k_s' = 0$, a process with no variability!

Another matter of interest is the general dependence of both first- and second-order moments in the process in this boundary, whereas the boundary in the conservative process involved only second-order moments.

We will find it convenient to note the parametric form of this boundary, analogous to (10), in terms of a parameter α, when $a > 0$, namely

$$k_i'^2 = \frac{1}{1-(1-\alpha)^2}\, a, \quad k_r'^2 = \frac{\alpha^2}{1-(1-\alpha)^2}\, a. \tag{22}$$

Theorem 5, in analogy to Theorem 2, shows how a class of truncated policies, closely related to the optimal class in the conservative case, is asymptotically optimal in the non-conservative case.

THEOREM 5. *Suppose, in a semi-conservative inventory model with independent identically distributed shipments, that the requisition policy*

$$r_1 = 0$$

$$r_t = \begin{cases} \alpha d_{t-1} + (1-\alpha) r_{t-1} & \text{if } i_t \le B \\ 0 & \text{if } i_t > B \end{cases} \tag{23}$$

such that $0 < \alpha \le 1$, $B \ge 0$ *determines a stable process. If in addition* $\sigma_i \le \bar{i}$ *in this process, then*

(a) $$k_r'^2 \le \frac{\alpha^2}{1-(1-\alpha)^2} + \frac{(\bar{d}-\bar{s})(\bar{d}+\bar{s})}{\sigma_d^2},$$

(b) $$k_i'^2 \le \frac{1}{1-(1-\alpha)^2}.$$

PROOF. Case (a). Let $T(t)$ be the least number of periods preceding t such that $r_{t-T(t)} = 0$; $T(t)$ is a random variable. Then, referring to (23) we find

$$r_t = \frac{\alpha}{1-\alpha} \sum_{j=1}^{T(t)} (1-\alpha)^j\, d_{t-j} \ge 0$$

whence

$$0 \le r_t \le \frac{\alpha}{1-\alpha} \sum_{j=1}^{\infty} (1-\alpha)^j\, d_{t-j}$$

where the d_u, $u < t - T(t)$ are taken to be independent replicates of the identically distributed random variable d_u, $u > 0$. Then

$$0 < (r_t - \bar{r}_t) + \bar{r}_t \le \frac{\alpha}{1-\alpha} \sum_{j=1}^{\infty} (1-\alpha)^j [(d_{t-j} - \bar{d}_{t-j}) + \bar{d}_{t-j}].$$

Square both expressions (preserving the inequality), take expectations and limits, to obtain

$$\sigma_r^2 + \bar{r}^2 \leq \frac{\alpha^2}{(1-\alpha)^2} \sum_{j=1}^{\infty} (1-\alpha)^{2j} \sigma_d^2 + \frac{\alpha^2}{(1-\alpha)^2} \left(\sum_{j=1}^{\infty} (1-\alpha)^j \bar{d} \right)^2$$

which becomes

$$\sigma_r^2 \leq \frac{\alpha^2}{1-(1-\alpha)^2} \sigma_d^2 + \bar{d}^2 - \bar{r}^2$$

or, since $\bar{r} = \bar{s}$,

$$k_r'^2 \leq \frac{\alpha^2}{1-(1-\alpha)^2} + \frac{(d-s)(d+s)}{\sigma_d^2}$$

as was to be proved for case (a).

Case (b). Let $S(t)$ be the least number of periods preceding t such that $i_{t-S(t)} = 0$; $S(t)$ is a random variable. Then using (1) and (23), we find

$$i_t = \sum_{k=1}^{S(t)} \left(-d_{t-k} + \frac{\alpha}{1-\alpha} \sum_{j=1}^{T(t-k)} (1-\alpha)^j d_{t-k-j} \right),$$

and

$$0 \leq i_t \leq \sum_{k=1}^{S(t)} \left(-d_{t-k} + \frac{\alpha}{1-\alpha} \sum_{j=1}^{\infty} (1-\alpha)^j d_{t-k-j} \right).$$

For each t, and $S = S(t)$, the right-hand expression can be rewritten as

$$- \sum_{k=1}^{S} d_{t-k} + \frac{\alpha}{1-\alpha} \sum_{k=1}^{S} \sum_{j=1}^{\infty} (1-\alpha)^j d_{t-k-j}$$

$$= - \sum_{h=1}^{S} d_{t-h} + \frac{\alpha}{1-\alpha} \sum_{h=2}^{S} \sum_{j=1}^{h-1} (1-\alpha)^j d_{t-h}$$

$$+ \frac{\alpha}{1-\alpha} \sum_{h=S+1}^{\infty} \sum_{k=1}^{S} (1-\alpha)^{h-k} d_{t-h}$$

(changing indices and orders of summation),

$$= -d_{t-1} + \sum_{h=2}^{S} \left[-1 + \frac{\alpha}{1-\alpha} \sum_{j=1}^{h-1} (1-\alpha)^j \right] d_{t-h}$$

$$+ \sum_{h=S+1}^{\infty} \left[\frac{\alpha}{1-\alpha} \sum_{k=1}^{S} (1-\alpha)^{h-k} \right] d_{t-h},$$

$$= -d_{t-1} - \sum_{h=2}^{S} (1-\alpha)^{h-1} d_{t-h} + \sum_{h=S+1}^{\infty} [(1-\alpha)^{-S} - 1](1-\alpha)^{h-1} d_{t-h}$$

(carrying out the inside summations).

Using this latter form, the inequality above can be rewritten as

$$0 \leq (i_t - \bar{i}_t) + \bar{i}_t \leq -\sum_{h=1}^{S}(1-\alpha)^{h-1}(d_{t-h} - \bar{d}_{t-h})$$
$$+ [(1-\alpha)^{-S} - 1] \sum_{h=S+1}^{\infty} (1-\alpha)^{h-1}(d_{t-h} - \bar{d}_{t-h}).$$

Square both non-trivial expressions (preserving the inequality), take expectations and limits, to obtain

$$\sigma_i^2 + \bar{i}^2 \leq \sum_{h=1}^{S}(1-\alpha)^{2h-2}\sigma_d^2 + [(1-\alpha)^{-S} - 1]^2 \sum_{h=S+1}^{\infty} (1-\alpha)^{2h-2}\sigma_d^2$$

which becomes

$$\sigma_i^2 + \bar{i}^2 \leq \left(\frac{1-(1-\alpha)^{2S}}{1-(1-\alpha)^2} + \frac{[1-(1-\alpha)^S]^2}{1-(1-\alpha)^2}\right)\sigma_d^2$$
$$= \frac{2[1-(1-\alpha)^S]}{1-(1-\alpha)^2}\sigma_d^2$$
$$\leq \frac{2}{1-(1-\alpha)^2}\sigma_d^2$$

for every S. Now, using the hypothesis

$$\sigma_i^2 \leq \bar{i}^2,$$

we have

$$2\sigma_i^2 \leq \sigma_i^2 + \bar{i}^2 \leq \frac{2}{1-(1-\alpha)^2}\sigma_d^2,$$

and the outside expressions provide the relation

$$k_i'^2 \leq \frac{1}{1-(1-\alpha)^2},$$

as was to be proved for case (b). This completes the proof of the theorem.

Theorems 4 and 5 combine to "box in" the point (k_i', k_r') induced by the policy of Theorem 5 as $\bar{s}$ approaches $\bar{d}$. To see this notice the three inequalities

$$k_i' \geq \frac{1}{2}\left(k_r' + \frac{a}{k_r'}\right), \quad a = 1 - \frac{(\bar{d}-\bar{s})(2b + 2\bar{i} - \bar{d} + \bar{s})}{\sigma_d^2}$$
$$k_r' \leq \frac{\alpha^2}{1-(1-\alpha)^2} + \frac{(\bar{d}-\bar{s})(\bar{d}+\bar{s})}{\sigma_d^2}$$
$$k_i' \leq \frac{1}{1-(1-\alpha)^2}$$

define a non-empty curvilinear triangle which degenerates to a point as $\bar{s}$ approaches $\bar{d}$, as diagrammed in Figure 4.

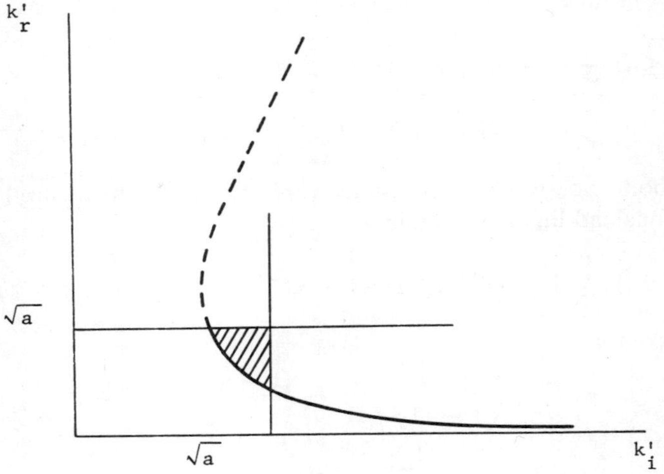

FIGURE 4

This point is the performance attained in the corresponding conservative case. Thus, smoothing limitations and optimal policies in the semi-conservative case approaches those of the conservative case as the operation fills a large fraction of the demands.

The effect of time delays, extended from Theorem 3, in information or requisition response, is similarly more complex, but approaches that of the conservative case under the same conditions.

REFERENCES

[1] Feller, William, *Probability Theorem and its Applications*, Vol. 1, 2nd ed. New York, Wiley, 1957.
[2] Mills, Harlan, "Smoothing in Servo Processes," *SIAM Review*, Vol. 3, pp. 131–139.
[3] Pinkham, Roger S., "An Approach to Linear Inventory Production Rules," *Operations Research*, Vol. 6 (1958), pp. 185–189.
[4] Simon, Herbert A., "On the Application of Servo-mechanism Theory in the Study of Production Control," *Econometrica*, Vol. 20 (1952).
[5] Vassian, Herbert J., "Application of Discrete Variable Servo Theory to Inventory Control," *Operations Research*, Vol. 3 (1955), p. 272.
[6] Wald, A., *Sequential Analysis*, New York, Wiley, 1947.

CHAPTER 12

A Bayesian Approach to Team Decision Problems

By KOICHI MIYASAWA*

1. TEAM DECISION PROBLEM

Let us consider the team T consisting of m members $1, 2, \ldots, m$ in which each member i is concerned with the state of nature θ_i and has to choose an act a_i against the state of nature θ_i, $i = 1, \ldots, m$. We assume that each state of nature θ_i and each act a_i are represented by real numbers and the set Θ_i of all possible values of the state of nature θ_i and the set A_i of all possible acts a_i for the member i are the sets of all real numbers, $i = 1, \ldots, m$. If each member i chooses an act a_i when the true state of nature for the member i is θ_i, $i = 1, \ldots, m$, then the team T will suffer the loss $L(\theta, a)$ which is given by the following quadratic function:

$$L(\theta, a) = \sum_{i=1}^{m} q_{ii}(\theta_i - a_i)^2 + 2\sum_{i<j} q_{ij}(\theta_i - a_i)(\theta_j - a_j) \tag{1.1}$$
$$= (\theta - a)Q(\theta - a)',$$

where $\theta = (\theta_1, \ldots, \theta_m)$, $a = (a_1, \ldots, a_m)$ and the notation "$'$" denotes the transposition of the matrix and $Q = \|q_{ij}\|$ is the given positive definite matrix which will be called the loss matrix of the team decision problem. The team T as a whole wants to suppress the loss $L(\theta, a)$ as much as possible by letting each member i choose an adequate act a_i, $i = 1, \ldots, m$. Each member i does not know the value of the true state of nature θ_i, but it is assumed that he knows that $\theta = (\theta_1, \ldots, \theta_m)$ is distributed on $\Theta = \Theta_1 \times \ldots \times \Theta_m$ according to the known prior probability density function (p.d.f.) $\varphi(\theta) \equiv \varphi(\theta_1, \ldots, \theta_m)$. Accordingly, the average loss, i.e., the risk $R(a)$ to the team T accruing from the choice of a set of acts $a = (a_1, \ldots, a_m)$ which will be called a team act is given by

$$R(a) = EL(\theta, a)$$
$$= \sum_{i} q_{ii}E(\theta_i - a_i)^2 + 2\sum_{i<j} q_{ij}E\{(\theta_i - a_i)(\theta_j - a_j)\} \tag{1.2}$$

Let $\bar{\theta}_i$ be the prior expected value of θ_i, i.e.,

$$\bar{\theta}_i = E\theta_i = \int \theta_i \varphi_i(\theta_i)\, d\theta_i, \tag{1.3}$$

where $\varphi_i(\theta_i)$ is the prior marginal p.d.f. of θ_i, i.e.,

$$\varphi_i(\theta_i) = \int \ldots \int \varphi(\theta_i, \ldots, \theta_m)\, d\theta_1 \ldots (i) \ldots d\theta_m \tag{1.4}$$

* University of Tokyo.

Then, in our case, it is clear that the optimal team act $\hat{a} = (\hat{a}_1, \ldots, \hat{a}_m)$ which gives the minimum risk $R(\hat{a})$ in the set of all team acts $A = A_1 \times \ldots \times A_m$ is determined as the solution of the following simultaneous equation:

$$\frac{1}{2}\frac{\partial R}{\partial a_i} = q_{ii}\bar{\theta}_i - q_{ii}\hat{a}_i - \sum_{j\neq i} q_{ij}(\hat{a}_j - \bar{\theta}_j) = 0, \qquad i = 1, \ldots, m. \quad (1.5)$$

That is

$$Q\hat{a} = Q\bar{\theta}, \quad (1.6)$$

where $\bar{\theta} = (\bar{\theta}_1, \ldots, \bar{\theta}_m)$.
From (1.6), we have

$$\hat{a}_i = \bar{\theta}_i, \qquad i = 1, \ldots, m. \quad (1.7)$$

Accordingly, from (1.2) and (1.7), we have

$$R(\hat{a}) = \sum_i q_{ii}(E\theta_i^2 - \bar{\theta}_i^2) + 2\sum_{i<j} q_{ij}(E\theta_i\theta_j - \bar{\theta}_i\bar{\theta}_j),$$

that is

$$\min_a R(a) = R(\hat{a}) = E\{\theta\, Q\, \theta'\} - \bar{\theta}\, Q\, \bar{\theta}'. \quad (1.8)$$

In this way, we have the following theorem.

THEOREM 1.1. *In the team decision problem, if each member i chooses his act a_i without having any additional information, then the optimal act $\hat{a}_i$ for the member i is given by* (1.7) *and the minimum risk $R(\hat{a})$ to the team is given by* (1.8).

REMARK. It should be mentioned that the optimal team act a does not depend on the loss matrix Q.

Now let us consider the team decision problem in which each member i is allowed to observe the value of the statistic x_i which is distributed on the space X_i according to the known p.d.f. $f_i(x_i \mid \theta_i)$ when the state of nature θ_i is given, $i = 1, \ldots, m$. It is also assumed that each statistic x_i, $i = 1, \ldots, m$ is distributed independently to each other when the state of nature $\theta = (\theta_1, \ldots, \theta_m)$ is given. Accordingly, the joint p.d.f. $f(x \mid \theta)$ of $x = (x_1, \ldots, x_m)$ under the condition that the state of nature $\theta = (\theta_1, \ldots, \theta_m)$ is given can be written as

$$f(x \mid \theta) = \prod_i f_i(x_i \mid \theta_i). \quad (1.9)$$

Now, it may be supposed that the communication system which may exist among the team members will be of help to the team in decreasing the risk to it. Then, according to the method of R. Radner [2], we shall characterize the communication system by the information function η_i, $i = 1, \ldots, m$, which is defined as follows. An information function η_i for the member i is a measurable function from $X = X_1 \times \ldots \times X_m$ to $Z_i = \{z_i\}$, where Z_i is defined as the set of all signals z_i on which the member i

can base his choice of act a_i in A_i, $i = 1, \ldots, m$. In other words, a signal z_i which is the value of an information function η_i of $x = (x_1, \ldots, x_m)$ provides the final information for his selection of an act a_i in A_i. Therefore, under the given information function η_i, we shall define a decision function α_i for the member i by a measurable function from Z_i to A_i, $i = 1, \ldots, m$. That is, $\eta_i(x) = z_i \in Z_i$ and $\alpha_i(z_i) = \alpha_i(\eta_i(x)) = a_i \in A_i$. The m-tuple $\eta = (\eta_1, \ldots, \eta_m)$ of the information functions will be called the information structure for the team T. If there is no ambiguity concerning the given information structure η, for simplicity $\alpha_i(\eta_i(x))$ will sometimes be written $\alpha_i(x)$. From the above consideration, in the team decision problem, the following questions naturally arise: (i) Under the given information structure η, what will be the optimal team decision function? (ii) What will be the optimal information structure? In the paper we shall mainly be concerned with question (i) and touch only slightly on question (ii) through the comparison of the values of complete communication and no communication which are defined below.

2. OPTIMAL TEAM DECISION FUNCTION AND VALUE OF INFORMATION STRUCTURE

Now, we assume that the information structure $\eta = (\eta_1, \ldots, \eta_m)$ of the team is given. Under this condition, at first, we shall find out the risk $R(\alpha)$ of any given team decision function $\alpha = (\alpha_1, \ldots, \alpha_m)$.

Let us write

$$L(\theta, \alpha(x)) = (\theta - \alpha(x))Q(\theta - \alpha(x))', \tag{2.1}$$

where

$$\alpha(x) = \big(\alpha_1(\eta_1(x)), \ldots, \alpha_m(\eta_m(x))\big).$$

Then we have

$$\begin{aligned} R(\alpha) &= \int \cdots \int L(\theta, \alpha(x)) f_1(x_1 \mid \theta_1) \cdots f_m(x_m \mid \theta_m) \\ &\quad \times \varphi(\theta_1, \ldots, \theta_m)\, dx_1 \ldots dx_m\, d\theta_1 \ldots d\theta_m \\ &= \int_\Theta \int_X L(\theta, \alpha(x)) f(x \mid \theta)\varphi(\theta)\, dx\, d\theta \end{aligned} \tag{2.2}$$

Now, the unconditional joint p.d.f. $f(x) = f(x_1, \ldots, x_m)$ of the set of statistics $x = (x_1, \ldots, x_m)$ is given by

$$f(x) = \int_\Theta f(x \mid \theta)\varphi(\theta)\, d\theta. \tag{2.3}$$

Using the unconditional joint p.d.f. $f(x)$, it can easily be seen that the risk $R(\alpha)$ can be expressed as follows:

$$R(\alpha) = \int_X \left[\int_\Theta L(\theta, \alpha(x))\varphi_x(\theta)\, d\theta\right] f(x)\, dx, \tag{2.4}$$

that is

$$R(\alpha) = E\left\{\int_{\Theta} L(\theta, \alpha(x))\varphi_x(\theta)\, d\theta\right\}, \tag{2.5}$$

where $\varphi_x(\theta)$ is the posterior joint p.d.f. of $\theta = (\theta_1, \ldots, \theta_m)$ given that the set of observed values of the statistics of all members is $x = (x_1, \ldots, x_m)$, that is

$$\varphi_x(\theta) = \varphi(\theta)\frac{f(x \mid \theta)}{f(x)} = \varphi(\theta_1, \ldots \theta_m)\frac{f_1(x_1 \mid \theta_1) \ldots f_m(x \mid \theta_m)}{f(x_1, \ldots, x_m)} \tag{2.6}$$

Now, let us define the function $\psi_\alpha(x)$ of x by

$$\begin{aligned}\psi_\alpha(x) &= \int_{\Theta} L(\theta, \alpha(x))\, \varphi_x(\theta)\, d\theta \\ &= \int_{\Theta}\left[\sum_{i=1}^{m} q_{ii}(\theta_i - \alpha_i(x))^2 + 2\sum_{i<j} q_{ij}(\theta_i - \alpha_i(x))(\theta_j - \alpha_j(x))\right]\varphi_x(\theta)\, d\theta.\end{aligned} \tag{2.7}$$

Then, from (2.5) and (2.7), we have

$$R(\alpha) = E\psi_\alpha(x). \tag{2.8}$$

Next, let us denote the posterior expected value of θ_i given the observed value $x = (x_1, \ldots, x_m)$ by $\bar{\theta}_i(x)$, that is

$$\begin{aligned}\bar{\theta}_i(x) &= \int_{\Theta} \theta_i\varphi_x(\theta)\, d\theta \\ &= \int \ldots \int \theta_i\varphi(\theta_1, \ldots, \theta_m)\frac{f_1(x_1 \mid \theta_1) \ldots f_m(x_m \mid \theta_m)}{f(x_1, \ldots, x_m)}\, d\theta_1 \ldots d\theta_m.\end{aligned} \tag{2.9}$$

Then it is clear that in the expression $\psi_\alpha(x)$ of (2.7) the part which contains $\alpha_i(x)$ has the following form:

$$\begin{aligned}&-2q_{ii}\alpha_i(x)\bar{\theta}_i(x) + q_{ii}\alpha_i(x)^2 \\ &-2\alpha_i(x)\sum_{j\neq i} q_{ij}\bar{\theta}_j(x) + 2\alpha_i(x)\sum_{j\neq i} q_{ij}\alpha_j(x).\end{aligned} \tag{2.10}$$

Now, let us define the set $\eta_i^{-1}(z_i)$ for any $z_i \in Z_i$ by

$$\eta_i^{-1}(z_i) = \{x\colon \eta_i(x) = z_i\}, \qquad i = 1, \ldots, m. \tag{2.11}$$

Then, from the definition of the information function η_i, any decision function α_i for the member i is constant on the set $\eta_i^{-1}(z_i)$. And, from (2.8) we can express $R(\alpha)$ as follows:

$$R(\alpha) = E\{\psi_\alpha(x) \mid \eta_i(x) = z_i\}. \tag{2.12}$$

Now, expressing the value of the decision function α_i on $\eta_i^{-1}(z_i)$ by a_i, the conditional expected value $E\{\psi_\alpha(x) \mid \eta_i(x) = z_i\}$ may be considered as a function of a_i as long as all members except the member i continue their decision functions, which will be written $\Phi_i(a_i \mid z_i)$. Let us define the decision function $\hat{\alpha}_i$ for the member i by letting it take as its constant value on a set $\eta_i^{-1}(z_i)$ the value of a_i which gives the minimum value of $\Phi_i(a_i \mid z_i)$ as the function of a_i, $i = 1, \ldots, m$. If we can determine the set of decision functions $(\hat{\alpha}_1, \ldots, \hat{\alpha}_m)$ for which the above conditions hold simultaneously for all members $i = 1, \ldots, m$, then in our case, as can be checked by reference to Radner [4], it is known that $\hat{\alpha} = (\hat{\alpha}_1, \ldots, \hat{\alpha}_m)$ is the optimal team decision function under the given information structure η. In our case, such a set of decision functions $(\hat{\alpha}_1, \ldots, \hat{\alpha}_m)$ will be obtained by defining their values $(\hat{\alpha}_1(x), \ldots, \hat{\alpha}_m(x))$ at x by the values of $(a_1, \ldots, a_m)$, which satisfies the following simultaneous equation:

$$\frac{\partial \Phi_i(a_i \mid z_i)}{\partial a_i} = 0, \qquad i = 1, \ldots, m, \tag{2.13}$$

where

$$z_i = \eta_i(x), \qquad i = 1, \ldots, m.$$

From (2.10), it is clear that the equation (2.13) has the following form:

$$\begin{aligned} &-q_{ii}E\{\bar{\theta}_i(x) \mid \eta_i(x) = z_i\} + q_{ii}\hat{\alpha}_i(x) \\ &- \sum_{j \neq i} q_{ij}E\{\bar{\theta}_j(x) \mid \eta_i(x) = z_i\} + \sum_{j \neq i} q_{ij}E\{\hat{\alpha}_j(x) \mid \eta_i(x) = z_i\} = 0, \qquad (2.14) \\ &\qquad\qquad i = 1, \ldots, m. \end{aligned}$$

Thus, we have the following fundamental theorem.

THEOREM 2.1. *In the team decision problem under the given information structure η, the value $\hat{\alpha}_i(x)$ of the optimal decision function $\hat{\alpha}_i$ at x, $i = 1, \ldots,$ m, is given by the solution of the following simultaneous equation*:

$$\begin{aligned} q_{ii}\hat{\alpha}_i(x) &+ \sum_{i \neq j} q_{ij}E\{\hat{\alpha}_j(x) \mid \eta_i(x) = z_i\} \\ &= \sum_{j=1}^{m} q_{ij}E\{\bar{\theta}_j(x) \mid \eta_i(x) = z_i\}, \qquad i = 1, \ldots, m. \end{aligned} \tag{2.15}$$

Next, we shall determine the minimum value of the risk $R(\alpha)$ with respect to α, that is $R(\hat{\alpha})$. From (2.5), we have

$$\begin{aligned} R(\hat{\alpha}) &= E\left\{\int_{\Theta} (\theta - \hat{\alpha}(x))Q(\theta - \hat{\alpha}(x))' \varphi_x(\theta)\, d\theta\right\} \\ &= E\{\theta\, Q\, \theta'\} - 2E\{\hat{\alpha}(x)Q\bar{\theta}(x)'\} + E\{\hat{\alpha}(x)Q\hat{\alpha}(x)'\}, \end{aligned} \tag{2.16}$$

where $\bar{\theta}(x) = (\bar{\theta}_1(x), \ldots, \bar{\theta}_m(x))$ and $\bar{\theta}_i(x)$ is given by (2.9). Now, since $\hat{\alpha}(x)$ satisfies the equation (2.15), multiplying both sides of the equation (2.15) by $\hat{\alpha}_i(x)$ and then taking the expected value of both sides, we have

$$\sum_j q_{ij}E\hat{\alpha}_i(x)\hat{\alpha}_j(x) = \sum_j q_{ij}E\hat{\alpha}_i(x)\bar{\theta}_j(x), \qquad i = 1, \ldots, m. \tag{2.17}$$

Summing (2.17) on i from 1 to m, we have

$$E\{\hat{a}(x)Q\hat{\alpha}(x)'\} = E\{\hat{\alpha}(x)Q\bar{\theta}(x)'\} \tag{2.18}$$

Accordingly, from (2.16) and (2.18), we have

THEOREM 2.2.

$$\min_{\alpha} R(\alpha) = R(\hat{\alpha}) = E\{\theta\ Q\ \theta'\} - E\{\hat{\alpha}(x)Q\hat{\alpha}(x)'\}, \tag{2.19}$$

Now, taking the expected value of the both sides of the equation (2.15), we have

$$QE\{\hat{\alpha}(x)\} = QE\{\bar{\theta}(x)\}. \tag{2.20}$$

And it is clear that we have

$$E\bar{\theta}_j(x) = E\left\{\int_{\Theta} \theta_j \varphi_x(\theta)\, d\theta\right\} = \bar{\theta}_j, \qquad i = 1, \ldots, m. \tag{2.21}$$

Then, from (2.20) and (2.21), it will be seen that we have

$$E\{\hat{\alpha}_i(x)\} = \bar{\theta}_i, \qquad i = 1, \ldots, m. \tag{2.22}$$

In other words, the expected value of the optimal decision function is equal to the optimal act in the case of no observation.

Next, the natural procedure will be to define the value $V(\eta)$ of an information structure η by

$$V(\eta) = \min_{a} R(a) - \min_{\alpha} R(\alpha). \tag{2.23}$$

Then, from (1.8) and (2.19), we have

THEOREM 2.3. *The value $V(\eta)$ of an information structure η is given by*

$$V(\eta) = E\{\hat{\alpha}(x)Q\hat{\alpha}(x)'\} - \bar{\theta}Q\bar{\theta}'. \tag{2.24}$$

Now, since it is assumed that the value of $\bar{\theta}_i$, $i = 1, \ldots, m$, is known to the team, without loss of generality, it will be assumed that

$$\bar{\theta}_i = 0, \qquad i = 1, \ldots, m. \tag{2.25}$$

Then, under the assumption (2.25), the value $V(\eta)$ is given by

$$V(\eta) = E\{\hat{\alpha}(x)Q\hat{\alpha}(x)'\}. \tag{2.26}$$

In the following sections, we shall examine the structure of the optimal team decision functions for the two special information structures which are called complete communication and no communication, as will be defined below.

3. COMPLETE COMMUNICATION AND NO COMMUNICATION

The communication system of the team in which each team member i conveys his observed value of the statistic x_i to all other members will be called complete communication η_c. That is the case in which the information function η_i is defined by

$$\eta_i(x) = z_i = x = (x_1, \ldots, x_m), \qquad i = 1, \ldots, m. \tag{3.1}$$

In this case, it is clear that the fundamental equation (2.15) which gives the optimal team decision function $\hat{\alpha}$ has the following form:

$$\sum_{j=1}^{m} q_{ij}\hat{\alpha}_j(x) = \sum_{j=1}^{m} q_{ij}\bar{\theta}_j(x), \qquad i = 1, \ldots, m,$$

that is

$$Q\hat{\alpha}(x)' = Q\bar{\theta}(x)' \tag{3.2}$$

Therefore we have the following theorem:

THEOREM 3.1. *In the case of complete communication η_c, the optimal team decision function $\hat{\alpha} = (\hat{\alpha}_1, \ldots, \hat{\alpha}_m)$ is given by*

$$\hat{\alpha}_i(x) = \bar{\theta}_i(x), \qquad i = 1, \ldots, m. \tag{3.3}$$

Accordingly, the value $V(\eta_c)$ of complete communication η_c is given by

$$V(\eta_c) = E\{\bar{\theta}(x)Q\bar{\theta}(x)\} - \bar{\theta}Q\bar{\theta}'. \tag{3.4}$$

REMARK. It should be mentioned that in the case of complete communication, the optimal decision for each member i is to choose the action which is equal to the posterior expected value $\bar{\theta}_i(x)$ of θ_i, given the information $x = (x_1, \ldots, x_m)$, and is independent of the loss matrix Q.

If each team member i does not communicate his observed value x_i to any other member and bases his choice of an act a_i just on his own observed value x_i, then it will be said that the team is acting under no communication. That is the case in which the information function η_i for each member i is given by

$$\eta_i(x) = z_i = x_i, \qquad i = 1, \ldots, m. \tag{3.5}$$

In this case, the fundamental equation (2.15) which gives the optimal team decision function $\hat{\alpha} = (\hat{\alpha}_1, \ldots, \hat{\alpha}_m)$ has the following form:

$$\begin{aligned} q_{ii}\hat{\alpha}_i(x) &+ \sum_{j \neq i} q_{ij}E\{\alpha_j(x_j) \mid x_i\} \\ &= \sum_{j=1}^{m} q_{ij}E\{\bar{\theta}_j(x) \mid x_i\}, \qquad i = 1, \ldots, m. \end{aligned} \tag{3.6}$$

It will be convenient, for the following calculation, to rewrite equation (3.6). At first, it can easily be seen that

$$\begin{aligned} E\{\bar{\theta}_i(x) \mid x_i\} &= \int \cdots \int \theta_i \varphi(\theta_1, \ldots, \theta_m) \frac{f_1(x_1 \mid \theta_1) \ldots f_m(x_m \mid \theta_m)}{f(x_1, \ldots, x_m)} \\ &\quad \times \frac{f(x_1, \ldots, x_m)}{f_i(x_i)} \, d\theta_1 \ldots d\theta_m \, dx_1 \ldots (i) \ldots dx_m \\ &= \int \theta_i \varphi_i(\theta_i) \frac{f_i(x_i \mid \theta_i)}{f_i(x_i)} \, d\theta_i, \end{aligned} \tag{3.7}$$

that is

$$E\{\bar{\theta}_i(x) \mid x_i\} = \bar{\theta}_i(x_i), \qquad i = 1, \ldots, m,$$

where $\bar{\theta}_i(x_i)$ is the posterior expected value of θ_i given that x_i is observed and is defined by the last expression in (3.7). If $j \neq i$, then we have

$$\begin{aligned} E\{\bar{\theta}_j(x) \mid x_i\} &= \int \cdots \int \theta_j \varphi(\theta_1, \ldots, \theta_m) \frac{f_1(x_1 \mid \theta_1) \ldots f_m(x_m \mid \theta_m)}{f(x_1, \ldots, x_m)} \\ &\quad \times \frac{f(x_1, \ldots, x_m)}{f_i(x_i)} d\theta_1 \ldots d\theta_m \, dx_1 \ldots (i) \ldots dx_m \qquad (3.9) \\ &= \iint \theta_j \varphi_{ij}(\theta_i, \theta_j) \frac{f_i(x_i \mid \theta_i)}{f_i(x_i)} d\theta_i \, d\theta_j, \end{aligned}$$

where $\varphi_{ij}(\theta_i, \theta_j)$ is the prior marginal p.d.f. of θ_i and θ_j, $j \neq i$, that is

$$\varphi_{ij}(\theta_i, \theta_j) = \int \cdots \int \varphi(\theta_1, \ldots, \theta_m) \, d\theta_1 \ldots (i) \ldots (j) \ldots d\theta_m. \qquad (3.10)$$

It is also clear that we have

$$E\{\alpha_j(x_j) \mid x_i\} = \int \alpha_j(x_j) \frac{f_{ij}(x_i, x_j)}{f_i(x_i)} dx_j, \qquad j \neq i, \qquad (3.11)$$

where $f_{ij}(x_i, x_j)$ is the unconditional marginal p.d.f. of x_i and x_j, that is

$$j_{ij}(x_i, x_j) = \int \cdots \int f((x_1, \ldots, x_m) \, dx_1 \ldots (i) \ldots (j) \ldots dx_m, \quad j \neq i. \qquad (3.12)$$

In this way, applying (3.8) — (3.12) to (3.6), we have the following theorem:

THEOREM 3.2. *In the case of no communication η_0, the optimal team decision function $\hat{\alpha} = (\hat{\alpha}_1, \ldots, \hat{\alpha}_m)$ is given by the solution of the following simultaneous equation with respect to $\hat{\alpha}_i(x_i)$, $i = 1 \ldots m$:*

$$\begin{aligned} q_{ii}\hat{\alpha}_i(x_i) &+ \sum_{j \neq i} q_{ij} \int \hat{\alpha}_j(x_j) \frac{f_{ij}(x_i, x_j)}{f_i(x_i)} dx_j \\ &= q_{ii} \int \theta_i \varphi_i(\theta_i) \frac{f_i(x_i \mid \theta_i)}{f_i(x_i)} d\theta_i \qquad (3.13) \\ &+ \sum_{j \neq i} q_{ij} \iint \theta_j \varphi_{ij}(\theta_i, \theta_j) \frac{f_i(x_i \mid \theta_i)}{f_i(x_i)} d\theta_i \, d\theta_j, \qquad i = 1, \ldots, m. \end{aligned}$$

Next, we shall take up a special case of the team decision problem in which the number of team members is $m = 2$ and the prior distribution $\varphi(\theta_1, \theta_2)$ of $\theta = (\theta_1, \theta_2)$ is normal. And for that case we shall make a comparative study of complete communication and no communication.

4. TWO-MEMBER NORMAL CASE

In the following, we shall examine the team decision problem in which the team T consists of two members $i = 1, 2$ and both of them know that the prior joint distribution $\varphi(\theta_1, \theta_2)$ of the states of nature θ_i, $i = 1, 2$, which the members $i = 1, 2$ are confronted with, is normal with means $(0, 0)$ and variances σ_1^2, σ_2^2 and correlation coefficient ρ, i.e.,

$$\varphi(\theta_1, \theta_2) = \frac{1}{2\pi\sigma_1\sigma_2\sqrt{1-\rho^2}} \exp\left\{-\frac{1}{2(1-\rho^2)}\left[\frac{\theta_1^2}{\sigma_1^2} - 2\rho\frac{\theta_1\theta_2}{\sigma_1\sigma_2} + \frac{\theta_2^2}{\sigma_2^2}\right]\right\}. \tag{4.1}$$

It is also assumed that the statistic x_i which the member i is allowed to observe is normally distributed with mean θ_i and variance d_i^2 when the true state of nature for him is θ_i, $i = 1, 2$.

That is

$$f_i(x_i \mid \theta_i) = \frac{1}{\sqrt{2\pi}\, d_i} \exp\left\{-\frac{1}{2\, d_i^2}(x_i - \theta_i)^2\right\}, \qquad i = 1, 2. \tag{4.2}$$

We shall call the team decision problem under these conditions a two-member normal case. It is clear that we have

$$\varphi_i(\theta_i) = \frac{1}{\sqrt{2\pi}\sigma_i} \exp\left\{-\frac{\theta_i^2}{2\sigma_i^2}\right\}, \qquad i = 1, 2. \tag{4.3}$$

For convenience in the following calculations, let us take up the simple integral formula as

LEMMA 4.1. *It holds*

$$\int \exp\{-\tfrac{1}{2}(\alpha x^2 - 2\beta x)\}\, dx = \frac{\sqrt{2\pi}}{\sqrt{\alpha}} \exp\left\{\frac{\beta^2}{2\alpha}\right\}, \tag{4.4}$$

$$\int x \exp\{-\tfrac{1}{2}(\alpha x^2 - 2\beta x)\}\, dx = \sqrt{2\pi}\,\frac{\beta}{\alpha^{\frac{3}{2}}} \exp\left\{\frac{\beta^2}{2\alpha}\right\}, \tag{4.5}$$

where α and β are constants and $\alpha > 0$.

We shall prove the following

LEMMA 4.2. *In the two-member normal case which is defined by* (4.1) *and* (4.2), *the unconditional marginal p.d.f. $f_i(x_i)$, $i = 1, 2$, and the unconditional joint p.d.f. $f(x_1, x_2)$ are given by the following expressions* (4.6) *and* (4.7) *respectively*:

$$f_i(x_i) = \frac{1}{\sqrt{2\pi}\ell_i\sigma_i} \exp\left\{-\frac{x_i^2}{2\ell_i^2\sigma_i^2}\right\}, \qquad i = 1, 2, \tag{4.6}$$

$$f(x_1, x_2) = \frac{1}{2\pi\sigma_1\sigma_2\sqrt{\ell_1^2\ell_2^2 - \rho^2}} \times \exp\left\{-\frac{1}{2(\ell_1^2\ell_2^2 - \rho^2)}\left[\frac{\ell_2^2}{\sigma_1^2}x_1^2 - 2\frac{\rho}{\sigma_1\sigma_2}x_1x_2 + \frac{\ell_1^2}{\sigma_2^2}x_2^2\right]\right\}, \tag{4.7}$$

where l_i^2 is defined by

$$d_i^2 = k_i^2\sigma_i^2, \qquad l_i^2 = k_i^2 + 1, \qquad i = 1, 2. \tag{4.8}$$

Furthermore, variance Var (x_i), $i = 1, 2$, *and covariance* Cov (x_1, x_2) *of the unconditional joint distribution of* x_1 *and* x_2 *are given by*

$$\text{Var}\,(x_i) = l_i^2\sigma_i^2, \qquad i = 1, 2, \tag{4.9}$$

$$\text{Cov}\,(x_1, x_2) = \rho\sigma_1\sigma_2. \tag{4.10}$$

PROOF. From (4.2) and (4.3), we have

$$\begin{aligned} f_i(x_i) &= \int f_i(x_i \mid \theta_i)\varphi_i(\theta_i)\, d\theta_i \\ &= \frac{1}{2\pi\, d_i\sigma_i^2} \exp\left\{-\frac{x_i^2}{2\, d_i^2}\right\} \int \exp\left\{-\tfrac{1}{2}\left[\left(\frac{1}{\sigma_i^2} + \frac{1}{d_i^2}\right)\theta_i^2 - 2\frac{x_i}{d_i}\theta_i\right]\right\} d\theta_i \end{aligned} \tag{4.11}$$

Applying Lemma 4.1 to the last integral in (4.11), it can easily be seen that (4.11) reduces to (4.6).

Next, from (4.1), (4.2), and the assumption that x_i, $i = 1, 2$ are independent given $\theta = (\theta_1, \theta_2)$, we have

$$\begin{aligned} f(x_1, x_2) &= \iint f_1(x_1 \mid \theta_1) f_2(x_2 \mid \theta_2)\varphi(\theta_1, \theta_2)\, d\theta_1\, d\theta_2 \\ &= \frac{1}{(2\pi)^2 k_1 k_2 \sigma_1^2\sigma_2^2\sqrt{1-\rho^2}} \\ &\quad \times \int \exp\left\{-\frac{1}{2k_1^2\sigma_1^2}(x_1 - \theta_1)^2 - \frac{1}{2k_2^2\sigma_2^2}x_2^2 - \frac{1}{2(1-\rho^2)\sigma_1^2}\theta_1^2\right\} \\ &\quad \times \int \exp\left\{-\tfrac{1}{2}(\alpha\theta_2^2 - 2\beta\theta_2)\right\} d\theta_2\, d\theta_1, \end{aligned} \tag{4.12}$$

where

$$\alpha = \frac{\ell_2^2 - \rho^2}{(1-\rho^2)k_2^2\sigma_2^2}, \qquad \beta = \frac{\rho k_2^2\sigma_2\theta_1 + (1-\rho^2)\sigma_1 x_2}{(1-\rho^2)k_2^2\sigma_1\sigma_2^2}.$$

Then applying Lemma 4.1 to the last integral with respect to θ_2 in (4.12) and then rearranging the terms, we have

$$\begin{aligned} f(x_1, x_2) &= \frac{1}{(2\pi)^{\frac{3}{2}} k_1\sigma_1^2\sigma_2\sqrt{\ell_2^2 - \rho^2}} \\ &\quad \times \exp\left\{-\frac{x_1^2}{2k_1^2\sigma_1^2} - \frac{\ell_2^2 - 1}{2(\ell_2^2 - \rho^2)k_2^2\sigma_2^2}x_2^2\right\} \\ &\quad \times \int \exp\left\{-\tfrac{1}{2}(\alpha\theta_1^2 - 2\beta\theta_1)\right\} d\theta_1, \end{aligned} \tag{4.13}$$

where

$$\alpha = \frac{\ell_1^2\ell_2^2 - \rho^2}{k_1^2\sigma_1^2(\ell_2^2 - \rho^2)}, \qquad \beta = \frac{x_1}{k_1^2\sigma_1^2} + \frac{\rho x_2}{(\ell_2^2 - \rho^2)\sigma_1\sigma_2}.$$

Therefore, applying Lemma 4.1 to the last integral with respect to θ_1 in (4.13) and then rearranging the terms we see that we have the expression (4.7) for $f(x_1, x_2)$. From (4.6) and (4.7), it is clear that variances and covariance of x_1 and x_2 are given by (4.9) and (4.10). (q.e.d.)

THEOREM 4.1. *In the two-member normal case of the team decision problem under complete communication η_c, the optimal team decision function $\hat{\alpha} = (\hat{\alpha}_1, \hat{\alpha}_2)$ is given by the following expressions:*

$$\hat{\alpha}_1(x) = \bar{\theta}_1(x) = \frac{1}{\sigma_2(\ell_1^2\ell_2^2 - \rho^2)}[\sigma_2(\ell_2^2 - \rho^2)x_1 + \rho\sigma_1(\ell_1^2 - 1)x_2], \tag{4.14}$$

$$\hat{\alpha}_2(x) = \bar{\theta}_2(x) = \frac{1}{\sigma_1(\ell_1^2\ell_2^2 - \rho^2)}[\rho\sigma_2(\ell_2^2 - 1)x_1 + \sigma_1(\ell_1^2 - \rho^2)x_2] \tag{4.15}$$

The value $V(\eta_c)$ of complete communication η_c is given by

$$\begin{aligned} V(\eta_c) = {} & \frac{1}{(\ell_1^2\ell_2^2 - \rho^2)^2} \\ & \times [q_{11}\sigma_1^2\{\ell_1^2\ell_2^4 + \ell_2^2(\ell_1^4 - 2\ell_1^2 - 1)\rho^2 + (2 - \ell_1^2)\rho^4\} \\ & + 2q_{12}\rho\sigma_1\sigma_2\{\ell_1^2\ell_2^2(\ell_1^2 + \ell_2^2 - 1) \\ & - (\ell_1^2\ell_2^2 + \ell_1^2 + \ell_2^2 - 1)\rho^2 + \rho^4\} \\ & + q_{22}\sigma_2^2\{\ell_1^4\ell_2^2 + \ell_1^2(\ell_2^4 - 2\ell_2^2 - 1)\rho^2 + (2 - \ell_2^2)\rho^4\}, \end{aligned} \tag{4.16}$$

where the loss matrix is

$$Q = \begin{Vmatrix} q_{11} & q_{12} \\ q_{12} & q_{22} \end{Vmatrix}.$$

PROOF. It has already been proved in Theorem 3.1 that the optimal team decision function $\hat{\alpha}$ under complete communication is given by $\hat{\alpha}_i(x) = \bar{\theta}_i(x)$, $i = 1, \ldots, m$. So, our concern here is to express it explicitly. First we shall make the following remark. Let $A_i(x)$ be defined by

$$A_i(x) = \iint \theta_i\varphi(\theta_1, \theta_2)f_1(x_1 \mid \theta_1)f_2(x_2 \mid \theta_2)\, d\theta_1\, d\theta_2, \quad i = 1, 2, \tag{4.17}$$

then it is clear that

$$\bar{\theta}_i(x) = A_i(x)/f(x_1, x_2), \qquad i = 1, 2, \tag{4.18}$$

$$E\bar{\theta}_i(x)^2 = \iint \frac{A_i(x)^2}{f(x_1, x_2)}\, dx_1\, dx_2, \qquad i = 1, 2, \tag{4.19}$$

and

$$E\bar{\theta}_1(x)\bar{\theta}_2(x) = \iint \frac{A_1(x)A_2(x)}{f(x_1, x_2)}\, dx_1\, dx_2. \tag{4.20}$$

For the computation of $A_1(x)$, we put

$$B_2(\theta_1, x_2) = \int \varphi(\theta_1, \theta_2) f_2(x_2 \mid \theta_2)\, d\theta_2, \tag{4.21}$$

then

$$A_1(x) = \int \theta_1 f_1(x_1 \mid \theta_1) B_2(\theta_1, x_2)\, d\theta_1. \tag{4.22}$$

In our case, from (4.1) and (4.2), $B_2(\theta_1, x_2)$ defined by (4.21) has the following form:

$$\begin{aligned} B_2(\theta_1, x_2) = {} & \frac{1}{(2\pi)^{\frac{3}{2}} k_2 \sigma_1 \sigma_2^2 \sqrt{1-\rho^2}} \\ & \times \exp\left\{-\frac{\theta_1^2}{2(1-\rho^2)\sigma_1^2} - \frac{x_2^2}{2k_2^2\sigma_2^2}\right\} \\ & \times \int \exp\{-\tfrac{1}{2}(\alpha\theta_2^2 - 2\beta\theta_2)\}\, d\theta_2, \end{aligned} \tag{4.23}$$

where

$$\alpha = \frac{\ell_2^2 - \rho^2}{(1-\rho^2)k_2^2\sigma_2^2}, \qquad \beta = \frac{\rho k_2^2 \sigma_2 \theta_1 + (1-\rho^2)\sigma_1 x_2}{(1-\rho^2)k_2^2\sigma_1\sigma_2^2}.$$

Therefore, applying Lemma 4.1 to the last integral with respect to θ_2 in (4.23), we have

$$\begin{aligned} B_2(\theta_1, x_2) = {} & \frac{1}{2\pi\sigma_1\sigma_2\sqrt{\ell_2^2 - \rho^2}} \\ & \times \exp\left\{-\tfrac{1}{2}\cdot\frac{\ell_2^2}{(\ell_2^2-\rho^2)\sigma_1^2}\theta_1^2 + \frac{\rho x_2}{(\ell_2^2-\rho^2)}\theta_1 - \tfrac{1}{2}\frac{x_2^2}{(\ell_2^2-\rho^2)\sigma_2^2}\right\} \end{aligned} \tag{4.24}$$

Then, from (4.22) and (4.24), we have

$$\begin{aligned} A_1(x) = {} & \frac{1}{(2\pi)^{\frac{3}{2}} k_1 \sigma_1^2 \sigma_2 \sqrt{\ell_2^2 - \rho^2}} \\ & \times \exp\left\{-\frac{x_1^2}{2k_1^2\sigma_1^2} - \frac{x_2^2}{2(\ell_2^2-\rho^2)\sigma_2^2}\right\} \\ & \times \int \theta_1 \exp\{-\tfrac{1}{2}(\alpha\theta_1^2 - 2\beta\theta_1)\}\, d\theta_1, \end{aligned} \tag{4.25}$$

where

$$\alpha = \frac{\ell_1^2\ell_2^2 - \rho^2}{k_1^2\sigma_1^2(\ell_2^2-\rho^2)}, \qquad \beta = \frac{(\ell_2^2-\rho^2)\sigma_2 x_1 + \rho k_1^2 \sigma_1 x_2}{k_1^2\sigma_1^2\sigma_2(\ell_2^2-\rho^2)}.$$

Applying Lemma 4.1 to the last integral in (4.25), we have finally the following expression for $A_1(x)$:

$$A_1(x) = \frac{1}{2\pi\sigma_1\sigma_2^2(\ell_1^2\ell_2^2 - \rho^2)^{\frac{3}{2}}}[\sigma_2(\ell_2^2 - \rho^2)x_1 + \rho k_1^2\sigma_1 x_2] \times \exp\left\{-\frac{1}{2(\ell_1^2\ell_2^2 - \rho^2)}\left[\frac{\ell_2^2}{\sigma_1^2}x_1^2 - 2\frac{\rho}{\sigma_1\sigma_2}x_1x_2 + \frac{\ell_1^2}{\sigma_2^2}x_2^2\right]\right\}. \tag{4.26}$$

Now, comparing the expression (4.7) for $f(x_1, x_2)$ with the expression (4.26) for $A_1(x)$, we see that $A_1(x)$ can be expressed as follows:

$$A_1(x) = \frac{1}{\sigma_2(\ell_1^2\ell_2^2 - \rho^2)}[\sigma_2(\ell_2^2 - \rho^2)x_1 + \rho k_1^2\sigma_1 x_2]f(x_1, x_2), \tag{4.27}$$

where $f(x_1, x_2)$ is the unconditional joint p.d.f. of x_1 and x_2 given by (4.7). It is clear that we similarly have the following expression for $A_2(x)$:

$$A_2(x) = \frac{1}{\sigma_1(\ell_1^2\ell_2^2 - \rho^2)}[\rho k_2^2\sigma_2 x_1 + \sigma_1(\ell_1^2 - \rho^2)x_2]f(x_1, x_2). \tag{4.28}$$

Then, from (4.18), (4.27), and (4.28), it is clear that $\bar{\theta}_i(x)$, i.e. $\hat{\alpha}_i(x)$, is given by (4.14) and (4.15).

Next, from (4.19) and (4.27), we have

$$E\bar{\theta}_1(x)^2 = \frac{1}{\sigma_2^2(\ell_1^2\ell_2^2 - \rho^2)^2}\iint[\sigma_2(\ell_2^2 - \rho^2)x_1 + \rho k_1^2\sigma_1 x_2]^2 f(x_1, x_2)\,dx_1\,dx_2$$

$$= \frac{1}{\sigma_2^2(\ell_1^2\ell_2^2 - \rho^2)^2}[\sigma_2^2(\ell_2^2 - \rho^2)^2 \operatorname{Var}(x_1) + 2\rho k_1^2\sigma_1\sigma_2(\ell_2^2 - \rho^2)\operatorname{Cov}(x_1, x_2) + \rho^2 k_1^4\sigma_1^2 \operatorname{Var}(x_2)].$$

Here, applying (4.9) and (4.10) in Lemma 4.2, and then rearranging the terms, we have

$$E\bar{\theta}_1(x)^2 = \frac{\sigma_1^2}{(\ell_1^2\ell_2^2 - \rho^2)^2}[\ell_1^2\ell_2^4 + \ell_2^2(\ell_1^4 - 2\ell_1^2 - 1)\rho^2 + (2 - \ell_1^2)\rho^4]. \tag{4.29}$$

Similarly, we have

$$E\bar{\theta}_2(x)^2 = \frac{\sigma_2^2}{(\ell_1^2\ell_2^2 - \rho^2)^2}[\ell_1^4\ell_2^2 + \ell_1^2(\ell_2^4 - 2\ell_2^2 - 1)\rho^2 + (2 - \ell_2^2)\rho^4]. \tag{4.30}$$

Applying (4.27) and (4.28) to (4.20), and using the results (4.9) and (4.10), we similarly have the following expression:

$$E\bar{\theta}_1(x)\bar{\theta}_2(x) = \frac{\rho\sigma_1\sigma_2}{(\ell_1^2\ell_2^2 - \rho^2)^2} \times [\ell_1^2\ell_2^2(\ell_1^2 + \ell_2^2 - 1) - (\ell_1^2\ell_2^2 + \ell_1^2 + \ell_2^2 - 1)\rho^2 + \rho^4]. \tag{4.31}$$

Now, from the result (3.4) in Theorem 3.1 and the fact that $\bar{\theta}_1 = \bar{\theta}_2 = 0$ in our case, we have

$$V(\eta_c) = q_{11}E\bar{\theta}_1(x)^2 + 2q_{12}E\bar{\theta}_1(x)\bar{\theta}_2(x) + q_{22}E\bar{\theta}_2(x)^2. \tag{4.32}$$

Applying (4.29) – (4.31) to (4.32), it is clear that we have the expression (4.16) for the value $V(\eta_c)$ as is required. q.e.d.

THEOREM 4.2. *In the two-member normal case of the team decision problem under no communication η_0, the optimal team decision function $\hat{\alpha} = (\hat{\alpha}_1, \hat{\alpha}_2)$ is given by*

$$\hat{\alpha}_i(x_i) = K_i\bar{\theta}_i(x_i) = K_i \frac{x_i}{\ell_i^2}, \qquad i = 1, 2, \tag{4.33}$$

where

$$K_1 = \frac{\ell_1^2[q_{11}q_{22}\ell_2^2\sigma_1 + q_{12}q_{22}(\ell_2^2 - 1)\sigma_2\rho - q_{12}^2\sigma_1\rho^2]}{\sigma_1(q_{11}q_{22}\ell_1^2\ell_2^2 - q_{12}^2\rho^2)}, \tag{4.34}$$

$$K_2 = \frac{\ell_2^2[q_{11}q_{22}\ell_1^2\sigma_2 + q_{12}q_{11}(\ell_1^2 - 1)\sigma_1\rho - q_{12}^2\sigma_2\rho^2]}{\sigma_2(q_{11}q_{22}\ell_1^2\ell_2^2 - q_{12}^2\rho^2)}. \tag{4.35}$$

And the value $V(\eta_0)$ of no communication is given by

$$V(\eta_0) = q_{11}\frac{K_1^2}{\ell_1^2}\sigma_1^2 + 2q_{12}K_1K_2 - \frac{\rho\sigma_1\sigma_2}{\ell_1^2\ell_2^2} + q_{22}\frac{K_2^2}{\ell_2^2}\sigma_2^2, \tag{4.36}$$

where K_1 and K_2 are the constants defined by (4.34) *and* (4.35) *respectively.*

REMARK. It might be supposed from the result of Theorem 4.1. that the optimal decision function $\hat{\alpha}_i$ for the member i under no communication is the posterior expected value of θ_i, given his observed value x_i. But this supposition is incorrect; Theorem 4.2 shows that in order to obtain the optimal decision function $\hat{\alpha}_i(x_i)$ under no communication, we should correct the posterior expected value $\bar{\theta}_1(x_i)$ by the correction coefficient K_i which depends on the loss matrix Q, as can be seen from (4.34) and (4.35).

PROOF. In order to apply the fundamental equation (3.13) in Theorem 3.2 to our case, we first calculate the following. From (4.2), (4.3), and (4.6), we have

$$\begin{aligned} E\{\bar{\theta}_1(x) \mid x_1\} &= \int \theta_1\varphi_1(\theta_1)\frac{f_1(x_1 \mid \theta_1)}{f_1(x_1)}\, d\theta_1 \\ &= \frac{\ell_1}{\sqrt{2\pi}k_1\sigma_1} \exp\left\{\frac{x_1^2}{2\ell_1^2\sigma_1^2} - \frac{x_1^2}{2k_1^2\sigma_1^2}\right\} \\ &\quad \times \int \theta_1 \exp\left\{-\tfrac{1}{2}\left[\left(\frac{1}{\sigma_1^2} + \frac{1}{k_1^2\sigma_1^2}\right)\theta_1^2 - 2\frac{x_1}{k_1^2\sigma_1^2}\theta_1\right]\right\} d\theta_1 \end{aligned}$$

Here, applying Lemma 4.1 to the last integral and then rearranging terms, it can be seen that we have

$$E\{\bar{\theta}_1(x) \mid x_1\} = \bar{\theta}_1(x_1) = \frac{x_1}{\ell_1^2} \tag{4.37}$$

Similarly, we have

$$E\{\bar{\theta}_2(x) \mid x_2\} = \bar{\theta}_2(x_2) = \frac{x_2}{\ell_2^2} \tag{4.38}$$

Next, from (4.1), (4.2), and (4.6), we have

$$\begin{aligned} E\{\bar{\theta}_1(x) \mid x_2\} &= \iint \theta_1 \varphi(\theta_1, \theta_2) \frac{f_2(x_2 \mid \theta_2)}{f_2(x_2)}\, d\theta_1, d\theta_2 \\ &= \frac{\ell_2}{2\pi k_2 \sigma_1 \sigma_2 \sqrt{1-\rho^2}} \exp\left\{\frac{x_2^2}{2\ell_2^2\sigma_2^2} - \frac{x_2^2}{2k_2^2\sigma_2^2}\right\} \qquad (4.39) \\ &\quad \times \int \theta_1 \exp\left\{-\frac{\theta_1^2}{2(1-\rho^2)\sigma_1^2}\right\} \cdot \int \exp\{-\tfrac{1}{2}(\alpha\theta_2^2 - 2\beta\theta_2)\}\, d\theta_2 \cdot d\theta_1, \end{aligned}$$

where

$$\alpha = \frac{1}{(1-\rho^2)\sigma_2^2} + \frac{1}{k_2^2\sigma_2^2}, \qquad \beta = \frac{\rho k_2^2 \sigma_2 \theta_1 + (1-\rho^2)\sigma_1 x_2}{(1-\rho^2)k_2^2\sigma_1\sigma_2^2}.$$

Therefore, applying Lemma 4.1 to the last integral with respect to θ_2 in (4.39), we see that the integral is equal to

$$\sqrt{2\pi}\, \frac{k_2\sigma_2\sqrt{1-\rho_2}}{\sqrt{\ell_2^2 - \rho^2}} \exp\left\{\frac{[\rho\sigma_2 k_2^2\theta_1 + (1-\rho^2)\sigma_1 x_2]^2}{2(\ell_2^2-\rho^2)(1-\rho^2)k_2^2\sigma_1^2\sigma_2^2}\right\}. \tag{4.40}$$

Substituting (4.40) in (4.39), we have

$$\begin{aligned} E\{\bar{\theta}_1(x) \mid x_2\} &= \frac{\ell_2}{\sigma_1\sqrt{2\pi}\sqrt{\ell_2^2-\rho^2}} \\ &\quad \times \exp\left\{-\frac{x_2^2}{2\ell_2^2k_2^2\sigma_2^2} + \frac{(1-\rho^2)x_2^2}{2(\ell_2^2-\rho^2)k_2^2\sigma_2^2}\right\} \qquad (4.41) \\ &\quad \times \int \theta_1 \exp\{-\tfrac{1}{2}(\alpha\theta_1^2 - 2\beta\theta_1)\}\, d\theta_1, \end{aligned}$$

where

$$\alpha = \frac{1}{(1-\rho^2)\sigma_1^2} - \frac{\rho^2k_2^2}{(\ell_2^2-\rho^2)(1-\rho^2)\sigma_1^2}, \qquad \beta = \frac{\rho x_2}{(\ell_2^2-\rho^2)\sigma_1\sigma_2}.$$

Again applying Lemma 4.1 to the last integral in (4.41), we see that the integral with respect to θ_1 in (4.41) is equal to

$$\sqrt{2\pi}\, \frac{\rho\sigma_1^2\sqrt{\ell_2^2-\rho^2}}{\ell_2^3\sigma_2} \cdot x_2 \cdot \exp\left\{\frac{\rho^2x_2^2}{2\ell_2^2\sigma_2^2(\ell_2^2-\rho^2)}\right\} \tag{4.42}$$

Then, substituting (4.42) in (4.41), we have

$$E\{\bar{\theta}_1(x) \mid x_2\} = \frac{\rho\sigma_1\sigma_2}{\ell_2^2\sigma_2^2} x_2. \tag{4.43}$$

Similarly, we have

$$E\{\bar{\theta}_2(x) \mid x_1\} = \frac{\rho\sigma_1\sigma_2}{\ell_1^2\sigma_1^2} x_1. \tag{4.44}$$

In this way, using (4.37), (4.38), and (4.43) and (4.44), we see that the fundamental equation (3.12) in Theorem 3.2 which gives the optimal team decision function $\hat{\alpha} = (\hat{\alpha}_1, \hat{\alpha}_2)$ under no communication can be expressed as follows:

$$\begin{aligned} q_{11}\hat{\alpha}_1(x_1) + q_{12}\int \hat{\alpha}_2(x_2)\frac{f(x_1, x_2)}{f_1(x_1)}\,dx_2 &= q_{11}\frac{x_1}{\ell_1^2} + q_{12}\frac{\rho\sigma_2}{\ell_1^2\sigma_1} x_1 \\ q_{22}\hat{\alpha}_2(x_2) + q_{12}\int \hat{\alpha}_1(x_1)\frac{f(x_1, x_2)}{f_2(x_2)}\,dx_1 &= q_{22}\frac{x_2}{\ell_2^2} + q_{12}\frac{\rho\sigma_1}{\ell_2^2\sigma_2} x_2 \end{aligned} \tag{4.45}$$

Now, from (4.6) and (4.7), $f(x_1, x_2)/f_i(x_i)$ has the following form in our case:

$$\begin{aligned} \frac{f(x_1, x_2)}{f_1(x_1)} = {} & \frac{\ell_1}{\sigma_2\sqrt{2\pi}\sqrt{\ell_1^2\ell_2^2 - \rho^2}} \\ & \times \exp\left\{-\frac{1}{2(\ell_1^2\ell_2^2 - \rho^2)}\left[\frac{\rho^2}{\sigma_1^2\ell_1^2} x_1^2 - 2\frac{\rho}{\sigma_1\sigma_2} x_1x_2 + \frac{\ell_1^2}{\sigma_2^2} x_2^2\right]\right\} \end{aligned} \tag{4.46}$$

Then, using Lemma 4.1 for the calculation of the integral

$$\int x_2 \frac{f(x_1, x_2)}{f_1(x_1)}\,dx_2$$

we have

$$\int x_2 \frac{f(x_1, x_2)}{f_1(x_1)}\,dx_2 = \frac{\rho\sigma_2}{\ell_1^2\sigma_1} x_1 (= E\{\bar{\theta}_2(x) \mid x_1\}) \tag{4.47}$$

Similarly, we have

$$\int x_1 \frac{f(x_1, x_2)}{f_2(x_2)}\,dx_1 = \frac{\rho\sigma_1}{\ell_2^2\sigma_2} x_2 (= E\{\bar{\theta}_1(x) \mid x_2\}) \tag{4.48}$$

Now, using unspecified constants K_i, $i = 1, 2$, let us tentatively define $\hat{\alpha}_i(x_i)$ by

$$\hat{\alpha}_i(x_i) = K_i\bar{\theta}_i(x_i) = K_i\frac{x_i}{\ell_i^2}, \qquad i = 1, 2, \tag{4.49}$$

and then try to determine the value of the constants K_i so that $\hat{\alpha}_i(x_i)$ defined by (4.49) will satisfy the fundamental equation (4.45). To do this we substitute (4.49) in (4.45) and then use (4.47) and (4.48); it can then be seen that the optimal team decision function $\hat{\alpha} = (\hat{\alpha}_1, \hat{\alpha}_2)$ under no communication is given by (4.49) with the coefficients K_i, $i = 1, 2$, which are the solution of the following simultaneous equation:

$$\begin{aligned} \frac{q_{11}}{\ell_1^2} K_1 + \frac{q_{12}\rho\sigma_2}{\ell_1^2\ell_2^2\sigma_1} K_2 &= \frac{q_{11}\sigma_1 + q_{12}\rho\sigma_2}{\ell_1^2\sigma_1} \\ \frac{q_{12}\rho\sigma_1}{\ell_1^2\ell_2^2\sigma_2} K_1 + \frac{q_{22}}{\ell_2^2} K_2 &= \frac{q_{22}\sigma_2 + q_{12}\rho\sigma_1}{\ell_2^2\sigma_2} \end{aligned} \tag{4.50}$$

A quick calculation shows that the solution K_i, $i = 1, 2$ of equation (4.50) is given by (4.34) and (4.35). Furthermore, using the fact that $\bar{\theta}_1 = \bar{\theta}_2 = 0$ in case and using Theorem 2.3 and equation (4.33), we can easily see that the value $V(\eta_0)$ of no communication is given by

$$\begin{aligned} V(\eta_0) &= q_{11}E\hat{\alpha}_1(x_1)^2 + 2q_{12}E\hat{\alpha}_1(x_1)\hat{\alpha}_2(x_2) + q_{22}E\hat{\alpha}_2(x_2)^2 \\ &= q_{11}\frac{K_1^2}{\ell_1^4}\operatorname{Var}(x_1) + 2q_{12}\frac{K_1K_2}{\ell_1^2\ell_2^2}\operatorname{Cov}(x_1, x_2) + q_{22}\frac{K_2^2}{\ell_2^4}\operatorname{Var}(x_2). \end{aligned} \tag{4.51}$$

If we now use (4.9) and (4.5) in Lemma 4.2, it is clear that we have the expression (4.36) for $V(\eta_0)$ as is required. (q.e.d.)

5. COMPARISON OF VALUES OF COMPLETE COMMUNICATION AND NO COMMUNICATION

In the case of complete communication η_c, a possible decision function α_i for the team member i is any measurable function of $x = (x_1, \ldots, x_m)$. Therefore, it is also allowed to be a function only of x_i. This means that any possible decision function for the member i under no communication can be viewed as a special decision function for him under complete communication. Accordingly, from the definition of the value of the information structure, it is clear that we have

$$V(\eta_c) \geq V(\eta_0). \tag{5.1}$$

Therefore, if we define the relative efficiency of complete communication with respect to no communication by

$$\nu = V(\eta_c)/V(\eta_0), \tag{5.2}$$

then we should certainly have $\nu \geq 1$.
In this section, we shall examine the explicit value of ν for the two-member normal case in which the loss matrix is given by

$$q_{11} = q_{22} = 1, \qquad q_{12} = q, \qquad -1 < q < 1. \tag{5.3}$$

Then, it can be seen, from (4.16) and (4.36), that the difference between $V(\eta_c)$ and $V(\eta_0)$ is given by

$$V(\eta_c) - V(\eta_0) = \frac{\rho^2(1-q^2)}{(\ell_1^2\ell_2^2 - \rho^2 q^2)(\ell_1^2\ell_2^2 - \rho^2)} \tag{5.4}$$
$$\times \{\sigma_1^2\ell_2^2(\ell_1^2 - 1)^2 - 2q\rho\sigma_1\sigma_2(\ell_1^2 - 1)(\ell_2^2 - 1) + \sigma_2^2\ell_1^2(\ell_2^2 - 1)^2\}$$

where

$$-1 < \rho, \quad q < 1 \quad \text{and} \quad \ell_i^2 > 1, \qquad i = 1, 2.$$

It can easily be seen that the discriminant D of the quadratic form in σ_1 and σ_2 within the parentheses in (5.4) is expressed as follows

$$D = (\ell_1^2 - 1)^2(\ell_2^2 - 1)(q^2\rho^2 - \ell_1^2\ell_2^2) < 0. \tag{5.5}$$

Accordingly, from (5.4) and (5.5), it is also clear that we surely have $V(\eta_c) \geq V(\eta_0)$.

Let us further specify our case by assuming that

$$\sigma_1^2 = \sigma_2^2 = \sigma^2, \qquad d_1^2 = d_2^2 = d^2. \tag{5.6}$$

Then

$$k_1^2 = k_2^2 = k^2, \qquad \ell_1^2 = \ell_2^2 = \ell^2 = k^2 + 1. \tag{5.7}$$

Now, let us define t by

$$\frac{1}{d^2} = t\frac{1}{\sigma^2}, \tag{5.8}$$

then $t > 0$ and t exhibits the relative magnitude of the precision of the statistic x_i which the member i is allowed to observe with respect to that of the prior information concerning the state of nature θ_i which the member i is confronted with. For convenience, let us call t the relative precision of the observation. It is clear that

$$\ell^2 = 1 + \frac{1}{t}. \tag{5.9}$$

Under these conditions, our concern here is to examine the variation of the relative efficiency v of complete communication with respect to no communication as a function of t for several pairs of values of ρ and q. From the results shown below it may be possible to obtain some idea of whether it is better to spend money to increase the accuracy of observation, or to spend it for better communication among the team members. It is a matter of course that for the final decision it is necessary to take into consideration the costs for observation and for communication. Now, from Theorems 4.1 and 4.2, under the assumptions (5.3) and (5.6), it is clear that we have the following results. At first, the optimal team decision function

$\hat{\alpha}^{(c)} = (\hat{\alpha}_1^{(c)}, \hat{\alpha}_2^{(c)})$ under complete communication η_c is given by

$$\hat{\alpha}_1^{(c)}(x) = \frac{1}{\ell^4 - \rho^2}\{(\ell^2 - \rho^2)x_1 + \rho(\ell^2 - 1)x_2\},$$
$$\hat{\alpha}_2^{(c)}(x) = \frac{1}{\ell^4 - \rho^2}\{\rho(\ell^2 - 1)x_1 + (\ell^2 - \rho^2)x_2\}. \tag{5.10}$$

Expression (5.10) shows how the optimal decision function assigns increasing weight to the other member's observation of which he is informed by complete communication as the absolute value of the correlation coefficient ρ of θ_1 and θ_2 increases. The value $V(\eta_c)$ of complete communication η_c is given by

$$V(\eta_c) = \frac{2\sigma^2}{\ell^4 - \rho^2}\{\ell^2 + (2\ell^2 - 1)q\rho + (\ell \; - 2)\rho^2 - q\rho^3\}, \tag{5.11}$$

or

$$V(\eta_c) = \frac{2t\sigma^2}{1 + (1 - \rho^2)t^2}\{1 + t + (2 + t)q\rho + (1 - t)\rho^2 - tq\rho^3\}. \tag{5.12}$$

Next, the optimal team decision function $\hat{\alpha}^{(o)} = (\hat{\alpha}_1^{(o)}, \hat{\alpha}_2^{(o)})$ under no communication η_0 is given by

$$\hat{\alpha}_i^{(0)}(x_i) = \frac{1 + \rho q}{\ell^2 + \rho q} x_i, \qquad i = 1, 2, \tag{5.13}$$

and the value $V(\eta_0)$ of no communication is given by

$$V(\eta_0) = \frac{2\sigma^2(1 + \rho q)^2}{\ell^2 + \rho q} \tag{5.14}$$

or

$$V(\eta_0) = \frac{2\sigma^2(1 + q\rho)^2 t}{1 + (1 + q\rho)t} \tag{5.15}$$

The variation of $V(\eta_c)$ given by (5.12) and that of $V(\eta_0)$ given by (5.15) are shown as functions of t in Figures 1 and 2 for several pairs of values of ρ and q.

From (5.12) and (5.15), it can be seen that the relative efficiency ν is given by

$$\nu = \frac{V(\eta_c)}{V(\eta_0)} = 1 + \frac{\rho^2(1 - q^2)}{(1 + q\rho)^2} \cdot \frac{1}{(1 - \rho^2)t^2 + 2t + 1} \tag{5.16}$$

The variation of ν given by (5.16) is shown as a function of t in Figure 3 for several pairs of values of ρ and q.

The following points are of interest:

1. The expression (5.10) for the optimal decision function $\hat{\alpha}_i^{(c)}$ shows

how much weight should be put on the observation which is conveyed from the other member according to the magnitude of ρ.

2. It naturally follows from (5.10) and (5.13) that

$$\lim_{\rho\to 0} \hat{\alpha}_i^{(c)}(x) = \lim_{\rho\to 0} \hat{\alpha}_i^{(0)}(x).$$

3. From (5.12) and (5.14) we have

$$\lim_{t\to 0} V(\eta_c) = \lim_{t\to 0} V(\eta_0) = 0.$$

This means that as the relative precision of the observations becomes very poor the contributions which the observations can render to the team become negligible.

4. But it should be seen from (5.16), that the relative efficiency ν increases as t decreases and we have

$$\lim_{t\to 0} \nu = 1 + \frac{\rho^2(1-q^2)}{(1+q\rho)^2}\,.$$

5. It is easily seen from (5.10), (5.13), and (5.16) that we have

$$\lim_{t\to\infty} \hat{\alpha}_i^{(c)}(x) = \lim_{t\to\infty} \hat{\alpha}_i^{(0)}(x) = x_i,$$

and

$$\lim_{t\to\infty} \nu = 1.$$

This means that as the relative precision of the observations increases, the information which will be conveyed from the other member contributes little to the team.

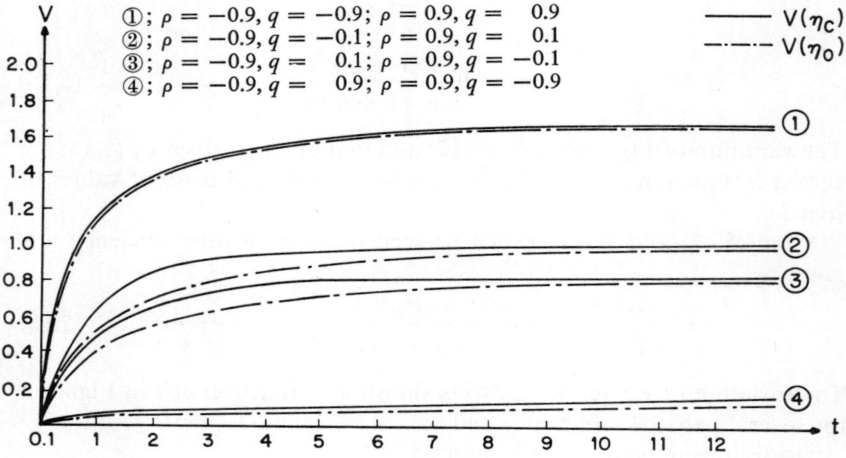

FIGURE 1

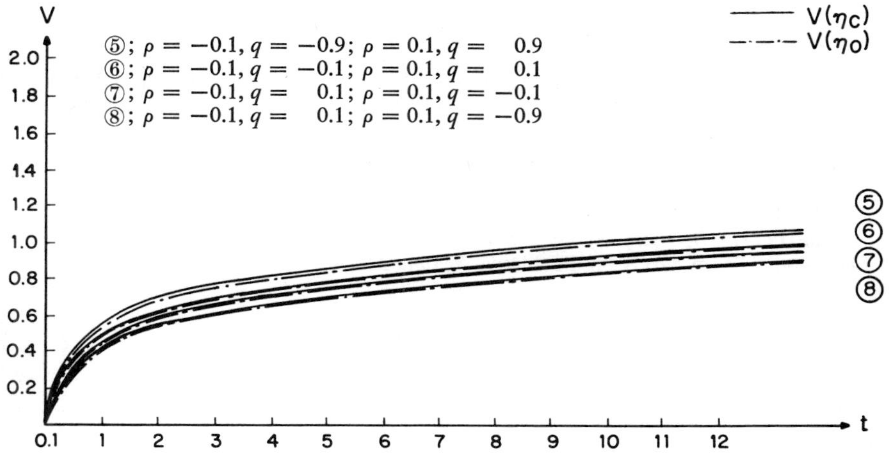

FIGURE 2

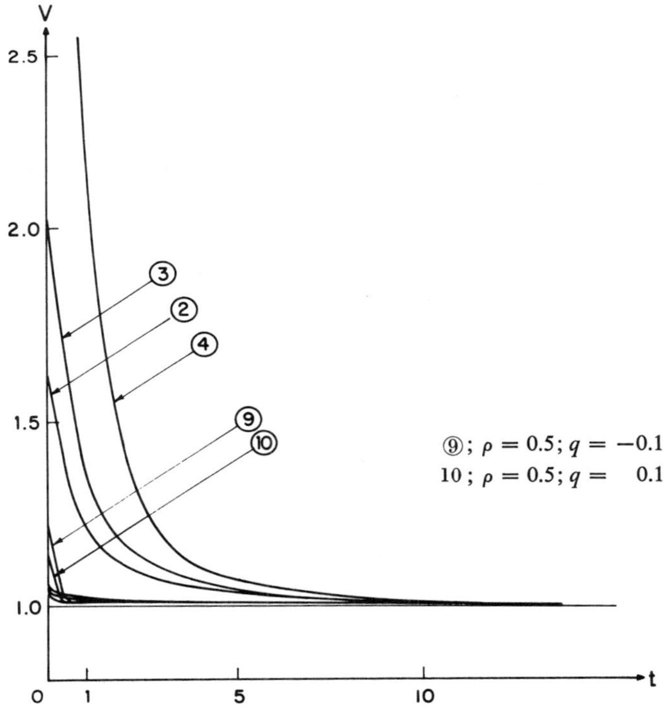

FIGURE 3

REFERENCES

[1] J. Marschak, "Towards an Economic Theory of Organization and Information," *Decision Processes*, ed. by R. M. Thrall, C. H. Cooms, and R. L. Davis (1954), pp. 187–220.

[2] ——, "Elements for a Theory of Teams," *Management Sci.*, Vol. 1 (1955), pp. 127–137.

[3] R. Radner, "The Evaluation of Information in Organizations," *Proc. of the Fourth Berkeley Symp. on Math. Stat. and Prob.* (1961), pp. 491–530.

[4] ——, "Team Decision Problems," *Ann. Math. Stat.*, Vol. 33 (1962), pp. 857–881.

CHAPTER 13

Capital Flexibility and Long Run Cost under Stationary Uncertainty

By DANIEL ORR*

1. INTRODUCTION

In the comparative static theory of the firm, a long-run cost curve is drawn to represent the available array of undominated short-run alternatives. The marginal conditions then identify the optimal scale (they locate a short-run cost function associated with the firm's best plant and equipment holding) and the rate at which this equipment should be operated. Beyond this, results on the interdependence between output rates and the firm's capital holdings are scattered and fragmentary. Capital theory stipulates that all decisions should aim at maximizing the present value of net cash flows; this, like the long-run marginal condition of the price theory text, does little more than restate the assumed motive of profit maximization.

The firm's investment problem has been treated from the standpoint of overall optimization: among significant postwar contributions can be listed the Lutz and Lutz volume [7], a portion of Haavelmo's book [3], and Manne's recent article [8]. This last work is of particular interest; the optimal rules for capacity expansion when market demand comprises a trend and a random term are analyzed. "Capacity" is treated as a homogeneous attribute; significant costs are encountered each time capacity is changed, regardless of the size of the change. If the demand for output is Gaussian with known parameters, and if input proportions are fixed, the capacity requirements obey a well-known stochastic pattern, the diffusion process.[1]

Manne's analysis has a crisp analytic focus, which stems from the assumption that capital has one unambiguous quantitative dimension, and no qualitative dimensions: capacity is a scalar. The present paper is concerned with the quantitative "capacity" dimension, but it also examines an essentially qualitative aspect of the investment decision: the "flexibility" of the equipment in which the firm invests.[2]

If demand contains a random component, fluctuations in output are

* University of California, San Diego.

[1] Cf. Feller [2], Ch. XIV. In an earlier paper I used the same model to generate optimal production decisions for the case where production rate changes are costly and the costs are at least in part independent of the size of the change [10].

[2] An analysis of the qualitative impact of flexible capital, particularly its effect on the shape of the firm's cost curves, is found in Stigler's early paper [12]. See also the recent paper by Marschak and Nelson [13].

necessary if infinite fluctuations in inventory are to be avoided. The degree of inventory and output fluctuation is determined by the firm's production control rules; the relative costs of fluctuation in these two variables affect the decision. Although the flexibility of the firm's capital affects the cost of output variation, it is not normally regarded as an attribute subject to the control of the firm in planning production schedules over time. In the following, a quantitative representation of flexibility is devised, and the tradeoff between long-run cost (incurred by varying the flexibility of capital) and short-run cost (incurred by varying the inventory-output control rules) will be studied.[3]

The way around the difficulty of quantifying a qualitative attribute taken here is to assume that the firm's cost function contains only linear and quadratic terms of the output rate and inventory level. This assumption, as it turns out, leads to an obvious quantitative representation of flexibility. While this approach makes it possible to detect whether a firm is insufficiently or overly flexible, the analysis contains a great gap from a normative standpoint. There is no way to pick the best plant and equipment inputs, given the desired degree of flexibility. Thus, the capital programming problem is untouched.

2. CAPITAL FLEXIBILITY: DEFINITION

Adaptability is the attribute of a capital structure that permits the firm to operate with a larger (smaller) quantity of the variable inputs, and thus to obtain significant increases (reductions) in output: the smaller the unit cost changes incurred from this kind of output change, the more adaptable the capital is. Divisibility is another capital attribute that enables operation over a wide range of output rates: the greater the degree of divisibility, the wider the range of output rates for which unit costs do not deviate from the minimum level by more than a given percentage. In a divisible plant, the portion of capital actually utilized is variable; as output varies, input proportions remain constant, and capital not used stands idle.[4]

Both adaptability and divisibility lead to flexibility. Flexible capital permits operation over a wide range of production rates with little change in unit costs, but it cannot accommodate the single most efficient input combination for any one output rate. Flexible capital as a response to fluctuating demand was illustrated by Stigler [12] as in Figure 1; if output fluctuates within the range 0_1–0_2, the firm opts for the less flexible investment and the cost curve *ATUC*, instead of *atuc*.

[3] Despite the long awareness of flexibility as an important attribute of the firm's capital choice, the notion has not been incorporated in the mainstream of price theory (if the mainstream flows through intermediate-level textbooks). This is due, perhaps, to the obscurity of the marginal implications of flexibility: it is difficult to evaluate its quantitative impact, although its qualitative importance is obvious.

[4] Adaptability and divisibility are defined by Stigler [12], pp. 122, 127–8.

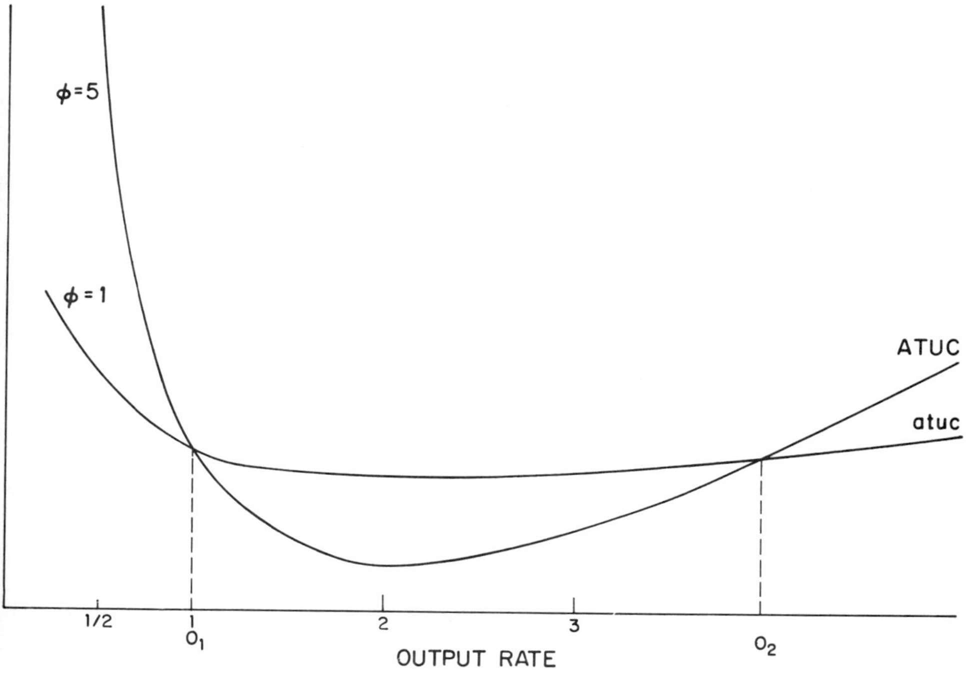

FIGURE 1
Average production-associated cost in two plants

Unfortunately, the choice is less clear cut than is implied in this diagrammatic analysis. As noted before, inventory-output policy determines the degree of output rate fluctuation the firm faces, and the flexibility of capital determines the cost of output fluctuation and thus partially shapes the production-inventory policy finally adopted. This interaction is lost in the diagrammatic analysis. The first objective of this paper is to explore these interactions. This objective will be approached via the "linear-servo" production-inventory policy, featured prominently in Mills' paper in this volume [9]. The demand and cost assumptions specified in the next section also lead to a natural definition of scale of plant. Unsurprisingly, the optimal degree of flexibility and the optimal scale of the operation turn out to be dependent on each other.

3. CAPITAL FLEXIBILITY: QUANTITATIVE REPRESENTATION

Consider a firm confronted by stationary, independent random demands. The cost of meeting demand fluctuations is, as usual, divided into two classes, one associated with the production rate, the other with the inventory level. The total production-associated costs (TPAC)[5] of operating at the rate x_t are

$$\text{TPAC} = \phi(x_t - x^o)^2 + (\Pi + \pi/\phi)x_t. \tag{1}$$

This at first may seem a non-traditional representation of total production cost, since the parabolic first term is more in keeping with our concept of average cost behavior. However, consider the form

$$\text{TPAC} = ax_t^2 + b \tag{2}$$

as a representation of total cost. This version yields a *U*-shaped average cost curve (as is seen by plotting $ax_t + b/x_t$) and an increasing marginal curve $2ax_t$. The second degree representation is advantageous to us because we propose to use Mills' results on the "linear-servo" policy. He shows that these policies cope effectively with mean square error, which suggests that they are appropriate for use with quadratic costs. If this were the only consideration, we would happily choose (2) instead of (1) as the form of production cost. However, our method of representing variations in flexibility will be to vary the cost coefficient of the second-degree term. If the value of a in (2) is increased, the sides of the unit cost curve become steeper, which is consistent with a lower degree of flexibility in the plant we are representing. However, the value of x_t at which unit cost is minimized does not remain constant as a varies, and as a increases, the minimum value of unit cost shifts upward, contrary to the pattern visualized in Figure 1, where greater flexibility is associated with higher values of unit cost at the minimum.

The relation (1), by contrast, avoids these objections and does not differ too radically from the appealing representation (2). The significant difference between (1) and (2) is the interval of decreasing total cost between $x_t = 0$ and $x_t = x^o - \Pi/2\phi - \pi/2\phi^2$ (which is not troublesome if ϕ is sufficiently small). The relation (1) is graphed in Figure 1; two values of $\phi(\frac{1}{2}$ and 5) are used, in conjunction with $\Pi = 1$, $\pi = 5$, and $x^o = 2$.

The second set of costs—those associated with positive and negative levels of period-ending inventory—are represented by

$$\text{TIAC} = \eta(i_t - i^o)^2 \tag{3}$$

where η is a positive constant, i^o is the "target" inventory,[6] and i_t is the

[5] These costs are the short-run costs expounded in the static theory of the firm, and the U shape of the unit cost function that may be derived from our total cost function (1) is rationalized in the same way as in that traditional theory.

[6] It may be felt that the most economical inventory level is zero: neither shortages nor holding costs are incurred at this level. However, we are using a discrete-time model in which the status of a stock over an entire interval is being appraised in terms of its value at the end of the interval. A terminal inventory level of zero will typically mean that the level during the period was positive or negative; if the latter, troublesome delays in response to orders may have occurred. If negative stock levels are more expensive than positive stock levels of the same absolute magnitude, the firm should regard x^o as positive.

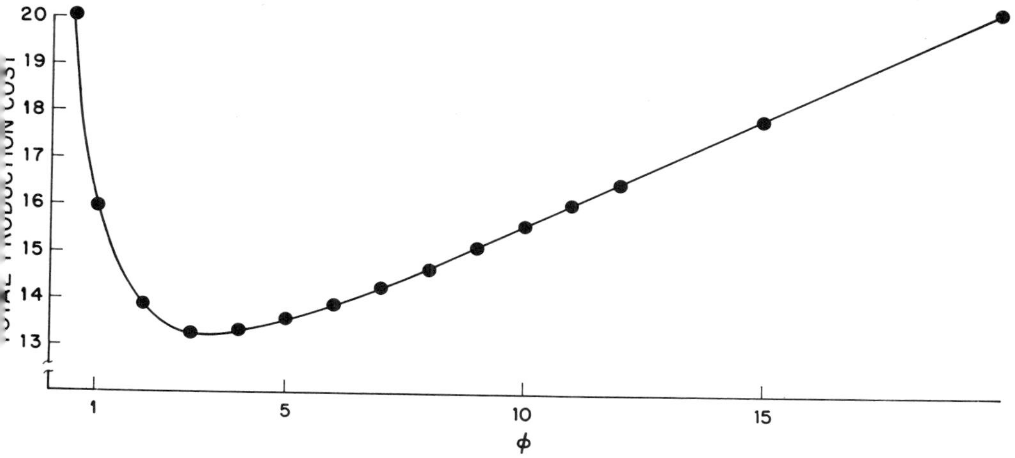

FIGURE 2
Sample tabulation of loss function

level of finished goods inventory in the firm at the end of the tth period.

The factors responsible for the costs impounded in (3) have been discussed at length elsewhere; therefore, it seems unnecessary to explore in detail the various ramifications of holding inventory, running out of stock, changing production rates, etc. The representation (3) implies that small positive stock levels are inexpensive but large accumulations are costly. The cost of shortages varies directly with both η and i^o.

The cost function

$$L = E[\phi(x_t - x^o)^2 + (\Pi + \pi/\phi)x_t + \eta(i_t - i^o)^2] \tag{4}$$

represents the expected losses in period t, conditional upon knowledge of the state of the firm at the beginning of that period; that is, the initial inventory level i_{t-1} is known at the time the output decision x_t is made. The optimal output decision x_t may be obtained from (4), using the relation

$$i_t = i_{t-1} + x_t - d_t \tag{5}$$

where d_t is the quantity demanded by customers during period t. Expectations are taken on this stochastic term.[7] The minimum cost condition is

$$2\phi(x_t - x^o) + \Pi + \pi/\phi + 2\eta(i_{t-1} + x_t - \bar{d} - i^o) = 0$$

from which we obtain the rule

$$x_t = x^o - \alpha(i_{t-1} - i^*) \tag{6}$$

[7] The derivative of the expectation in (4) is equal to the expectation of the derivative, since x_t does not appear in the bounds over which d_t is integrated.

where

$$\alpha = \eta/(\phi + \eta) \tag{7}$$

$$i^* = \bar{d} - x^o + i^o - \left(\frac{\phi\Pi + \pi}{2\eta\phi}\right). \tag{8}$$

The cost parameters, expected demand, and initial inventory completely determine the optimal output rate, defined by the system (6-8).

The loss function (4) is straightforward, yields rules of appealingly simple form, and may be satisfactory as a representation of the implications of a decision x_t in period t. Its main virtue for the present paper is that it permits the simplest representation of flexibility imaginable. We may view flexibility as an instrument variable of the firm, controlled by intermittent selection of a value ϕ.

Precedents for use of a quadratic loss function, and some possibly severe shortcomings of it as a representation of costs confronting the firm, are explored in the Appendix.

4. FLEXIBILITY AND PRODUCTION RESPONSE

Given the loss function (4), the production control rule (6-8) is optimal because it minimizes the expected costs of operation in each period, on the basis of information available at the beginning of the period. Mills' results for policies of the class (6), which he obtains under the assumption of stationary, random, independent demands, permit further derivations of interest to us. He shows that under the posited demand conditions and supply rules, the means and variances of sales and inventories have these values:

$$\begin{aligned} \sigma_x^2 &= \alpha^2\sigma_d^2/[1 - (1 - \alpha)^2] & \bar{x} &= \bar{d} \\ \sigma_i^2 &= \sigma_d^2/[1 - (1 - \alpha)^2] & i &= i^* + \frac{1}{\alpha}(x^o - \bar{d}) \end{aligned} \tag{9}$$

where $\bar{d}$ is the mean demand, i^*, x^o, and α have the values given in (7-9), and σ_i^2, σ_x^2, and σ_d^2 are the variances of inventory, output, and demand.

These moments, which depend only on the demand assumptions and the production rule, permit easy evaluation of the one-period expected loss associated with the policy (6-8). The function (4) evaluates cost performance in a specific future period t; the policy rule (6) is obtained by optimizing with respect to the exact value of i_{t-1} and the expected value of d_t. To determine the mean costs that will be encountered if the demand assumptions hold and the policy (6-8) is used, the expected value of (4) is taken with respect to the cost-generating variables, x_t^2, i_t^2, x_t, and i_t. This function is

$$E(L) = \lim_{t\to\infty} E[\phi(x_t - x^o)^2 + (\Pi + \pi/\phi)x_t + \eta(i_t - i^o)^2] \tag{10}$$

or

$$E(L) = \phi\sigma_x^2 + \phi(\bar{x} - x^o)^2 + (\Pi + \pi/\phi)\bar{x} + \eta\sigma_i^2 + \eta(\bar{i} - i^o)^2. \quad (11)$$

We are interested in determining how ϕ (the flexibility parameter) is related to the values established for other cost parameters, and how sensitive cost performance is to deviations from the optimum in this instrument variable. It is also of passing interest to investigate the optimal scale of capital.

Measure scale by the value of x^o, the output rate that yields minimum average production-associated cost. The firm can optimize the scale, as well as the flexibility of its equipment, by choosing x^o and ϕ to minimize (11). The terms of (11) containing x^o are (substituting wherever appropriate from (7), (8), and (9))

$$\phi(\bar{x} - x^o)^2 + \eta\left[-\frac{\phi}{\eta}(\bar{x} - x^o) - \frac{\phi\Pi + \pi}{2\eta\phi}\right]^2. \quad (12)$$

The minimum of (12)—and hence of (11) with respect to x^o—is given by

$$x^o = \bar{x} + \frac{\phi\Pi + \pi}{2\phi(\eta + \phi)} \quad (13)$$

With the operating scale set at this value, (11) becomes

$$E(L) = \frac{[\eta\phi + (\eta + \phi)^2]\sigma_d^2}{2\phi + \eta} + \frac{(\phi\Pi + \pi)^2}{4\phi^2(\eta + \phi)} + \frac{(\phi\Pi + \pi)}{\phi}\bar{d}$$

which may be maximized with respect to ϕ. Optimal values of ϕ are presented in Table 1, a different array of values for the cost and demand parameters being assumed for each.[8]

We are also interested in deviations away from the optimal ϕ for a given set of cost and demand parameters. In Figure 2 these costs are plotted for two different $(\eta, \Pi, \pi, \sigma_d^2, \bar{x})$ sets.

Note that (13) stipulates that the firm install excess capacity. The minimum average cost output x^o should exceed mean output $\bar{x}$, and hence from (9), should also exceed mean demand $\bar{d}$. Thus, this dynamic and stochastic model with quadratic costs offers an interesting addendum to the excess-capacity controversy of monopolistic competition.

Apparently, the factors responsible for the presence of this excess capacity are the costs of holding inventory—see (12). However, inventory-associated costs cannot by themselves lead to excess capacity: If the term $(\Pi + \pi/\phi)x_t$ were not in (1), the optimal x^o would be $\bar{x}$, and there would be no excess capacity. Because of this cost term, increments of output change above x^o are more expensive than decrements below x^o, and hence the prescription $\bar{x} < x^o$. If inventory costs were zero, the firm would set

[8] I am grateful to Hodson Thornber for these tables.

x^o equal to $\bar{x}$, and produce x^o in every period, without regard for unfilled demands or accumulated finished goods.

Thus, the stationary random demand assumption in effect states that at a given price, marginal revenue equals expected marginal cost. Cost shifts are generated from period to period by purely random demand

TABLE 1

The Loss Function (11)

(All entries use $\Pi = 1$, $\sigma_d^2 = 1$)

Example no.	$\bar{d}$	π	η	Optimal ϕ	Cost per period
1	0.25	1	0.003	1.657	1.620
2	0.50	1	0.003	1.674	2.017
3	1	1	0.003	1.973	2.784
4	2	1	0.003	2.368	4.246
5	5	1	0.003	3.351	8.297
6	10	1	0.003	4.583	14.56
7	1	0.05	0.003	1.522	2.383
8	1	2	0.003	2.625	3.373
9	1	5	0.003	3.932	4.569
10	1	10	0.003	5.413	5.932
11	5	0.05	0.003	2.443	7.397
12	5	2	0.003	4.651	9.589
13	5	5	0.003	7.252	12.18
14	5	10	0.003	10.19	15.10
15	1	1	0.010	1.971	2.792
16	1	1	0.025	1.966	2.809
17	2	1	0.010	2.367	4.254
18	2	1	0.025	2.364	4.271
19	1	2	0.010	2.623	3.381
20	1	2	0.025	2.619	3.398
21	2	5	0.010	4.907	6.711
22	2	5	0.025	4.906	6.729
23	5	2	0.010	4.651	9.598
24	5	2	0.025	4.650	9.616

fluctuations, which are attributable to factors other than price; the excess capacity stipulation we encounter is a figment of cost minimization, rather than of profit maximization.

5. THE NUMBER OF PLANTS AND THE DEPENDENCE OF OPTIMAL FLEXIBILITY ON SIZE

Three questions are entertained in this section: (1) Does a one-plant firm alter the flexibility of its operations if the mean and variance of demand increase in proportion, or are shifts in the coefficient of variation of demand ($\sigma_d/\bar{d}$) necessary to evoke a flexibility change? (2) Under a straightforward hypothesis about costs of multi-plant operation, is the firm better off with several optimally designed but identical small plants, or with an optimally designed single large plant? (3) Can a multi-plant firm realize gains by varying the flexibility and scale of different plants, e.g., by letting some plants with inflexible capital produce at a steady rate, while demand fluctuations are met by flexible plants with low mean output?[9]

These questions could in principle be answered directly by solving (11) for ϕ and differentiatiating with respect to $\bar{x}$. This is not practical, however, so we seek information from numerical examples. In Table 1, results are presented on optimal values of ϕ, given specific values of the other parameters.

It is seen in Table 1 that optimal ϕ increases as the mean demand increases, all other parameters held constant (examples 1-6). A similar effect is observed as π increases (examples 7-10, and 11-14). On the other hand, optimal ϕ decreases as inventory holding cost η becomes higher (examples 15-24). In all cases, the increase of a parameter value leads to higher operating costs.

Table 1 shows that as the ratio $\sigma_d/\bar{d}$ decreases, the optimal plant becomes less flexible. (In effect, Table 1 lets σ_d establish the output scale and Π establish the cost scale of the firm's cost function.) Table 2 investigates whether this ratio is an appropriate and reliable demand characteristic for predicting optimal flexibility. Will a ratio of 10/5 give the same costs and optimal flexibility as one of 2/1?

Apparently not. Table 2 shows that as mean and variance increase in proportion, the firm's optimal flexibility also increases. Thus, if increases in size are comprised of independent, homogeneous demand "units," the single-plant firm's optimal flexibility also increases (examples 1, 4, 9). Table 2 also suggests that a group of identical, independently operated small plants, each confronted by an equal fraction of total demand, is less costly than a single large plant of optimal flexibility (cases 1, 4, 9; cases 3, 6; any cases x and y where the mean and variance of x are the same fraction of their counterparts in y).

These examples suggest that cost advantages may accrue from multiplant operation. The third question asks if further gains can be realized by apportioning the variability of the firm's demand unequally among the plants, i.e., by making some plants specialists in absorbing mean demand and making other plants specialists in absorbing demand variations.

Table 3 shows a range of possible combinations open to a two-plant firm, confronted by random demands, with mean 5 and variance 2. The "optimal" two-plant configuration (calculated to the nearest 0.5 in σ_d^2 and $\bar{d}$) is fifteen to twenty per cent cheaper than the optimal single-plant

[9] This third question is prompted by a remark of Stigler's: ". . . when the firm has several plants . . . it is possible, within certain limits, to concentrate the fluctuations of output in certain plants, and to build a maximum of flexibility into these plants only" ([12], footnote 20, p. 131). We are curious as to whether this policy of building plants of differing flexibility is preferable, given our cost function (4), to a single plant or to several identical plants.

TABLE 2

The Loss Function (11), with Shifts in σ_d^2
(All examples use $\Pi = 1$, $\eta = 0.01$)

Example no.	π	σ_d^2	$\bar{d}$	Optimal ϕ	Loss	Loss/d
1	0.5	1	1	1.519	2.390	2.390
2	0.5	1	2	1.780	3.693	1.847
3	0.5	1	5	2.442	7.405	1.481
4	0.5	2	2	0.984	5.606	2.803
5	0.5	2	5	1.274	9.938	1.988
6	0.5	2	10	1.675	16.64	1.664
7	0.5	2	25	2.547	35.19	1.408
8	0.5	5	2	0.489	12.52	6.259
9	0.5	5	5	0.576	18.34	3.668
10	0.5	5	10	0.710	27.24	2.724
11	0.5	5	25	1.035	50.85	2.034
12	1	2	2	1.310	6.786	3.393
13	1	2	5	1.741	11.76	2.352
14	1	2	10	2.325	19.22	1.922
15	1	2	25	3.577	39.31	1.572
16	1	5	2	0.659	15.95	7.976
17	1	5	5	0.784	23.12	4.624
18	1	5	10	0.981	33.80	3.380
19	1	5	25	1.450	61.17	2.447
20	2	2	2	1.776	8.487	4.243
21	2	2	5	2.407	14.37	2.873
22	2	2	10	3.251	22.90	2.290
23	2	2	25	5.038	45.15	1.806
24	2	5	2	0.899	20.86	10.43
25	2	5	5	1.081	29.94	5.987
26	2	5	10	1.365	43.13	4.313
27	2	5	25	2.038	75.80	3.032

TABLE 3

Flexibility Allocation in Multi-Plant Operations[a]

Example no.	${}^1\sigma_d^2$	${}^2\sigma_d^2$	${}^1\bar{d}$	${}^2\bar{d}$	ϕ_1	ϕ_2	Loss
1	0.50	1.50	0	5	1.590	2.013	8.637
2	0.50	1.50	1	4	2.036	1.851	8.655
3	0.50	1.50	2	3	2.436	1.677	8.595
4	0.50	1.50	2.5	2.5	2.619	1.585	8.541
5	0.50	1.50	3	2	2.793	1.489	8.470
6	0.50	1.50	4	1	3.116	1.287	8.279
7	1	1	0	5	1.238	2.442	8.432
8	1	1	1	4	1.519	2.239	8.582
9	1	1	2	3	1.780	2.019	8.650
10	1	1	2.5	2.5	1.902	1.902	8.658
11	0	2	0	5	0	1.274	9.938

[a] Uses values:

σ_d (total demand variance) = 2
$\bar{d}$ (total mean demand) = 5
$\Pi = 1$
$\eta = 0.01$
$\pi = 0.5$

Production and inventory associated costs are assumed identical for the two plants.

firm, and about five per cent cheaper than a firm comprising two identical plants.[10] The lowest cost arrangement tabulated calls for one inflexible plant to supply 4/5 of mean demand, while accommodating only 1/4 of the demand variance; the remaining demand variability would be accommodated in the second flexible plant.

Thus, divisibility is apparently a more economical way of achieving flexibility of operation than is adaptability, but a mixture of divisibility and adaptability is still superior.

The preference for divisibility may in part be due to a bias introduced in the loss function (4): no cost of changing the labor force is incorporated in that function.[11] Because of the near fixity of input proportions in a divisible plant, the variance of the labor force will be approximately equal to the variance of the production rate; in an adaptable plant, this variance might well be lower. Consequently, if the labor force is taken explicitly into account, the preference for divisibility over adaptability might well disappear.

6. MULTI-PLANT OPERATIONS AND LONG-RUN COST

The questions and answers of Section 5 may be relevant to recent discussions of long-run cost and the size distribution of firms.

If Gibrat's "law of proportionate effect" held exactly, the growth rate of firms would be independent of their base size: the probability of a ten per cent gain would be the same for a small firm as for a large firm, and the mean and variance of growth rate computed for a group of small firms should be the same as for a group of large firms.

Inferences about the shape of the long-run cost function have been drawn from the empirical behavior of size changes. Simon and Bonini [11] observe that the law of proportionate effect holds approximately, and conclude that long-run average cost does not vary with output.[12] Hymer and Pashigian [5] disagree. They find that the mean percentage size change

[10] It should be remembered that producing in two differently structured plants is likely to entail higher transportation costs than producing in two identical plants would entail.

[11] These costs play an important part in shaping the analysis of Holt, Modigliani, Muth, and Simon [4], and in determining the form of the production control policy analyzed in an earlier paper of mine [10]. Ignoring changes in the rate of output is a representational shortcoming of the function (4) that is discussed more fully in the Appendix of the present paper.

[12] They assert that "... the characteristic cost curve for the firm shows virtually constant returns to scale for sizes above some critical minimum ... we postulate that size has no effect upon the percentage growth of a firm ... if, as we have postulated, there exists approximately constant returns to scale (above a critical minimum size of firm) it is natural to expect the firms in each size-class to have the same chance on the average of increasing or decreasing in size in proportion to their present size" ([11], p. 609).

is invariant with base size (a finding consistent with proportionate effect) but that the variance of percentage size changes is smaller for large firms than for small firms. From these findings, they conclude that the long-run average cost function has a negative slope. In support of this contention, they first rule out a positively sloped long-run average cost function because it is inconsistent with the significant industrial growth which was accomplished without pronounced reduction in business concentration; if there were long-run diseconomies of size, large firms would grow less rapidly than small firms, and concentration would decline. This is unassailable, unless there have been significant shifts in long-run cost that favor growth over the period during which the measurements were made.

The remaining possibilities are constant long-run cost, as advocated by Simon and Bonini, or pervasively declining long-run cost. Hymer and Pashigian vote for the latter on these grounds:

Suppose that in a given industry the firms of the largest quartile are z times as large as the firms of the smallest quartile. The large firm may be regarded as a conglomeration of either *z independent* small firms, or *z interdependent* small firms. In the first case, the standard deviation of the growth rate of a large firm should be $1/\sqrt{z}$ times that of a small firm.[13] This assumes that the growth rate is a random variable and that the growth rates of different firms are mutually independent.

They find that the growth rate standard deviation for large firms is smaller than for small firms, but not as much smaller as is predicted by the mode of large firms as independent amalgams of small firms. They conclude that large firms are not an amalgam of z independent divisions, each equivalent to a single small firm. They state that ". . . large firms are . . . a collection . . . of related small firms. The divisions of a large firm may be related through economies of scale: the large firm is not subdivided into more units because unit costs rise. Then, the assumption that unit costs are constant beyond the critical minimum size must be rejected . . ."[14]

By "independence," Hymer and Pashigian must mean that neither the costs nor the demands of the several plants interact. A firm might achieve cost independence by decentralization, but demand rates confronting the plants would be correlated as long as the plants produce approximately the same goods. Conversely, if demand independence were achievable, flexibility-related considerations might still imply that the large firm should

[13] Let x_t be the size of the small firm in period t, and zx_t be the size of the large firm. Then, if the large firm were an amalgam of *z independently operated* small firms, the respective mean sizes in period $t + 1$ would be μ_{t+1} and $z\mu_{t+1}$, and the variances of the period $t + 1$ size distribution would be σ^2 and $z\sigma^2$. Then the mean growth rates would be equal (μ_{t+1}/x_t and $z\mu_{t+1}/zx_t$), but the standard deviations of the growth rates would be σ_{t+1}/x_t and $\sqrt{z}\sigma_{t+1}/zx_t$, i.e. the large firms would have a growth rate standard deviation $1/\sqrt{z}$ times that of the small firm.

[14] [5a], p. 176.

vary its asset holdings in response to demand shifts at a different rate than does the small firm. This second eventuality is explored in this section; we are concerned both with why it should arise, and with its implications for the presence of scale economies.

The response of capital holdings to demand increases can follow any of three patterns: (1) The firm could add a plant of minimum efficient size every time existing capacity became fully utilized[15] in anticipation of future shifts in demand. (2) It could add to an existing plant or plants in response to, or in anticipation of, demand shifts. (3) It could pass up growth opportunities and not alter capacity. The third alternative makes very little sense. The second alternative is preferable to the first, and provided demand increments are sufficiently small so that the alteration of an existing plant is more economical than construction of a new plant, the rate of change in asset holdings will differ between small firms and large. In Section 5 it was seen that as size increases, optimal flexibility changes, even with a homogeneous demand process. This increase in flexibility with size will dictate relatively larger and less frequent asset shifts by large firms, and hence a higher variability of size changes in the large size category than would be predicted by a model of the large firm as an independent collection of small firms. Because the large firm is prepared economically to meet larger transitory demand fluctuations than is the small firm, it can tolerate larger permanent fluctuations as well, and consequently need not respond by altering capital as frequently as does the small firm.

Thus, deviations from the pattern of asset change that are consistent with supply independence can crop up because of flexibility-related considerations. What does this mean with regard to whether the long-run average cost function is negatively sloped?

The observed deviations from "independence" mean that the large firms expand either by making increases in a single plant, or by specializing some plants to absorb demand fluctuations while specializing others to absorb mean demands. The analysis of Section 5 strongly suggests that the latter course is the most economical way to handle a given demand load; why should the possibility of expanding a single plant even be entertained? Appropriate changes in a single plant might be made in response to demand shifts, even over long periods of time, because the resulting higher cost of meeting the demand load may not completely outweight the lower costs of

[15] The plants need not be exactly of critical minimum size; however, to generate a size change variance that is proportional to existing size, plant additions must be of a standard size for all firms in the industry, and each plant addition must be operated independently of already existing plants in the same firm. The most feasible way of attaining these possibly far fetched conditions is to assume that each asset increment is self contained; and the smaller, the better. Hence, the assumption in the text that the increments are of the minimum efficient size.

adjusting the firm's capital to changes in demand. If the firm has a single plant well adapted to existing demand, and demand happens to change slightly, it may be extremely uneconomical to respond by adding a new plant.

Thus, for several reasons, very little can be concluded regarding long-run cost from the behavior of firm sizes, particularly along the lines followed by Hymer and Pashigian. First, considerations about demand are not adequately handled in their reasoning: firms could be aggregated (independent division of minimum economical size), but if customers of the independent divisions responded together, the size change rate would differ in variance between large and small firms. Second, there may be conflicts between long-run static cost considerations (the only costs to which the question of scale economies or diseconomies is pertinent) and the costs of responding to growth opportunities. The pattern of change in the firm's asset holdings may be dictated more by the latter (dynamic) cost category than by the former (static) cost category. Moreover, these costs of response to growth opportunities may vary significantly between large and small firms (a possibility that is suggested by our finding that flexibility should increase with size in a one-plant firm, even if demand variance, and demand mean, change proportionately with firm size).

APPENDIX

The results we arrive at in this paper are strongly affected by the selection of a quadratic cost function, a choice with ample precedent. Justification for its use in the present paper is on the ground that it (a) captures the response of cost to output and inventory levels and (b) offers an exceedingly convenient analytical framework. This convenience largely stems from the fact that quadratic criteria, like (4), lead to linear decision rules, like (6). In models with more complicated cost structures involving a greater number of state and decision variables, this can be a great benefit. This is seen best, perhaps, in the work of Holt, Modigliani, Muth, and Simon [4], where the size of the work force is taken as a separate decision variable.

This argument favoring a seemingly arbitrary model structure on grounds of heuristic efficiency is familiar; both linear and dynamic programming analyses are largely based on these considerations.

The quadratic representation also possesses the property of certainty equivalence (whereby its mean serves as a sufficient statistic to describe any stochastic exogenous variable) which may be of substantial analytic benefit. For the present paper, certainty equivalence facilitates determination of the optimal production policy, but the optimal degree of flexibility requires additional information about the demand distribution. To serve as a proper certainty equivalent, the mean estimate should be unbiased;

the forecasted variable must be truly exogenous, and not affected by values assigned to instrument variables.

Apparently, there are two major obstacles to treating the quadratic utility function as a paradigm for decision problems; these are (a) "kinky" and (b) "lumpy" costs. A pair of examples may clarify the critical effects of kinkiness on the validity of a quadratic cost approximation. First, consider the familiar "newsboy" or "Christmas tree" problem [6] in which the cost of period-ending inventory is incontrovertably V-shaped, with the kink at the origin. The left segment of the V, associated with shortages of available stock, represents the profit on lost sales opportunity. The other segment represents losses on stock invested in, left unsold, and doomed to obsolesence. If both "arms" of the cost function have slope equal in absolute value, the newsboy (or tree vendor) is indifferent between shortages and surpluses; his only concern is to keep the stock level close to zero. However if the cost of surpluses is lower than the cost of shortages, there is incentive to avoid negative terminal stock levels; some premium is placed on finishing with positive stocks. A quadratic approximation to this asymmetric function must be carefully designed to discriminate in favor of positive terminal inventories, and is unlikely to lead to the optimal level of initial inventory.

A second example is contained in a paper by Beckmann [1]. He posits a production-inventory system in which the costs of changing the production rate are proportional to the size of the change, but differ between production increases and production decreases. With linear costs of inventory holding, stockout, and production, the firm finds that avoidance of production rate changes will be profitable. Consequently, two bounds are defined in the space of inventory levels and past output rates. If the (past output inventory) pair lies between these bounds, current output is set equal to past output. If the state variables describe a point above the upper bound, current output is reduced to a value such that the (current output inventory) pair lies on the upper bound. A symmetric policy is pursued if the state variables lie below the lower bound.

This response discontinuity stems directly from the kink in the production change cost curve. In direct analogy to the Sweezy theory of oligopoly demand, the derivative of the cost function has a "jump" where current output equals past output. This leads to a similar policy prescription: Don't change the value of the instrument variable unless conditions have really changed significantly.

If a quadratic approximation to the production change cost function is used, the derivative becomes continuous and the policy response discontinuity is eradicated.

However, in any operational situation, there must be some question as to whether the relevant cost data are generated by a pair of linear functions or whether they arise from a system with a differentiable "bottom" as in the quadratic approximation. In the newsboy case, the underlying basis of cost break is sufficiently clear to recommend the kinky construction. Beckmann, on the other hand, accepts on faith the far-reaching policy implications of a hypothesis that change in output is proportionately as costly as any other change in the same direction. This hypothesis has no particular prior claim to validity; it is certainly not true in general that the cost of output rate change is V-shaped instead of U-shaped.

The other type of cost that leads to trouble if a quadratic representation is used is cost "lumpiness." (A lumpy cost is zero if some instrument variable takes a zero value, but if the instrument variable takes any positive value, cost jumps to some high level.) The characteristic response to lumpy costs is observed in the well-known "*s*, *S*" or two-bin inventory control rule. Because the cost of ordering is lumpy, orders are not placed at regular intervals (e.g. once per period), nor is the inventory position continuously adjusted toward some "target" level. Instead, orders are placed only when inventory deviates sufficiently from the target level so that the expected shortage costs of *not* ordering exceed the cost of placing an order.

If a quadratic is fitted to a lumpy cost function, the prescribed policy may differ quite sharply from the optimal *s*, *S* configuration. With the quadratic representation, the marginal cost of ordering will continuously decline instead of being constant for all units above the first. The marginal benefit from ordering, which stems from a reduction in the likelihood of stockout, is, under reasonable assumptions about the demand distribution, a continuous function. It is always possible to equate marginal cost and marginal benefit under the quadratic approximation since both functions are continuous. With the lumpy representation, marginal benefits must be greater than the lump sum cost of placing an order; otherwise, no order is placed.

In both cases for which the quadratic offers a dubious approximation, discontinuity of response is desirable. "Kinky" cost functions demand special treatment to keep most realizations of the cost-inducing variable on the cheap side of the kink. "Lumpy" cost functions require special treatment to keep the cost-inducing variable at a zero level as much as possible. Quadratic costs yield linear responses, and linear responses are contrary to the best interests of the firm if costs are lumpy or kinky.

REFERENCES

[1] Beckmann, Martin J., "Production Smoothing and Inventory Control," *Operations Research*, 9 (October 1961), pp. 456–467.

[2] Feller, William, *An Introduction to Probability Theory and Its Applications*, Vol. 1, 2nd ed., New York, John Wiley and Sons, 1957.

[3] Haavelmo, Trygvie, *A Study in the Theory of Investment*, Chicago, University of Chicago Press, 1960.
[4] Holt, Charles C., Modigliani, Franco, Muth, John F. and Simon, Herbert, *Planning Production, Inventories, and Work Force*, Englewood Cliffs, N.J., Prentice-Hall, Inc., 1960.
[5] Hymer, Stephen and Pashigian, Peter, "The Size and Growth of Firms," *Journals of Political Economy*, 70 (December 1962), pp. 556–569.
[5a] ——, "The Size and Growth of Firms" (abstract), *Journal of Economic Abstracts*, 1 (April 1963), p. 176.
[6] Laderman, Jack, Littauer, Sebastian, and Weiss, Lionel, "The Inventory Problem," *Journal of the American Statistical Association*, 48 (December 1953), pp. 717–732.
[7] Lutz, Friedrich, and Lutz, Vera, *The Theory of Investment of the Firm*, Princeton, N.J., Princeton University Press, 1951.
[8] Manne, Alan S., "Capacity Expansion and Probabilistic Growth," *Econometrica*, 29 (October 1962), pp. 632–649.
[9] Mills, Harlan D., "Smoothing in Inventory Processes," Chapter 11, this volume.
[10] Orr, Daniel, "A Random Walk Production-Inventory Policy," *Management Science*, 9 (October 1962), pp. 108–122.
[11] Simon, Herbert, and Bonini, Charles, "The Size Distribution of Business Firms," *American Economic Review*, 48 (September 1958), pp. 607–617.
[12] Stigler, George J., "Production and Distribution in the Short Run," in *A.E.A. Readings in the Theory of Income Distribution* (Fellner and Haley, eds.) Philadelphia, The Blakiston Co., 1951.
[13] Marschak, Thomas, and Nelson, Richard R., "Flexibility, Uncertainty, and Economic Theory," *Metroeconomica*, 14 (January 1962), pp. 42–58.

PART IV

Economic Theory

CHAPTER 14

The Ricardo Effect in the Point Input–Point Output Case

By WILLIAM J. BAUMOL*

I have argued elsewhere[1] that in the general case, the Ricardo effect does not follow from the assumptions which are alleged to lead to that result. In this note I hope to demonstrate that even where the firm's capital is effectively rationed and it is operating in a point input–point output situation, the Ricardo effect need not follow, though it will necessarily hold in two subcases often encountered in the literature.

The Ricardo effect amounts to the following assertion: In the firm whose capital is effectively rationed, an increase in the price of its output, all input prices remaining unchanged, will make profitable a decrease in the roundaboutness (or capital intensity or average period of investment) of the productive process. This is certainly plausible for the case where the entrepreneur has inelastic expectations, for then a rise in current output price will lead to a less than proportionate rise in the future prices expected by him. Hence it will be profitable for him to decrease his employment of durable inputs, i.e., inputs which are relatively future-output producing. However, the Ricardo effect is supposed to hold also for unit, and even somewhat more elastic expectations. In this paper I shall concentrate on the case of unit elasticity of expectations, relying on a continuity assumption to carry the result over to situations involving more elastic expectations.

The point input–point output case is very convenient for the analysis of our problem, for in this circumstance and in no other, there is a well-defined turnover period given by the amount of time the input is permitted to "ripen." In this situation we may then define the Ricardo effect as follows: With a *ceteris paribus* increase in output prices, even with elasticity of expectations equal to or slightly greater than unity, it will pay the firm with rationed capital to reduce its turnover period.

I employ the Hicksian device which treats outputs of the same commodity at different dates as outputs of different goods. This is particularly convenient here since a lengthening of the turnover period may affect the quality as well as the quantity of the product; i.e., it may give us a "better" (or after a while a "worse") tree and not just "more" tree and, even more obviously, better wine rather than more wine. This device also permits us to measure the output of the same good in different units at different dates,

* Princeton University.

[1] "The Analogy Between Producer and Consumer Equilibrium Analysis," *Economica*, Vol. XVII (n.s.), February 1950, pp. 75–80.

and I shall then be able without loss of generality to choose the unit of output at every date to be such that its initial price is exactly p.[2]

1. THE ZERO TIME DISCOUNT CASE

The case considered by Professor Hayek[3] to argue the plausibility of his hypothesis assumes that a return of six per cent in six months is worth the same as twelve per cent in one year. This procedure dodges the difficult problem of comparing a dollar now with a dollar in six months, for with capital rationing there is no interest rate with which to discount returns expected in the future.[4] We can only allow for the opportunity cost of not having future profits to invest now and for psychic time preference cost. Hayek's approach seems to imply that the entrepreneur is determined not to invest profits since otherwise there would be profit on the reinvested profit, making the total return on two consecutive investments more than twice that on the first investment alone. The Hayekian "simple interest case" is thus a peculiar case to include under the head of effective capital rationing when resources will surely not be kept idle, and I discuss it only because the analysis here is so simple and because it has been used in the literature.

Let $f(I, T)$ represent the maximum output obtainable in a period, with I the money value of input, and T the number of turnover periods in a year, which is assumed to be an integer throughout the paper ($1/T$ is the length of one turnover period). Then Hayek's approach involves an examination of the effect of maximization of what we may call the annual equivalent of the returns on each turnover, that is to say, the return on each turnover multiplied by the number of turnover periods in a year. Thus, since the return on each turnover is the total revenue $pf(I, T)$ minus total cost I, the annual equivalent net return is

$$T[pf(I, T) - I] \tag{1}$$

which is to be maximized by an appropriate choice of length of turnover period subject to the constant value of I imposed by capital rationing. In the differentiable case, if this is to be a true maximum, the second derivative of (1) with respect to T must be negative, i.e., we must have

$$2pf' + Tpf'' < 0. \tag{2}$$

[2] I am assuming that the output resulting at any given date from our single initial input is homogeneous.

[3] See *Profits Interest and Investment* (London: Routledge, 1939), p. 9, and Section 2 of "The Ricardo Effect," *Economica*, Vol. IX (N.S.), May 1942, reprinted in *Individualism and Economic Order* (London: Routledge, 1949).

[4] So, strictly speaking, it is not correct to refer to Professor Hayek's procedure as the neglect of compound interest, as does Professor Haberler in *Prosperity and Depression* (New York: United Nations, 1946), p. 485.

Maximization of (1) also requires

$$pf + Tpf' - I = 0. \tag{3}$$

We can now investigate the Ricardo effect in this case by seeing how the system adjusts to a change in price p. If (3) is to continue to hold when price changes, we must have (by total differentiation)

$$[2pf' + Tpf''] \, dT + [f + Tf'] \, dp = 0$$

or

$$dT/dp = -[f + Tf']/[2pf' + Tpf''] \tag{4}$$

where the denominator of the right-hand side is negative, by (2), and the numerator will normally be positive since it is equal to I/p, by (3). Thus dT/pd will be positive, so that the number of turnover periods in a year will increase in response to a *ceteris paribus* rise in output price; i.e., we will have a Ricardo effect.

2. REINVESTMENT IN THE FIRM: PRODUCTION FUNCTION LINEAR IN I

Let us now leave the simple Hayekian "non-compounding" case. Effective capital rationing would, more plausibly, call for reinvestment of revenues, living costs of the entrepreneur being treated as costs in the same way, say, as payments to workers. Here period analysis is called for and we get the first-order difference equation

$$I(t) = pf[I(t - 1), T] \tag{5}$$

relating investment $I(t)$ in the tth turnover period to that in the preceding period, where $f[I(t - 1), T]$ is the output in period $t - 1$ resulting from an investment of $I(t - 1)$ and a turnover period T. The equation merely states that gross returns in each period will be invested in the next period. The object is again to examine the effects of maximization of annual returns, i.e., maximization of

$$I(T) = pf[I(t - 1), T], \tag{6}$$

where the magnitude of $I(0)$, that is, the initial investment, is the quantity of capital rationed to the firm.

In the special case where the production function is linear and homogeneous in I alone we obtain

$$f[I, T] \equiv a(T)I$$

where $a(T)$ is independent of the value of I. In this case, (5) becomes the first-order homogeneous linear difference equation with constant coefficients

$$I(t) = Pa(T)I(t - 1).$$

This, as is well known, has the solution

$$I(t) = [Pa(T)]^t I(0),$$

i.e., investment will increase geometrically from its initial level, $I(0)$. We are interested in the amount accumulated at the end of the first year, i.e., after the passage of T periods. Thus substituting $t = T$ in the last equation we obtain

$$I(T) = [pa(T)]^T I(0) \tag{7}$$

and we desire to choose that turnover period T that makes this quantity as large as possible. Expression (7) will be at a maximum only where

$$\log I(T) = T \log p + T \log a(T) + \log I(0) \tag{8}$$

is. But this in turn requires

$$-\log p = d[T \log a(T)]/dT \tag{9}$$

which differentiated totally yields

$$-dp/p = [d^2 T \log a(T)/dT^2]\, dT$$

where the bracketed expression is the second derivative of (8) and so must be negative for a true maximum. This at once gives the required result: $dT/dp > 0$. The Ricardo effect holds here.

3. REINVESTMENT IN THE FIRM: PRODUCTION FUNCTION NON-LINEAR IN I

Where the production function is not linear in I the Ricardo effect need not follow. In particular if the production function corresponding to processes involving longer turnover periods yields relatively sharply increasing returns to I, or if the production function corresponding to processes involving shorter turnover periods yields relatively sharply diminishing returns to I, an increase in output price may increase annual profits on long-turnover-period processes more than in proportion to the increase in short-turnover-period profits. This is, of course, the reverse of the Ricardo effect.

To prove this consider the following two counterexamples:

EXAMPLE 1 (Sharply increasing returns on the more roundabout process). Let $f[I(t-1), 2]$ (the two-turnover period per year, i.e., the 6-month turnover production function) be $[I(t-1)]^3$. Then successive substitution into (5) gives the firm's assets at the end of the year as $p^4 I(0)^9$. Similarly, let $f[I(t-1), 3]$ (the four-month turnover production function) be, say, $3I(t-1)$. Then with this process the firm's year-end assets will be $27p^3 I(0)$. Clearly, a rise in p will raise profits on the six-month turnover period more than in proportion to that on the four-month process, contrary to the Ricardo effect assertion.

EXAMPLE 2 (Sharply diminishing returns on the less roundabout process).

Here it is convenient to employ $S(t)$ to denote investment in the tth period in a (shorter) four-month process ($T = 3$), and $L(t)$ to denote the corresponding item for a (longer) six-month process ($T = 2$). Now let $f[L(t-1), 2]$ be $30L(t-1)$ and let $f[S(t-1), 3]$ be given by $10S(t-1)$ for $0 \leq S \leq 100$, and by $1000 + 1.1[S(t-1) - 100]$ for $S > 100$. This is a kinked but continuous function giving positive marginal profits throughout, despite the sharp decrease in marginal returns to investment S after $S = 100$.

Assume that price is initially equal to unity and the firm considers employing a unit of investment on one of these processes for one year. The six-month process would yield by successive substitution into (5), $L(1) = 30$ and $L(2) = 900$, the firm's assets at the end of the year. The four-month process offers $S(1) = 10$, $S(2) = 100$, and $S(3) = 1000$. Suppose now that the unit product price goes up to two. Then we obtain the figures $L(1) = 60$, $L(2) = 3600$, $S(1) = 20$, $S(2) = 400$, and $S(3) = 2660$. Clearly the annual gross returns on the longer process will have increased fourfold (from 900 to 3600) in response to the price rise, whereas the returns to the shorter process will have been multiplied by only 2.66 (having risen from 1000 to 2660). Indeed, the rise in final product price will make it profitable to substitute the longer process for the shorter, which is the reverse of what the Ricardo effect would have us expect.

4. INTERPRETATION

It may be helpful to offer an intuitively plausible explanation for the efficacy of the Ricardo effect in the more simple cases and its breakdown in the presence of certain types of non-proportional returns.

If we examine the expressions (1) and (7), which represent the end-of-year yield in the two cases where the effect holds, it is not too difficult to see what is going on. In the compounding linear homogeneous production function case (7), the most simple-minded view of the matter turns out to be valid. In (7) price has been raised to the Tth power, meaning that it has had T periods to be reinvested and compound itself as a yield to the firm. Hence a given price rise will add more to profits the greater the value of T, the exponent of p^T. That is, a rise in price will be more lucrative to management if its investment is turned over more times in a year.

A somewhat similar though less obvious explanation applies to (1). Roughly speaking, a process with more turnover periods gives a specified price increase more turns to add up during the course of a year. More specifically, let us divide the net yield of the process $pTf - IT$ into its total revenue pTf and its total investment cost IT. Then a rise in p will increase both the total revenue and the marginal revenue, $dpTf/dT = p(f + Tf')$ of an increase in the number of turnover periods. However, the price rise will leave the total and marginal investment cost (I) unchanged. Hence,[5] with an increased marginal yield to T and an unchanged marginal cost, it will pay to employ a larger value of T.

But why can relatively increasing returns to capital intensive (long-turnover-period) processes cause trouble in the compounding case? The answer is fairly straightforward. A rise in price increases the scale of the entire operation after the first period; i.e., it offers the firm more capital to invest in subsequent periods. Hence, a process which offers greater returns to large-scale operation is given an advantage. If this advantage goes to the long-turnover-period process, it can, if it is of sufficient magnitude, overcome the fact that this process does not give the price rise as many opportunities to compound as does one with a more frequent turnover.

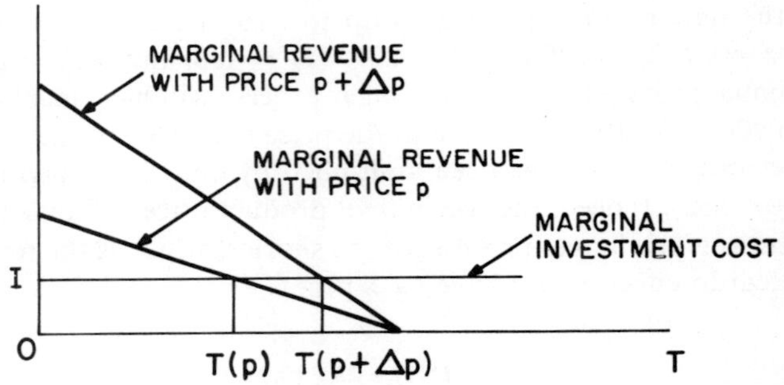

FIGURE 1

[5] More precisely, we will have the situation shown in Figure 1 where the marginal investment cost curve and the marginal revenue curves before and after a price rise are depicted. In the linear case, when price increases from p to $p + \Delta p$, the marginal revenue curve is rotated upward about its old intersection with the T axis. That is because marginal revenue is multiplied throughout by the same constant factor $(p + \Delta p)/p$, i.e., it goes up from $p(f + Tf')$ to $(p + \Delta p)(f + Tf')$. Since the marginal investment cost curve is a horizontal line and the marginal revenue curve must cut it from above (it must have a negative slope) by the second-order maximum conditions, the optimal number of turnovers will *increase* from $T(p)$ to $T(p + \Delta p)$.

CHAPTER 15

The Economics of Uncertainty

By KARL BORCH*

1. INTRODUCTION

1.1 In this paper we shall discuss a few of the new and rather unexpected problems which we encounter when we try to introduce uncertainty into some of the classical economic models. In Section 2 we shall study some simple numerical examples which may help the reader to get an intuitive grasp of the problems involved. In Section 3 we shall discuss some practical implications indicated by some slight generalization of our numerical examples. In Section 4 we shall give a heuristic survey of the mathematical problems involved in a full generalization of the simple models developed in the preceding sections.

1.2 The key to the economics of uncertainty appears to be Bernoulli's utility principle, or the "expected utility hypothesis." This principle was first proposed by Daniel Bernoulli [3] in a paper published in 1738, and was applied occasionally and reluctantly by a few economists and statisticians during the following two centuries. One of these was Barrois [2], who published a fairly complete theory of insurance based on the Bernoulli principle as early as 1834. In general, however, the principle was ignored until 1947, when it was made respectable and even fashionable by Von Neumann and Morgenstern [12].

Von Neumann and Morgenstern proved that the Bernoulli principle can be derived as a theorem from a few simple assumptions as to how rational people make their decisions under uncertainty. There has been a considerable amount of discussion, often confused, over the general validity of these assumptions. We shall not take up these questions in any detail in the present paper, but we shall discuss the subject briefly in Section 4.

2. TWO NUMERICAL EXAMPLES

2.1 We shall consider two persons and assume that each of them owns a business.

Person 1 owns Business 1, which will give a profit of either 1 or zero with equal probability. We shall write $(\frac{1}{2}, 1)$ for this business.

Person 2 owns Business 2, $(\frac{1}{4}, 2)$, which will give a profit of 2 with probability $\frac{1}{4}$, and nothing with probability $\frac{3}{4}$.

* The Norwegian School of Economics and Business Administration.

[1] The research described in this paper was supported partially by The Carnegie Foundation of New York, and partially by The Rockefeller Foundation.

According to the Bernoulli principle, the preferences of a person can be represented by a function, which it is convenient to call "utility of money."

If the utility of money to Person 1 is $u_1(x) = 8x - x^2$ he will, according to the Bernoulli principle, assign the following utility to Business 1:

$$U_1(1, 0) = \tfrac{1}{2}u_1(0) + \tfrac{1}{2}u_1(1) = 3.5$$

Assume further that the utility of money to Person 2 is $u_2(x) = 8x - \tfrac{1}{2}x^2$. The utility which he assigns to Business 2, i.e., to his own business, is then given by

$$U_2(0, 1) = \tfrac{3}{4}u_2(0) + \tfrac{1}{4}u_2(2) = 3.5$$

2.2 We now assume that the two persons can exchange shares in their businesses. It is easy to see that they can both increase their utility by such transactions. For instance, if the outcome of an exchange is that Person 1 owns 60% of Business 1 and 40% of Business 2, the utilities of the two persons will be (assuming statistical independence):

$$U_1(0.6, 0.4) = \tfrac{3}{8}u_1(0) + \tfrac{3}{8}u_1(0.6) + \tfrac{1}{8}u_1(0.8) + \tfrac{1}{8}u_1(1.4) = 3.54$$

and

$$U_2(0.4, 0.6) = \tfrac{3}{8}u_2(0) + \tfrac{3}{8}u_2(0.4) + \tfrac{1}{8}u_2(1.2) + \tfrac{1}{8}u_2(1.6) = 3.72$$

This exchange implies that Person 1 has received a 40% interest in Business 2 in exchange for a 40% interest in Business 1. This obviously means that shares in the two businesses have been traded at a price ratio of 1:1.

The exchange in this example gives both persons a higher utility than they have in the initial situation. If they both act rationally, they may well agree on this exchange arrangement.

If the two persons agree to split even, i.e., agree that both should hold a 50% interest in each business, the utilities will be

$$U_1(0.5, 0.5) = 3.50 \quad \text{and} \quad U_2(0.5, 0.5) = 3.75$$

This means that only Person 2 benefits from the exchange since the utility of Person 1 remains unchanged. If Person 1 acts rationally, he may be indifferent as to whether he should take part in this exchange or not, or he may suggest another exchange arrangement which will give him a part of the gain.

If the outcome of the exchange is that Person 1 owns 75% of Business 1 and 50% of Business 2, i.e., that shares have been traded at the price ratio 1:2, the utilities become:

$$U_1(0.75, 0.5) = 4.28 \quad \text{and} \quad U_2(0.25, 0.5) = 2.83$$

If Person 2 acts rationally, he will not accept this exchange since it gives him a lower utility than he has in the initial situation.

2.3 The examples we have considered show that both persons can increase their utility by a suitable exchange of shares. If the two persons act rationally, we would expect them to reach agreement on an exchange arrangement which will give them both a "fair" increase in utility.

Assume now that the two persons have agreed on an exchange (x, y) such that Person 1 holds $100x\%$ and $100y\%$ interests in the two businesses. If our two persons behave rationally, they will agree on the exchange (x, y) only if there exists no exchange (x_0, y_0) which gives both persons a higher utility. If we further assume that a rational person will not agree to an exchange which gives him a lower utility than he has in the initial situation, we arrive at the following result:

If our two persons behave rationally, they will agree on an exchange (x, y) which satisfies the conditions:

(i) There exists no exchange (x_0, y_0) such that:

$$U_1(x_0, y_0) > U_1(x, y)$$
$$U_2(1 - x_0, 1 - y_0) > U_2(1 - x, 1 - y)$$

(ii)

$$U_1(x, y) > U_1(1, 0)$$
$$U_2(1 - x, 1 - y) > U_2(0, 1)$$

2.4 Condition (i) is obviously satisfied if there exists an exchange (x, y) such that the total differentials

$$dU_1 = \frac{\partial U_1(x, y)}{\partial x} dx + \frac{\partial U_1(x, y)}{\partial y} dy$$

$$dU_2 = \frac{\partial U_2(1 - x, 1 - y)}{\partial x} dx + \frac{\partial U_2(1 - x, 1 - y)}{\partial y} dy$$

have opposite signs for any value of dx and dy. This condition is obviously satisfied if there exists a positive constant k such that

$$\frac{\partial U_1(x, y)}{\partial x} = -k \frac{\partial U_2(1 - x, 1 - y)}{\partial x}$$

$$\frac{\partial U_1(x, y)}{\partial y} = -k \frac{\partial U_2(1 - x, 1 - y)}{\partial y}$$

Hence the solution of these equations will give exchanges (x, y) which satisfy condition (i) in Par. 2.3. It follows from a theorem by Kuhn and Tucker [9] that this also holds for more general cases than our particular example.

2.5 If we apply this result to our numerical example, we obtain the conditions

$$16 - 2(2x + y) = k[13 + (2x + y)]$$
$$16 - 2(x + 4y) = k[11 + (x + 4y)]$$

Solving these equations we obtain

$$x = \frac{48 - 41k}{7(k + 2)} \quad \text{and} \quad y = \frac{16 - 9k}{7(k + 2)}$$

These expressions together with condition (ii) in Par. 2.3 determine the optimal exchanges (x, y), i.e., the set of all exchanges which two rational persons can agree upon.

Table 2 shows some of the optimal arrangements which can be reached by an exchange of shares.

2.6 We shall now assume that instead of exchanging shares, our two persons consider a more general arrangement of the following kind:

(i) If Business 1 succeeds, and Business 2 fails, the profit shall be split so that Person 1 receives x_1 and Person 2 receives $1 - x_1$

(ii) If Business 1 fails, and Business 2 succeeds, the profit of 2 shall be split as follows:

Person 1: x_2, Person 2: $2 - x_2$

(iii) If both businesses succeed, total profits, i.e., $1 + 2$ shall be split:

Person 1: x_3, Person 2: $3 - x_3$

This means that our two persons realize that there are three "states of the world," and that they need an arrangement as to how profits should be divided for each of these states. It is easy to see that an arrangement of this kind is equivalent to an exchange of shares only if $x_1 + x_2 = x_3$.

2.7 Applying the same reasoning as in Par. 2.3, we find that a general arrangement (x_1, x_2, x_3) is optimal only if there exists a constant h such that

$$\frac{\partial U_1}{\partial x_1} = h\frac{\partial U_2}{\partial x_1}$$
$$\frac{\partial U_1}{\partial x_2} = h\frac{\partial U_2}{\partial x_2}$$
$$\frac{\partial U_1}{\partial x_3} = h\frac{\partial U_2}{\partial x_3}$$

Applied to our numerical example these conditions give:

$$4 - x_1 = h(7 + x_1)$$
$$4 - x_2 = h(6 + x_2)$$
$$4 - x_3 = h(5 + x_3)$$

Solving with respect to x_1, x_2, and x_3, we obtain

$$x_1 = 4 - \frac{11h}{1+h}, \quad x_2 = 4 - \frac{10h}{1+h}, \quad x_3 = 4 - \frac{9h}{1+h}$$

Table 1 shows some of the optimal situations which can be reached by such general profit-split arrangements.

TABLE 1

Optimal Arrangements

		Profits received by Person 1		
U_1	U_2	State 1	State 2	State 3
3.500	3.775	0.600	0.908	1.217
3.550	3.719	0.612	0.920	1.228
3.600	3.665	0.624	0.931	1.238
3.650	3.610	0.637	0.943	1.249
3.700	3.555	0.651	0.955	1.260
3.750	3.501	0.663	0.966	1.269

2.8 Comparison of Table 1 and Table 2 shows that an exchange of shares between the two persons inevitably will lead to a sub-optimal situation. This means that for any exchange (x, y) that appears optimal under the rules of Para. 2.3, it will be possible to find a number of general arrangements (x_1, x_2, x_3) that will give both persons a higher utility.

TABLE 2

Sub-optimal Arrangements Which Can Be Reached by an Exchange of Shares

		Shares held by Person 1	
U_1	U_2	Business 1	Business 2
3.500	3.764	0.588	0.400
3.550	3.707	0.600	0.403
3.600	3.650	0.613	0.406
3.650	3.593	0.626	0.410
3.700	3.536	0.638	0.413
3.750	3.479	0.651	0.416

2.9 The two tables do not look very different. The fact that the utility differences appear small has little significance, since the scale for measuring utility can be chosen arbitrarily. However, the arrangements which lead to a general optimum differ considerably from some exchange of shares. This means that a rearrangement of some importance is required in a move from a sub-optimal to an optimal situation.

These rearrangements become even more important if the utility functions of the two persons are essentially different. To illustrate this point, let us consider a second example where

$$u_1(x) = x^{\frac{1}{2}} \quad \text{and} \quad u_2(x) = x^{\frac{3}{4}}$$

The conditions for an optimal general arrangement are

$$x_1^{-\frac{1}{2}} = h(1 - x_1)^{-\frac{1}{4}}$$
$$x_2^{-\frac{1}{2}} = h(2 - x_2)^{-\frac{1}{4}}$$
$$x_3^{-\frac{1}{2}} = h(3 - x_3)^{-\frac{1}{4}}$$

Solving with respect to x_i we obtain

$$x_i = (h_0^2 + 2h_0 i)^{\frac{1}{2}} - h_0 \qquad (i = 1, 2, 3)$$

where h_0 is an arbitrary constant. Some optimal arrangements for this example are given in Table 3.

TABLE 3

Optimal Arrangements for the Second Example

		Profits received by Person 1		
U_1	U_2	State 1	State 2	State 3
4.005	4.626	0.500	0.781	1.000
4.136	4.456	0.531	0.836	1.075
4.247	4.307	0.557	0.884	1.414
4.342	4.174	0.580	0.927	1.200
4.443	4.054	0.600	0.965	1.254
4.500	3.945	0.618	1.000	1.303
4.566	3.844	0.634	1.032	1.348
4.626	3.751	0.649	1.061	1.390
4.730	3.584	0,675	1.114	1.466
4.776	3.508	0.686	1.137	1.500
4.812	3.437	0.697	1.160	1.532
4.858	3.370	0.706	1.180	1.563

3. ECONOMIC INTERPRETATION OF THE MODEL

3.1 The two examples which we have discussed are obviously too simple to have any economic significance. We shall discuss the various possibilities of generalizing the model in some detail in Section 4. For the time being we shall just note that there is no serious difficulty involved in extending the model to *n* persons. If we consider *n* persons, each owning a business, we can study the various arrangements they can make in order to reach an optimal distribution of potential profits. It is easy to see that this situation will be essentially the same as in the two-person case.

In the *n*-person case the number of "states of the world" will be

$$n + \binom{n}{2} + \ldots + \binom{n}{n} = 2^n - 1$$

A general arrangement will then be determined by $n(2^n - 1)$ numbers:

$$x_i^{(j)} = \text{profits payable to person } i \text{ if state } j \text{ occurs.}$$

These numbers must obviously satisfy the equations

$$\sum_{i=1}^{n} x_i^{(j)} = x^{(j)} \quad \text{for all} \quad j$$

where $x^{(j)}$ is the total amount of profits available for distribution in state j.

An exchange of shares is determined by n^2 numbers. x_{ij} = the share person i holds in business j. These numbers must satisfy the conditions

$$\sum_{i=1}^{n} x_{ij} = 1 \quad \text{for all} \quad j.$$

3.2 From the considerations in the preceding paragraph it follows that if our n persons seek a general arrangement, they have to agree on $n(2^n - 1)$ positive numbers, subject to $2^n - 1$ linear constraints. This means that their optimizing problem has $(n - 1)(2^n - 1)$ "degrees of freedom."

If our persons only consider arrangements that can be reached by exchanging shares, they will have to agree on n^2 positive numbers subject to n linear constraints. It is clear that this in general will lead to a sub-optimal situation since the optimizing is done with only $n(n - 1)$ degrees of freedom.

3.3 Let us now assume that these n persons create a stock exchange where they can buy and sell shares in their businesses. If all persons behave in accordance with the usual assumptions of classical economic theory, each share will find its *equilibrium price* and the market as a whole will reach a *competitive equilibrium*. This competitive equilibrium will be stable in the sense that no further exchange of shares is possible without reducing the utility of at least one of the persons who take part in the transaction. This means that no further transactions will take place, if we make the usual assumptions of rational behavior.

However, we found above that an exchange of shares could, in general, only lead to a sub-optimal situation. This means that if the market has reached a competitive equilibrium by trading of shares, it may still be possible to make more general arrangements that will increase the utility of all participants.

3.4 It is evident from the considerations in the preceding paragraph that there will be an inherent element of instability in the stock exchange created by our n persons. It is always possible that some person will discover that the competitive equilibrium is a sub-optimal situation, and that the market can be brought closer to the general optimum by the creation of new securities such as preferred shares, premium bonds, and investment trusts with different leverage.

If the general optimum is determined by some complicated expression as the one we found in Par. 2.9, it is clear that the optimum cannot be reached by creating a finite number of new securities. However, there is no limit to the number of different securities that can be created by ingenious brokers. This means that if the managers of our stock exchange have sufficient ingenuity, there will be a steady flow of new securities, and by trading in these new securities the market will gradually approach, but never reach, a general optimum.

3.5 It is evident, even to the most casual observer, that stock markets in the real world are never in a state of stable equilibrium. Such markets usually seem to be moving all the time in one direction or another. One can explain this by introducing dynamic consideration, and by assuming that the market never has time to reach an equilibrium before external conditions or personal probabilities change. One may also explain the fluctuations in the market as the "tatonnements" of the static Walras model of general equilibrium.

Our model offers another explanation, which may or may not be considered simpler and more plausible. Heuristically we can formulate our explanation as follows: A competitive equilibrium in our model of a stock market will in general be sub-optimal as long as the number of different securities is finite.

3.6 The considerations in the preceding paragraphs suggest that the natural elements or units in our model may be not the businesses but the "states of the world." This approach has been explored by Arrow [1], who suggested that the price should be attached to the "state of the world." He introduces a price concept defined as:

> q_i = the amount one has to pay in order to be assured of receiving one unit of money if state i occurs.

Arrow shows that with this price concept and the classical behavior assumptions, there will exist a competitive equilibrium, which also is Pareto optimal.

With this ingenious device, Arrow is able to save the classical equilibrium model and to make it work also under uncertainty. However, it is not so easy to accept his underlying behavior assumption that people in the market take a system of Arrow-prices as given and buy and sell until their utility is maximized. There are certainly a number of people who believe that common shares have a market price and act accordingly.

3.7 Stock exchanges offer the most obvious real-life example of markets where uncertainty is an essential element. A less familiar example is the reinsurance market. If we reverse the signs in our numerical examples and assume that the persons seek an arrangement which will lead to an optimal distribution of losses, the model can be interpreted as a reinsurance market. Instead of persons owning businesses we will then have to consider insurance companies holding portfolios of insurance contracts. This interpretation of the model has been discussed in some detail in another paper [6], so we shall not pursue the subject any further.

It is, however, worth noting the interesting growth of the so-called non-proportional reinsurance during the last decade. The essence of such reinsurance arrangements is that the reinsurer is called to pay specified amounts only if certain combinations of events (or states of the world) occur. The older forms of "proportional" reinsurance correspond to the arrangements which we referred to as "exchange of shares," and as such they will lead to sub-optimal situations. It is possible that practical insurance men have discovered this, and that they have found that the market could be brought closer to a general optimum through the introduction of non-proportional reinsurance treaties. If this is so, we have a very instructive example of how a businessman's hunch can be well ahead of economic theory.

3.8 As long as our n persons consider the exchange of shares and other familiar securities (such as cash) as the only way to reach an optimal situation, it seems natural to analyze their problem within the framework of classical economic theory. Concepts like "price," "supply," and "demand" are easily defined, and the usual equilibrium analysis can be carried through without any particular difficulty. The difficulty comes only at the end when we realize that this approach leads to a sub-optimal situation.

In order to reach a general optimum our persons have to reach a profit-split agreement covering every state of the world. It seems natural to assume that an agreement of this kind is reached through a bargaining process rather than by the help of some semi-automatic price mechanism. This again should indicate that the theory of n-person games is the appropriate tool for analyzing the problem. It has been shown in some other papers, [4] and [5], that this theory can give a fairly realistic explanation of the transactions that take place in the international reinsurance markets.

Brokers seem to play an important part in markets where uncertainty is an essential element. This can be taken as evidence that the classical price mechanism does not work in such markets, since if it did, brokers would be unnecessary middle-men. If, however, these markets are analyzed in terms of game theory, it becomes clear that brokers fulfill an essential function by helping the people in the market to negotiate their way toward a general optimum.

4. GENERALIZATION OF THE MODEL

4.1 In Section 3 we have drawn a number of rather far-reaching conclusions from a mathematical model which is so simple that it must be considered just a little more than a child's toy. It is hardly possible, without a considerable amount of goodwill, to see that this simple model contains any hints as to how important economic phenomena can be explained. In the following we shall indicate in a heuristic way how the model can be generalized to the extent that may make it useful in serious economic analysis.

We have already mentioned in Par. 3.1 that generalization to an arbitrary number of persons does not present any difficulty, and we shall not discuss this question any further.

4.2 In order to obtain a more general concept of a "business" than the one introduced in Par. 2.1, we can assume that the profit is X, where X is a stochastic variable determined by a probability distribution $F(x)$. This means that a business is completely described by a probability distribution. If persons owning such businesses shall be able to make rational decisions, we must assume that they have a consistent preference ordering over the set of all probability distributions. This preference ordering can be represented by a functional which assigns the utility $U(F)$ to the distribution $F(x)$. The Bernoulli principle states that there exists a function $u(x)$ such that

$$U(F) = \int_{-\infty}^{+\infty} u(x)\, dF(x)$$

This representation must hold also in the degenerate case when $F(x) = \epsilon(x - a)$, i.e., when profit is equal to a with probability 1. From this it follows that $u(x)$ can be interpreted as the utility assigned to profits which are certain, or in classical terminology, as the "utility of money."

4.3 It is obviously not very satisfactory to consider a business as defined just by a simple probability distribution. A business is usually a continuing affair, so if we want a realistic model we must also bring in a time element. The obvious way of doing this is to assume that a business can be completely described by a *stochastic process* $X_1, X_2, \ldots X_t, \ldots$, where the stochastic variable X_t is the profit which the business will give in period t.

If the stochastic process is finite, it is completely determined by a joint probability distribution $F(x_1, \ldots, x_n)$. We can then follow the argument used by von Neumann and Morgenstern and assume that a rational person has a consistent preference ordering over the set of all joint probability distributions with a finite number of variables.

It then follows that there exists a function $u(x_1 \ldots x_n)$ so that

$$U(F) = \int_{-\infty}^{+\infty} u(x_1 \ldots x_n)\, dF(x_1 \ldots x_n)$$

There are no mathematical difficulties involved on this point. If the preference ordering is in some sense continuous, such an integral representation will exist.

If we want to remove the finite horizon, and let n go to infinity, we run into mathematical problems of some complexity, which we shall return to in Par. 4.5.

4.4 The immediate difficulty is to determine the shape of the function $u(x_1 \ldots x_n)$ for a rational person. This function obviously expresses the "timing preference," i.e., a preference ordering over sequences of profit payments which are all considered as certain. Most economists who have worked with this concept have assumed that there is a preference for earlier payments, i.e., that a sequence such as

$$(3, 2, 2, 1, 1) \quad \text{is preferred to} \quad (2, 2, 2, 2, 2)$$

In many economic situations it seems natural to assume that such an "impatience element" (Böhm-Bawerk's "Minderschätzung") exists. If, however, a steady or steadily increasing payment sequence is preferred to a fluctuating one, the preferences above must be reversed. This will often be a reasonable assumption and will not present any serious mathematical difficulties as long as n is finite.

The whole subject of timing preference under uncertainty is still relatively unexplored, and it is by no means certain that functions $u(x_1, \ldots, x_n)$ which seem acceptable on intuitive reasons will correspond to "reasonable" preference orderings over the set of distributions $F(x_1 \ldots x_n)$—and vice versa.

4.5 It is obviously desirable to remove the finite horizon in our model. This problem has recently been studied by Koopmans [8].

Koopmans defines a utility function $u(x_1, x_2, \ldots, x_t, \ldots)$ over an infinite sequence of payments. From some innocent looking assumptions about this function he proves that we must have

$$u(x_1, \ldots x_t + a, \ldots x_{t+i}, \ldots) > u(x_1, \ldots x_t, \ldots x_{t+i} + a, \ldots)$$

for any positive i and a. This means that the existence of an impatience element can be derived as a mathematical consequence of some more basic assumptions about timing preferences.

The most critical of Koopmans' assumptions appears to be his Postulate 3, which in our terms can be stated as follows:

> For any positive a and for any payment sequence $x_1, x_2, \ldots\ldots\ldots x_t, \ldots$, the following inequality must hold
>
> $$u(x_1 + a, x_2, \ldots x_t, \ldots) > u(x_1, x_2, \ldots x_t, \ldots)$$

The postulate is essentially one of inter-period independence, which has mathematical convenience as its main justification.

4.6 Koopmans [8] has pointed out that it follows from an earlier result of Debreu [7] that a slight strengthening of the independence assumption implies that the utility function must be of the form

$$u(x_1, \ldots x_t, \ldots) = \sum_{t=1}^{\infty} r^{t-1} u(x_t)$$

where r is a positive constant.

If we write ${}_3x$ for the infinite vector $\{x_3, x_4, \ldots\}$, the stronger independence assumption can be formulated as follows:

$$u(x_1, x_2, {}_3x) > u(y_1, y_2, {}_3x) \quad \text{implies} \quad u(x_1, x_2, {}_3y) > u(y_1, y_2, {}_3y)$$

and

$$u(x_1, x_2, {}_3x) > u(y_1, x_2, {}_3y) \quad \text{implies} \quad u(x_1, y_2, {}_3x) > u(y_1, y_2, {}_3y)$$

hold for all x_1, x_2, ${}_3x$, y_1, y_2, and ${}_3y$.

4.7 It is tempting also to assume full independence in the stochastic process, i.e., that $X_1, \ldots X_t \ldots$ are stochastically independent. This means if $F(x, t)$ is the probability distribution of X_t, the utility assigned to the infinite stochastic process $X_1 \ldots X_t \ldots$ will be given by

$$U(X_t) = \sum_{t=1}^{\infty} r^{t-1} \int_{-\infty}^{+\infty} u(x) \, d_x F(x, t)$$

If we introduce $r = e^a$, and write the sum as a Stieltjes integral, the expression becomes

$$U(X_t) = \int u(x) \, d\{e^{at} F(x, t)\}$$

where integration is over the whole $x\ t$ space.

This is in many ways a suggestive and interesting formula. It seems to indicate that uncertainty and the time element occur in an almost symmetrical manner in a general economic model. This has been suggested, although necessarily in vague terms, by some economists. The most explicit seems to be Morgenstern [10] and [11].

4.8 We reached the result in the preceding paragraph by making strong independence assumptions. In order to construct a realistic model, we must clearly allow for inter-period dependence both in the utility function and in the stochastic process which determines business profits. This will, as far as we can see at present, lead to mathematical problems of a really formidable nature.

If we accept the independence assumptions, it should be possible to determine the optimal states of an n-person market. The optimality conditions of Par. 2.3 can be generalized without any difficulties if we use the Kuhn-Tucker theorem [9] instead of the primitive method of solution which was adequate for our numerical examples.

4.9 Some readers may have doubts as to the usefulness of pursuing general equilibrium analysis to such generality as that of the preceding paragraphs. In order to illustrate the usefulness of a very general theoretical framework, we shall briefly consider the simple and "practical" problem of formulating the objectives of a firm.

Apparently no modern economist dares to assume that a firm seeks to maximize simple short-run profits. The literature is full of ingenious suggestions as to what firms want to maximize, such as market share, net worth, probability that profits shall stay above a certain threshold, etc., etc.

Instead of making arbitrary, and possibly contradictory, assumptions of this kind, we can assume that the manager of the firm realizes that future profits can only be described by a stochastic process. If the manager by his decisions can alter this stochastic process, his job will obviously be to make the decisions that will give the firm the "best" attainable process. This means that if the manager shall be able to make intelligent decisions, he must have a preference ordering over the set of all attainable processes.

If the firm has a finite time horizon, the stochastic processes can be represented by joint probability distributions of the type $F(x_1 \ldots x_n)$. The preference ordering can then, by the Bernoulli principle, be represented by a function $u(x_1 \ldots x_n)$.

This means that the general objective of the firm will be to maximize an expression of the form

$$U(F) = \int_{-\infty}^{+\infty} u(x_1, \ldots x_n)\, dF(x_1, \ldots x_n)$$

This may be a difficult problem, and a practical businessman may seek some "rule of thumb" which gives an approximate solution. For instance, a high market share this year may not give a high profit this same year, but it may increase the probability of high profits in following years. Hence to maximize the market share in the short run may be a good working rule for solving the general maximizing problem.

The assumptions we have made mean essentially that people are in business for the sake of profits, and that businessmen are more sophisticated about profits than most economists seem to be. These assumptions do not appear entirely unreasonable.

5. CONCLUSION

5.1 In this paper we have used some mathematics which may be considered "advanced," and we have indicated that mathematics of a far more advanced nature may be required in order to develop a complete theory for the economics of uncertainty. It may be useful to conclude with an attempt to explain why such mathematics is essential, owing to the very nature of the problem we set out to study.

5.2 In classical commodity markets people trade "commodity bundles" which can be thought of as vectors with a finite number of elements, or as points in an n-dimensional space. In the markets we have considered, the commodities traded are probability distributions or stochastic processes. A probability distribution can also be thought of as a vector, but this vector will in general have an infinite number of elements and it can be represented as a point only in a Hilbert space.

The introduction of uncertainty in the classical economic theory means essentially that we go from the finite to the infinite. We cannot expect this to be an easy step and we should not be surprised that mathematics of a quite different level is required.

5.3 In classical models it is to a large extent a matter of taste whether we take the underlying market structure to be given in the form of indifference curves, demand and supply functions, substitution rates, or utility functions. However, of these practically equivalent concepts, utility is the only one that can be generalized to the infinite case. It is for this reason that the Bernoulli utility concept plays such an important part in the theory we have tried to outline in this paper.

REFERENCES

[1] Arrow, K. J., "Le rôle de valeur boursières pour la répartition la meilleure des risques," *Colloques Internationaux du CNRS* 40, pp. 41–48, Paris, 1953.

[2] Barrois, T., *Essai sur l'application du calcul des probabilités aux assurances contre l'incendie*, Lille, 1834.

[3] Bernoulli, D., "Exposition of a new theory on the measurement of risk," *Econometrica* 1954, pp. 23–46. (Translation of, "Specimen Theoriae Novae de Mensura Sortis," St. Petersburg, 1738.)

[4] Borch, K., "Reciprocal Reinsurance Treaties seen as a Two-Person Cooperative Game," *Skandinavisk Aktuarietidskrift* 1960, pp. 29–58.

[5] Borch, K., "The Safety Loading of Reinsurance Premiums," *Skandinavisk Aktuarietidskrift* 1960, pp. 163–184.

[6] Borch, K., "Equilibrium in a Reinsurance Market," *Econometrica* 1962, pp. 424–444.

[7] Debreu, G., "Topological Methods in Cardinal Utility Theory," *Mathematical Methods in the Social Sciences*, pp. 16–26, Stanford 1960.

[8] Koopmans, T. C., "Stationary Ordinal Utility and Impatience," *Econometrica* 1960, pp. 287–309.

[9] Kuhn, H. W. and Tucker, A. W., "Nonlinear Programming," *Proceedings of the Second Berkeley Symposium*, pp. 481–492, Berkeley 1951.

[10] Morgenstern, O., "Das Zeitmoment in der Wertlehre," *Zeitschrift für Nationalökonomie*, 1934, pp. 433–458.

[11] Morgenstern, O., "Vollkommene Voraussicht und wirtschaftliches Gleichgewicht," *Zeitschrift für Nationalökonomie*, 1935, pp. 337–357.

[12] Neumann, J. von and O. Morgenstern, *Theory of Games and Economic Behavior*, 2nd edition, Princeton 1947.

CHAPTER 16

The Role of Uncertainty in Economics

By KARL MENGER*

1. THE PETERSBURG GAME

In the theory of probability, the "Petersburg Game" designates the following two-person game.[1] A coin is flipped. If heads turns up, individual B receives one dollar from A, and the game is over. If, however, the coin shows tails, it is flipped again. If the second throw turns up heads, then B receives two dollars from A, and the game ends. If the coin shows tails again, it is flipped a third time. If the third throw shows heads, then B gets four dollars from A and the game is over, but if the third throw produces another tails, then the coin is flipped a fourth time, and so on. The coin is flipped until heads appears for the first time. If this happens at the nth throw (that is, if the first $n - 1$ throws all come up tails, but the nth throw is heads), then B receives the sum of 2^{n-1} dollars from A and the game ends. Depending on whether $n = 1, 2, 3, 4, \ldots$, the size of B's winnings is $2^0 = 1$, $2^1 = 2$, $2^2 = 4$, $2^3 = 8, \ldots$ dollars.

What is B's mathematical expectation when playing this game?[2] The

* Illinois Institute of Technology.

Translated from the *Zeitschrift für Nationaloekonomie*, Band V, Heft 4, 1934, pp. 459–485. English translation by Wolfgang Schoellkopf, with the assistance of W. Giles Mellon.

As Morgenstern reports (cf. *Bulletin Amer. Math. Soc.*, 64 (1958) Number 3, Part 2, p. 108), this paper played a primary role in persuading J. von Neumann to undertake a formal treatment of utility, which was incorporated in the second and third editions of von Neumann and Morgenstern's *Theory of Games and Economic Behavior*, Princeton University Press, 1947 and 1953. The paper brings out some basic facts of "a much more refined system of psychology than the one now available for purposes of economics," which, according to von Neumann and Morgenstern (loc. cit. p. 28) would be necessary to take utility of gambling into account. Essential ideas concerning the so-called Petersburg Paradox propounded in this paper (especially, the theorem of the unsolubility of the paradox by bounded evaluation functions) have recently been restated by P. A. Samuelson, *Internat. Econom. Review*, vol. 1, 1960, pp. 30–37.

The essential features of this essay were completed in 1923 and presented to the Economic Society of Vienna in 1927. See also, *Ergebnisse eines mathematischen Kolloquiums*, Wien 1935, Heft 6, p. 26. At that time, the author was unaware of F. P. Ramsey's paper "Truth and Probability," reprinted in *The Foundations of Mathematics and Other Logical Essays*, New York, 1931.

[1] This game was related by Nikolaus Bernoulli to the mathematician Montmort, who published it in his *Essai d'analyse sur les jeux de hazard*, Paris, 1713. It became better known through the treatise of Daniel Bernoulli, "Specimen theoriae novae de mensura sortis" in the *Commentarii Academia Petropol.*, V, 1738, p. 175. The game received its name because of the connection of the Bernoulli family with the City of St. Petersburg.

[2] Mathematical expectation is defined as the amount of pS units of money, when the probability is p that an amount S is won. In a game where two possible amounts can be won with different probabilities, so that there exists a probability p_1 of winning S_1,

probability that the first throw produces heads is $\frac{1}{2}$, and since B in this case will win one dollar, this possibility contributes $\frac{1}{2}$ dollar to his expectation. The probability that the second throw is heads so that B receives two dollars is $\frac{1}{4}$, which means that this possibility contributes $\frac{1}{4}2 = \frac{1}{2}$ dollar. In general, for any natural number *n*, the probability that heads appears on the nth throw is $1/2^n$ and then B collects 2^{n-1} dollars. This contributes the amount of $(1/2^n)2^{n-1} = \frac{1}{2}$ to B's mathematical expectation, which therefore totals $\frac{1}{2} + \frac{1}{2} + \frac{1}{2} \ldots + \frac{1}{2} + \ldots$, i.e. is infinitely large.

Ever since the Petersburg Game was invented, and the infinite expected value calculated, it has been regarded as a paradox, which numerous investigators from various fields have attempted to resolve.

2. THE APPARENT PARADOX AND THE ACTUAL PROBLEM

Even the very formulation of the problem is a matter of some dispute. What usually is considered paradoxical or at least requiring explanation is the fact that the mathematical expectation of the person playing the game is not a finite sum, but infinite. In the present paper, however, the position will be taken that this infinitude in itself is not paradoxical. If B, who has an infinitely large expectation, would be willing to pay any amount of money for the privilege of playing the game, then his behavior would conform to his infinitely large mathematical expectation—at least as far as possible, considering the fact that there are only finite amounts of money. Thus, in this case, despite the infinite mathematical expectation, nothing would seem to need further explanation.

Actually, however, the situation is quite different. Not only will B not pay an infinitely high price to play the game, since this is impossible, but he will not even pay a very high price that he could afford. In any case, he will not, if he is sane, risk all or even a considerable portion of his wealth in a Petersburg Game. Now the fact that B's actual behavior differs from that indicated by his mathematical expectation is not paradoxical in any logical sense of the word. The discrepancy is a matter of observation (namely, the observation of normal human behavior); and a discrepancy

---

and also a probability p_2 of winning S_2, the mathematical expectation of the risk-taker is defined as $p_1S_1 + p_2S_2$ units of money. Assume, for example, that a game has the following rules. A die is thrown once; the gambler receives six dollars when a six shows, and three dollars for a five, but nothing for a 1, 2, 3, or 4. In this case, since the probability is 1/6 that six, and 1/6 that three, dollars will be won, his mathematical expectation is $1/6(6) + 1/6(3) = 3/2$ dollar. If the dice are "honest" (that is if thrown, for example, 120 times, each of the point numbers 1, 2, 3, 4, 5, and 6 will appear in about 1/6 of the throws, or about 20 times), then winning and losses will be about equal, provided that the game is played a large number of times and that in each game the gambler risks his mathematical expectation of 3/2 dollar.

between an observation and a previously calculated formula never constitutes a logical paradox, just as conformity of an observation to a previously stated formula never is a matter of logical necessity. Still, the discrepancy under discussion is very peculiar, and it is understandable that many investigators have wanted to find an explanation. The view, taken in this paper is that the explanation of a fact consists in its classification within a larger group of facts. Since the question herein analyzed deals with the economic behavior of normal people, one should expect from the start that the factual classification used for its explanation would be primarily of a psychological-economic nature, despite the mathematical interest which the problem may claim on historical grounds.

Summarizing, we can state for orientation purposes the following: What is, if not paradoxical, at least very peculiar about the Petersburg Game is not the discrepancy between mathematical expectation and possible behavior—a discrepancy that is inevitable because all sums of money are finite—but the discrepancy between the behavior indicated by the mathematical expectation and that of normal individuals.

3. THE ALLEGED INCONSISTENCY OF THE GAME

Disregarding the mathematically unsound considerations of the game,[3] we first consider an attempted solution that makes the formulation of the problem responsible for the so-called paradox.[4] It is being stressed that A has only a finite fortune; yet according to the rules of the game he may find himself in a position where he would have to pay extremely high amounts. These cases, which in their totality produce B's mathematical expectation, are thus, in fact, illusory. In actuality, B cannot expect to win more than A's entire estate, which means that B's expectation is finite. To this may be added the fact that because each throw requires a finite amount of time, the game would, of necessity, have to be stopped after some finite (though very high) number of throws even before heads turns up.

These and similar arguments, however, change the conditions of the

[3] We mention here the remarks of d'Alembert, "Doutes et questions sur le calcul des probabilités," *Mélange de Littérature, d'Histoire, et de Philosophie*, V, 1773; and of Béguelin, *Histoire de l'Acad. de Sciences et Belles-Lettres*, Berlin, 1767. These were already contradicted by Lichtenberg, *Physik. und Mathem. Schriften*, Vol. 4, 1806. For the whole discussion, compare Czuber, *Archiv d. Math. u. Phys.*, 67, 1882, p. 3 sq.

[4] The concepts described in this section were originated by Poisson, *Recherches s.l. probabilité des Jugements en Matière Criminelle et en Matière Civile*, 1837, and are supported partially by Czuber, *op. cit.*, p. 19 sq. (despite the critical remarks on p. 13), and especially by Pringsheim in his notes to the German edition of the essay by D. Bernoulli (mentioned in footnote 2), "Versuch einer neuen Theorie der Wertbestimmung von Gluecksfaellen," Leipzig, 1896.

game by introducing factors that are outside the problem, such as A's wealth and the length of time needed to make a throw. Some maximal numbers (of paid-out amounts and of throws) are introduced which are not in the definition of the game. Furthermore, the purpose of such explanations is simply to eliminate the infinitude of B's mathematical expectation, and therefore, to eliminate the alleged paradox mentioned in Section 2. But these solutions are unable to explain the remarkable discrepancy between mathematical expectation and actual behavior; and this is the more critical consideration, since, as will be shown below (Sections 7 and 8), this discrepancy is also to be found in cases where the mathematical expectation is not infinite but finite, and where one cannot possibly fall back on any inconsistency in the game.

4. A PROBABILITY-THEORETICAL SOLUTION

Another solution of the problem that has been proposed is this:[5] Only if a game is repeated many times is it reasonable to pay the expected value for each game. Indeed, only a repeated replaying of the game would result in the approximate equality (mentioned in footnote 3 in the case of throwing dice) between the sum paid out in winnings and the sum of the prices paid for playing, either one being equal to the mathematical expectation. Applied to a single game, however, especially to a single Petersburg Game, the concept of mathematical expectation loses its practicality. The real meaning of the statement that B possesses an infinitely large mathematical expectation, according to this view, is the following: Even if S is an extremely high amount, B will probably have an advantage over A (that is, B will win more than he has paid in) provided that A and B play the game not only once, but sufficiently often, and that B risks the amount S in every single game. This is true because in a sufficiently long series of Petersburg Games A will probably have to pay to B such large amounts that no matter how high an amount B risks in any single game, the result will be advantageous to B in the long run.

This unassailable argument is itself sufficient proof that the problem has no paradoxical elements. The statement that B's mathematical expectation is infinite is the source of statements similar to those which are usually made about the game of dice. The aspects of the problem, however, which are responsible for the (if not paradoxical, at least strange) discrepancy between expected value and the behavior of normal persons in the case of a single game, are of a psychological-economic nature and are not even touched upon in the above argument; e.g., the fact that a normal

[5] This solution was given by Fries, *Versuch einer Kritik der Prinzipien der Wahrscheinlichkeitsrechnung*, Braunschweig, 1842, p. 116, and also by Czuber, *op. cit.*, p. 15 sq.

person playing a single game of chance may be willing to pay a larger or only a smaller amount than his mathematical expectation, remains uninterpreted.

5. THE SOLUTION WITH UNBOUNDED VALUE FUNCTIONS

Other attempts at a solution have been based on the assumption that the gambler B evaluates his potential winnings not according to their absolute amount, but according to the utility which they have for him. This idea[6] has been used to analyze the Petersburg problem in two ways.

The better known version,[7] which we will discuss first, begins with the observation that B possesses a certain fortune when the game begins so that whatever the game may bring in, only constitutes an addition to this wealth. The way in which one evaluates additions to one's fortune, however, depends on one's total wealth. A person who owns a great fortune will evaluate a given sum less highly than one who owns less. The fact that the winnings have a value smaller than their absolute amounts results in a reduction of B's expectation; and this supposedly explains that, while his mathematical expectation is infinite, his actual expectation is finite.

[6] It seems to have appeared for the first time in the writings of the mathematician Cramer (whose contributions to the theory of linear equations are widely known). In 1728 he wrote a letter addressed to N. Bernoulli and published by D. Bernoulli about the question of the origin of the difference between mathematical calculation and normal estimation: "I think it stems from the fact that (in the theory) mathematicians evaluate money simply by its quantity, while reasonable people (in the real world) evaluate it according to its utility." D. Bernoulli formulated this idea independently (op. cit.) and also expressed it publicly for the first time, defining it in a way to be discussed footnote 8. "Valor non est aestimandus ex pretio rei, sed ex emolumento quod unusquisque inde capessit. Pretium ex re ipsa aestimatur (these words show incidentally that the basic idea of modern economic theory, which explains prices through the interplay of the valuations by various individuals in barter, was not known to Bernoulli in any form), omnibusque idem est, emolumentum ex conditione personae. Ita procul dubio pauperis magis refert lucrum facere mille ducatorum, quam divitis, etsi pretium utrique idem sit." Buffon also supported this idea in his "Essai d'Arithmetique Morale" (supplement to Volume IV of the Histoire Naturelle, 1777, p. 72 sq.). "L'argent ne doit pas être estimé par sa quantité numérique . . . . Il s'en faut bien que les advantages qu'on tire de l'argent soient en juste proportion avec sa quantité; un homme riche à 100.000 écus de rente, n'est pas dix fois plus heureux que l'homme qui n'a que 10,000 écus," and (l.c., p. 89): "L'avare est comme le Mathématicien; tous deux éstimant l'argent par sa quantité numérique." This idea was then emphasized by Laplace, Théorie Analytique des Probabilités, Paris 1812 (Oeuvres VII, Paris 1886, p. XIX sq.). "On doit distinguer dans le bien espéré sa valeur relative de sa valeur absolue: celle-ci est indépendente des motifs qui le font désirer, au lieu que la première croît avec les motifs," and this was repeatedly emphasized by F. A. Lange (e.g. "Die Arbeiterfrage" 1865). In theoretical economics, the discovery of this idea in the 1870's introduced a new era. Good source material about the predecessors of the marginal utility theory can be found in Weinberger, "Die Grenznutzentheorie," 1926, where, however, the theory itself is not completely presented and, in my opinion, unfairly criticized.

[7] D. Bernoulli, *op. cit.*

This idea has also been defined quantitatively.[8] To each addition to wealth was assigned a "subjective" value directly proportional to the addition and inversely proportional to the total wealth of the evaluator. It follows from this assumption that an addition D to wealth W is given the subjective value of $c \log (1 + D/W)$, where c is a number which depends only on the evaluator (but is independent of D) and log indicates the natural logarithm. According to this formula, the subjective value of the addition kD to kW is $c \log (1 + kD/kW) = c \log (1 + D/W)$, or equal to the subjective value of the addition D to W.

If the logarithmic formula of subjective evaluation is introduced into the Petersburg problem, the following results are obtained. For each number $n = 1, 2, 3, \ldots$ the probability is $1/2^n$ that B adds to his initial wealth W the amount 2^{n-1}. This amount has for B a subjective value of $c \log (1 + 2^{n-1}/W)$. The sum B's subjective expectations is therefore

$$\frac{c}{2} \log \frac{W+1}{W} + \frac{c}{4} \log \frac{W+2}{W} + \frac{c}{8} \log \frac{W+4}{W} + \ldots + \frac{c}{2^n} \log \frac{W+2^{n-1}}{W} + \ldots.$$

As is easily seen, this series converges when $W > 0$, i.e., when B starts the game with any wealth at all.[9] Hence B's subjective expectation is finite.

[8] At first probably by Cramer, who considered the valuation formula $C\sqrt{W}$, according to which a gain of 4W is evaluated only twice as high as W, and a gain of 9W, three times as high. The famous definition which leads to the logarithmic valuation function was introduced under the title, *Mensura sortis*, by D. Bernoulli, *op. cit.* "Ita vero valde probabile est, lucrum quod vis semper emolumentum aferre summae bonorum reciprocae proportionale." It was used by Laplace. "On ne peut donner de règle générale pour apprécier cette valeur relative; cependant il est naturel de supposer la valeur absolue, en raison inverse au bien total de la personne interessée." Following the latter, one generally designates the expression $c \log (W + D)/W$ as the moral value of the increase D to wealth W, and the product of the moral value of a gain and its probability as its moral expectation. Oettinger (*Crelles Journ. f.d.r.u. angew. Math.*, 36, 1848, p. 26 and p. 300 sq.), proposes the terms subjective value and subjective expectation that are used in the present article. Later the Weber-Fechner Law of Psychology proceeded generally from the assumption that the sensation of an increase d to a stimulus s is directly proportional to the increase d and inversely proportional to the stimulus s and is, therefore, proportional to $\log (s + d)/s$. Fechner, *Elemente der Psychophysik*, 1860, I, p. 236, II. p. 10, states that this formulation encompasses the Bernoulli valuation function, if one treats the size of wealth as the stimulus, and its value as the sensation. *Sensations* can never (and the expressions of sensations can only rarely) be added and consequently, measured and compared on a numerical basis; at best they can be arranged according to their intensity. In other words, sensations cannot be mapped congruently, but at best topologically, on real numbers. The *expressions* of valuation "sensations," however, (that is, the behavior in exchanging commodities) can be determined quantitatively. For example, one can determine, for various numbers x, the amount of money that a man will spend in order to acquire x units of a commodity. Marginal utility theory, if carefully formulated, can, therefore, well stand up to logical critique.

[9] In order to make sure that this assumption is always satisfied, D. Bernoulli uses a particular definition of wealth including the ability to earn something through work or even begging. Hence $W = 0$ only for a person deprived of even these opportunities and starving.

The basic idea of this attempted solution—that the value of an addition to wealth depends not only on its size, but on the size of the wealth of the evaluator—is, of course, excellent and has played a great role in theoretical economics since its formulation by the founders of the subjective theory of value. But as far as the discrepancy between mathematical expectation and the behavior of normal people towards games of chance is concerned, that basic idea fails, as will now be shown, to offer a satisfactory explanation.

We begin by demonstrating that the solution of the Petersburg paradox according to the logarithmic formula for subjective value (contrary to the views expressed for more than one hundred years in the textbooks on probability theory) is unsatisfactory on formal grounds. (We shall not discuss in this context whether the logarithmic value function is suitable for describing the actual evaluations of normal individuals, though this also might be questioned.) That the formula does not explain the paradox is most easily shown by considering the *ad hoc* character of the solution. It is true, the assumption that the addition D to wealth W has the subjective value $c \log (1 + D/W)$ implies that the subjective expectation in the Petersburg Game is finite, but by changing the game slightly, one can stipulate a similar game for which not only the mathematical but also the subjective expectation based on the logarithmic value function is infinite and yet no one in his right mind would risk a substantial amount. Consider, for example, the following modified Petersburg Game. A coin is flipped until heads appears for the first time. If this occurs at the nth throw, then B wins from A the amount $We^{2n} - W$, where W is B's initial fortune. For all large n, perhaps even for all n, this amount exceeds 2^{n-1}, which B would receive in the Petersburg Game. In this modified Petersburg Game, for each number n = 1, 2, . . . to infinity, *B* has the probability $\frac{1}{2}/2^n$ to win $We^{2n} - W$ and his mathematical expectation is

$$\tfrac{1}{2}(We^2 - W) + \tfrac{1}{4}(We^4 - W) + \tfrac{1}{8}(We^8 - W) + \ldots + \frac{1}{2^n}(We^{2n} - W) + \ldots,$$

thus infinite. If we now calculate B's subjective expectation using the logarithmic value function, we obtain the following. For each natural number n, B has the probability $1/2^n$ of adding $We^{2n} - W$ to his wealth W, which has a subjective value $c \log (W + We^{2n} - W)/W = c \log e^{2n} = 2^n c$, and which contributes for each whole number n the amount $1/2^n c 2^n = c$ to his subjective expectation. In total, therefore, the subjective expectation is $\frac{1}{2}2c + \frac{1}{4}4c + \frac{1}{8}8c + \ldots$, that is, infinite. We therefore see that in this slightly modified Petersburg Game, B has not only an infinite mathematical, but also an infinite subjective expectation on the basis of the logarithmic value function. At the same time it is obvious that, even in the modified Petersburg Game, no normal person would risk his total fortune or a substantial amount. Thus, the solution of the Petersburg game through the use of the logarithmic formula for subjective evaluation is only an apparent one.

Furthermore, we can show the following. The fact that the (correct) idea of evaluations depending on the wealth of the evaluators is unsuitable as an explanation of the discrepancy between mathematical expectation and actual gambling behavior, is not due to the specific casting of the idea in the form of the logarithmic function (that is, of a subjective value proportional to the size of the addition and inversely proportional to the fortune of the evaluator). Instead, we consider, more generally, the value $f(\mathrm{D})$ that an individual B attributes to the addition D to his fortune, where f is any function (dependent on *B*'s wealth and perhaps other parameters) that is unbounded; that is to say a function which, for sufficiently large D, assumes arbitrarily large values. This is the case, for example, for the functions $\sqrt{\mathrm{D}}$ and $\mathrm{c} \log \log (1 + \mathrm{D}/\mathrm{W})$, but also, for example, for the linear function $\mathrm{cW} + \mathrm{c}_1$, where $\mathrm{c} > 0$, and for the function $\mathrm{c} \log (1 + \mathrm{D}/\mathrm{W})$. We now can state the following general theorem.

For any evaluation of additions to a fortune by an unbounded function, there exists a game related to the Petersburg Game in which the subjective expectation of the risk-taker on the basis of that value function is infinite.

Assume that D_1 is an addition whose subjective value $f(\mathrm{D}_1)$ is at least $2^0 = 1$ dollar, D_2 an addition whose subjective value $f(\mathrm{D}_2)$ is at least $2^1 = 2$ dollars, and for each natural number n, D_n is an addition whose subjective value is at least $2^{\mathrm{n}-1}$. It follows from the assumed unboundedness of the function f, that for each n such a D_n exists. Having determined such an addition D_n for each natural number n, consider the following modified Petersburg Game. A coin is flipped until heads appears for the first time. If this happens on the nth toss, then B wins D_n dollars from A and the game ends. B's subjective expectation, given the valuation function f, clearly is

$$\tfrac{1}{2}f(\mathrm{d}_1) + \tfrac{1}{4}f(\mathrm{D}_2) + \tfrac{1}{8}f(\mathrm{D}_3) + \ldots,$$

which is at least equal to $\tfrac{1}{2}1 + \tfrac{1}{4}2 + \tfrac{1}{8}4 + \ldots$, thus infinite.

While, as was shown in Section 2, the essence of the Petersburg problem is by no means the replacement of an infinite expectation by a finite amount, the above theorem shows that the introduction of an unbounded valuation function does not in all cases achieve even that replacement.

6. THE SOLUTION WITH BOUNDED VALUE FUNCTIONS

The idea that the value of an amount of money does not only depend on its size, has also been used in a second, less widely known form. The assumption has been made that the function f describing the subjective value be bounded;[10] that is to say that there is a number W* such that for every W, the value f(W) is less than W*. Under this assumption, B's expectation in the Petersburg Game, as well as in all Petersburg-type games considered in the previous section, is finite. This is true since B's subjective expectation is a number that is smaller than $W^*/2 + W^*/4 + W^*/8 + \ldots + W^*/2^n + \ldots$, and this sum is equal to W*.

Yet this solution cannot be regarded as satisfactory. (Note that the critical arguments advanced in what follows can also be raised against the unbounded valuation functions, which in the preceding section were found not even to entail finite expectations.) First, simple introspection shows that if we are unwilling to risk a substantial amount on the Petersburg Game, this cannot be explained or at least not completely explained, by the idea that the possible large winnings do not have a substantial value to us. More important, as introspection shows, is the consideration that it is highly unlikely that we shall win one of those very large, subjectively valuable prizes. Second, in connection with this line of reasoning, we can cite games where a man under normal circumstances would not even be willing to risk that amount which corresponds to his subjective expectation on the basis of a bounded value function. Examples will be given in Section 7. Thus even the introduction of bounded value functions does not eliminate the discrepancy between calculated and actual behavior. We can summarize the results in Sections 5 and 6 by stating that the correct and

[10] Cramer, before considering the valuation function $\sqrt{W}$, suggested loc.cit. that the value of money beyond a certain very high amount (he arbitrarily assumed a figure of $10 million), no longer increases. He then based his calculations on the following valuation function. The subjective value f(W), of an amount of money W, is equal to W, if W is smaller than $10 million, and is equal to $10 million whenever W exceeds that amount. Now it is clear that this sudden change in the function at $10 million finds no support from actual experience, nor does the linear increase up to the $10 million limit, which the assumption of a logarithmic valuation function has replaced by a slower increase that corresponds more closely to reality. In what follows, we shall, however, disregard the faults of this special valuation function and deal with the general question of bounded valuation functions. Buffon also states, *op.cit.* p. 72: ". . . c'est que l'argent, dès qu'on passe de certaines bornes, n'a presque plus de valeur réelle; et ne peut augmenter le bien de celui qui le possede; un homme qui decouvriroit une montagne d'or, ne seroit pas plus riche que celui qui n'en trouveroit qu'une toise cube." An example of a steadily increasing bounded function which approaches asymptotically with increasing W a certain amount and which, therefore, combines the advantages of the logarithmic and the bounded functions, is, as was stated by Timerding in his essay on Bernoulli's theory of value (*Zeitschrift f. Math. u. Phys.* 47, 1902, p. 337), the function $f(W) = a + b/(W + c)$, where a, b, c, are numbers or constant functions. If this function is (1) to assume the value 0 for $W = 0$, (2) to approach with increasing W asymptotically the value W*, and (3) to have the derivative 1 for $W = 0$, then one arrives at the hyperbolic function

$$f(W) = \frac{W^*W}{W^* + W}\cdot$$

The following graphs represent the logarithmic function of Bernoulli, the parabolic function $\sqrt{W}$ of Cramer, and the hyperbolic function mentioned above.

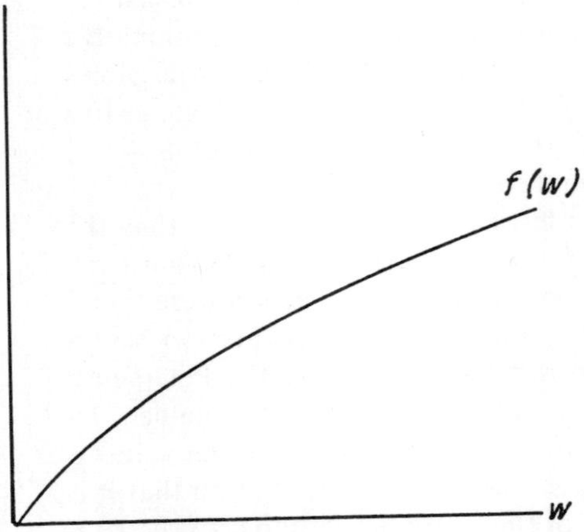

$f(w) = C \log \frac{v+w}{v}$

For $C = v = 1$

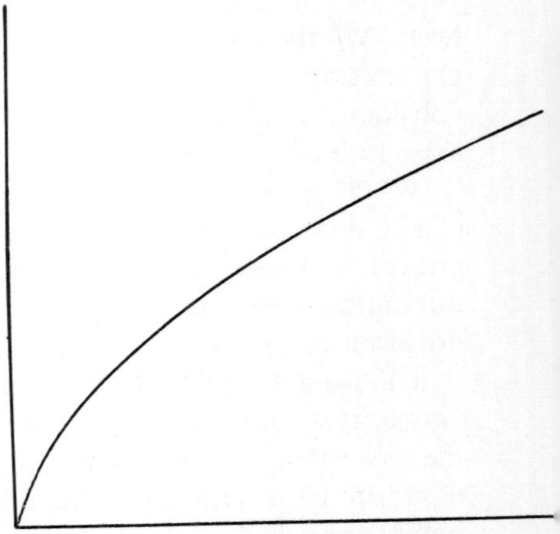

$f(w) = \sqrt{w}$

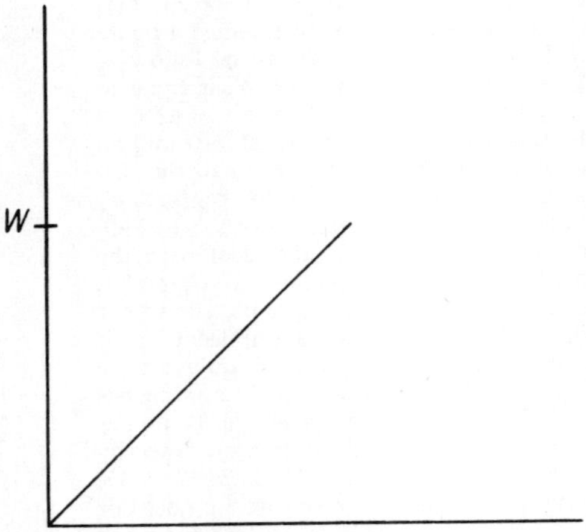

$f(w) = \begin{cases} w & \text{for } w < W \\ W & \text{for } w > W \end{cases}$

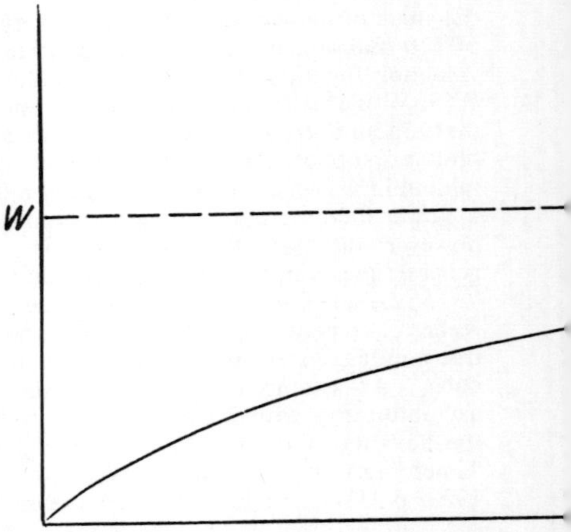

$f(w) = \frac{Ww}{w+W}$

important idea of the value of goods not depending solely on their quantities, is insufficient to explain gambling behavior.

7. THE SOLUTION WHICH DISREGARDS SMALL PROBABILITIES

Another attempt at a solution begins with the observation that events whose occurrence has a very small probability are, for practical purposes, completely disregarded.[11] Consider, for example, a game where a single winning ticket of one billion dollars is drawn from a total of 100 million tickets. The expected value to the owner of each ticket is ten dollars. Yet would many among 100 million people risk ten dollars in a lottery in which a single one of them is to become a billionaire?

Incidentally, this example shows that even in some games with completely finite conditions, normal behavior differs from that indicated by mathematical expectation. Hence the elements of infinity in the conditions of the Petersburg Game cannot be essential in explaining that discrepancy. Moreover, this line of thought contributes to the criticism, started in Section 6, of the attempt to solve the Petersburg problem on the basis of a bounded value function. Consider, for example, W^* as the bound of the valuation function f, where W^* has to be a rather large sum. (At the beginning of the 18th century,[12] a sum equivalent to $10 million was mentioned.) Most people would refuse to risk one dollar in order to obtain a probability of 1/10,000,000 of winning an amount which has even a subjective value of $10 million. Hence also the introduction of bounded value functions is insufficient to remove the discrepancy between mathematical expectation and the actual behavior of gamblers.

It is clear that by setting small probabilities equal to zero, all but a finite number of the infinitely many terms that make up B's infinite expectation in the Petersburg Game are reduced to zero. Hence the resulting expectation is finite, which removes the element that some writers have regarded as paradoxical in the Petersburg Game. A complete description

[11] Buffon was probably the first (*op.cit.* p. 48 ff.) to use this fact in the analysis of the Petersburg Game, separately from the other fact that the subjective valuation of sums of money does not only depend on their size. Apparently none of the later commentators has sufficiently emphasized this circumstance. Buffon distinguishes between mathematical certainty and moral certainty, mathematical impossibility and moral impossibility. Moral certainty is already determined in the case of a very large probability. "On sent bien que c'est un certain nombre de probabilités qui fait la certitude morale, mais quel est ce nombre?" He discusses in this connection the most important of fears, that of death. The probability that a fifty year old man may die within the next twenty-four hours is 1 in 10189. This probability is, according to Buffon, for practical purposes equal to 0. A smaller expectation fails to cause any emotion and is not worth a thought. Without discussing the specific limit below which Buffon proposes to disregard probabilities—a limit which, incidentally, was criticized by Bernoulli—we shall consider the general assumption that probabilities below a certain level are set equal to zero.

[12] Compare footnote 10.

of the behavior of normal people with regard to games of chance however, cannot be achieved by simply setting all small probabilities equal to zero. This may be seen by comparing this attempt at a solution with the one presented in the following section.

8. THE FUNDAMENTALS OF BEHAVIOR IN GAMBLING GAMES

In the following pages, we present an explanation of the discrepancy between the behavior according to mathematical expectation and the actual behavior in the Petersburg Game. The explanation consists of a general description of normal behavior in all types of games of chance.

Such a description cannot disregard the fact that a risk-taker in evaluating a good, takes into account, in addition to the mere quantity of that good, the size of his other wealth; nor can it disregard the fact that he refuses to risk money if the probability of winning is extraordinarily small. To this latter point one must add, however, that experience does not indicate a definite lower bound such that all smaller probabilities are equated to zero, while all greater probabilities are treated according to their values. Such a probability limit cannot even be observed in the case of a single individual, and it would be utterly impossible to indicate a common probability limit, valid for everyone. Instead, the actual behavior of individuals reveals a steadily increasing undervaluation of small probabilities. Similarly, the lower valuation of additions to wealth generally does not begin, as has sometimes been assumed, at a certain limit (so that additions beyond this limit are valued at zero, while additions to all lower levels of wealth below the limiting value are valued according to their size); and what one observes is rather a steadily decreasing valuation of larger additions relative to their size.[13]

Before we go on to state the most important facts about normal behavior in games of chance, we must note the following. How the chance pD (that is, the probability p to add an amount D to wealth W) is to be evaluated is not an unequivocal question. One may ask either: "What is the largest amount that a man will pay in order to buy the chance pD?" or "What sure addition to his wealth W does he equate in value to the chance pD?" For example, he must ask himself the first question when he has to purchase the chance pD, but he must answer the second question when he can choose between two gifts; of a certain sum or of a chance pD. These two amounts, which both correspond to the chance pD (though each in a different sense) need not be equal, and, in fact, the first is usually smaller than the second. In the following discussion we shall concern ourselves only with the first question (despite the fact that it involves

[13] This obviously is a direct consequence of the fact that the valuation function $f(W)$ grows monotonically with increasing W, but also has a steadily declining rate of growth.

some complications which do not arise in the second question), since in nearly all cases this is the basic question one has to ask oneself in daily life before entering into any gambling situation. We now list a number of important features which can be regularly observed in the valuation (according to the first definition) of chances.

A first type of characteristic stems from the fact that the purchase of a chance presupposes a payment which is valued in proportion to the wealth W of the evaluator. Taken by themselves, the laws of subjective value theory (which deal with the estimation of certainty), when applied to the price that one has to pay to participate in the game, state that given the same probability, a wealthier person may risk more and perhaps even a greater proportion of his wealth than someone who owns less.

It follows in particular that in the case of very high probabilities (of winning) nothing will be risked.[14] This statement seems peculiar only at first glance; a closer look shows that it is confirmed by experience. In general, one would not risk a substantial part W′ of one's wealth in order to obtain a probability 99/100 of winning 100/99W′ (i.e., about 1.01W′). Between the behavior indicated by expected value on the one hand, and actual behavior on the other, there exists a considerable discrepancy in the case of larger values of p (i.e., values close to 1). According to the formula pW, in the example just described, one would have to risk the amount W′. In fact, however, one is unlikely to risk the amount W′ if this constitutes a substantial portion of one's wealth, even if one could get a 99/100 chance of winning 1.1W′ or 1.2W′. This is true because, although the chance (1/100) of losing the risked amount W′ is small, if this should happen one would suffer a considerable loss. Conversely, while the chances of winning are very high, the winnings would be very small, even if they amounted to 1.1W′ or 1.2W′, since after subtracting the risked amount, the gain would only be 1/10 W′ or 1/5 W′ respectively.

Of course, the theorem of the undervaluation of chances with high probabilities, which thus can be deduced from the basic theorems about the valuation of sure events in subjective value theory and can also be observed in the real world, should not be interpreted as meaning that a businessman prefers relatively uncertain to almost certain transactions. While we have been dealing with the characteristics of the purchases of chances in gambling where one either wins or loses, the circumstances surrounding a businessman's decisions are usually such that, even if he does not win, he will get back some part of his original investment. In such cases, the larger the part of the original investment that he can be sure to recover, the more he will prefer deals with a high probability of gain. In the sense of the theory of games of chance, one can regard as risked only that part of the investment which cannot be expected to be regained

[14] The essence of this remark is already contained in the work of D. Bernoulli.

with certainty. If, however, he faces transactions where in the event of not winning the businessman will lose a large part of his investment (and where, therefore, a large risk—in the sense of the theory of games—exists), then the businessman (conforming to the rules about the behavior in gambling) will considerably undervalue a very likely gain as compared to its expected value (which, in the cases of high probabilities, is nearly identical with the value of the gain itself).

We shall see presently how these theorems are modified because of other considerations related to the wealth of the evaluator.

Second, it is clear that a normal man will risk only a limited part of his total wealth to buy chances in games. Call W_M that part of his wealth which a man M is ready to spend on such chances. Then this number, which is dependent on M—i.e., varies from person to person—lies between 0 and 1, usually much closer to 0 than to 1. There exist, however, cases where $W_M = 1$, that is, where M is ready to risk his whole fortune on a single card. For example, in order to continue living he may need an amount exactly six times his whole estate, say, for an expensive medical treatment or to settle a debt of honor, while life under his present circumstances may not appear worth living. In this case, he may well risk all his money in a game of dice, since he feels that if he wins he will have the opportunity to go on living, while if he loses he is not any worse off than before the game when he had already given up the idea of living. An analysis of this example leads to a more exact description of the following characteristics. Apart from M's wealth, let us also consider a number U which measures the mimimum wealth which he desires if he is to go on living. If his wealth W is considerably smaller than U, then $W_M = 1$. If, on the other hand, W is larger than U (and this is true in particular when $U = 0$, which is normally the case, since continuation of life is not tied to ownership of a certain amount of money, while W is greater than 0, since M has at least some money), then we can consider the amount U' which M needs in order to continue his present standard of living, and the amount Z which he desires to own. If W is between U' and Z, then M will only risk amounts short of $W - U'$; that is to say, W_M will not exceed $(W - U')/W = 1 - U'/W$. Furthermore, one can say the following. The smaller M's chances to obtain the desired difference $Z - W$ through work and saving, the closer W_M approaches its maximum value, $1 - U'/W$. At the same time, M will prefer to assume risks in which the possible gain is close to the difference $Z - W$.

The third group of important characteristics describing the behavior of individuals in taking a chance pD is concerned with the dependence of this behavior on the probability p. The first basic rule here is that people, in general, tend to undervalue very small probabilities as we remarked in Section 7. We can assign each individual M a probability $p_M(R)$ such that

M will not risk an amount R whenever the probability is smaller than $p_M(R)$. When the probability p exceeds the value of $p_M(R)$ only slightly, then M will undervalue p inasmuch as for such p he will risk an amount R to obtain a chance pD only if pD is larger than R, or, in other words, when the potential gain D is larger than R/p. At the same time, the difference $D_M(p) - R/p$ between the gain $D_M(p)$ which M, given the probability p, demands in order to risk an amount R, and the expected value, R/p, becomes greater as $p - p_M(R)$ decreases. The relation of $p_M(R)$ to R is usually one of monotonic increase, that is, in order for *M* to risk larger amounts it is usually necessary to provide a higher minimal probability than when he risks smaller amounts. Hence, the smaller p, the larger the difference between actual behavior and that conforming to the expected value of the chance pD. Let us designate, as above, W_M the maximum share of his wealth that M is willing to risk, so that $W_M W$ is the maximum amount of money that he will risk. Then $p_M(W_M W)$ is the largest value which the function $p_M(R)$ can assume with respect to R.

If we add to the fact of the undervaluation of chances where the probabilities are very small (i.e. close to 0) the undervaluation, discussed above, of chances where the probabilities are very high (i.e. close to 1), we see that only chances with medium probabilities are valued in a way which begins to correspond to expected values. Furthermore, chances with certain medium probabilities will be overvalued. The confirmation of this statement can be found in the fact that in lotteries, roulette, etc. risk-takers will knowingly and with pleasure, bet more than the expected value of the game. The difference is usually in the expenses and profits of the organizer of the lottery or the owner of the casino.

The probabilities at which the maximum overvaluation takes place depend on the potential gain, on the wealth of the risk-taker, on the relationship of his fortune to the quantities U′ and Z discussed above, and on other personal circumstances.

As far as the Petersburg game is concerned, the difference between behavior according to mathematical expectation and actual behavior has now been completely explained (in the sense that it is included in a general description of behavior in game situations). Let us define, for every natural number k, a kth approximation game of the Petersburg Game. In this game, a coin is flipped until heads appears, but with the initial stipulation that at most k tosses will be made. B receives from A 2^{n-1} dollars when heads shows for the first time at the nth throw, but if heads does not show up in k tosses, then B receives 2^{k-1} dollars. B's mathematical expectation is thus

$$1/2 + 1/4(2) + 1/2^{n-1}(2^n) + \ldots + 1/2^{k-1}(2^{k-2}) + 1/2^k(2^{k-1}) = (k + 1)/2 \text{ dollars.}$$

In the mth approximating game, B has the mathematical expectation $(m + 1)/2$ and thus a greater expectation than in the kth game if $m > k$. The amount that under normal circumstances B is willing to risk in the mth game, however, is not in the same proportion greater than the amount he is willing to risk in the kth game. In fact, the actual amounts B is willing to risk for the nth game stay, however large n may be, below a certain bound that is not very high, while $(n + 1)/2$ increases with n beyond all bounds. The larger n, the larger, therefore, the discrepancy between mathematical expectation and actual behavior. This is why for the Petersburg Game itself, which is something like a limit of the approximating games, the discrepancy becomes infinite, in that one does not want to risk more than a rather small sum while the mathematical expectation is infinite.

9. THE DESCRIPTION OF PERSONAL TRAITS

It has been repeatedly emphasized in what precedes that the magnitude of the constants and the course of the functions involved in the description of behavior vary from person to person. It may now be pointed out, briefly, that those magnitudes and courses are characteristic of certain circumstances of the evaluator, in particular, of some personal traits of his.

It has been shown, for example, that risks in which the probability is either very small or very large are underestimated while risks in which the probability belongs to a certain middle bracket are overestimated compared with their mathematical expectation. As a result, the actual evaluation of a chance pD (apart from modifications due to other circumstances) corresponds to a multiplication of pD by a factor $\varphi(p)$ which depends upon p in the following way: $\varphi(p) = 0$ for all p less than a certain p_0. Then $\varphi(p)$ continuously increases up to a maximum assumed, say, for p_1. This maximum $\varphi(p_1)$ is greater than 1 (corresponding to the overestimation of the medium probabilities). For large p (that is, for p close to 1), $\varphi(p)$ decreases to 0.

Consider a man M for whom the probability p_0 (above which he is willing to take risks) is smaller than for most other people, while from p_0 on, $\varphi(p)$ increases faster than for other people in similar circumstances. M will be called a *gambler*. It is clear that, in a similar way, rather general statements about the magnitude of some constants and the course of some functions, as considered in the preceding section, characterize some forms of imprudence, of cautiousness, of pettiness, and the like.

10. METHODOLOGICAL REMARKS

This paper has primarily dealt with the Petersburg Game. Even the very formulation of the problem is somewhat different from the traditional one. The fact that the mathematical expectation is infinite has, in itself, not been considered here as paradoxical; and the fact that despite the infinite expectation, no one would risk more than a very small amount in this game, has been considered merely as a striking symptom for a rather general discrepancy between the behavior dictated by mathematical expectation and the actual behavior in games of chance. This subordination of the phenomena occurring in connection with the Petersburg Game to general observations concerning the behavior in evaluating chances and risks, is what has been considered here as an explanation of those phenomena. Since those general statements concern actual behavior, they are purely empirical. Yet it has been shown that one can in this way ascertain some regularities—not *a priori* truths nor logical necessities or logically provable statements, or even comprehensive empirical statements that observation would verify without any exception—still, laws of about the same character and the same degree of certainty as those studied in psychology. While primarily dealing with the Petersburg Game, we clearly realize that what has been done here for this special problem—the way it has been formulated, what has been considered as its explanation, and the general type of treatment it has received—could be useful in dealing with many as yet unsolved problems or confusedly treated questions of social science.

In particular, one cannot overemphasize the importance of the fact that all the results established in Section 8 are purely empirical. For it is in this point that our theory differs, with regard to its claims, from many other theories—even from some that take subjective elements into consideration and have been developed by mathematicians; for those theories stipulate certain formulae, e.g., the logarithmic evaluation formula, as *norms* for behavior, or they *recommend* only behavior according to those formulae. An incidental consequence of those views which should be pointed out, is the fact that no games exist that can be recommended to both partners. For even if the conditions of a game are formulated in such a way that the rules make it advisable for one player to accept it, the rules would stipulate that the other player should not offer the game. Our simple description of the observed behavior with regard to games of chance, without arrogating any normative validity, takes into account the undeniable facts that under certain circumstances both partners gladly enter a game, that people like to take part in keno and in (especially European-type, State-sponsored) lotteries and the like, while the sponsors gladly offer the games since they certainly will make a profit. The fact that, within certain probability brackets and in a certain relation to their fortune, people evaluate risks beyond their mathematical expectation, makes it possible that games of chance, lotteries, and the like (i.e., the exchange of bets for chances) materialize and lead to that satisfaction of both parties which results from every voluntary exchange of commodities.

It should further be stressed that in formulating the general regularities in the behavior with regard to games of chance, we have confined ourselves to rather qualitative statements about increase and decrease, the places of maxima, and the like; and that we have refrained from defining specific functions possessing those qualitative properties. This has three reasons: 1. The observed qualitative properties are shared by many, and even by many simple, functions so that there may be no reason to prefer a particular one to the others. 2. An exact description of actual evaluations is not supplied by any of these functions. 3. As shown in Section 8, the magnitude of constants and the nature of functions involved in the description vary from person to person. A comprehensive function, therefore, would have to include numerous parameters which, for the description of the behavior of a particular person, would have to be replaced by specific numbers. In order to synthesize the most important material, the number of parameters would have to be so large, and the dependence of the functions on these parameters would be so complicated, that the general function would lack any transparency. And what would be gained by setting up such formulae? Could more than general remarks corresponding to the qualitative observations (ascertained directly from the empirical material) be predicated about the functions? Not unless one incorporated in their definition accessories contradicting experience.

We thus see: the mere assertion that a functional connection of some kind exists does not transcend the general qualitative statements that can be made without the introduction of a function symbol. Consider, for example, the statement that the quantity of a commodity G_2 that one is willing to spend for the acquisition of an addition D to his wealth W in a commodity G_1, is a function $f(D, W)$ which, if W is kept fixed, increases with increasing D and, if D is kept fixed, decreases with increasing W, so that, if the function is assumed to have partial derivatives,

$$\partial f(D, W)/\partial W > 0 \quad \text{and} \quad \partial f(D, W)/\partial W < 0.$$

This statement does not in any way transcend the simple verbal statement that a larger addition to a fixed wealth is preferred to a smaller one, while the same addition is more appreciated by the poorer than by the richer man. Those who believe that by writing $f(D, W)$ they introduce a mathematical formulation superior to verbal expressions, are under a misapprehension, just as are those who when seeing the symbol $f(D, W)$ suspect the use of mathematics, which because of ignorance or on the ground of general prejudices, they reject in the social sciences. Special functions, however, as the example of Bernoulli's logarithmic evaluation demonstrates, are not in the theory of value borne out by experience as they are in physics. In Section 8 we therefore have refrained from setting up such formulae and have followed rather the method of the Austrian school of economists—hopefully without being suspected of underestimating mathematics.

It may be appropriate to consider more closely the parallelism between the statements about the evaluation of chances, as developed in Section 8, and the propositions about the evaluation of commodities about which there is no uncertainty. Both express empirical regularities that can be ascertained in one of two ways: by observing the behavior of individuals or by collecting statements of theirs as to how they would act under certain circumstances. In order to obtain data concerning the evaluation of sure commodities one asks[15] a man what quantity, y, of a commodity G_2 he would be willing to spend in exchange for a quantity, x, of a commodity G_1. Comparing the information received from various people, despite many differences in detail, one discovers similarities in the general structure of the data, e.g. monotonic increase and a decreasing rate of increase for most individuals. Interpolating the finite set of data obtained from one person by a curve that is as smooth as possible, one arrives at a twice differentiable function $y = f(x)$ for which $f'(x) > 0$ and $f''(x) < 0$ for each x (without adding by this interpolation procedure to the immediate observations[16]). If one wishes to assemble corresponding material concerning risks one has to ask a man how much he is willing to spend for the acquisition of a chance, of course varying, in so doing, W and D. One can interpolate a function of two variables or, graphically, embed the data in a surface. Here again one observes similarities between the data obtained from various people as to their general course. These similarities can be expressed as general qualitative regularities, some of which have been mentioned in Section 8. One observes occasional exceptions from the regularities concerning the evaluation of the uncertain as well as of the certain. These exceptions are reflected in wrinkles of the surfaces and curves. The peculiarities of an individual in evaluating risks (as well as sure commodities) lend themselves to the description of personal traits. If a man is a collector of a certain type of commodity, this fact is reflected in his demand curve for that commodity in that his evaluation of each new acquisition is relatively high and his evaluation curve, mentioned above, is steeper than that of noncollectors. The miser evaluates

[15] Compare D. Bernoulli, *op.cit.*, and Oettinger *op.cit.*

[16] One asks naturally only for the values of y corresponding to a few values of x and thereby obtains a finite number of points in the plane, which then can be connected by a curve that is as smooth as possible, or simply by a polygon. Even if the interviewed person reports a function $y = f(x)$ or if he constructs a graphical relationship which associates with every x a corresponding value of y, this is naturally the result of a finite number of prior decisions by him to which he applied interpolation. When asking him for the values of y corresponding to a finite number of values of x one should choose these values of x efficiently. That is, one should not necessarily distribute them evenly, but especially in the case of monotonic functions, should ask more questions in those ranges where the values of y may show greater changes.

each addition to his fortune according to its amount.[17] Similarly, the evaluation curves and surfaces for risks reflect imprudence or cautiousness or other traits of the evaluator. This supplies us with a means to endow those vague terms with a certain degree of precision.

Quantitative precision, even in a restricted way, for the dependence of the evaluation of a chance upon its probability and the potential gain can be achieved only in the case of games of chance. In other domains of economic actions, uncertainty also plays a very important role indeed but is rarely capable of being made numerically precise. This is most obvious with regard to general economic and political uncertainty, however important their influence on economic actions may be. If some piece of real estate lends itself only to a special use, say, in connection with a luxury hotel or an armament factory, then its evaluation will largely depend upon the evaluator's views on the economic development of the country or the prospect of war—thus on his views about uncertain circumstances. But even if he can make precise his personal judgment of the likelihood (e.g. by saying: I bet 2:1 that within a year the price index in country A will rise by 10%, or: I bet 1:2 that within two years the countries B and C will be at war with one another) one cannot speak of a probability in a stricter sense. Yet, even in some domains in which numerical probabilities cannot be formulated, one can compare the behavior patterns of various people with each other and with their average. One can for example, observe whether a man is willing to risk a larger part of his fortune for a certain chance than are others in similar circumstances. Two different aspects must be clearly distinguished: a. his evaluation of the uncertainty; b. the risk he is willing to take on the basis of that evaluation. If these two aspects can be separated at all, the first can be described by statements of the form: I bet a:b that the event E will occur; the second, by statements: My betting quotient for obtaining a gain G being a:b, I am willing to risk W′. If, of two men, A estimates the likelihood of a gain almost always higher than B does in similar situations, then A may be called more *optimistic* than B. If A is willing in most cases to risk more than B, who lives in similar circumstances and has the same betting quotients, then A will be called more *daring* or *venturesome* than B. More generally, in cases other than games of chance, the probability p is replaced by the betting quotient a:b, which varies from person to person.[18]

In conclusion, it may be appropriate to point out the connection of the

[17] The empirical character of all these laws about the satisfaction of needs is evident from the fact that a world is imaginable in which most needs would increase with increasing satisfaction. In some cases, this relation may be observed even in our actual world, for example, with regard to narcotics.

[18] Compare the remarks of Buffon, *op.cit.*, that the mathematicians calculate like misers because they set the expectation proportional to the size of the gains. To this

preceding discussion of value theory with studies of the will, as contained in treatises on ethics, psychology, and general philosophy. Those philosophical discussions of the motivation of human actions that I know (especially in studies of the freedom of will) neglect or at least underemphasize the problem of uncertainty; e.g., the statement that actions always follow the course of the strongest motive certainly is false unless in speaking of the stronger motive one takes the probability of the realization into account. In general, however, this point is, at least explicitly, insufficiently stressed. Suppose someone lets me choose between some ticket offering a winning of $1,000 with a probability of 1/1,000 and a ticket which offers a winning of $2,000 with a probability of 1/4,000. The stronger motive certainly is the larger amount. Yet, apart from exceptional cases discussed in Section 8 (where I absolutely need the acquisition of $2,000), I shall choose the ticket promising $1,000 with a probability of 1/1,000. Of importance to our actions, beside the strength of the motive, is thus the probability of achieving the aim and, in cases where this probability cannot be numerically described, our evaluation of the likelihood, as expressed in betting quotients.[19] The second element is altogether different from the first unless in the idea of motivation one explicitly includes the idea of uncertainty.[20]

---

one may add, in a similar vein, that mathematicians calculate in some ranges of probability like daredevils and are cautious in others because they set the expectation proportional to the probability. (This, however, should not be construed as a major criticism of the concept of mathematical expectation which, in any case, does not describe actual behavior in the case of single gains but has a different purpose.)

[19] The general questions concerning the economic behavior under uncertainty are naturally more complicated than the simple games of chance, to which the discussion of this paper is limited. The general question is not one of a single gain with a certain betting quotient, but of various (often continuously varying) gains with different betting quotients. This, incidentally, is also the problem in somewhat complicated games of chance like lotteries.

[20] In another field, however, will-theoretical considerations of philosophers may be useful for the analysis of problems in value theory. Schlick correctly points out in his "Questions Concerning Ethics," Vienna, 1930, that the statement "the actual human behavior occurs always in the direction of the strongest motivation" is not a tautology but an empirical statement and that this point is illustrated by the fact that one may imagine a world where behavior is determined by those motives which are least satisfying. In several discussions about the theory of value, these aspects are presented as if the statement "we prefer the more valuable to the less valuable" were a tautology since the fact that "something is more valuable to us" can only be explicated by the fact that we prefer it. This, in some respects, extreme behavioristic conception (which also underlies, in a more or less vague form, the rejection of value theory as "metaphysics of price theory" as well as the criticism of a supposedly redundant doubling of concepts, etc.), I regard as incorrect. Through self-observations we arrive at the theorems of subjective value theory and, furthermore, establish a rather regular connection between these introspective experiences and the actual behavior that leads to the formation of prices.

CHAPTER 17

Changing Utility Functions

By MAURICE H. PESTON*

Most of the theory of the individual consumer in economics assumes that he has fixed preferences. A question which is rarely discussed is: How are the preferences formed?[1] If it is agreed, however, that preferences do in fact change, this question cannot be ignored. It would seem to be important for positive economics and even more so for welfare economics. In particular, a tax or subsidy may affect not only the constraints confronting a consumer, but also his marginal rate of substitution between commodities. A government may, therefore, wish to know not so much the initial preferences of consumers, but the final ones after a policy has had its full effect.

In dealing with the question of changing preferences, one possibility is to bring into the picture the experience of the consumer. We may regard him as experimenting, learning, and possibly even deteriorating as a decision-making mechanism. The consumer with stable preferences, capable of translating these into relevant choice, may perhaps be interpreted as being in a sort of psychological equilibrium. We may enquire, therefore, what are the conditions for the existence of such an equilibrium. Moreover, moving from the static to the more explicitly dynamic, we may consider a dynamic equilibrium of the consumer who is able to translate new phenomena into stable preferences and choices, and contrast this with an appropriate notion of dynamic disequilibrium.

In this paper I present an extremely simple and specific model of the behavior of the individual consumer. It illustrates the type of theorem that can be proved, and possibly indicates the direction in which further analysis might proceed.

In each period the consumer is assumed, for purposes of illustration, to have a utility function of the Cobb-Douglas variety.

$$U_t = Ax_t^{a_t}y_t^{b_t} \tag{1}$$

Note that the exponents of this function have time subscripts but that the form of the function does not change over time.

The effect of his consumption experience is translated in each period by the consumer into new exponents in the utility function. These

* Queen Mary College, University of London.

[1] Oskar Morgenstern was one of the first economists to raise this as a serious issue. *vid.* "Demand Theory Reconsidered," *Quarterly Journal of Economics*, Feb. 1948 pp. 165–201.

experiences translating functions are assumed to be homogeneous of degree zero in x and y.

$$a_t = f\left(\frac{x_{t-1}}{y_{t-1}}\right) \tag{2}$$

$$b_t = g\left(\frac{x_{t-1}}{y_{t-1}}\right) \tag{3}$$

It is also assumed that those functions are (a) single valued, (b) continuous, and (c) positive and finite for x/y non-negative. It follows that a/b will be a continuous, positive, single-valued function of x/y.

In every period, utility is maximized relative to the constraint

$$p_x x_t + p_y y_t = Z. \tag{4}$$

The first-order condition for a maximum (ignoring the possibility of a corner point) is then

$$\frac{a_t}{b_t} = \frac{x_t}{y_t} \frac{p_x}{p_y}. \tag{5}$$

It is now easy to see that there exists at least one equilibrium ratio a/b in the sense that the x/y obtained by maximizing U subject to the constraint is the same x/y that generates a/b in equations (2) and (3). In other words, there is an equilibrium utility function.

From equation (5), a/b is a homogeneous increasing linear function of x/y. From equations (2) and (3), a/b is also a finite positive function of x/y. These two functions must intersect for finite positive a/b and x/y. They may, of course, intersect more than once so that the equilibrium utility function is not necessarily unique. It may be noted that variations in income z do not affect any of equations (2), (3), or (5). It follows that the equilibrium utility function is independent of income. A shift in relative prices will affect equation (5) so that the equilibrium utility function is not independent of relative prices. This is indicated in Figure 1.

The initial relative prices are given by $P_{x(1)}/P_{y(1)}$, the later relative prices by $P_{x(2)}/P_{y(2)}$. The initial equilibrium positions are e_1 and E_1. The later equilibrium conditions are given by e_2 and E_2. It is impossible to say without investigating the dynamic possibilities whether any increase in P_x/P_y will lead to an increase or decrease in x/y. In Figure 1, a/b increases in both cases. That this is not necessary may be seen from Figure 2.

A tax on commodity x will cause a change in the equilibrium utility function. According to both his initial and final utility functions he is worse off as a result of the tax. The income change which will compensate him for the tax change will vary with the utility function, however. In general, the exactly compensating variation for the original utility function may be too much or too little for the new one.

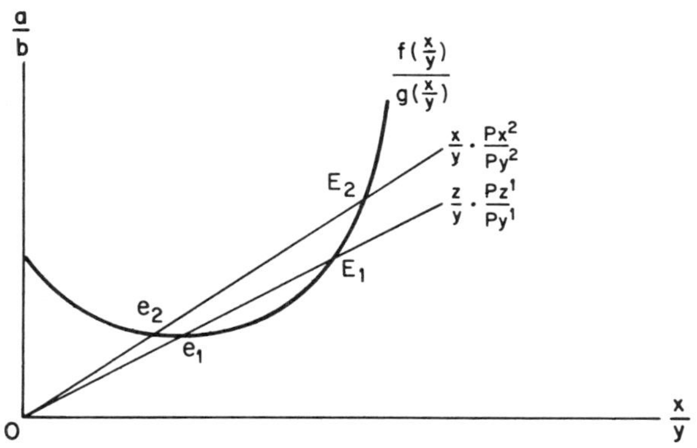

FIGURE 1

Let us now examine briefly the simple dynamics of this model. From equations (2), (3), and (5) we can see that we are dealing with a first order difference equation in x_t/y_t.

$$\frac{x_t}{y_t} = \frac{f\left(\frac{x_{t-1}}{y_{t-1}}\right)}{g\left(\frac{x_{t-1}}{y_{t-1}}\right)} \bigg/ \frac{P_x}{P_y} \tag{6}$$

For stability in the neighborhood of equilibrium we have the usual condition of the cobweb cycle namely that the absolute value of the slope of

$$f\left(\frac{x_{t-1}}{y_{t-1}}\right) \Big/ g\left(\frac{x_{t-1}}{y_{t-1}}\right)$$

must be less than P_x/P_y. Thus e_1, in Figure 1 and e_2 in Figure 2 are stable

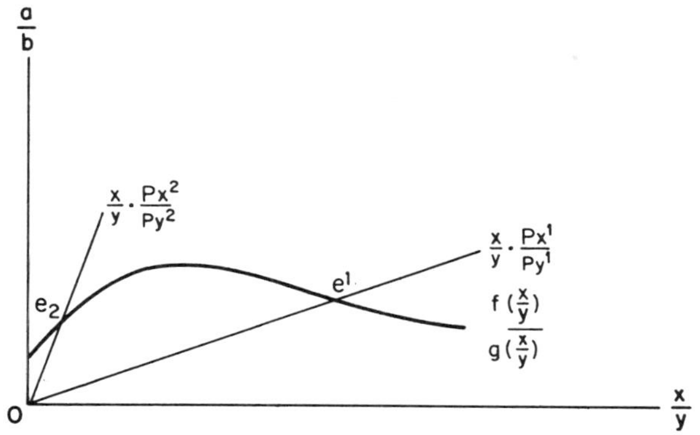

FIGURE 2

while E_1 and E_2 in Figure 1 and e_1 in Figure 2 are unstable. Figure 3 illustrates the simple dynamics of Figure 1.

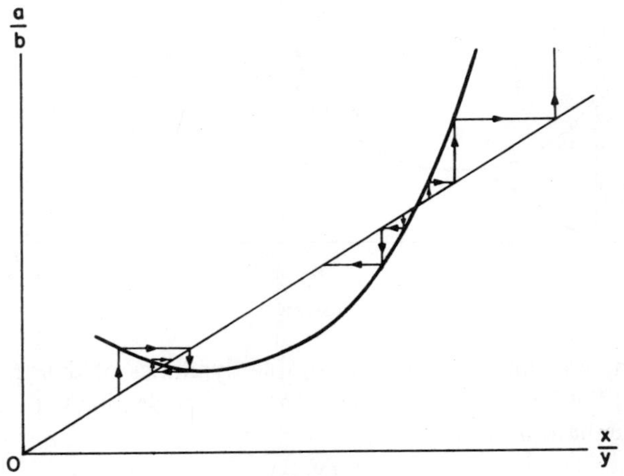

FIGURE 3

It will be noted that the case that arose in Figure 1 where an increase in P_x/P_y leads to increases in a/b and x/y is unstable.

In conclusion, it is clear that many of the results of this model follow from the special assumptions made about the form of equations (1), (2), and (3). Nevertheless, it does illustrate how one may deal with certain obvious questions that arise when changing utility functions are considered. In particular, it emphasizes the possibilities of unstable preferences arising in the small at least, which may be contrasted with the given preference functions of orthodox theory.

CHAPTER 18

Subjective Probability Derived from the Morgenstern–von Neumann Utility Concept

By J. PFANZAGL*

INTRODUCTION

The utility concept developed by Morgenstern–von Neumann in [16] was criticized by several authors because of the assumption that the objective probabilities (in the sense of the frequency interpretation) are relevant for the evaluation of wagers. This attitude is, for example, taken by Suppes and Winet [15, p. 259]: "The interaction between probability and utility makes it difficult to make unequivocal measurements of either one or the other. The recent Mosteller and Nogee experiments [8] may be interpreted as measuring utility if objective probabilities are assumed or as measuring subjective probabilities if utility is assumed linear in money."

There is an obvious way out of this dilemma: to determine one of the two scales independently and to use the evaluation of wagers for the determination of the other scale. So, for example, Davidson and Suppes [6] start with a utility scale based on utility differences and use the evaluation of wagers for the determination of subjective probability. L. J. Savage [13], on the other hand, develops first a measure of subjective (personal) probability, determining utility afterwards.

A rather unusual attempt to measure utility and subjective probability simultaneously was invented by Luce [7].

In [9] and [10] the author presented a theory yielding a cardinal scale of utility and a unique measure of subjective probability *uno actu*. This theory is based on the evaluation of wagers depending on an uncertain event, P (say). In this approach, the subjective probability, relevant for the evaluations of the wager, is however *not* assumed to be identical with the objective probability of P. To be more accurate, the wagers may also be based on events for which an objective probability does not even exist.

The result of this theory was that under rather plausible assumptions, given an uncertain event P, there exists a utility function $U(\cdot)$, monotone increasing and unique up to linear transformations, such that

$$U(xPy) = pU(x) + (1 - p)U(y),$$

xPy being the wager which leads to outcome x, if P obtains, and to

* University of Cologne.

The author wants to thank Mr. R. Borges for his valuable comments and suggestions on earlier versions of this paper.

outcome y, if P does not obtain ($\bar{P}$ obtains). The quantity p is a constant (independent of x and y), uniquely determined by the event P. In [9] and [10] the interpretation of p as "subjective probability of P" was suggested. In the following this suggestion will be outlined in detail: It will be shown that p, regarded as a function of P, has all the properties usually required for probabilities.

This approach can be considered as a generalization of the Morgenstern-von Neumann approach, in so far as our assumptions are less restrictive than the assumptions made by Morgenstern–von Neumann. The essential relaxations are: allowing for a divergence between objective and subjective probability, allowing for dependence between the events, and not requiring the range of the probabilities to be a continuum. For a more detailed discussion of the relation to the Morgenstern–von Neumann approach see [9, pp. 34–35 and p. 53, Punkt 23].

This approach, however, can equally be interpreted as a generalization of the Ramsey approach [11] too. The most important relaxation compared to the Ramsey approach is: admitting arbitrary events and not only events of subjective probability $\frac{1}{2}$, as Ramsey does. Ramsey does not confine himself to measuring utility: Though his procedure of measuring utility uses only events of subjective probability $\frac{1}{2}$, after having a utility scale he defines subjective probability (actually, degree of belief) for arbitrary events, and even for conditional events (pp. 179–180) and shows that probability defined in this way has the usual properties (pp. 181–182). Therefore, the approach taken here can be mainly regarded as an elimination of the special role played by events of subjective probability $\frac{1}{2}$ in the Ramsey approach.

THE ALGEBRA OF EVENTS[1]

In the following, the concept of an "event" will be of fundamental importance. Formally, this concept will be defined by stating the rules for the operations to be performed with events. A possible intuitive interpretation of this concept is that of a proposition which can be either true or wrong.

Let $\mathscr{E}$ be a set of events. The events themselves will be denoted by capital letters like P, Q, ... By $P \cap Q$ (or PQ) we mean the conjunction of the events P and Q, i.e. an event which obtains if both P and Q obtain. By $P \cup Q$ we mean the disjunction of the events P and Q, i.e., an event which obtains if at least one of the events P and Q obtains.

[1] This section may be omitted by readers mainly interested in the measurement of subjective probability and utility and who do *not* insist on a formally rigorous treatment. For the basic concepts used here see Birkhoff [3], especially p. 152 ff. or Sikorski [14].

Both operations $\cap$ and $\cup$ are idempotent, commutative, associative, and mutually distributive.

By $P \subset Q$ we mean that P implies Q, i.e., Q obtains as soon as P obtains. This relation can be formally defined by either of the operations $\cap$ or $\cup$ as follows: $P \subset Q$ if and only if $P \cap Q = P$ ($P \cup Q = Q$). The relation "$\subset$" induces a partial order between the events.

Furthermore, we assume the existence of an impossible event O, defined by $O \subset P$ for any $P \in \mathscr{E}$, and the existence of a sure event E, defined by $P \subset E$ for any $P \in \mathscr{E}$.

Finally, we assume that to each event $P \in \mathscr{E}$ there exists an event $\bar{P} \in \mathscr{E}$, called "negation" or "complement" of P. $\bar{P}$ is the event which obtains if P does not obtain and vice versa. Formally, $\bar{P}$ is characterized by $P \cap \bar{P} = O$ and $P \cup \bar{P} = E$. It can easily be shown that $\bar{\bar{P}} = P$, $\overline{P \cap Q} = \bar{P} \cup \bar{Q}$ and $\overline{P \cup Q} = \bar{P} \cap \bar{Q}$. The mapping $P \to \bar{P}$ induces a dual automorphism of $\mathscr{E}$.

A set $\mathscr{E}$ with the properties stated above is called a Boolean algebra.

A subset $\mathscr{T} \subset \mathscr{E}$ is called an ideal, if it is closed under the operation $\cup$ (i.e., P, $Q \in \mathscr{T}$ implies $P \cup Q \in \mathscr{T}$) and if $P \cap Q \in \mathscr{T}$ for $P \in \mathscr{T}$ and $Q \in \mathscr{E}$.

To any ideal $\mathscr{T}$ there corresponds a congruence relation defined as follows: $P \equiv Q(\mathscr{T})$ if there exists an $I \in \mathscr{T}$ such that $P \cap I = Q \cap I$. As is well known, this congruence relation is a reflexive, symmetric, and transitive relation and has the so called substitution property: $P \equiv Q$ implies $P \cap R \equiv Q \cap R$ and $P \cup R \equiv Q \cup R$. (Then, necessarily, $\bar{P} \equiv \bar{Q}$.) Conversely, in a Boolean algebra any congruence relation with these properties can be defined by an ideal, namely the ideal of the elements congruent O (Birkhoff [3] p. 159).

We now define the concept of a conditional algebra (or implicative algebra, as Copeland [5] says). If event P obtains, a congruence relation in $\mathscr{E}$ is induced. Two events Q', Q'' are congruent, if

$$Q' \cap P = Q'' \cap P, \tag{1}$$

i.e., if—given P obtains—the occurrence of Q' leads to the same event as the occurrence of Q''.

This congruence is induced by the ideal $\mathscr{T}_P$ of events congruent O: $\mathscr{T}_P = \{Q \in \mathscr{E}\colon Q \cap P = O\}$. Obviously $\mathscr{T}_P$ is an ideal (and therefore the relation defined by (1) really a congruence relation).

The conditional event $Q \mid P$ is formally defined as the class of events congruent Q modulo $\mathscr{T}_P$. The conditional algebra $\mathscr{E} \mid P$ is defined as the algebra of congruence classes. $\mathscr{E} \mid P$ is (with respect to all operations) homomorphic to $\mathscr{E}$ with the homomorphism defined by the mapping $\mathscr{T}_P \to O$. Especially $\mathscr{T}_E = \{O\}$ is the ideal inducing an isomorphism from $\mathscr{E} \mid E$ to $\mathscr{E}$.

From the above, we immediately obtain relations like

$$\overline{Q \mid P} = \bar{Q} \mid P \tag{2}$$

$$Q' \mid P \cap Q'' \mid P = (Q' \cap Q'') \mid P \tag{3}$$

as used by Copeland [5] to define conditional events. There is, however, a fundamental difference between this and Copeland's approach in so far as Copeland regards conditional events as elements of $\mathscr{E}$ rather than classes of elements of $\mathscr{E}$.

We now consider conditioning in $\mathscr{E} \mid P$. If $Q \mid P$ obtains, a congruence relation between the elements of $\mathscr{E} \mid P$ is induced which reduces the conditional algebra $\mathscr{E} \mid P$ further to an algebra of equivalence classes denoted by $(\mathscr{E} \mid P) \mid (Q \mid P)$. Let $\mathscr{T}_{Q \mid P} = \{R \mid P \in \mathscr{E} \mid P\colon (R \mid P) \cap (Q \mid P) = O \mid P\}$ be the ideal in $\mathscr{E} \mid P$ generating this congruence relation. If $\mathscr{T}_{Q \mid P}$ is not regarded as a collection of equivalence classes, but as a collection of elements of $\mathscr{E}$, $\mathscr{T}_{Q \mid P}$ forms an ideal in $\mathscr{E}$ which is

$$\mathscr{T}_{Q \cap P} = \{R \in \mathscr{E}\colon R \cap (Q \cap P) = O\}.$$

In this understanding we can write

$$\mathscr{T}_{Q \mid P} = \mathscr{T}_{Q \cap P} \tag{4}$$

The conditional algebra $\mathscr{E} \mid (Q \cap P)$ induced in $\mathscr{E}$ by the ideal $\mathscr{T}_{Q \cap P}$ is isomorphic to the algebra $(\mathscr{E} \mid P) \mid (Q \mid P)$ which is induced in $\mathscr{E} \mid P$ by the ideal $\mathscr{T}_{Q \mid P}$:

$$(\mathscr{E} \mid P) \mid (Q \mid P) = \mathscr{E} \mid (Q \cap P).$$

This relation corresponds to Copeland's axiom $(R \mid P) \mid (Q \mid P) = R \mid (Q \cap P)$.

The proof of (4) is straightforward: For $R \mid P \in \mathscr{T}_{Q \mid P}$, we have $(R \mid P) \cap (Q \mid P) = O \mid P$. On account of (3) we obtain: $(R \cap Q) \mid P = O \mid P$, i.e., $R \cap Q \cap P = O$, so that $\mathscr{T}_{Q \mid P} \subset \mathscr{T}_{Q \cap P}$. By an inversion of the sequence we obtain the opposite inclusion, whence the equality follows.

A special consequence of (4) is that for $Q \subset P$: $\mathscr{T}_{Q \mid P} = \mathscr{T}_{Q}$

THE SPACE OF WAGERS

In order to fix our ideas, we restrict our considerations to a set $\mathfrak{M}$, consisting of different quantities of a specific commodity (e.g., money). Formally, $\mathfrak{M}$ will be assumed to be order-isomorphic to a connected subset of the real numbers i.e., an interval. Wagers of the type xPy, where the outcomes x and y are elements of $\mathfrak{M}$, are called simple wagers. In addition, we consider wagers with simple wagers as outcomes. This set of wagers will be denoted by $\mathfrak{M}(\mathscr{E})$. If the wagers are based on conditional events, given P, we will denote the space of wagers by $\mathfrak{M}(\mathscr{E} \mid P)$.

ORDER-AXIOM. The set $\mathfrak{M}(\mathscr{E} \mid P)$ is completely ordered for any $P \in \mathscr{E}$. (Especially for $P = E$: $\mathfrak{M}(\mathscr{E})$ is completely ordered.)

According to the order-axiom for any two elements of $\mathfrak{M}(\mathscr{E})$ there holds exactly one of the relations $\gtrless$. The relation $\sim$ is reflexive, symmetric, and transitive; the relations $\succ$ and $\prec$ are irreflexive, asymmetric, and transitive. The intuitive interpretation of this order is an order according to utility. The relation $\sim$ is to be interpreted as an equivalence relation (in the sense of equal utility), not as an equality. Such an order relation can, for example, be defined by means of the (objective) probability with which one element of $\mathfrak{M}(\mathscr{E})$ is preferred to another, equivalence (preference) being the case in which this probability equals (is greater than) 1/2. (We remark that we have cases of imperfect as well as cases of perfect discrimination between the wagers: whereas the discrimination between xPy and $x'Py'$ is perfect in case $x < x'$, $y < y'$, it is imperfect in case $x < x', y > y'$.)

The order-axiom has to be supplemented by two other axioms, ensuring that 1) the order between the wagers is connected with the order between the outcomes in a natural way, and 2) that the order induced by the order of $\mathfrak{M}(\mathscr{E})$ in the algebra of events $\mathscr{E}$ is consistent (in the sense that it does not depend on the specific outcomes).

The connection between the order of the wagers and the order of the outcomes is constituted by the so-called:

SURE-THING PRINCIPLE. (a) If the outcome of wager 1) in case P obtains is equivalent to the outcome of wager 2) in case P obtains, and if also the outcome of wager 1) in case $\bar{P}$ obtains is equivalent to the outcome of wager 2) in case $\bar{P}$ obtains, then wager 1) is equivalent to wager 2).

(b) Moreover, if the outcome of wager 1) in case P obtains is preferred to the outcome of wager 2) in case P obtains and if also the outcome of wager 1) in case $\bar{P}$ obtains is preferred (or equivalent) to the outcome of wager 2) in case $\bar{P}$ obtains, then wager 1) is preferred to wager 2), provided P is not regarded as virtually impossible.

The essential content of this axiom is, that (a) the utility of a wager is not changed if an outcome is substituted by an equivalent outcome, (b) the utility of a wager is increased by substituting an outcome by an outcome which is preferred.

The reader will realize, that the above formulation is essentially that of Savage [13, pp. 21–22]. It is somewhat stronger than Samuelson's "strong-independence axiom" [12, p. 672], as it is not confined to even-chance wagers.

The order relation between the wagers xPy and xQy introduces an order relation between the events P, Q by $P \gtrless Q$ according to whether

$xPy \gtrless xQy$, if $x > y$. We have to assume that this order-relation is uniquely determined, i.e., that it does not depend on the specific amounts x and y involved (see Savage, [13], p. 31, P4).

UNIQUENESS-AXIOM. If there exist quantities $x_0 > y_0$ such that $x_0Py_0 \gtrless x_0Qy_0$, then $xPy \gtrless xQy$ for any $x, y \in \mathfrak{M}$, $x > y$.

A special role is played by the events equivalent O, called "almost impossible." (At the end of this paper we will show that these events form an ideal in $\mathscr{E}$, so that we can restrict ourselves to the algebra of equivalence classes modulo this ideal.) The events equivalent to E are called "almost sure." The complement of an almost impossible event is almost sure and vice versa.

This definition of "almost impossible" implies that for any almost impossible event O^* we have $x'O^*y \sim x''O^*y$ whatever x', x'' and y might be. Assume that $x'O^*y \prec x''O^*y$ if $x' < x''$ though O^* is almost impossible. Then we would have a non-archimedean order for wagers based on almost impossible events (and correspondingly for almost sure events), because $x'O^*y' \prec x''O^*y''$ if $y' < y''$ regardless of what x' and x'' might be. The map of $\mathfrak{M}(\mathscr{E})$ onto $\mathfrak{M}$ does not allow for a representation of this order; all elements xO^*y (for any x) are mapped into the same element y. The reader interested in the representation of non-archimedean preferences is referred to the paper by Chipman [4].

$\mathfrak{M}(\mathscr{E})$ contains a subset similar to $\mathfrak{M}$, namely the subset of the wagers xPx, $x \in \mathfrak{M}$. This subset will be denoted by the same letter $\mathfrak{M}$. We require:

DENSITY-AXIOM. $\mathfrak{M}$ is dense in $\mathfrak{M}(\mathscr{E})$, i.e., between any two elements xPy and uQv which are not equivalent, there is an element of $\mathfrak{M}$.

From the density-axiom we immediately obtain that there exists an order preserving mapping of $\mathfrak{M}(\mathscr{E})$ onto $\mathfrak{M}$, which is defined by $xPy \rightarrow m_P(x, y) = \sup\{m\colon m \in \mathfrak{M}, m \precsim xPy\}$. It is easy to show that

$$m_P(x, y) \sim xPy.$$

This means that each equivalence class in $\mathfrak{M}(\mathscr{E})$ contains exactly one element of $\mathfrak{M}$. The element of $\mathfrak{M}$ contained in the equivalence class of xPy is denoted by $m_P(x, y)$.

CONTINUITY-AXIOM. For any event P we have: If $xPy \prec m$, then there exists $x' > x$ such that $x'Py \prec m$.

From these axioms we obtain the following properties of $m_P(x, y)$:

P1: $m_P(x, x) = x$.

This follows immediately from the fact that $xPx = x$.

P2: $m_{\bar{P}}(x, y) = m_P(y, x)$.

This follows immediately from the fact that both wagers $x\bar{P}y$ and yPx are virtually identical: $P \rightarrow y$, $\bar{P} \rightarrow x$ for both wagers.

P3: $m_P(x, y)$ is strictly increasing in both variables, except the case that P is an almost sure or an almost impossible event.

P4: $m_P(x, y)$ is continuous in both variables.

P5: $m_E(x, y) = x$.

P6: If $m_P(x, y) = m_Q(x, y)$ for a special pair $x \neq y$, then equality holds for any $x, y \in \mathfrak{M}$.

Beside wagers based on events of $\mathscr{E}$, we will also consider wagers based on conditional events, i.e., on elements of $\mathscr{E} \mid Q$. By $x(P \mid Q)y$ we denote the wager leading to outcome x if P obtains and to outcome y if P does not obtain, all this under the condition that the event Q obtains. By $m_{P \mid Q}(x, y)$ we mean the amount equivalent to $x(P \mid Q)y$. We remember, that $P \mid Q$ denotes a class of congruent elements. The function $m_{P \mid Q}(x, y)$ is the same for congruent elements, or, to express it in another way: if $\mathscr{E}$ is regarded as the parameter space of $m_P(x, y)$, then the parameter space of $m_{P \mid Q}(x, y)$ is $\mathscr{E} \mid Q$. It immediately follows that $m_{P \mid Q}(x, y)$ has the properties P1–P6 stated above.

By a combined wager such as $(x(P \mid Q)y)Q(u(P \mid \bar{Q})v)$ we mean a wager leading to $x(P \mid Q)y$ if Q obtains, leading to $u(P \mid \bar{Q})v$ if $\bar{Q}$ obtains. The function $m_{P \mid Q}$ enables us to reduce combined wagers to simple wagers:

P7: $(x(P \mid Q)y)Qz \sim m_{P \mid Q}(x, y)Qz$.

The equivalence of these expressions can be seen as follows: If Q obtains, then the first wager leads to $x(P \mid Q)y$, which is by definition in this case equivalent to $m_{P \mid Q}(x, y)$. If $\bar{Q}$ obtains, both wagers lead to z. As both wagers lead to equivalent outcomes, regardless of whether Q or $\bar{Q}$ obtains, they are equivalent according to the sure-thing principle.

As this relation is basic for the results obtained in this paper, it seems fair to draw the reader's attention to the objections against the plausibility of this relation as raised by Allais [2, pp. 524–530] and which can be illustrated by the following example, which is similar to one by Allais.

Let Q be an event of low (subjective) probability, say 0.1 and P an event which obtains with high probability (say 0.9), given Q obtains. Despite the high probability of $P \mid Q$, there will be many persons who, in case Q obtains, prefer the amount 4 to the wager $5(P \mid Q)0$, so that $m_{P \mid Q}(5, 0) < 4$. On the other hand, some of these persons might prefer the wager $5(PQ)0$ to the wager $4Q0$, as PQ has a subjective probability of 0.09, differing only very little from the subjective probability of Q, being 0.1. But this contradicts the sure-thing principle which implies

$$5(PQ)0 \sim (5(P \mid Q)0)Q0 \sim m_{P \mid Q}(5, 0)Q0 \succ 4Q0.$$

(A detailed discussion of Allais' example is given by Savage [13, pp. 101–103].)

Though examples like this are rather plausible, we think that the sure-thing principle must necessarily be adopted by any "rational" person, so that anyone having preferences in contradiction to the sure-thing principle is bound to a revision of his preferences, just as anyone who is intuitively convinced of the truth of a specific mathematical theorem, is bound to a revision of this conviction if a careful logical analysis shows this theorem to be wrong. Both the "sure-thing principle" and the order axioms are to be regarded as normative principles.

P8: $m_Q(m_{P\mid Q}(x, y), m_{P\mid \bar{Q}}(u, v)) = m_P(m_{Q\mid P}(x, u), m_{Q\mid \bar{P}}(y, v))$.

This follows from the identity of the wagers $(x(P \mid Q)y)Q(u(P \mid \bar{Q})v)$ and $(x(Q \mid P)u)P(y(Q \mid \bar{P})v)$. In both cases, $PQ \to x$, $\bar{P}Q \to y$, $P\bar{Q} \to u$, $\bar{P}\bar{Q} \to v$.

P9: $m_Q(m_{P\mid Q}(x, y), y) = m_{PQ}(x, y)$.

This equivalence relation follows from the identity of the wagers

$(x(P \mid Q)y)Qy$ and $xPQy$: In both cases $PQ \to x$, $\bar{P}Q \to y$, $P\bar{Q} \to y$, $\bar{P}\bar{Q} \to y$.

Properties 8 and 9 can also be regarded as specific consequences of the general (normative) principle that there is no illusion due to the way in which a wager is presented, which means that virtually identical wagers presented in different ways are nevertheless judged to be equivalent.

Properties 1–9, which are consequences of the four axioms stated above, form a sufficient basis for the following theorems. Therefore, another possibility would be to use properties 1–9 themselves as axioms.

TWO PRELIMINARY LEMMAS

For the sake of brevity, we will use the following notation: Two functions $U_0(x)$ and $U_1(x)$ are "congruent" in symbols $U_0(x) \backsimeq U_1(x)$, if they differ only by an additive constant.

LEMMA 1. *The representation of a function in the form* $U_0^{-1}(V_0(x) + W_0(y) + k_0)$ *with strictly monotone and continuous functions* U_0, V_0, W_0 *and a constant* k_0 *is essentially unique: If there exists another representation of the same type with functions* U_1, V_1, W_1 *and a constant* k_1, *then:*

$$U_1(x) \backsimeq aU_0(x)$$
$$V_1(x) \backsimeq aV_0(x)$$
$$W_1(x) \backsimeq aW_0(x).$$

PROOF. From

$$U_0^{-1}(V_0(x) + W_0(y) + k_0) = U_1^{-1}(V_1(x) + W_1(y) + k_1)$$

we obtain

$$U_1 U_0^{-1}(\xi + \eta) = V_1 V_0^{-1}(\xi) + W_1{}^* W_0^{*-1}(\eta) \tag{5}$$

with $W_i{}^*(y) = W_i(y) + k_i$, $\xi = V_0(x)$, $\eta = W_0{}^*(y)$. The functional equation (5) is a generalization of the well-known functional equation of Cauchy. As the functions involved in (5) are strictly monotone and continuous, the general solution is (see Aczél [1, pp. 116–117]):

$$\begin{aligned} U_1 U_0^{-1}(t) &= at + b + c \\ V_1 V_0^{-1}(t) &= at + b \\ W_1{}^* W_0^{*-1}(t) &= at + c, \end{aligned}$$

whence the lemma follows immediately.

LEMMA 2. *Conditions* P1–P9 *imply the existence of functions* $U_R(x)$, *defined for all* $R \in \mathscr{E}$, *strictly increasing and continuous for all* $x \in \mathfrak{M}$, *such that:*

$$U_P(m_{P\mid Q}(x, y)) \cong U_{PQ}(x) + U_{P\bar{Q}}(y). \tag{6}$$

The functions U_R *are unique up to linear transformations. If one of the functions, e.g.,* U_E, *is fixed, all the other functions are unique up to an additive constant.*

COROLLARY 1. *This lemma comprises as a special case for* $P = E$

$$U_E(m_Q(x, y)) \cong U_Q(x) + U_{\bar{Q}}(y). \tag{7}$$

PROOF OF LEMMA 2. Since $m_R(x, y)$ is strictly monotone and continuous in x and y, we obtain from P8 according to Aczél [1, pp. 215–216] the existence of strictly monotone and continuous functions U, U_P, $\bar{U}_P$, U_Q, $\bar{U}_Q$, $U_{Q\mid P}$, $\bar{U}_{Q\mid \bar{P}}$ such that:

$$U(m_P(x, y)) = U_P(x) + \bar{U}_P(y) \tag{8}$$

$$U(m_Q(x, y)) = U_Q(x) + \bar{U}_Q(y) \tag{9}$$

$$\left.\begin{aligned} U_P(m_{Q\mid P}(x, y)) &= U_{Q\mid P}(x) + \bar{U}_{Q\mid P}(y) \\ \bar{U}_P(m_{Q\mid \bar{P}}(x, y)) &= U_{Q\mid \bar{P}}(x) + \bar{U}_{Q\mid \bar{P}}(y) \\ U_Q(m_{P\mid Q}(x, y)) &= U_{Q\mid P}(x) + U_{Q\mid \bar{P}}(y) \\ \bar{U}_Q(m_{P\mid \bar{Q}}(x, y)) &= \bar{U}_{Q\mid P}(x) + \bar{U}_{Q\mid \bar{P}}(y) \end{aligned}\right\} \tag{10}$$

We have to face the problem that, in principle, all functions U, U_P, $\bar{U}_P$, U_Q, $\bar{U}_Q$, depend on P and Q. We shall, however, show that the functions U_P, $\bar{U}_P$ can be chosen independent of Q, and U can be chosen independent of both P and Q. This follows from the fact that representations of the form (8)–(10) are essentially unique (Lemma 1).

If we take an event Q^* instead of Q we obtain instead of (8).

$$U^*(m_P(x, y)) = U_P{}^*(x) + \bar{U}_P{}^*(y). \tag{11}$$

Because of Lemma 1, the relations (8) and (11) imply

$$U^*(x) \lesseqgtr aU(x)$$
$$U_P{}^*(x) \lesseqgtr aU_P(x)$$
$$\bar{U}_P{}^*(x) \lesseqgtr a\bar{U}_P(x).$$

Therefore, for given P the functions U, U_P, and $\bar{U}_P$ are unique up to linear transformations. Hence, we can choose U, U_P, and $\bar{U}_P$ independent of Q; that is, a representation (8) based on a specific event Q^* is valid for any event Q.

By applying the same argument to equation (9) we see that for given Q, the functions U, U_Q, and $\bar{U}_Q$ are unique up to linear transformations. Hence we can choose U independent of P and Q.

If U is fixed, all the other functions occurring in (8)–(10) are uniquely determined up to an additive constant, as follows from Lemma 1.

Starting from (10)

$$U_P(m_{Q\,|\,P}(x, y)) = U_{Q\,|\,P}(x) + \bar{U}_{Q\,|\,P}(y),$$

we obtain from $m_{\bar{Q}\,|\,P}(x, y) = m_{Q\,|\,P}(y, x)$ (P2):

$$\bar{U}_{Q\,|\,P}(y) \lesseqgtr U_{\bar{Q}\,|\,P}(y),$$

so that

$$U_P(m_{Q\,|\,P}(x, y)) \lesseqgtr U_{Q\,|\,P}(x) + U_{\bar{Q}\,|\,P}(y).$$

Furthermore, from P9 we obtain:

$$U_{Q\,|\,P}(x) \lesseqgtr U_{PQ}(x),$$

whence relation (6) follows.

For $P = E$ in (6), we obtain:

$$U_E(m_Q(x, y)) \lesseqgtr U_Q(x) + U_{\bar{Q}}(y).$$

Therefore,

$$U(x) \lesseqgtr U_E(x).$$

SOME THEOREMS ON UTILITY AND SUBJECTIVE PROBABILITY

Up to now, conditional events belonging to different conditional algebras, say $\mathscr{E} \mid P$ and $\mathscr{E} \mid P'$, were not assumed to be comparable. The comparison of wagers based on events of one and the same (conditional) algebra was sufficient for the construction of a scale. As soon as scales are given, these can be used for a comparison of events belonging to different conditional algebras:

We define $Q \mid P \gtreqless Q' \mid P'$ if

$$m_{Q\,|\,P}(x, y) \gtreqless m_{Q'\,|\,P'}(x, y) \text{ for all } x, y \in \mathfrak{M},\ x > y.$$

Now we can define independence as follows:

DEFINITION. Q is called independent of P, if $Q \mid P \sim Q \mid \bar{P}$, i.e., if the two congruence classes are of equal probability.

(We note that, in general, to any two congruence classes $Q' \mid P$ and $Q'' \mid \bar{P}$, there exists a uniquely determined common element, namely $Q = (Q' \cap P) \cup (Q'' \cap \bar{P})$. If $Q' \mid P \sim Q'' \mid \bar{P}$, then Q is independent of P.)

Therefore we have

$$m_{Q \mid P}(x; y) = m_{Q \mid \bar{P}}(x, y) \quad \text{for every} \quad x, y \in \mathfrak{M} \tag{12}$$

in the case of independence.

The essential content of this definition is that in the case of independence, the knowledge of whether P or $\bar{P}$ obtains is irrelevant for the evaluation of the wager xQy.

DECOMPOSITION-THEOREM. *If to an event $P \in \mathscr{E}$ there exists an event $Q^* \in \mathscr{E}$ ($Q^* \nsim O, E$) which is independent of P, we have*

$$U_P(x) \mathrel{\substack{\scriptstyle <\\[-2pt] \scriptstyle =\\[-2pt] \scriptstyle >}} s(P)U_E(x). \tag{13}$$

PROOF. By the use of (6), we obtain from (12):

$$U_P{}^{-1}(U_{PQ^*}(x) + U_{P\bar{Q}^*}(y) + k_P) = U_{\bar{P}}{}^{-1}(U_{\bar{P}Q^*}(x) + U_{\bar{P}\bar{Q}^*}(y) + k_{\bar{P}}).$$

Hence, from Lemma 1 follows

$$U_{\bar{P}}(x) \mathrel{\substack{\scriptstyle <\\[-2pt] \scriptstyle =\\[-2pt] \scriptstyle >}} aU_P(x)$$

with the constant a depending on P. Inserting this into (7) (written with P instead of Q), we obtain (13)

COROLLARY 2. *For $P = E$, we obtain from* (13)

$$s(E) = 1 \tag{14}$$

REPRESENTATION-THEOREM. *If* (13) *holds for the events P, PQ and $P\bar{Q}$, we have*

$$U_E(m_{Q \mid P}(x, y)) = \frac{s(PQ)}{s(P)}\, U_E(x) + \left(1 - \frac{s(PQ)}{s(P)}\right) U_E(y). \tag{15}$$

COROLLARY 3. *For $P = E$, we obtain from* (15)

$$U_E(m_Q(x, y)) = s(Q)U_E(x) + (1 - s(Q))U_E(y). \tag{16}$$

PROOF. Inserting (13) and the analogous relations for PQ and $P\bar{Q}$ in (6) we obtain

$$s(P)U_E(m_{Q \mid P}(x, y)) \mathrel{\substack{\scriptstyle <\\[-2pt] \scriptstyle =\\[-2pt] \scriptstyle >}} s(PQ)U_E(x) + s(P\bar{Q})U_E(y). \tag{17}$$

For $x = y$ we obtain from P1:

$$s(P)U_E(x) \mathrel{\substack{\scriptstyle <\\[-2pt] \scriptstyle =\\[-2pt] \scriptstyle >}} [s(PQ) + s(P\bar{Q})]U_E(x).$$

As this congruence relation is necessarily an equation, we obtain that (17) is also an equation and that furthermore

$$s(P) = s(PQ) + s(P\bar{Q}), \tag{18}$$

whence (15) follows.

It suggests itself to interpret the factor $s(Q)$ occurring in (16) as the subjective probability of the event Q. Accordingly, $s(PQ)/s(P)$ ought to be interpreted as the subjective probability of the conditional event $Q \mid P$. Introducing the symbol $s(Q \mid P)$ for this probability, we obtain the common formula

$$s(Q \mid P) = s(PQ)/s(P). \tag{19}$$

The interpretation of $s(Q)$ as the subjective probability of the event Q requires us to show that $s(P)$ has the properties usually required of probabilities.

By taking $P = E$ in (18), we obtain from (14):

$$s(Q) + s(\bar{Q}) = 1. \tag{20}$$

Together with (14), (20) yields:

$$s(O) = 0. \tag{21}$$

Since $m_Q(x, y)$ is strictly increasing in both variables according to P3 except in the case of an almost sure or almost impossible event, we obtain from (16)

$$0 < s(Q) < 1 \quad \text{if} \quad Q \nsim O, E. \tag{22}$$

If P is less probable than Q, according to the definition of P8 we have for $x > y$

$$m_P(x, y) < m_Q(x, y). \tag{23}$$

Together with (16), (23) implies

$$s(P) < s(Q). \tag{24}$$

As mentioned in the introduction, in Ramsey's approach a special role is played by events H, for which x_0 and y_0 $(x_0 \neq y_0)$ exist, such that $x_0Hy_0 \sim y_0Hx_0$, i.e., $m_H(x, y) = m_H(y, x)$. For these events, $s(H) = s(\bar{H})$, so that according to (20): $s(H) = \frac{1}{2}$.

Going back to the definition of independence by (12), we obtain from (15) and (19) a characterization of independence in terms of subjective probability, namely

$$s(Q \mid P) = s(Q \mid \bar{P}). \tag{25}$$

Relations (19) and (25) together with the relation obtained from (18) by interchanging P and Q yield the celebrated

MULTIPLICATION THEOREM FOR INDEPENDENT EVENTS:

$$s(P \cap Q) = s(P)s(Q). \tag{26}$$

This shows clearly that independence is a symmetric property: If Q is independent of P, then also P is independent of Q.

From relations (18) and (20) we can easily obtain

$$s(P \cap Q) + s(P \cup Q) = s(P) + s(Q). \tag{27}$$

If $P \cap Q = O$, (27) reduces to the *addition theorem* on account of (21).

In order to prove (27), we need the relation obtained from (18) by substituting $\bar{Q}$ for P and $\bar{P}$ for Q. This yields:

$$s(\overline{\bar{P} \cap \bar{Q}}) = 1 - s(\bar{P} \cap \bar{Q}) = 1 - s(\bar{Q}) + s(P \cap \bar{Q}) = s(Q) + s(P \cap \bar{Q}).$$

Adding $s(P \cap Q)$ on both sides we obtain (27) by use of (18).

An immediate consequence of the addition theorem is that, if $P \subset Q$, then $s(P) \leq s(Q)$, i.e., P is not of a higher subjective probability than Q. This property together with the addition theorem yields that the events with subjective probability zero form an ideal in $\mathscr{E}$.

SOME CONCLUDING REMARKS

The theory of measuring utility and subjective probability developed here is based on a set of axioms. These axioms are to be considered as normative principles defining the concept of rational behavior rather than as a description of actual behavior. The most important of these axioms are: (1) the sure-thing principle; (2) the principle that there is no illusion due to the way in which an option is presented; (3) the principle that the subjective probability of an event is independent of the outcome of a wager based on this event.

Starting from these axioms, we obtain as a mathematical theorem that there exist two functions, U (defined in the domain $\mathfrak{M}$) and s (defined in the domain $\mathscr{E}$) such that

$$xPy \sim s(P)U(x) + (1 - s(P))U(y).$$

The interpretation of U and s as *utility* and *subjective probability* respectively is strongly suggested by the fact that U is strictly increasing and s has all the properties usually associated with probabilities. Therefore, rational behavior in situations involving risk can be described by means of utility and subjective probability. The "rational person" behaves as if he would evaluate situations involving risk according to their expected utility (the expectation being based on subjective probabilities).

The author desires to stress that the subjective probabilities are a means for describing rational behavior. Nothing more! They can not be used as estimates of the objective probability of an event, or credibility of a statement, or corraboration of a theory.

Another rule for behavior under uncertainty is the so-called minimax principle. What is the relation between the minimax principle and the result obtained above? Actually, the minimax principle is here excluded in advance. Its application is ruled out by the axioms defining the concept of rational behavior. To be more specific, a person acting according to the minimax principle would violate the uniqueness axiom. He would behave as if his subjective probability associated with a specific event (which is neither sure nor impossible) would depend on the outcomes. Each fixed amount $a > 0$ would be preferred to $1P0$, so that a person applying the minimax principle would act as if he would consider P as an almost impossible event. On the other hand, the same event connected with the outcomes 0 and 1, i.e., $0P1$, will again be judged inferior to each fixed amount $a > 0$; so that now a person applying the minimax principle would act as if he would consider P as an almost sure event, thereby violating the uniqueness axiom.

The result that rational behavior can be described by means of expectation might suggest that this theory is of relevance only for decisions related to events which are repeated independently a great number of times. However, this is not the case. The theory is exclusively designed to deal with unique events. (It can be applied to a combination of n independent repetitions by considering this combination as a single event: If $P_1, P_2, \ldots, P_n$ are n independent repetitions of an event P, this can be considered as one single event with the 2^n possible outcomes $P_1' P_2' \ldots P_n'$, where P_i' is either P_i or $\bar{P}_i$.)

Let us finally consider the question whether there is any relationship between subjective probability and objective probability for those events which have objective probabilities associated with them. If we assume that subjective probability is uniquely determined by objective probability (i.e., that events with the same objective probability also have the same subjective probability), and if in addition objectively independent events are also subjectively independent, then subjective probability is equal to objective probability (see [9, p. 35]). We let the sense in which objective probability exists be undetermined. In order to establish the equality of subjective and objective probability, we only need to show that both concepts of probability fulfill the following two conditions: (1) the multiplication theorem holds and (2) the probabilities of complementary events add to unity.

Therefore, under the conditions stated above, subjective probability would merely be an extension of the concept of probability to events for which no objective probability exists. Very important and deep questions remain undiscussed—Why should a rational person have subjective probabilities uniquely determined by objective probabilities? If we consider unique events only, then what are the disadvantages of having subjective probabilities different from the objective probabilities?

REFERENCES

[1] Aczél, J., *Vorlesungen über Funktionalgleichungen und ihre Anwendungen*, Basel, Birkhäuser-Verlag, 1961.
[2] Allais, M., "Le comportement de l'homme rationnel devant le risque: Critique des postulats et axiomes de l'école Americaine," *Econometrica*, Vol. 21, 1953, pp. 503–546.
[3] Birkhoff, G., *Lattice Theory*, New York, Amer. Math. Soc., 1948.
[4] Chipman, J. S., "The foundations of Utility," *Econometrica*, Vol. 28 1960, pp. 193–224.
[5] Copeland, A. H. Sr., "Probabilities, Observations and Predictions," *Proc. 3rd Berkeley Symposium*, Vol. II pp. 41–47.
[6] Davidson, D. and Suppes, P., "A finitistic Axiomatization of Subjective Probability and Utility," *Econometrica*, Vol. 24 (1956), pp. 264–275.
[7] Luce, R. D., "A Probabilistic Theory of Utility," *Econometrica*, Vol. 26 (1958), pp. 193–222.
[8] Mosteller, F. and Nogee, P., "An Experimental Measurement of Utility," *J. of Political Economy*, Vol. 59 (1951), pp. 371–404.
[9] Pfanzagl, J., *Die axiomatischen Grundlagen einer allgemeinen Theorie des Messens*, Schriftenreihe des Stat. Inst. Univ. Wien, Vol. 1, Würzburg Physica-Verlag, 1959.
[10] Pfanzagl, J., "A general Theory of Measurement—Applications to Utility," *Naval Research Logistics Quarterly*, Office of Naval Research, Vol. 6 (1959), pp. 283–294. Reprinted in: *Readings in Mathematical Psychology* Vol. 2, p. 492–502, New York, Wiley, 1965.
[11] Ramsey, F. P., *The Foundation of Mathematics and Other Logical Essays*, New York, Harcourt, Brace & Co, 1931.
[12] Samuelson, P. A., "Probability, Utility and the Independence Axiom," *Econometrica*, Vol. 20 (1952), pp. 670–678.
[13] Savage, L. J., *The Foundations of Statistics*, New York, Wiley, 1954.
[14] Sikorski, R.: *Boolean Algebras*, 2nd ed., Berlin, Göttingen, Heidelberg, New York, Springer-Verlag, 1964.
[15] Suppes, P. and Winet, M., "An Axiomatization of Utility Based on the Notion of Utility Differences," *Management Science*, Vol. 1 (1955), pp. 259–270.
[16] von Neumann, J. and Morgenstern, O., *The Theory of Games and Economic Behavior*, Princeton, Princeton Univ. Press, 3rd edition, 1953.

PART V

Management Science

CHAPTER 19

Some Notes on Oligopoly Theory and Experiments

By DAVID H. STERN*

The market situation in which a small number of sellers (two or more) are competing with each other in selling to a public consisting of many buyers is called oligopoly. In the three sections of this paper we analyze the structure of a certain class of oligopoly markets, catalog several of the hypotheses that have been suggested for the behavior of firms in such markets, and describe the procedure and results of several laboratory experiments aimed at providing evidence concerning the truth of these hypotheses.

The first attempt to analyze oligopoly markets was made by Augustin Cournot in 1838;[1] in this paper we will be concerned with a market model only slightly more general than his. Our model involves five kinds of variables: number of business firms; market price; and, for each firm, quantities produced, unit cost of production, and profits. Other variables, such as advertising, investment in plant and equipment, market power, research and development, and personality differences between businessmen, are covered by a general *ceteris paribus* assumption. Of the included variables, only the quantities produced by each firm are decision variables; hence we call our model the Cournot quantity oligopoly model.[2]

The symbols we will need to specify this model are:

n		Number of sellers.
q_i	$(i = 1, \ldots, n)$	Quantity of product produced and supplied to the market by seller i.
$Q \equiv \sum_{i=1}^{n} q_i$		Total quantity of product produced and supplied to the market by all sellers.

* Unaffiliated, residing in San Diego, California.

This work was supported partly by the Office of Naval Research under Task 047–003 (Management Sciences Research Project), and partly by the Western Management Science Institute under a grant from the Ford Foundation. Reproduction in whole or in part is permitted for any purpose of the United States Government.

The author wishes to thank professors Jacob Marschak and James R. Jackson for their thoughtful suggestions and helpful criticism.

[1] *Researches into the Mathematical Principles of the Theory of Wealth*, translated by N. T. Bacon (New York: Macmillan, 1897).

[2] Good descriptions of the model which emphasize the underlying economic assumptions may be found in Fritz Machlup, *The Economics of Sellers' Competition* (Baltimore: Johns Hopkins, 1952), Chapter 12; E. H. Chamberlin, *The Theory of Monopolistic Competition* (6th edition; Cambridge, Massachusetts: Harvard University Press, 1950), Chapter 3; and William Fellner, *Competition Among the Few* (New York: Alfred A. Knopf, 1949).

$c_i = c_i(q_i)$	$(i = 1, \ldots, n)$	Cost to seller i of producing one unit of product: determined by the quantity he produces. The functions $c_i(q_i)$ are called the cost functions.
$p = p(Q)$		Market price at which one unit of product is sold: determined by the total quantity supplied to the market. The function $p(Q)$ is called the demand function.
$r_i \equiv (p - c_i)q_i$	$(i = 1, \ldots, n)$	Net reward to seller i, defined as his total revenue pq_i less his total cost c_iq_i. We will refer to r_i as i's profit or payoff.
$R \equiv \sum_{i=1}^{n} r_i$		Total profits of all sellers.

We shall speak of both price and cost per unit as measures of *value* per unit of product.

It is assumed here that the demand function and cost functions remain constant over time. We may imagine the market to function thus: In each time period each firm decides how much to produce. Communication with other firms is specifically prohibited, so that the decisions are independent. At the start of any period each firm has this information: the exact nature of the demand function and of all cost functions, its own past quantities produced, its own past profits, past quantities produced by all other firms taken together, past profits of all other firms taken together, and past market prices. Hence, it has at the start of a particular time period t the following five time series for all time periods through $t - 1$: q_i, r_i, $Q - q_i$, $R - r_i$, p. Each firm sends its q_i to market, where, after all firms' product has been received, the total amount Q is presented to the buyers, who are willing to pay p per unit for it. This price p is such that the market is exactly cleared: inventories at the start and end of a period are always zero. An amount r_i as defined above accrues to the firm at the end of the period, after which a new period commences.

PART I

With this background we proceed to our first topic, the analysis of the structure of a certain class of duopoly situations. Duopoly is the special case of oligopoly in which only two firms supply the market. Following in the footsteps of Nash, Mayberry, and Shubik,[3] we shall consider the class of duopoly markets in which the demand function and both cost

[3] John Nash, J. Mayberry, and Martin Shubik, "A Comparison of Treatments of a Duopoly Situation," *Econometrica*, XXI (1953), pp. 141–154.

functions are linear:

$p = a + bQ$	Market price (demand function)
$c_1 = d + eq_1,\ c_2 = f + gq_2$	Unit costs (cost functions)

We shall call the six constants, a, b, d, e, f, and g, *market coefficients*. The profits for the two firms are defined to be

$$r_1 = (p - c_1)q_1,\ r_2 = (p - c_2)q_2.$$

In order for the model to make economic sense we must place these restrictions on the ranges of the variables and the market coefficients:

1. $p \geq 0$	Market price must be greater than or equal to zero.
2. $c_1 \geq 0$ $c_2 \geq 0$	Unit costs must be greater than or equal to zero.
3. $q_1 \geq 0$ $q_2 \geq 0$	Quantities produced must be greater than or equal to zero.
4. $a > d$ $a > f$	In the quantity–value-per-unit plane, the demand curve must intersect the value axis farther from the origin than either of the cost curves.
5. $d > 0$ $f > 0$	The value-intercepts of the cost curves must be positive.
6. $e > b$ $g > b$	The cost curves must have a greater arithmetic slope than the demand curve.
7. $b < 0$	The demand curve slopes down to the right.
8. $ae - bd > 0$, $ag - bf > 0$.	The cost curves intersect the demand curve in the upper right quadrant of the quantity–value plane.

Cournot quantity duopoly can also be regarded as a two-person non-zero-sum game in which the players are the firms, the payoffs are the profits, and choosing a pure strategy means deciding how much to produce. The payoff to Player 1 is

$$\begin{aligned} r_1 &= (p - c_1)q_1 \\ &= [a + b(q_1 + q_2) - (d + eq_1)]q_1 \\ &= (a - d)q_1 + (b - e)q_1^2 + bq_1q_2. \end{aligned} \tag{1}$$

Similarly, Player 2's payoff is

$$r_2 = (a - f)q_2 + (b - g)q_2^2 + bq_1q_2. \tag{2}$$

If the values of the market coefficients are fixed, and if a particular value is chosen for q_1 in equations (1) and (2), then by allowing q_2 to vary we may locate points on a curve in the r_1–r_2 plane or "payoff space" which show the possible pairs of payoffs, given Player 1's quantity strategy. In Figure 1, where the values of a, b, d, e, f, and g are given, the curve X

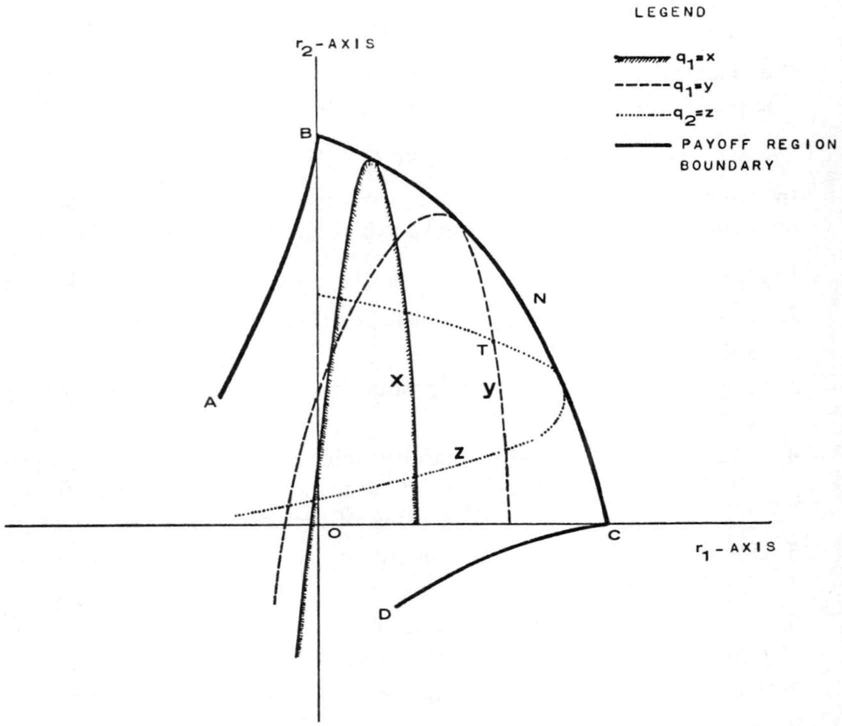

FIGURE 1
Strategy curves and payoff region boundary for an overall-increasing-cost linear duopoly game

shows the payoff combinations possible if Player 1 produces x units of product; Y shows the payoffs possible if he produces y units, where $y > x$. Similarly, by setting $q_2 = z$ and allowing q_1 to vary we get the curve Z. One of the intersections of Y with Z, namely, the one labeled T, indicates the payoffs to the two players if Player 1 produces y units and Player 2 produces z units.

In language drawn from duopoly theory these are iso-quantity curves for Firms 1 and 2 respectively; using language more appropriate to game theory we shall call them *strategy curves*. The set of all strategy curves defined by equations (1) and (2) with the market parameters fixed we shall call the *strategic structure* of duopoly game $D(a, b, d, e, f, g)$.

Next we shall show that the strategic structure of a duopoly game can be specified uniquely by four suitably chosen linear functions of the six market coefficients—that is, that there are two-parameter families of duopoly games which are *strategically equivalent*. Let h and k be arbitrary constants and define:

$$\begin{array}{lll} a' = a/h + k & b' = b/h^2 & q_1' = hq_1 \\ d' = d/h + k & e' = e/h^2 & q_2' = hq_2 \\ f' = f/h + k & g' = g/h^2 & \end{array}$$

To prove $D'(a', b', d', e', f', g')$ strategically equivalent to $D(a, b, d, e, f, g)$ we must show that $r_1' = r_1$ and $r_2' = r_2$. Thus, using equation (1),

$$\begin{aligned} r_1' &= (a' - d')q_1' + (b' - e')q_1'^2 + bq_1'q_2' \\ &= \left(\frac{a}{h} + k - \frac{d}{h} - k\right)hq_1 + \left(\frac{b}{h^2} - \frac{e}{h^2}\right)h^2q_1^2 + \frac{b}{h^2}hq_1hq_2 \\ &= (a - d)q_1 + (b - e)q_1^2 + bq_1q_2 \\ &= r_1. \end{aligned}$$

That $r_2' = r_2$ can be similarly proved. Any four independent functions of the market coefficients which remain invariant under the *h-k* transformation suffice to specify uniquely the strategic structure of a duopoly game. The economic interpretation is most prosaic: If a fixed amount k is added to total cost and to total revenue, nothing is changed; h refers to the units in which quantities are measured—if q_1 is measured in pounds and q_1' in ounces, $h = 16$.

Given the strategic structure, one may determine which payoff pairs (r_1, r_2) are attainable and which are not. The boundary of the *payoff region* is the locus of points satisfying the following Jacobian equation, which defines the envelope of the strategy curves in payoff space:

$$\begin{vmatrix} \dfrac{\partial r_1}{\partial q_1} & \dfrac{\partial r_2}{\partial q_1} \\[2ex] \dfrac{\partial r_1}{\partial q_2} & \dfrac{\partial r_2}{\partial q_2} \end{vmatrix} = 0,$$

and the conditions: $q_1 \geq 0$ and $q_2 \geq 0$.[4]
When our symbols are substituted in the equation and the indicated differentiation is performed the equation becomes

$$\begin{aligned} &2b(b - e)q_1^2 + 4(b - e)(b - g)q_1q_2 + 2b(b - g)q_2^2 \\ &+ [2(a - f)(b - e) + b(a - d)]q_1 + [2(a - d)(b - g) + b(a - f)]q_2 \\ &+ (a - d)(a - f) = 0. \end{aligned}$$

[4] Only in duopoly games with "overall decreasing cost," as defined below, do the requirements that $q_1 \geq 0$ and $q_2 \geq 0$ restrict the payoff region in an important way beyond that defined by the Jacobian alone.

This may be solved for either q_1 or q_2 in terms of the other by using the quadratic formula with the minus sign.[5] For example:

$$\begin{aligned} q_2 = [&-4(b-e)(b-g) - 2(a-d)(b-g) - b(a-f) \\ &- (\{4(b-e)(b-g) + 2(a-d)(b-g) + b(a-f)\}^2 \\ &- 8b(b-g)\{2b(b-g)q_1^2 + [2(a-f)(b-d) + b(a-d)]q_1 \\ &+ (a-d)(a-f)\})^{\frac{1}{2}}]\left[\frac{1}{4b(b-g)}\right]. \end{aligned} \tag{3}$$

With the aid of a computer, as many points on this boundary can be found as are desired by solving equation (3) for various values of q_1 and substituting the pairs of values so obtained in equations (1) and (2). Theoretically, one should be able to express the equation of the boundary in terms of r_1, r_2, and the market parameters by simultaneously solving equations (1)–(3); in practice this is very difficult to do, for the resulting equation is generally of the eighth degree! This is a fine example of how even the simplest problems in economics can become entangled in Gordian knots of mathematical red tape.

In Figure 1, the boundary of the payoff region is labeled *ABNCD*; this curve is convex in the first quadrant and illustrates a typical game with "overall increasing cost" (a term to be defined). Figure 2 shows the boundary for an overall-decreasing-cost game. While the envelope of the strategy curves is *ABCDNEFGH*, the payoffs in the darkened regions *BCD* and *EFG* correspond to negative q_2 and q_1 respectively. The payoff region boundary therefore is *ABDNEGH*. (See footnote 4 above.) The portion of the payoff region boundary which lies in the first quadrant, including any portion which lies on the axes, is called in economics the Edgeworth contract curve. In Figure 1 this is curve *BNC*; in Figure 2 it is *BDNEG*.

What follows is a conjecture which we cannot prove, though it is supported by numerous computer calculations. It seems that not four but three properly chosen functions of the market coefficients suffice to specify uniquely the payoff region of a duopoly game. In other words, there seem to be one-parameter families of duopoly games which have identical payoff regions but different strategic structures. One way to show this analytically would be to show how to express the payoff region boundary in terms of r_1, r_2, and three functions of the market coefficients; but we have found this at least as difficult as expressing the boundary

[5] If the plus sign is used in front of the expression in large parentheses the equation is that of the "threat curve" defined in John Nash, "Two Person Cooperative Games," *Econometrica*, XXI (1953), pp. 128–140, and used by Nash, Mayberry, and Shubik, *op.cit.*, in defining several solutions to duopoly games in which threats and/or side-payments are allowed.

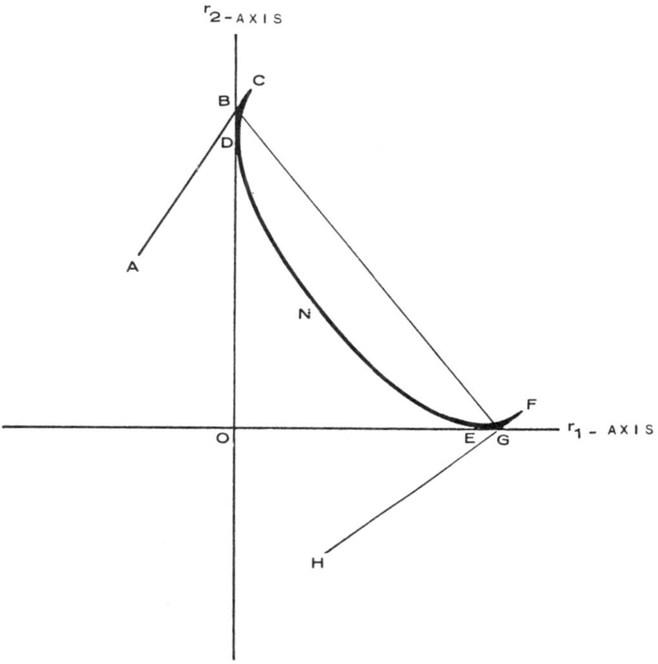

FIGURE 2
Payoff region for an overall-decreasing-cost linear duopoly game

equation in terms of r_1, r_2, and the market coefficients themselves. Anyhow, three functions which seem to be satisfactory are:

$$J = [(b - e)(b - g)]/b^2$$
$$M_1 = [(a - d)^2/[-4(b - e)]$$
$$M_2 = [(a - f)^2]/[-4(b - g)]$$

If the values of these three functions of the market coefficients are fixed, the payoff region is apparently fixed also. One may then specify a particular strategic structure consistent with this payoff region by setting a value for some fourth independent function invariant under the *h-k* transformation, such as e/b. Player 1 may be made to have increasing, constant, or decreasing unit costs according as e/b is greater than, equal to, or less than zero. Once this is done, however, the slope of Player 2's cost curve is determined by the value of J. If $J > 1$, all associated duopoly games possess a property we may call *overall increasing cost*. This means that it is *possible* to choose e/b in such a way that both players have increasing costs: both $e > 0$ and $g > 0$. If $J = 1$, the associated families of duopoly games have *overall constant cost*, which means that whenever $e = 0$, $g = 0$ also. If $J > 1$, the associated duopoly games have *overall decreasing cost*, which means that both players can (though need not) have decreasing costs: if e/b is properly chosen, both $e < 0$ and $g < 0$. In payoff space the Edgeworth contract curve of overall increasing cost games is convex, as in Figure 1; it is a straight line if the game is one of overall constant cost; and it is concave as in Figure 2 if the game has overall decreasing cost.

M_1 and M_2 have simple interpretations, they are the monopoly payoffs of the two players: if Player 2 produces nothing, Player 1's maximum profit is M_1, which he will obtain if he chooses $q_1 = (a - d)/[-2(b - e)]$. Similarly, if Player 1 produces zero, Player 2 can receive M_2 if he chooses $q_2 = (a - f)/[-2(b - e)]$. M_1 is shown as point C in Figure 1, point G in Figure 2; M_2 is shown as point B in both figures.

It is possible to choose a different set of three functions of the six market coefficients to specify the payoff region and a different fourth function to specify the strategic structure. The choice in a particular situation will depend on the purpose of the analysis. It is suggested here that separate specification of payoff regions and strategic structures, given the payoff region, will be useful for many kinds of analysis of market structure as well as for construction of experiments which make use of the technique of gaming.

The duopoly model just examined is at once more and less general than that used in the experiments described in Section III. There we permitted n to be other than two but considered only markets in which the costs were constant and equal for all the sellers:

$$c_i(q_i) = c \qquad (c \geq 0;\ i = 1, \ldots, n)$$
$$p = a + bQ/n \qquad (a > c;\ b < 0).^{6}$$

The inequalities are necessary for the model to make economic sense. We obtain:

$$r_i = (a + bQ/n - c)q_i \qquad (a > c \geq 0 > b;\ i = 1, \ldots, n)$$

The four parameters, a, b, c, and n, completely determine the structure of such oligopoly games.

To simplify analysis one may transform any oligopoly game $\Gamma(a, b, c, n)$ into what we shall call the normalized quantity oligopoly game $\Gamma^*(a^*, b^*, c^*, n)$ where:

$$a^* = 2$$
$$b^* = -1$$
$$c^* = 0$$

[6] This b is not identical with the b in the above duopoly analysis.

The necessary transformations are all linear:

$$q_i^* = -2bq_i/(a - c)$$
$$p^* = 2(p - c)/(a - c)$$
$$r_i^* = -4br_i/(a - c)^2$$

PART II

We are now ready to ask how the players will behave, given that the market structure has been specified. A variety of "solutions" or hypotheses about the quantities the firms will produce have been suggested in the 125 years since Cournot first presented his model. The following outline is not meant be to exhaustive but describes the best-known of these hypotheses; the subordination of hypothesis X to hypothesis Y means that X is a subset of Y. Table I summarizes the solutions for the game Γ^*.

TABLE I

Solutions for the Game Γ^*

Name of Solution	p	q_i	r_i	Q	R
I. Static					
A. Contract curve	1	indeterminate	indeterminate	n	n
1. Cooperative	1	1	1	n	n
2. Maximum market profit	1	indeterminate	indeterminate	n	n
3. Obvious points					
a. Equalizing	1	1	1	n	n
b. Monopoly:	1			n	n
Monopolist		n	n		
Others		0	0		
B. Non-contract curve					
1. No-trade	0	0	0	0	0
2. Competitive	0	2	0	$2n$	0
3. Non-cooperative	$\frac{2}{n+1}$	$\frac{2n}{n+1}$	$\frac{4n}{(n+1)^2}$	$\frac{2n^2}{n+1}$	$\frac{4n^2}{(n+1)^2}$
4. Cournot leadership:	$\frac{1}{n}$			$2n - 1$	$\frac{2n-1}{n}$
Leader		n	1		
Followers		1	$1/n$		
II. Dynamic:					

A. Cournot

$$p_t = 2 - n + \frac{n-1}{2n} Q_{t-1}$$
$$q_{it} = n - \tfrac{1}{2}(Q_{t-1} - q_{i,t-1})$$
$$Q_t = n^2 - \frac{n-1}{2} Q_{t-1}$$

B. Retaliation

$$p_t = 2 - 2n + \frac{n-1}{n} Q_{t-1}$$
$$q_{it} = 2n - (Q_{t-1} - q_{i,t-1})$$
$$Q_t = 2n^2 - (n-1)Q_{t-1}$$

I. STATIC HYPOTHESES. If the game is played repeatedly, the same strategies will be observed each period: the history of the market does not affect the firm's current behavior. All the points in payoff space corresponding to these hypotheses are shown in Figure 3, as is indicated below.

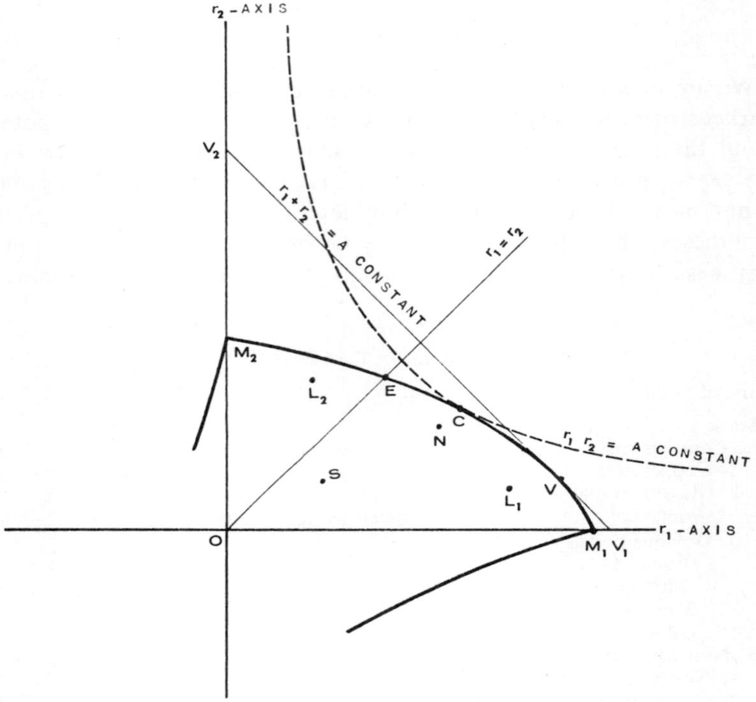

FIGURE 3
Some static solutions for an overall-increasing-cost linear duopoly game

A. *Contract curve solutions.* The contract curve consists of all payoff combinations in which each player is guaranteed at least his "security level" (see I-B-1 below), and such that it is impossible to increase one player's payoff without decreasing that of at least one other player.[7] In Figure 3, the line M_2CVEM_1 is the contract curve.

[7] The term "Pareto-optimal" is used in welfare economics to denote states which satisfy a similar criterion for the society under consideration—states such that no member of society's position can be improved without worsening that of another member. In the present context the contract curve solutions are Pareto-optimal if we think of the sellers alone as constituting the considered society, but if the buyers are also part of society it is a conclusion of welfare economics that the competitive solution (I-B-2 below) is under certain conditions Pareto-optimal.

1. *Cooperative solution.* This maximizes the product of the payoffs;[8] it is shown in Figure 3 as point *C*, where a rectangular hyperbola is tangent to the contract curve, so that r_1r_2 is maximized.

2. *Maximum market profit solution.* This maximizes the sum of the payoffs; von Neumann and Morgenstern originally defined this solution for games in which side-payments among the players were permitted, but in our model these are forbidden. If they were permitted, any payoff combination on Line V_2V_1 could be attained; but in our model the solution is point V only, where V_2V_1, with its slope of -1, is tangent to the contract curve, so that $r_1 + r_2$ is maximized.

3. "*Obvious points.*" Thomas Schelling has suggested that players may look for outcomes which stand out from the general run; they may be suggested by the game structure, by the environment in which the game is played, or by social custom. In the quantity oligopoly game, two solutions of this type are:

a. *Equalizing solution.* This is the contract curve solution which divides the total payoff equally among the players. In Figure 3 it is point *E*, where a line with slope $+1$ and defined by the relation $r_1 = r_2$ intersects the contract curve.

b. *Monopoly solutions.* There are *n* solutions, one for each player, such that one player (the monopolist) maximizes his own profits and all other players produce nothing and receive zero payoff. In Figure 3, M_1 and M_2 are the monopoly solutions for Firms 1 and 2 respectively. These are important in games with overall decreasing cost, for if by using correlated mixed strategies the firms can take turns at being monopolists, they can raise average profits over the long run above those experienced when all firms remain in production. By doing this the firms are in effect enlarging and "convex-izing" the concave payoff region; in Figure 2, the payoff region has been extended to include all points up to and including the straight line *BG*.

B. *Non-contract curve solutions.*

1. *No-trade point.* This is the solution where no firm produces and all payoffs are zero; it is the origin, point 0, in Figure 3. In a more general game-theory context, the no-trade point is defined by the *security levels* of the respective players. A player's security level is the amount he can guarantee himself, no matter what strategies the other players use. In all oligopoly games with zero fixed costs, a player can guarantee himself a payoff of zero by producing nothing.

[8] In general, the cooperative solution is calculated after the payoff axes have been translated so that the no-trade point (see solution I-B-1 below) is at the origin; in oligopoly games this point is already the origin, so that no translation is necessary.

2. *Competitive solution.* Each firm produces according to the rule, "Let marginal cost equal price (average revenue)." In Figure 3 this is point S. In effect the players are assuming that the industry demand curve is not relevant to their production planning; instead they hypothesize that at some price which they expect to obtain in the market they will be able to sell as much as they can produce, and they produce up to the point where marginal cost equals marginal revenue expected under their belief about market demand. (See also Footnote 7.)

3. *Non-cooperative solution.* The players produce according to the rule, "Let marginal cost equal marginal revenue." This is the equilibrium approached after many time periods by all the non-leadership reaction-function hypotheses in which firms attempt to maximize profits; the only specific example given here is the Cournot hypothesis (II-A-1 below). It is unique among the solutions mentioned here in that for any firm, if all the other firms are using their non-cooperative solution strategies, this firm will maximize its own payoff by producing its non-cooperative solution quantity. Thus, once this solution is attained, no firm has any immediate incentive to initiate any unilateral change, even though concerted change could increase everyone's payoff. The solution is point N in Figure 3.

4. *Cournot leadership solutions.* If one player (the leader) knows that the others (the followers) have Cournot reaction functions (see below), he can maximize his profits in the light of this information. In Figure 3, L_1 and L_2 are the leadership solutions where Players 1 and 2 respectively, are the leaders.

II. DYNAMIC HYPOTHESES. In oligopoly theory dynamic hypotheses must consist of two parts: an assertion of how each player uses information available from previous periods to form expectations about current-period strategies of other players, and a statement of the goal the player pursues when he chooses his own strategy, given his expectations.

A. *Reaction function hypotheses.* This class of hypotheses asserts that each player i chooses a quantity q_{it} in period t which maximizes his expected profit $\hat{r}_{it}$ that period. (The hat reminds us that we are considering the amount i expects, not the amount he actually receives.) For him to calculate the expected profit associated with a particular strategy, he must be able to estimate price, $\hat{p}_t$, which depends on his estimate of total market quantity $\hat{Q}_t$, which in turn depends on the amount he expects others to produce, $\widehat{\sum_{j\neq i} q_{jt}}$—for it is clear that $\hat{Q}_t = \widehat{\sum_{j\neq i} q_{jt}} + q_{it}$. Reaction function hypotheses assert that $\widehat{\sum_{j\neq i} q_{jt}}$ is a function of quantities produced by the firms during previous periods.

1. *Cournot hypothesis.* The simplest reaction function is $\hat{\sum_{j \neq i}} q_{it} = \hat{\sum_{j \neq i}} q_{j,t-1}$—Player i assumes that the total quantity the others will produce will be the same this period as last period, and he chooses q_{it} to maximize expected profits.

B. *Retaliatory hypothesis.* This hypothesis asserts that the player who is retaliating does not attempt to maximize his own profits but pursues a different goal. He too assumes that the total quantity the others will produce will be the same this period as last period, but he chooses q_{it} so that no player (with the possible exception of himself) will experience a profit larger than zero.

It is worth repeating that not all dynamic hypotheses assume that firms attempt to maximize expected profits; but that reaction function hypotheses, as used in oligopoly theory, are a subset of those dynamic hypotheses which do assume expected profit maximization. The reader is again reminded that the above classification is not meant to be exhaustive.

PART III

While the *raison d'etre* of oligopoly theory is to explain real-world oligopolistic markets, the models we have just discussed are obviously highly abstract. They assume that no variables enter the market environment or affect firms' behavior except those specifically mentioned; but in the real world demand and cost conditions change, goals quite unrelated to profit maximization or anything else in the model are pursued, variables other than production quantity are subject to decision makers' control. In myriad ways the Cournot quantity oligopoly model's blanket *ceteris paribus* assumption is violated.

Perhaps it is possible in the laboratory to screen out some of the real world's noise; and though the laboratory artificially introduces several variables which in the real economic world are of minor significance, the large degree of control the researcher can exercise in the laboratory may give experimentation an advantage over real world empiricism as a means for testing the alternative behavior hypotheses suggested as solutions to the quantity oligopoly game, The rest of the paper describes several such experiments. They follow a tradition of recent origin in economics; but the area, becoming known as "experimental economics," is attracting more and more followers.[9]

[9] The following four papers report exploratory laboratory studies in oligopoly: Edward Hastings Chamberlin, "An Experimental Imperfect Market," *Journal of Political Economy*, LVI, 2 (April, 1948), pp. 95–108; Austin C. Hoggatt, "An Experimental Business Game," *Behavioral Science*, IV, 3 (July, 1959), pp. 81–96; Heinz Sauermann and Reinhard Selten, "An Experiment in Oligopoly," *General Systems*

Our method was essentially to realize as closely as possible in the laboratory the market conditions and procedure outlined in the first part of this paper. Subjects gathered in a large room were given the experiment instructions to read silently. The experimenter reviewed the instructions with them and the subjects asked clarifying questions. They were then forbidden to communicate with each other. They were paired by a random process into duopoly markets, but a subject did not know the identity of his opponent. Each subject was given his own and his opponent's cost schedules and the market demand schedule. However, to eliminate errors which might result from having subjects compute expected profits themselves from the basic economic data, he was also given the profits matrix (payoff matrix) showing for a variety of his own and his opponent's quantity strategies the profits he and his opponent would receive. At the start of a period he wrote his production quantity for that period on a "Decision and Results Form." A monitor collected the forms and on each subject's form wrote his opponent's quantity strategy, along with market price and both profits figures. The forms were returned and the next period commenced. The number of periods included in a particular experiment was not told in advance in order to avoid "end effects," which result when subjects, expecting the conclusion of an experiment, try during the final periods a wild strategy as a "last fling."

At this point, in order to give more of the flavor of these experiments, we present the complete instructions received by subjects who played the game $\Gamma(12, -1, 10, 2)$, which has the same solutions as Γ^* with $n = 2$ (see Table 1). Afterwards we shall explain how the instructions assisted in creating conditions conducive to valid testing. We note here only that the language of the instructions, which is more technical than that which would be used with persons totally unfamiliar with economics, was directed at the population from which the subjects were drawn—juniors, seniors, and graduate students of the School of Business Administration at the University of California, Los Angeles.

Yearbook, V (1960), pp. 85–114; and Vernon L. Smith, "An Experimental Study of Competitive Market Behavior," *Journal of Political Economy*, LXX, 2 (April, 1962), pp. 111–137.

The most complete series of studies has been carried out by Lawrence E. Fouraker and the late Sidney Siegel; they include these three monographs: S. Siegel and L. Fouraker, *Bargaining and Group Decision Making—Experiments in Bilateral Monopoly* (New York: McGraw-Hill, 1960); L. E. Fouraker, S. Siegel, and Donald L. Harnett, *Bargaining Behavior: I—The Uses of Information and Threat by Bilateral Monopolists of Unequal Strength* (University Park: Pennsylvania State University, 1961); and L. E. Fouraker and S. Siegel, *Bargaining Behavior: II—Experiments in Oligopoly* (University Park: Pennsylvania State University, 1961).

The present author wishes to call the reader's attention to his own monograph in a related field: David H. Stern, *Bargaining Experiments: An Exploratory Study*, Management Sciences Research Project Research Report No. 66 (Los Angeles: University of California, 1960).

INSTRUCTIONS FOR EXPERIMENT SUBJECTS

1. In these experiments in which you have been asked to take part, we are trying to examine the behavior of businessmen in markets. You are going to be in the role of a businessman, and we are going to provide you with a simulated business environment in which you will be able to make certain business decisions. We are trying to test certain theories which predict the sort of decisions you will make. These theories apply only to certain rather abstract kinds of business worlds -- abstract in the sense that many things which you could do in the real business world you cannot do in this simulated business world. In fact, we are eliminating all but its most essential features.

2. Our technique here is closely akin to a physicist's in conducting his experiments. A physicist tries to exclude or hold constant all the factors which might vary and distort the results of his experiment.[10] In particular, he tries to hold constant all of the factors which his theory does not specifically account for. In the same way we are excluding from our experiment all the factors which our theories do not account for: advertising, investment in plant and equipment, assets of the firms and their distribution in various ways, research and development, and psychological interaction between different businessmen with different personalities -- to name but a few. This experiment includes directly only the following variables:
    1. Time
    2. You (the businessman) and your competitor
    3. Market price
    4. Unit costs of production for each businessman
    5. Quantities produced by each businessman
    6. Profits of each businessman

3. In this experiment, you and one competitor are producers competing with each other in the same market for several consecutive time periods which we shall call "weeks". The product you and your competitor produce is called "X" and we shall measure it in tons. Costs of production are the same for you and your competitor: ten dollars per ton. There is a production capacity limit on both you and your competitor: neither you nor he can produce more than 3.6 tons per week.

4. The X you produce and the X he produces are identical in every respect; there is no brand preference or customer loyalty on the part of the X-buying public. Also you both produce X for an "isolated market" in the sense that there is no other product, either in existence now or in potential existence, which will compete with X or which will stimulate its sales. (That is, if X were butter, there is neither such a product as margarine nor one such as bread.)

5. The market demand for X is fixed and it is known to both you and your competitor. That means that the size of the market neither grows nor shrinks from week to week, and that the public's tastes remain the same throughout the weeks included in the experiment. Note that there is no demand for your X or for your competitor's X, as such; there is a fixed market demand for X. (You see that we are using the phrase "fixed market demand" the way economists use it -- as a fixed relationship between market price and market quantity.) The market demand to be used in this experiment is shown in Table II; graphically it appears as in Figure 4. The table and figure also show the profit per ton associated with each total market production quantity.

[10] We felt that this was not the place to discuss randomization!

Table II
Market Price and Unit Profit or Loss
for Various Total Productions

TOTAL NUMBER OF TONS OF X SUPPLIED TO THE MARKET BY YOU AND YOUR COMPETITOR	PRICE THE PUBLIC WILL PAY FOR IT, SO THAT THE MARKET IS EXACTLY CLEARED	PROFIT PER TON (OR LOSS WITH MINUS SIGN) *
0	$ 12.00	$ 2.00
.2	11.90	1.90
.4	11.80	1.80
.6	11.70	1.70
.8	11.60	1.60
1.0	11.50	1.50
1.2	11.40	1.40
1.4	11.30	1.30
1.6	11.20	1.20
1.8	11.10	1.10
2.0	11.00	1.00
2.2	10.90	.90
2.4	10.80	.80
2.6	10.70	.70
2.8	10.60	.60
3.0	10.50	.50
3.2	10.40	.40
3.4	10.30	.30
3.6	10.20	.20
3.8	10.10	.10
4.0	10.00	0
4.2	9.90	- .10
4.4	9.80	- .20
4.6	9.70	- .30
4.8	9.60	- .40
5.0	9.50	- .50
5.2	9.40	- .60
5.4	9.30	- .70
5.6	9.20	- .80
5.8	9.10	- .90
6.0	9.00	-1.00
6.2	8.90	-1.10
6.4	8.80	-1.20
6.6	8.70	-1.30
6.8	8.60	-1.40
7.0	8.50	-1.50
7.2	8.40	-1.60

* Production cost is $10.00 per ton for you and for your competitor

6. You will be able to decide each week how much you wish to produce that week, and your competitor will decide how much he wishes to produce. You will both make your decisions independently, without any communication. In fact, you will not even know which of the other subjects is your competitor, and he will not know who you are. You will write your decision

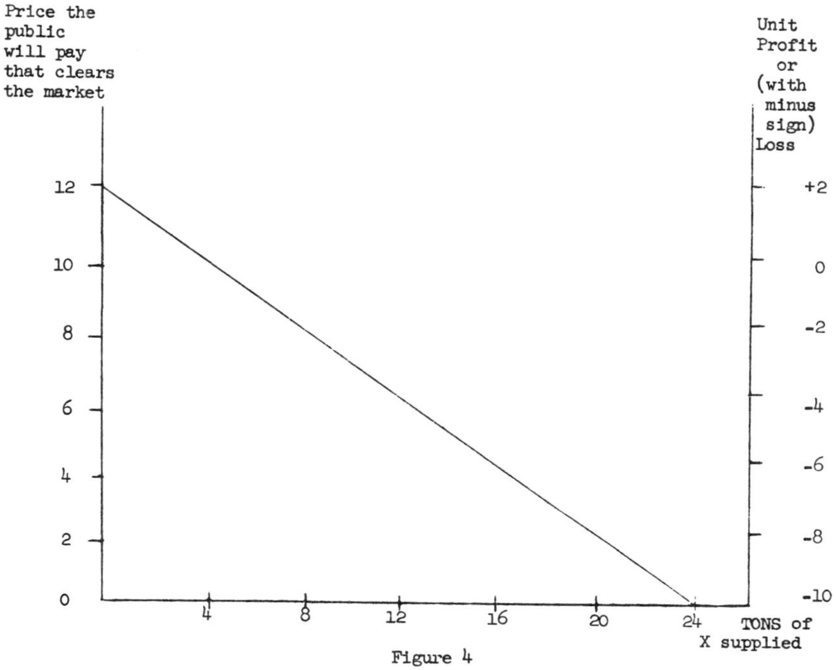

Figure 4

in column 4 of DECISION AND RESULTS FORM like that shown in Figure 5. You and your competitor will then give these forms to a laboratory assistant who will write your competitor's production decision in column 6 and return the form to you. You will then be able to fill in column 2 (market price), column 3 (profit per ton), and columns 5 and 7 (your own and your competitor's total profit) by referring to Table II. The laboratory assistant will inform your competitor of your production decision and he will do likewise. Then you will both repeat the procedure for the next week. The experiment will simulate many weeks, though not necessisarily 25 (as the DECISION AND RESULTS FORM might lead you to believe). Once everyone is familiar with the procedure, and experiment "week" should require between two and five real minutes.

7. An example may make the procedure clearer. Suppose in the first week you decide to produce 1.8 tons of X and your competitor independently decides to produce 2.5 tons. Your DECISIONS AND RESULTS FORM will look like this when you hand it to the laboratory assistant:

(1) WEEK	(2) MARKET PRICE	(3) UNIT PROFIT	(4) TONS OF X YOU PRODUCE	(5) YOUR TOTAL PROFIT	(6) TONS OF X YOUR COMPETITOR PRODUCES	(7) COMPETITOR'S TOTAL PROFIT
1			1.8			
2						

The laboratory assistant writes "2.5" in column 5 and returns the form to you. You now know that total production was 4.3 tons. Table II does not give the exact price, but you may interpolate to find that the market

SUBJECT______________

DECISIONS AND RESULTS FORM

(1) WEEK	(2) MARKET PRICE	(3) UNIT PROFIT	(4) TONS OF X YOU PRODUCE	(5) YOUR TOTAL PROFIT	(6) TONS OF X YOUR COMPETITOR PRODUCES	(7) COMPETITOR'S TOTAL PROFIT
1						
2						
3						
4						
5						
6						
7						
8						
9						
10						
11						
12						
13						
14						
15						
16						
17						
18						
19						
20						
21						
22						
23						
24						
25						

Figure 5

price is $9.85 and the per ton loss is fifteen cents. Your total profit is 1.8 times - $.15, or $ -.27; while your competitor's total profit is 2.5 times - $.15, or - $.38 (rounded off to the nearest cent). After you have entered this information on your DECISION AND RESULTS FORM its first line will look like this:

(1) WEEK	(2) MARKET PRICE	(3) UNIT PROFIT	(4) TONS OF X YOU PRODUCE	(5) YOUR TOTAL PROFIT	(6) TONS OF X YOUR COMPETITOR PRODUCES	(7) COMPETITOR'S TOTAL PROFIT
1	9.85	-.15	1.8	-.27	2.5	-.38

Your competitor will also receive information about your decision, and when he is finished, the first line of his DECISION AND RESULTS FORM will look like this:

1	9.85	-.15	2.5	-.38	1.8	-.27

Then the next week will commence: you will record your production decision in line 2, column 4, hand it to the laboratory assistant, and so forth.

8. To save you time in making production decisions, we have prepared a COMPUTING TABLE (Table III) which shows directly the profits resulting from many of the possible combinations of production quantities of yourself and your competitor. To find the profits you and your competitor will make if you produce 2.2 tons and he produces .8 tons, read across in the row numbered 2.2 and down in the column numbered .8. Your profit in cents is the upper number and his is the lower. The row numbered 2.2 shows pairs of profits for several possible outputs of your competitor ranging from 0 to 4 tons, given that you produce 2.2 tons. The column numbered .8 shows pairs of profits for possible outputs of yours between 0 and 4 tons, given that your opponent produces .8 tons. In this example, where your production is 2.2 and his .8, your profit is $1.10 and his is $.40.

9. Remember that your profit is the top number and your competitor's the bottom; and that the numbers in a row show profits with your production held constant, while those in a column show profits with your competitor's production held constant. The purpose of the COMPUTING TABLE is to save you much calculating involving quantities, prices, and costs, which you would need to do to make informed decisions.

10. An essential requirement in the design of this experiment, as has been pointed out, is that there be no communication between subjects during the experiment. Even talking about the experiment, or making unplanned exclamations to no one in particular may inadvertently affect the results Ideally, this experiment would be conducted with each subject in a separate cubicle in which he could neither see nor hear other subjects. Since facilities are not available for this, we meet in a large room and you sit far away from the other subjects. The rule about talking is that you must not talk with anyone about the experiment, either during the sessions or outside, and you should make a special point of refraining from exclamations and outbursts during the experiment.

11. On the application which you filled out, the specification of pay was left vague. This was done on purpose. We intend to relate your real money earnings as a subject to the profits which you succeed in making in our experiments. The "exchange rate" of real earnings for experiment profits will be announced in advance of each experiment. In the first experiment, if you are an average businessman, you may expect to earn about $1.50 per hour. However, if you are a very successful businessman you may come away with much more than $1.50 per hour; while if you are not, you may leave with less. But the amount you will earn will indeed depend on your success in the business experiments. Note especially that the amount you earn does not depend in any way on your rating with respect to the other subjects, but that it is determined by an absolute exchange rate. Thus it pays you (literally) to make your experiment profits as large as possible. Payment will be made after all the experiments are completed.

TABLE III

COMPUTING TABLE

AMOUNT YOU PRODUCE	AMOUNT YOUR COMPETITOR PRODUCES: 0	.2	.4	.6	.8	1.0	1.2	1.4	1.6	1.8	2.0	2.2	2.4	2.6	2.8	3.0	3.2	3.4	3.6
0	0 0	0 38	0 72	0 102	0 128	0 150	0 168	0 182	0 192	0 198	0 200	0 198	0 192	0 182	0 168	0 150	0 128	0 102	0 72
.2	38 0	36 36	34 68	32 96	30 120	28 140	26 156	24 168	22 176	20 180	18 180	16 176	14 168	12 156	10 140	8 120	6 96	4 68	2 36
.4	72 0	68 34	64 64	60 90	56 112	52 130	48 144	44 154	40 160	36 162	32 160	28 154	24 144	20 130	16 112	12 90	8 64	4 34	0 0
.6	102 0	96 32	90 60	84 84	78 104	72 120	66 132	60 140	54 144	48 144	42 140	36 132	30 120	24 104	28 84	12 60	6 32	0 0	-6 -36
.8	128 0	120 30	112 56	104 78	96 96	88 110	80 120	72 126	64 128	56 126	48 120	40 110	32 96	24 78	16 56	8 30	0 0	-8 -34	-16 -72
1.0	150 0	140 28	130 52	120 72	110 88	100 100	90 108	80 112	70 112	60 108	50 100	40 88	30 72	20 52	10 28	0 0	-10 -32	-20 -68	-30 -108
1.2	168 0	156 26	144 48	132 66	120 80	108 90	96 96	84 98	72 96	60 90	48 80	36 66	24 48	12 26	0 0	-12 -30	-24 -64	-36 -102	-48 -144
1.4	182 0	168 24	154 44	140 60	126 72	112 80	98 84	84 84	70 80	56 72	42 60	28 44	14 24	0 0	-14 -28	-28 -60	-42 -96	-56 -136	-70 -180
1.6	192 0	176 22	160 40	144 54	128 64	112 70	96 72	80 70	64 64	48 54	32 40	16 22	0 0	-16 -26	-32 -56	-48 -90	-64 -128	-80 -170	-96 -216
1.8	198 0	180 20	162 36	144 48	126 56	108 60	90 60	72 56	54 48	36 36	18 20	0 0	-18 -24	-36 -52	-54 -84	-72 -120	-90 -160	-108 -204	-126 -252
2.0	200 0	180 18	160 32	140 42	120 48	100 50	80 48	60 42	40 32	20 18	0 0	-20 -22	-40 -48	-60 -78	-80 -112	-100 -150	-120 -192	-140 -238	-160 -288
2.2	198 0	176 16	154 28	132 36	110 40	88 40	66 36	44 28	22 16	0 0	-22 -20	-44 -44	-66 -72	-88 -104	-110 -140	-132 -180	-154 -224	-176 -272	-198 -324
2.4	192 0	168 14	144 24	120 30	96 32	72 30	48 24	24 14	0 0	-24 -18	-48 -40	-72 -66	-96 -96	-120 -130	-144 -168	-168 -210	-192 -256	-216 -306	-240 -360
2.6	182 0	156 12	130 20	104 24	78 24	52 20	26 12	0 0	-26 -16	-52 -36	-78 -60	-104 -88	-130 -120	-156 -156	-182 -196	-208 -240	-234 -288	-260 -340	-286 -396
2.8	168 0	140 10	112 16	84 18	56 16	28 10	0 0	-28 -14	-56 -32	-84 -54	-112 -80	-140 -110	-168 -144	-196 -182	-224 -224	-252 -270	-280 -320	-308 -374	-336 -432
3.0	150 0	120 8	90 12	60 12	30 8	0 0	-30 -12	-60 -28	-90 -48	-120 -72	-150 -100	-180 -132	-210 -168	-240 -208	-270 -252	-300 -300	-330 -352	-360 -408	-390 -468
3.2	128 0	96 6	64 8	32 6	0 0	-32 -10	-64 -24	-96 -42	-128 -64	-160 -90	-192 -120	-224 -154	-256 -192	-288 -234	-320 -280	-352 -330	-384 -384	-416 -442	-448 -504
3.4	102 0	68 4	34 4	0 0	-34 -8	-68 -20	-102 -36	-136 -56	-170 -80	-204 -108	-238 -140	-272 -176	-306 -216	-340 -260	-374 -308	-408 -360	-442 -416	-476 -476	-510 -540
3.6	72 0	36 2	0 0	-36 -6	-72 -16	-108 -30	-144 -48	-180 -70	-216 -96	-252 -126	-288 -160	-324 -198	-360 -240	-396 -286	-432 -336	-468 -390	-504 -448	-540 -510	-576 -576

12. The experiment director will review these instructions with you after you have read them in order to be sure all is clear. If there is anything you do not understand about the experiment or the method of payment, he will answer your questions at that time.

Real-world oligopoly situations offer the players large monetary rewards for doing well; one would expect that even if an individual did not attempt to maximize profits, he would weight profits heavily in his criterion function. Since on a normal experiment budget it is possible to offer subjects only token payment, other aspects of the laboratory environment are used to aid in reinforcing the profit motive. Efforts in this direction commence with the first sentence of the instructions, where the subject's role as an entrepreneur is emphasized in order to assist in creating a basic attitude of being in business, for it is felt that to the extent that a subject thinks of himself as being in business, pursuit of profits should seem a reasonable goal: whereas if subjects feel they are "only playing a game" or "only taking a test," they might more easily be tempted to adopt alternative goals which in the real world are probably of less relative importance—goals such as beating the opponent, stalemating the opponent, or, worst of all, confounding the experimenter.

However, the chief device used for profit-motive reinforcement was the payment plan described in paragraph 11 of the Instructions. Subjects were told in advance that they would be paid for their services in direct proportion to their experiment profits, and the exchange rate between game money and real money was also announced in advance. Had the exchange rate not been announced, subjects might have concluded that there was a fixed amount of money to be distributed to them according to their relative scores. A subject who believes this finds beating his opponent a reasonable goal, for diminishing the latter's share of the pie increases the subject's own share. It was to discourage subjects from adopting this belief that it was strongly emphasized that absolute and not relative profits would determine the final payment. In the last analysis, though, if it is a truth that "money talks," it is equally true that lack of money is silent; subjects' comments confirm our suspicion that the desire to strive for profits was very much less in the laboratory than one would expect to find in the real world.

The mechanics for simulating the market procedure described above are explained in paragraphs 5–7 of the Instructions. Note that many real-world complications are eliminated in the laboratory merely by experimenter's edict. (See paragraphs 1–4.) For example, brand preference on the part of the public, customer loyalty to sellers, product substitution and complementarity are disposed of in paragraph 3; changes in consumer preferences and growth of the market in paragraph 4.

More conscious effort was required in meeting the condition that no collusion or communication be permitted. As mentioned earlier, subjects were assigned membership in a market in advance of the experiment by means of a random selection procedure. However, it was important—especially since all subjects were fellow students—that no one knew which market he was in, for it was felt that subjects who were friends might tend to cooperate if they knew they were competing in the same market. In this respect the laboratory conditions were less than ideal because the subjects were seated in the same large room instead of individual cubicles. For this reason it was essential to prohibit talking of any kind and to caution particularly against sudden exclamations which might reveal to a clever subject who his opponent was. Paragraphs 6 and 10 of the Instructions were aimed at enlisting the cooperation of the subjects themselves to counter this physical disadvantage in the laboratory conditions. In a questionaire administered after some of the experiments, subjects were unable to identify correctly their opponents with greater accuracy than chance would permit; this is circumstantial evidence that the non-communication assumption was not violated.

As was pointed out earlier, there may be *ceteris paribus* assumptions that do not figure prominently in the traditional presentations of oligopoly theory but which become important in the laboratory environment, an example being the use of the profits matrix to ensure that subjects be equally able to compute the profits associated with various quantity strategies—for choosing an optimal strategy implies knowing all possible consequences of all possible strategies. The use of this matrix (Table 3) is explained to subjects in paragraphs 8–9 of the Instructions.

A few words are in order now about the sample and the population from which it was drawn. Ideally the experiment would have used businessmen as subjects, for it is they whose behavior in oligopoly situations is the subject of theory.[11] Businessmen were not available, but—as is common in much psychological experimentation—students were. It was felt that of the group of all students, business administration majors would most closely approximate the desired population. The sample size was very small, ranging from eight to twelve in several experiment sessions. Because the sample was so small and because it was taken from a population other than businessmen, we caution the reader that we regard the results as exploratory and suggestive rather than conclusive. Also because of the small sample size, it was felt that it would not be

[11] For a different interpretation see Sauermann and Selten, *op. cit.*

practical to attempt to conduct auxiliary measurements in order to correlate observed oligopoly behavior with differences between subjects with respect to intelligence, personality factors, previous experience in business, economic status (which might be related to the subject's need to maximize profits), or other subject variables.

We turn at last to the results. The most striking duopoly data were produced in a pilot study by four subjects who made output decisions in two duopoly markets for twenty-five periods ("weeks"). The particular game they played was $\Gamma(120, -3.158, 60, 2)$; for this game the cooperative output is $q_i = 9.5$, the non-competitive output is $q_i = 12.667$, and the competitive output is $q_i = 19$.

TABLE IV

Number of times q_i was chosen in each of five five-period groups

Periods	5	8	9	10	11	12	13	14	q_i 15	16	17	18	19	22	25	Number of observations in group
1–5	1	2	2	0	0	2	1	1	1	1	1	6	0	1	1	20
6–10	0	0	0	0	1	1	4	3	3	1	2	5	0	0	0	20
11–15	0	0	1	2	1	1	5	2	2	1	0	4	1	0	0	20
16–20	0	1	0	10	1	5	2	0	0	0	0	1	0	0	0	20
21–25	0	0	0	4	0	2	9	2	3	0	0	0	0	0	0	20
1–25	1	3	3	16	3	11	21	8	9	3	3	16	1	1	1	100

The bottom line of Table IV shows the frequency distribution of the 100 quantities produced by the four subjects in these twenty-five time periods; it is a trimodal distribution. Furthermore, the first two modes approximate outputs of theoretical significance. No quantity choice is closer to the cooperative output ($q_i = 9.5$) than the 10-unit strategy which captured 16 percent of all observed decisions; and 12 and 13, the outputs on either side of the non-cooperative output ($q_i = 12.667$), acquired 32 percent of the choices. The mode at $q_i = 18$ results from a baby-blunder in the construction of the pilot study—the computing table (comparable to Table III) showed integral outputs from zero up to only 18 units; the absence of pre-calculated profits for these larger quantity strategies almost certainly was a psychological barrier to their more frequent use. In effect, the range of responses was truncated at 18: a decision to produce 18 units may therefore have represented the desire to produce some larger amount. Two static solutions—the competitive and the Cournot leadership—involve a firm's producing 19 units, and the dynamic "retaliation" hypothesis would frequently dictate outputs in excess of 18. The mode at 18, we suggest, is due to thwarted retaliatory, competitive, and leadership motives; had the computing table extended to higher outputs, there might have been a mode at 19, or the choices might have spread out over a wider range—we cannot know which.

The rest of Table IV shows observed frequency of quantity choices by successive groups of five time periods each. Here is demonstrated another interesting behavior pattern, namely, that subjects learn to avoid competing and retaliating. In periods 1–5, eight of the 20 choices (40%) were competitive or retaliatory (18 units or more), while none were at the cooperative output and only three at the non-cooperative output. By the last five periods the competitive and retaliatory outputs were not observed at all, and instead 75 percent of the choices were either cooperative (four choices) or non-cooperative (eleven choices). In the previous five periods even more were cooperative (ten) than non-cooperative (seven), with only three being anything else. Figure 6 illustrates this clearly. The dashed line shows percentage frequency of observed q_i (upper scale) or q_i^* (lower scale) for periods 1–10, the dotted line for periods 16–25. Table 4 shows that the shifting took place gradually, as subjects discovered it was in

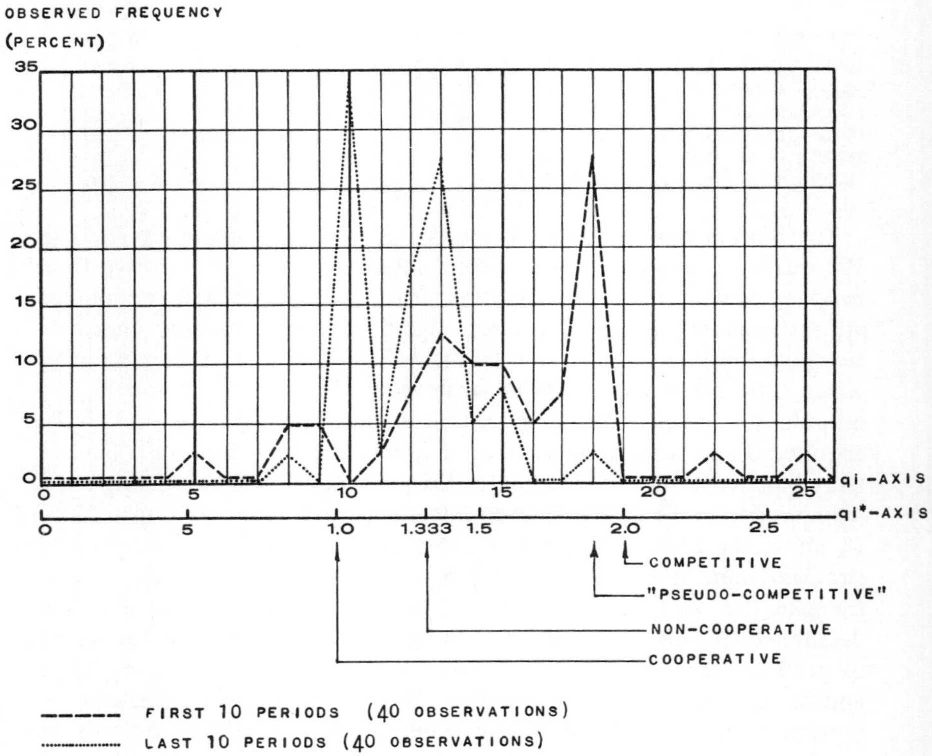

FIGURE 6
Observed frequency of quantity strategies in a duopoly pilot study

their interest to restrict production. Both the desire to pursue long-run rather than short-run profit maximization, and the decreased need to overproduce in period t in order to retaliate against an opponent for overproducing in period t-1, probably contributed to the observed secular decrease in output.

In the main set of experiments, modes were noted at the non-cooperative and cooperative solutions, but there were very few competitive or retaliatory responses, even in early periods—that is, learning behavior was absent. Perhaps improvements in the wording of the experiment instructions over those used in the pilot study made the structure of the duopoly situation so clear to the subjects that they discovered almost immediately the virtues of the cooperative and non-cooperative solutions.

We did not, in this series of experiments, test the dynamic hypotheses, but a few words about them are in order at this point. One possible test which, given sufficient data, would indicate whether an individual obeyed any simple reaction rule of the type $q_{it} = f(Q_{it} - q_{i,t-1})$ would be a chi-square test for the orthogonality of a contingency table in which the rows represent the subject's quantity chosen in period t, and the columns his opponent's quantity in period t-1. The explanatory value of any particular dynamic hypothesis specified in advance, such as Cournot, retaliatory, etc., could be discovered by testing the best-estimated regression coefficients against those specified by the hypothesis. Applying such statistical methods to short time series involves us in problems concerning sampling, the normal distribution assumption, autocorrelation, and other difficult matters discussed in the literature of econometrics.

Of several other experiments, the most interesting series was one in which the number of firms per market was permitted to vary. This required several experiment sessions. At one session in which nine subjects showed up, they were combined into a nine-firm market. Another day when nine appeared, they were split into three three-firm markets for a while and then another nine-firm market test was run. On a day when eight attended, two six-firm markets and one four-firm market were run, with each subject simultaneously in two of these three markets—by then the students had had enough experience, we felt, to be able to react independently yet simultaneously to two different market situations.

The results of these experiments are shown in Table 5 and in Figure 7, with all data transformed to the game Γ^*. The cooperative solution price is 1; for the competitive solution it is 0; and the non-cooperative solution price is indicated by the dotted line in Figure 7. The unweighted average observed price—averaged over all "weeks" and all markets of size n—is represented by the solid line as a function of n. The chief conclusion is that price does decrease with market size but not as fast as the non-cooperative solution would suggest.

TABLE 5

Summary of results in oligopoly experiment with n variable

Market Size (n)	Number of Time Periods	Average p^* in Each Market	Unweighted Average p^*, All Markets	Non-cooperative p^*	Z
2	6	1.017	0.983	0.667	0.950
	6	1.017			
	6	1.000			
	6	0.900			
3	7	1.000	0.705	0.500	0.409
	7	0.632			
	7	0.483			
4	25	0.640	0.640	0.400	0.400
6	25	0.674	0.569	0.286	0.396
	25	0.464			
9	16	0.564	0.493	0.200	0.366
	6	0.422			

Cooperative $p^* = 1$.
Competitive $p^* = 0$.
Non-cooperative $p^* = 1$ when $n = 1$; it approaches 0 as n grows large.

Another interesting feature of this experiment series is that the relative position of the average observed price with respect to the cooperative and non-cooperative prices seems to be very close to constant, for market sizes larger than 2. In the last column of Table 5 is a statistic Z, defined as the ratio of the difference between the observed average price and the non-cooperative price to the difference between the cooperative and non-cooperative prices. That is, $Z =$ (Observed p^* − Non-cooperative p^*)/(1 − Non-cooperative p^*). For $n > 2$, Z varies within a very small range (0.366 to 0.409). There is no behavioral theory within the framework of economics to account for this empirical regularity.

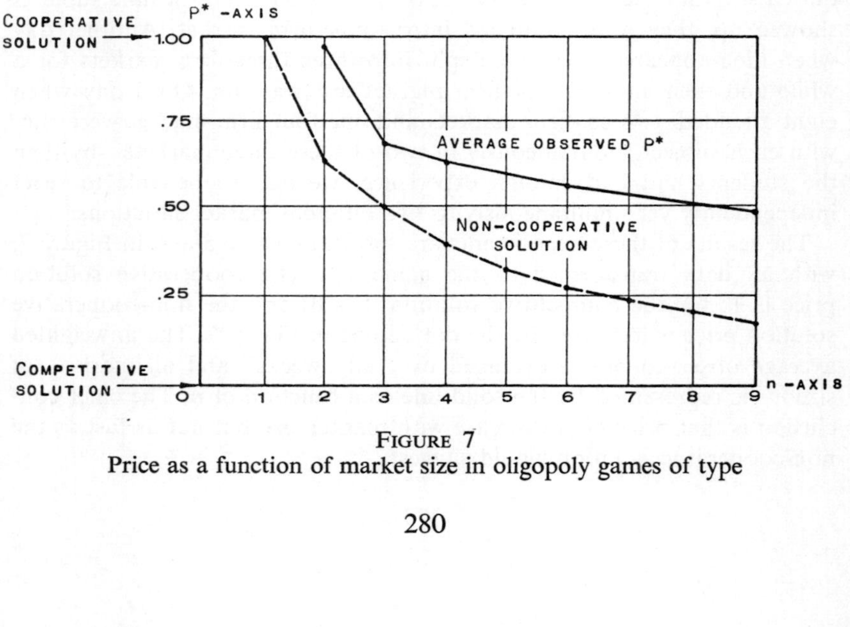

FIGURE 7
Price as a function of market size in oligopoly games of type

It is a milestone in experimental economics to find supported in the laboratory, the general conclusion—reached already in many real-world empirical studies, incorporated in many economic theories, and generally believed in lay folklore about economic affairs—that increasing the number of firms supplying a market brings about competitive pressure expressed in the form of a lowering of the price level.

CHAPTER 20

The Role of Economics in Management Science

By T. M. WHITIN*

INTRODUCTION

In the world of reality, the simplest principles of economic analysis are frequently (one might perhaps say typically) ignored. The following quotation from an unpublished memorandum of an economist working for a large government agency is illustrative:

> The fundamental principles of economics are typically ignored and sometimes even explicitly denied in [name of agency] papers. . . . Perhaps the worst abuse is the failure to accept the validity of marginal or "incremental" cost analysis. . . . Another serious abuse is treating dollars received or spent at different points of time as being identical, i.e., failing to "discount" costs and revenues. At a 3% discount rate a dollar, thirty years from today, is worth only 41 cents, yet staff papers add these dollars as though they were equal. Costs calculated by the [name of agency] and presented to congressional committees are sometimes off by a factor of two because of the failure to discount. A third fundamental principle that is sometimes ignored is that it is unwise to have an identical material priced at two different prices at the same time.

Nevertheless, economists are far from unanimous in believing that it is a legitimate function of an economist to try to remedy some of these abuses. For example, Pigou has written, "It is not the business of an economist to try to teach woollen manufacturers how to make or sell wool, or brewers to make and sell beer, or any other businessmen how to do their job." [1] Also, consider the statement that "the final aim of both cost theory and measurement is, of course, and could not be anything else but a better understanding of entrepreneurs' behavior."[2] Much discussion exists in the literature concerning whether or not entrepreneurs utilize or attempt to utilize marginal analysis.[3] There is little consideration

* Wesleyan University.

[1] "Empty Economic Boxes: A Reply," *Economic Journal* 1922, pp. 458–465, reprinted in *A. E. A. Readings in Price Theory* (editors G. J. Stigler and K. E. Boulding), p. 137.

[2] Staehle, H., "The Measurement of Statistical Cost Functions," *American Economic Review*, 1942, pp. 321–333, reprinted in *A. E. A. Readings in Price Theory*, p. 275.

[3] Lester, R. A., "Shortcomings of Marginal Analysis for Wage-Employment Problems," *American Economic Review* March 1946, pp. 63–82; Machlup F., "Marginal Analysis and Empirical Research," *American Economic Review*, September 1946, pp. 519–554; Hall, R. J., and Hitch, C. J., "Price Theory and Business Behavior," *Oxford Economic Papers II* (1959), pp. 12–45.

of the possibility that economists could *influence* businessmen to use marginal analysis or that an accurate understanding of business behavior may be acquired in this manner. The result is often a futile search for rationalization of business behavior which is not inconsistent with the economic theory of rational entrepreneurial behavior. Clearly in the many instances where entrepreneurial behavior and the reason for it are entirely illogical, such rationalizations cannot be realistic or useful. It would therefore seem sensible to make a frontal attack on the problem by scrutinizing the manner in which decisions are in fact made. In government decisions, which may involve gross errors of the type mentioned above, the direct approach certainly provides better clues concerning the causes of the behavior than would be possible by a more circuitous one. If such errors were restricted to government agencies, the economist could derive pleasure from the thought that competition would lead to the rapid extermination of firms making such elementary blunders. However, businesses seem to survive for extended periods of time under foolish management. If foolish management is assumed to exist, then the economist can say little beyond platitudes to the effect that the entrepreneur might like to maximize his profits.[4] Cost curves under inefficient management may bear little or no resemblance to the minimum cost curves of the theory of the firm.

Some economists are not concerned with entrepreneurial behavior, considering their role to be that of constructing *theories* of rational behavior. E. S. Mason has pointed out the pitfalls of this approach.[5]

> Some theorists, pursuing their analysis on a high plane, refer to their work as "tool making" rather than "tool using." A "tool maker," however, who constructs tools which no "tool user" can use is making a contribution of limited significance. Some knowledge of the use of tools is probably indispensable to their effective fabrication.

It has been contended that the assumptions of static equilibrium theory are such as to make its theoretical constructions "irrelevant to the real problems."[6] If such is the case, many would argue that the theory needs modification:

> What we have a right to ask of a conceptual model is that it seize on the strategic relationships that control the model it describes and

[4] It has been contended that entrepreneurs frequently attempt to maximize sales measured by gross revenue; see Baumol, W. J., "On the Theory of Oligopoly." *Economica*, August, 1958.

[5] Mason, E. S., "Price or Production Policies of Large-Scale Enterprise," *American Economic Review Supplement*, 1939, reprinted in *A. E. A. Readings in Industrial Organization and Public Policy*, (Heflebower and Stocking, editors), Homewood, Illinois, 1958, p. 191n.

[6] *Ibid.*, p. 193.

> that it thereby permit us to manipulate, i.e., to think about the situation.[7]
>
> It is incumbent on us . . . to set forth as explicitly as we can in just what way the tendencies work out and how our statements of them need to be modified in view of the complexities and disturbances of actual life.[8]

Only a very small portion of economic research involves any attempt to inject realism into the models or to narrow or bridge the wide gap between theory and practice. While it is legitimate for economists to be primarily interested in theories of market equilibrium or macro rather than micro economics, it is quite likely that improved micro theory has important implications for macro theory which the pure macro economist may ignore.[9] In the following pages, an attempt will be made to point out several areas where conventional theory bears little resemblance to reality. It will then be contended that there exist areas where economists can construct and implement fairly simple models, which, although falling far short of a general theory of business behavior, lead to considerable improvement over existing business practice as well as extending the static framework of the economic theory of the firm. It is recognized that there is no such thing as a "typical economist" and that specific counterexamples to the various points mentioned may be found. However, an attempt has been made to create an economist with widely held views rather than an absurd caricature.

DEFICIENCIES IN THE CONVENTIONAL THEORY OF THE FIRM

Monoperiodic analysis. As usually presented, the static theory of the firm is concerned with the economic analysis of one time period. Typically, there is no discussion concerning the manner in which the time period under analysis is selected or its length. Almost all realistic business problems involve more than one time period. A single period of substantial length (say a year) is inappropriate because of seasonal variations of demand (or costs) within the period. A single short period, on the other hand, encounters other types of difficulties. Once one admits the possibility of producing in one period for sales in later periods, the use of monoperiodic analysis seems inappropriate, for one becomes immediately involved in multi-period problems. This possiblity does not appear

[7] Samuelson, P., Solow, R., and Dorfman, R., *Linear Programming and Economic Analysis*, 1958, p. 9.

[8] Taussig, F. W., "Is Market Price Determinate?" *Quarterly Journal of Economics*, May 1921, p. 411.

[9] See for example Modigliani, F., and Cohen, K. J., "The Significance and Uses of Ex Ante Data," in *Expectations, Uncertainty, and Business Behavior* (editor M. J. Bowman), New York, 1958, pp. 151–164.

unreasonable either from an empirical or a theoretical point of view. It is seldom that a firm produces only for sale in the current (short) period, for several factors may make it profitable for the firm to produce for sales in subsequent periods. Among these are changes in costs or demand, distribution of set-up costs over a large number of items, and costs of varying the rate of production or labor force. An entrepreneur is all but certain to encounter some of these phenomena and hence is likely to produce for stock. Indeed, even under the artificial assumptions of a known, steady rate of demand and constant marginal costs of production, it is sometimes impossible to find a single period on which to base one's analysis of the firm. The relationships between the times when production of various items is started will shift in time, giving an endless number of possible sequences and periods, similar to those observed in a "vertical sequence" in art, where a very few shapes and colors, each with different time sequences, can give rise to a constantly changing picture for several weeks. Thus it seems necessary to use multi-period analysis if the analysis is to bear even a superficial resemblance to reality. Nevertheless, the static theory of the firm continues to receive much emphasis in modern books on economics.

Where attempts have been made to introduce multi-period analysis, these have often been of extremely limited scope. For example, in the final pages of Carlson[10] the analysis did not include any consideration of producing goods in one period for sales in later periods, i.e., no storage of final products was considered. Although Carlson admittedly passed over "many interesting details of the poly-periodic theory in order that the main relationships which we must consider may gain in lucidity,"[11] one may question whether or not he has left out some of the "main" relationships. Other economists have applied spatial price discrimination techniques to inventory problems.[12] Additional attempts to introduce inventories into economic models of the firm have been discussed elsewhere.[13] In spite of these few attempts, multi-period models have not yet been incorporated into the traditional economic theory of the firm. The usual reasons for the persistent neglect of what seem to be considerations of undeniable and obvious relevance are of two varieties.[14] One is typified by S. Weintraub's contention that the multi-period approach "can be charged with an even higher degree of sterility than traditional

[10] Carlson, Sune, *A Study on the Pure Theory of Production*, New York, 1956, pp. 109–125.

[11] *Ibid.*, p. 111.

[12] Lutz, Friedrich and Vera, *Theory of Investment of the Firm*, Princeton, 1951, pp. 94–100; Weintraub, S., *Price Theory*, New York, 1949, pp. 389–393.

[13] Whitin, T. M., *Theory of Inventory Management*, Princeton, 1957, Ch. IV.

[14] A third type of reason is the conceptual difficulty that the "last" period may affect behavior in the first, as the selection of a "last" period is itself arbitrary.

methods . . . [and] there seem to be few occasions, therefore, to erect elaborate models of equilibrium in time on the thought that they may have some empirical counterpart in the world we know."[15] The other reason is the contention that the static theory of the firm at least points out the important factors that influence business decisions and their interaction. Concerning the first reason, one may point out that there are already instances where multi-period models are applicable to business practice.[16] Concerning the second reason, it will be argued below that many relevant factors are not included in the conventional analysis, the omitted factors perhaps being of as much importance as those typically considered.

IMMANENT CRITICISM OF STATIC THEORY. A large portion of the relevant problem is avoided by the economist's usual assumption that the engineers possess and provide him a complete description of the physical production function, i.e., that the engineers can provide data on the output levels that result from all relevant combinations of the input factors. In practice, such knowledge of the production function is rarely found. The traditional economist, on the one hand, denies any responsibility for obtaining the information, and the engineer, on the other hand, "has never really grasped the theoretical concept of the production function and its usefulness in economizing decisions."[17] The very definition of the production function solves the purely "technical" maximization problem, since each isoquant is assumed to be a locus of points of "maximum product obtainable from the [factor] combination at the existing state of technical knowledge."[18] As mentioned, knowledge of these functions is largely non-existent. Even the definition itself is far from clear. Interperiod considerations creep back into the analysis, since the level of output for a given factor combination in one period affects the output of the same factor combination in future periods. Other difficulties stem from the assumption of "constant technical knowledge." In the first place, as Carlson wrote:

> A definition of constant technical knowledge is exceedingly difficult to give in any actual production process since the maximal technical organizations for the whole range of possible service combinations

[15] *Op.cit.*, p. 393. In spite of this contention, Weintraub himself has developed interesting dynamic models.

[16] See, for example, Holt, C. C., Modigliani, F., and Simon, H. A., "A Linear Decision Rule for Production and Employment Scheduling," *Management Science*, October 1955, pp. 1–30; Bowman, E. H., "Production Scheduling by the Transportation Method", *Operations Research*, February 1956, pp. 100–103; Wagner, H. M., and Whitin, T. M., "Dynamic Version of the Economic Lot Size Model, *Management Science*, October 1958, pp. 89–96.

[17] Smith, Vernon L., "The Theory of Investment and Production," *Quarterly Journal of Economics*, February 1959, p. 63n.

[18] Carlson, S., *op.cit.*, pp. 14–15.

> are seldom known in advance but have to be found out by practical experience.[19]

To the extent that the production function itself is not known until the factor combinations are chosen, the traditional analysis is somewhat circular, since one must know the production function to make rational decisions concerning factor combinations. The situation is made worse by the likelihood that the production function is also affected by the initial choice of factors. Actually it is only realistic to assume that some improvement in performance is obtained from experience with particular factor combinations. This learning would change the isoquants in the direction of a complementary relationship between the factors. An isoquant phenomenon is thus created which is somewhat similar to the "kink" in demand curves.[20] The "kink" would occur for the factor proportions initially selected. This may partially explain the observed lack of sensitivity of factor proportions to changes in the costs of the factors of production. A body of literature on learning is now in existence.[21] Should economists continue to neglect it in their theoretical formulations by merely assuming it to be non-existent?

In fact, technological change frequently worms its way back into economic analysis. The following quotation clearly is inconsistent with the usual assumption of constant technical knowledge.[22]

> Management is, because of lack of competition, able to operate in such a way that the actual cost for the output produced substantially exceeds the minimum possible total cost for that output . . . . Management may simply bestir itself and increase the technical efficiency of the firm, thereby raising the marginal physical productivity curves of some (or perhaps all) factors.

Another complication that realism introduces into the analysis is that factor price movements and technical knowledge are in fact not independent. Bela Gold has written:[23]

> Specific differentiation of technical change from factor price adjustments may represent so serious an underestimation of the extent of

[19] *Ibid.*, p. 16n.

[20] Sweezy, Paul M., "Demand Under Conditions of Oligopoly," *Journal of Political Economy*, August 1939, pp. 568–573; Hall and Hitch, *op.cit.*

[21] Andress, F. J., "The Learning Curve as a Production Tool," *Harvard Business Review*, January–February 1954, pp. 87–97; Mosteller F., and Bush, R. R., *Stochastic Models for Learning*, New York, 1955.

[22] Reder, M. W., "Marginal Productivity Theory Reconsidered," *Journal of Political Economy*, October 1947, p. 454. It is assumed that changes in the state of information constitute technical change.

[23] *Foundations of Productivity Analysis*, Pittsburgh, 1955, pp. 7–8. Economists have in fact considered the effect of technical changes on factor prices in their general equilibrium formulations. The interaction in the other direction, i.e., factor prices on technical change, is more typically ignored.

interaction between these categories of change as to limit the general applicability of the results . . . [in] a wide range of instances in which technical changes have affected factor prices.

Another type of difficulty plagues traditional economic analysis, namely that of decreasing average costs. Since this difficulty has been discussed at length in the literature, it will not be labored here.[24] There is "no substantial evidence that the long-run cost function for plants does not continue to decline up to the largest sizes found in practice, if factor prices are held constant (a generalization apparently justified by the evidence)."[25] Such a decline in average costs has serious consequences for economic theory:

> It is, I believe, only possible to save anything from this wreck—and it must be remembered that the threatened wreckage is that of the greater part of general equilibrium theory—if we can assume that the markets confronting most of the firms with which we shall be dealing do not differ greatly from competitive markets.[26]

Nevertheless, the writings on the theory of the competitive firm under static conditions continue to pervade the literature. One simple solution to the problem of what firms have minimum costs was suggested by Milton Friedman:[27]

> Surely the obvious answer is: firms of existing size. We can hardly expect to get better answers to this question than a host of firms, each of which has much more intimate knowledge about its activities than we as outside observers can have and each of which has a much stronger and immediate incentive to find the right answer . . . .

Here again is the typical expression of faith that entrepreneurs are

[24] Clapham, J. H., "Of Empty Economics Boxes," September 1922, pp. 305–314; Pigou, A. C., "Empty Economic Boxes: A Reply," December 1922, pp. 458–465; Robertson, D. H., "Those Empty Boxes," March 1924, pp. 16–30; Sraffa, P., "The Laws of Returns Under Competitive Conditions." December 1926, pp. 535–550; Pigou, A. C., "The Laws of Diminishing and Increasing Cost," June 1927, pp. 188–197; Robbins, L., "The Representative Firm," September 1928, pp. 387–404; Young, A. A., "Increasing Returns and Economic Progress," December 1928, pp. 527–542; Robertson, D. H., Shove, G. F., and Sraffa, P., "Increasing Returns and the Representative Firm: A Symposium" March 1930, pp. 79–116; Shove, G. F., "The Imperfection of the Market. A Further Note," March 1933, pp. 113–124; Kaldor, N., "The Equilibrium of the Firm," March 1934, pp. 60–76; Robinson, E. A. G., "The Problem of Management and the Size of Firms," June 1934, pp. 242–257—all the above in the *Economic Journal;* Chamberlin, E. H., "Proportionality, Divisibility and Economics of Scale," *Quarterly Journal of Economics*, February 1948, pp. 229–262; "Comments" by A. N. McLeod and F. H. Hahn, "Reply" by E. H. Chamberlin, *ibid.*, February 1949, pp. 128–143.

[25] Smith, Caleb A., "Survey of the Empirical Evidence on Economics of Scale," *Business Concentration and Price Policy*, National Bureau of Economic Research, Princeton, 1955, p. 230.

[26] Hicks, J. R., *Value and Capital* (second edition), Oxford, 1946, p. 84.

[27] "Comment on Economics of Scale," In *Business Concentration and Price Policy*, National Bureau of Economic Research, Princeton, 1955, p. 237.

indeed, in some way, behaving in optimal fashion. This expression of faith, in the absence of any positive evidence, appears somewhat naive.

Furthermore, as indicated above, the economist allegedly turns to the engineer for data concerning the production function. The engineering studies frequently demonstrate economies of scale, as evidenced by the ".6 rule:"[28]

> Briefly stated the rule says that the increase in cost is given by the increase in capacity raised to the .6 power . . . . The rule has been adduced from the fact that for some items of equipment as tanks, gas holders, columns, compressors, etc., the cost is determined by the amount of materials used in enclosing a given volume, i.e., cost is a function of surface area; while capacity is directly related to the volume of the container . . . . Cost varies as capacity to the 2/3 power. If the container is cylindrical . . . ; cost varies as capacity to the .5 power . . . . From a consideration of these factors the .6 rule has been developed.

Other economies of scale based on economic and probability considerations have been set forth elsewhere.[29]

It seems rather peculiar that the economist delegates the task of providing production function data to the engineer and then ignores the inconvenient fact that the engineer frequently finds increasing returns to scale present.

In summary, it has been argued that the static model of conventional economic theory contains several difficulties typically ignored. The theory is oversimplified by assuming known production functions and a state of constant technical (and entrepreneurial) knowledge. It also ignores difficulties due to increasing returns to scale. In the following section other objections are raised which perhaps have even more serious theoretical implications.

SCHEDULING AND SEQUENCING PROBLEMS. Other problems that are usually ignored in the conventional theory of the firm are the whole range of problems concerned with production scheduling and sequencing. These problems are implicit in the analysis only to the extent that it is assumed that the entrepreneur minimizes his costs, and hence in some obscure way has solved these problems. The variance of this assumption with the facts was mentioned in a recent article by M. H. Peston.[30]

> It is worth noting that recent operations research seems to indicate the need for modification of another part of the theory of the firm,

[28] Moore, F. T., "Economics of Scale: Some Statistical Evidence," *Quarterly Journal of Economics*, May 1959, pp. 234–235. This reference provides several references to the engineering literature.

[29] Whitin, T. M., and Peston, M. H., "Random Variations, Risk, and Returns to Scale," *Quarterly Journal of Economics*, November 1954, pp. 603–612.

[30] "On the Sales Maximization Hypothesis," *Economica*, May 1959, p. 131n.

> that concerned with cost functions. The assumption that the firm's factor combinations are optimal so that its cost curve is a minimum seems to be equally at variance with the facts . . . .

The emphasis on factor proportion *per se* has led to the almost total neglect of other considerations which may well be of equal importance from the standpoint of cost minimization. The following quotations indicate the potential importance of production scheduling:

> Due to scheduling, delays, avoidable mechanical difficulties, and so forth, that he was only producing 75 per cent of what he should have been producing.[31]

> In two months' time output has been increased 13 per cent. We expect to increase this figure to approximately 25 per cent, with practically no expense. In other words, detailed study and knowledge of the operating difficulties presented a program by which we are making unexpected gains . . . . The production line is more or less of a continuous operation and the problems to be corrected were those of management.[32]

> The decision rule with the moving average forecasts saved $173,000 annually against factory performance . . . . In the 1952–54 period actual factory costs exceeded costs under the decision rule by 8.5% or $51,000 per year on the average.[33]

> As can be seen, excess hours have been reduced by more than half, resulting in substantial savings in out-of-pocket costs. Another way of interpreting these economies is to say that the capacity of the machines has been increased by about 10 per cent.[34]

> In one company, scheduling setups by stock size and thread roll saved approximately 100 to 150 hours a week. Addition of the other three setup characteristics to setup sequencing doubled the savings. In the case of one machine shop, the only setup characteristic that was scheduled was die number . . . . Use of a comprehensive list of setup characteristics in the machine shop showed that several hundred setup hours a week could be saved.[35]

> Experience to date indicates that, in general, the Wilson Plan will

[31] Drucker, Ned, "Increasing Productivity Through a Proper Scheduling System," *Increasing Factory Output Through Better Production Control*, American Management Association Production Series No. 129, p. 15.

[32] Oberg, Harry U., "Increasing Capacity on a Production Line," *Increasing Factory Output Through Better Production Control*, American Management Association Production Series, No. 129, p. 12.

[33] Anshen, M., and others, "Mathematics for Production Scheduling," *Harvard Business Review*, March–April 1958, p. 56.

[34] Lohmiller, G. G., "Linear Programming: Its Application to a Typical Repetitive Scheduling Problem" in *How to Reduce Production Costs*, American Management Association Production Series No. 222, 1956, pp. 59–66.

[35] Reinfeld, N., and Vogel, W. R., *Mathematical Programming*, New York, 1958, p. 166.

> either reduce inventories by at least one-third with no decrease in stockroom service or reduce the present numbers of stock-outs per year to one-third of the present number with no increase in inventory, thus making a tremendous improvement in customer service without operating costs.[36]

The number of examples could be expanded considerably. It seems entirely possible that scheduling problems are of more importance than the choice of optimal factor combinations. There seems to be little reason for assuming that the entrepreneur optimizes the scheduling decisions. If the entrepreneur does not understand simple concepts, such as marginal analysis and present value, it is highly unlikely that he has successfully solved these more difficult problems. Data showing small differences between operations research results and management practice are based on a strongly biased sample.

Some scheduling problems have been solved by management science or operations research techniques, utilizing such tools as linear programming, linear decision rules, queuing theory, and probability analysis. The optimizing decisions are typically used for cost minimization rather than profit maximization. However, there seems to be substantial agreement among economists that the estimation of demand functions for the firm is an extremely difficult if not impossible task:

> It seems extremely unlikely that economists will be able by independent investigations to ascertain this [demand] shape except by the roughest sort of deduction from other data.[37]
>
> Little progress was made, however, in the methodology of statistical analysis of demand in the direction of bringing the statistically derived demand curve closer to the demand curve used by the economic theorist in his analysis. It has to be recognized that the difficulties in the way of attaining such an objective may simply be insoluble.[38]
>
> In view of the theoretical and statistical problems involved, it is not surprising that there are practically no empirical studies of demand curves or similar relationships for individual sellers.[39]

[36] Wilson, R. H., "Control of Inventories," reprinted from the *Bulletin* of Robert Morris Associates, February 1949.

[37] Haley, B. F., "Value and Distribution," in *A Survey of Contemporary Economics*, American Economic Association (editor H. S. Ellis), Philadelphia, 1948, p. 10. In spite of these opinions, a typical function of the economist in business is estimating sales for the firm.

[38] Mason, E. S., "Price and Production Policies of Large Scale Enterprise," *American Economic Review* (Supplement), 1939, pp. 61–74, reprinted in *A. E. A. Readings in Industrial Organization and Public Policy* (editors R. B. Heflebower and G. W. Stocking), Homewood, Ill., 1958, p. 193n.

[39] Bain, J. S., "Price and Production Policies," in *A Survey of Contemporary Economics*, p. 138.

In view of these difficulties, cost minimization is a fairly reasonable type of behavior. One may at least argue that cost minimization within various specific contexts is of more use to the entrepreneur than various statements and theorems concerning profit maximization which cannot be translated into the operational context of the firm. In any event, the fact that application of scientific techniques to cost minimization problems results in significant cost reductions demonstrates that businesses have operated at costs far from the minimum cost locus, which point theory typically assumes is known.

In focusing attention on optimal *quantities* of factor inputs, economists have neglected the scheduling problem. One cannot solve the problem of optimal factor combinations independently of the scheduling problem, which is usually the more difficult of the two problems. Solutions that are both optimal and practicable do not yet exist for most scheduling problems. When we turn to "sequencing" problems, i.e., problems involving the order in which units are processed or serviced, it is far more difficult to ascertain the minimum cost sequence. Problems involving as few as ten jobs and five machines involve such a number of possible sequences[40] that the fastest electronic computer could consider only a small fraction of them in a billion centuries. In some special cases the number of computations may be reduced significantly. For example, S. Johnson[41] has devised an ingenious method for scheduling jobs which must be processed by two machines in sequence. Given the set-up time and running time for each job, one could solve for minimum total production time for 100 jobs in a few minutes by hand, whereas there are 100! (9.3326×10^{157}) possible sequences. Without Johnson's method the chances of having the optimal sequence would be negligible. Other special cases involving more than two processes may be solved by Johnson's method.[42] Another technique that reduces the number of possible sequences to be considered has been evolved by S. B. Akers and J. Friedman.[43] Their technique uses symbolic logic to eliminate sequences that are technologically unfeasible because the various jobs require different routing. Another technique used in sequencing problems is simulation. Some preliminary work of James R. Jackson indicates that significant savings may be obtainable through the use of simulation and fairly simple decision rules.[44] Aside from

[40] $(10!)^5 \cong (3.6288 \times 10^6)^5 \cong 6.3 \times 10^{32}$.

[41] "Optimal Two and Three-Stage Production Schedules with Setup Time Included," *Naval Research Logistics Quarterly*, March 1954, pp. 61–68.

[42] *Ibid.*, pp. 65–68.

[43] Akers, S. B., and Friedman, J., "A Non-Numerical Approach to Production Scheduling Problems," *Journal of the Operations Research Society of America*, November 1955, pp. 429–442.

[44] Jackson, J. R., "Simulation Research on Job Shop Production," *Naval Research Logistics Quarterly*, December 1957, pp. 287–295.

these isolated cases, only trial and error methods, visual aids, etc., are available to aid in solving the sequencing problems. Even under stationary conditions, the entrepreneur could not solve these problems within a relevant time horizon. Thus, one can conclude that, for practical purposes, the minimum production cost for typical multi-stage processes is unascertainable in the present state of knowledge.

It is difficult to find explicit mention in conventional economic theory of the problems involved in scheduling or sequencing jobs. Except for the relatively new literature on linear programming, multi-process problems are frequently ignored or oversimplified. For example, Chenery claims much greater generality for the economist than for the engineer because the former considers all possible combinations of inputs.[45] On the other hand, he writes the following:

> Fortunately, the necessity of combining all these processes in a coordinated unit reduces the final number of variables.[46]

This would seem to imply that the economist does not in fact consider all different input combinations. In fact, in a broad sense, it is physically impossible to give explicit consideration to all possible sequences, even in fairly simple problems, as indicated above.

SUMMARY AND CONCLUSION

Most of the above has been destructive criticism of conventional economic theory. Much of this theory is not (and was not intended to be) useful in management science. The typical economist refrains from trying to influence business behavior, attempting rather to find rationalizations of this behavior which are consistent with economic theory. It is argued above that entrepreneurial behavior is frequently irrational to such an extent that it is useless to attempt to rationalize it in terms of economic models. The direct approach of influencing the entrepreneur to utilize economic theory is suggested as an alternative approach.

Several broad areas exist where fairly general principles of economic analysis may be applied. These have indeed been applied in many instances, although a surprising number of instances of their being ignored also exist. One of these general principles is discounting future values to achieve valid intertemporal comparisons. Another principle is that sunk costs should not affect current decisions, or that marginal (or incremental) analysis should be used. Most of the progress in the direction of using marginal analysis has been effected by accountants' recommending the use of direct costing methods rather than as the result of economic

[45] Chenery, H. B., "Engineering Production Functions," *Quarterly Journal o Economics*, November 1949, p. 509.

[46] *Ibid.*, p. 530.

analysis. Potential applications of static marginal analysis exist[47] although, as mentioned above, multi-period considerations are present in most problems. Applications of economic analysis in the inventory management area are possible.[48] Linear programming applications are becoming quite common both in scheduling problems and in transportation problems.[49] In the area of capital theory, applications to equipment replacement problems are common.

In short, there already exist many areas where economic analysis is used in management. There is need for the economic theoreticians to examine these areas and to attempt to incorporate more realistic features in their models. The improved models may then lead to further applications leading to a narrowing of the immense gap between economic theory and business practice. In particular, management science would benefit from economic theorists' abandoning such concepts as the "representative firm" and general equilibrium analysis and focusing instead on multi-period models, production scheduling and sequencing models, learning models, and other models incorporating many more realistic considerations in the analysis than is customary at the present time.

[47] Karr, H. W., and Geisler, M. A., "A Fruitful Application of Static Marginal Analysis," *Management Science*, July 1956, pp. 313–326.

[48] Eagle, Alan R., "Distribution of Seasonal Inventory of The Hawaiian Pineapple Company," *Operations Research*, June 1957, pp. 382–396; Holt, C. C., Modigliani, F., and Simon, H. A., "A Linear Decision Rule for Production and Employment Scheduling," *Management Science*, October 1955, pp. 1–30; Naddor, E., "Some Models of Inventory and an Application," *Management Science*, July 1956, pp. 299–312; Whitin, T. M., *Theory of Inventory Management* (second edition), Princeton, 1957.

[49] Bowman, E. H., "Production Scheduling by the Transportation Method of Linear Programming," *Operations Research*, February 1956, pp. 100–102; Charnes, A., Cooper, W. W., and Farr, D., "Linear Programming and Profit Preference Scheduling for a Manufacturing Firm," *Operations Research*, May 1953, pp. 114–129; Jewell, W. S., "Warehousing and Distribution of a Seasonal Product," *Naval Research Logistics Quarterly*, March 1957, pp. 29–34.

PART VI

International Trade

CHAPTER 21

Competition of American and Japanese Textiles in the World Market: An Empirical Test of the Theory of Comparative Cost

By ANTHONY Y. C. KOO*

I. INTRODUCTION

The theory of comparative advantage is the cornerstone of the theory of interregional and international trade. According to that theory, each country or region will specialize and export those goods whose costs of production per unit of output are lower than those of others. However, the importance of the theory of comparative advantage in explaining the composition of foreign trade has been questioned [3]. It is argued that there are important short-run aspects of the commodity trade pattern of a given moment that cannot be explained by a long-run construct of static economics. The theory of comparative cost explains what the commodity pattern would eventually be if tastes, resources, and production technique remained unchanged and if there were no serious restriction of free competition in foreign or domestic commerce by government or by large concentrations of economic power. But in the real world, neither of the above conditions is fully satisfied. It is, therefore, incumbent upon the students of international trade theory to demonstrate the extent to which the new dynamic factors such as product differentiation, trade restrictions, etc., tend to offset the international and interregional division of labor based on comparative cost.

One approach, by Sir Donald MacDougall, has been used to make an empirical test of the theory of comparative cost [4]. He compared the relative United States and United Kingdom outputs per worker in several industries with the relative total exports by each in the world market. In contrast, the present procedure will be to select the output of a single industry, namely, textiles, and study the effectiveness of the cost disparity between American and Japanese producers in the international market. The main difference between the present approach and Sir MacDougall's should be noted. First, the outputs of certain industries with the same general designation in each country are only more or less of similar character and do not necessarily provide good substitutes for each other in the world market. For example, Sir MacDougall included in his list such industries as motor cars, wireless sets, and machinery. It is well

* The University of Michigan.
The author was on a Ford Foundation Faculty Research Fellowship when this paper was written and wishes to thank the Ford Foundation for making the research possible.

known that in the three industries mentioned above considerations such as design and services may play just as important a role in international competition as the price. Textiles are by no means a homogeneous commodity but are a standardized one. A study of international competition in textiles should provide a more clear-cut test of the theory of comparative cost.

Second, competition in the "world market" as measured in terms of the total exports of a country is an ambiguous notion. As a matter of fact, each country has its own sphere of trade influence, and the exports of the similar goods of the countries do not compete on an equal footing everywhere in the world. In order to uncover the competitiveness of each country's exports, it is necessary to examine their respective shares in each sub-market.

Third, the cost data for international comparison are generally unavailable. The only direct cost information concerns the industrial wage rates in various countries. It is only in those industries where wage payments constitute a significant cost element that wage rates are a good reflection of the cost condition. The textile industry seems to be one in which the wage element is a dominant factor of the cost [1].

Fourth, the textile industry is relatively less concentrated in the hands of a few producers than, say, the steel and motor cars industries. The absence of a large measure of monopolistic elements among the producers of the industry tends to make cost a determining factor in setting the price of textile goods.

Our analysis proceeds as follows: Section II examines the components of cost data of textiles in the United States and Japan. Section III reviews the sales performance of the products from each country in various markets of the world and the elasticity of substitution between the two. A rough estimate of the effect of the change of cost components on exports to various markets will be made, thus bringing together a more direct verification of the theory of comparative costs. Some of the more important conclusions are summarized in Section IV.

SECTION II

There are two types of data on the cost of production of textiles. One is derived from overall employment, man-hours, and production figures in the textile industry. It is an aggregative, time-series approach to productivity measurement in terms of square yard per man-hour. These figures have not been related to particular fabrics in such a way as to bring out their full significance for comparative purposes; but they have often been cited in comparative cost studies. The other type is based on survey data. These data are obtained from individual producers in each country and are based on actual cost experience in the production of a specific fabric. Such cross-section information, though more useful for our purpose, is usually unavailable. Fortunately, the Department of Commerce, through the Survey and Research Corporation, undertook to compare the costs of producing a limited number of fabrics in several countries [5]. In the survey, exactly specified fabrics were selected as a

TABLE 1

Selected Cotton Fabrics: Major Elements of Average Production Costs in the United States and Japan, 1960[a]

(In U.S. cents per linear yard)

Item of cost	*Sheeting*[b]		*Print Cloth*[c]		*Broadcloth*[d]		*Gingham*[e]	
	U.S.	Japan	U.S.	Japan	U.S.	Japan	U.S.	Japan
Total cost of fabric	14.60	11.39	18.40	14.55	26.16	18.35	33.55	22.13
Net cotton cost	8.40	7.81	8.71	8.23	11.80	10.29	10.33	9.05
Costs over cotton	6.20	3.58	9.69	6.32	14.35	8.06	23.22	13.08
Labor	3.92	1.72	5.58	2.46	9.09	3.18	11.99	4.57
All other	2.28	1.86	4.11	3.86	5.26	4.88	11.23	8.51

[a] U.S. Department of Commerce, *Comparative Production Costs in the United States and Four Other Countries*, 1961.
[b] Sheeting (gray)
Construction (warp ends and filling threads per square inch): 44 × 40; Yarns: carded white, warp 20 single, filling 18 single; Width: 40 inches; Weight: 4.25 linear yards per pound.
[c] Print cloth (grey)
Construction: 80 × 80; Yarns: carded white, warp 40 single, filling 40 single; Width: 39 inches; Weight: 4.00 linear yards per pound.
[d] Broadcloth (gray)
Construction: 136 × 60; Yarns: combed white, warp 40 single, filling 40 single; Width: 40 inches; Weight: 3.65 linear yards per pound.
[e] Gingham (gray)
Construction: 90 × 60; Yarns: combed, 50% colored, warp 40 single, filling 40 single; Width: 47 inches; Weight: 4.37 linear yards per pound.

basis for comparison, because a difference in any physical characteristics involves a difference in production costs. However, it can be reasonably assumed, according to the authors of the study, that "the findings with respect to the relative costs in the several countries of production of the specified fabrics are indicative of the relative costs of producing other fabrics of the same type" [5, p. 3]. This observation is mostly true in the case of sheeting and print cloth but slightly less true in the case of fine cloth fabrics, since the latter category includes a greater variety of somewhat more differentiated fabrics. The data are summarized in Table 1. Of the three categories of cost, cotton, labor and others, it is the labor cost in which the disparity between the United States and Japan is the widest. In every type of fabric, the labor cost in the United States is more than twice that in Japan, and the difference explains in substantial part the total cost differential.

Thanks to the availability for the first time of such detailed comparative cost figures, it is possible to seek answers to questions about (1) the impact of the U.S. cotton export subsidy and (2) the effect of change of wage rates on the cost of production.

During the past two decades, the U.S. Government has supported the domestic price of cotton at a level higher than the world prices. The U.S. exports of cotton were largely assisted through cash subsidies, export differential payments, special grants, donations, loans, foreign currency sales and barter programs. The various programs often run concurrently. The net effect has been, and is, that the U.S. textile producers pay as much as eight cents more per pound for cotton than the foreign producers. The recent Presidential order to examine an equalization duty on imported textiles equivalent to the export subsidy of cotton has highlighted that portion of textile production cost.

Suppose the American textile producers pay the world market price for cotton: To what extent would the average production cost in the United States be reduced for sheeting, print cloth, broadcloth, and gingham? In order to answer this question, we first determine the average price per pound of cotton. It was 32.36¢ for 1960, when the survey of the cost of production was made.[1] The rates of payments to exporters were on the average 7¢ per pound.[2] This means that the domestic textile producers pay approximately 21.63% (7/32.36) more than the world price If we reduce the net cotton cost for sheeting, print cloth, broadcloth, and gingham by the above percentage, we find that the cost of sheeting will be reduced from 14.60¢ to 12.78¢ per linear yard, or 12%; the cost of sheeting will be reduced from 18.40¢ to 16.32¢, or 11%; that of broadcloth from 26.15¢ to 23.60¢, or 9.80%; and of gingham from 33.55¢ to 31.32¢, or 6.6%. In short, if it were not for the higher price brought about by the cotton support program, the average cost of production of American textiles would have been lower by 6–12%.

With productivity of labor assumed as given in the short run, to what extent will a change in wage rate affect the average production cost in the U.S.? To put the question in more concrete terms, suppose that the cost of U.S. textiles must be held at the level of the average cost of production of Japanese textiles, what will be the wage rate commensurate with the

[1] It was based on monthly average prices (Jan.–Dec., 1960) for four territory growth, even running lots, prompt shipments, delivered at Group 201 (Group B) mill points including landing costs and brokerage prices for the average quality of cotton used in each kind of cloth. *The Cotton Situation*, March, 1961, Table 21, p. 30, Agricultural Marketing Service, United States Department of Agriculture.

[2] Cotton: "Rates of Payments to Exporters, Export Differentials by C.C.C., United States, 1956–61." *Ibid.*, May, 1961, p. 21, Table 15.

given cost level? The answer can be derived in two steps. From Table 2, we know the man-hour content per 100 yards of selected fabrics. For example, it takes 0.0233 man-hour to produce one yard of sheeting in the United States and 0.0474 man-hour in Japan. Since we also know from Table 2 the labor costs per hour (including fringe benefits) of producing various kinds of fabrics both in the United States and in Japan, it is possible to calculate the wage cost of one yard of a selected fabric by multiplying labor cost per hour and man-hour necessary for one yard of this fabric. Again take sheeting as an illustration. Suppose the hourly wage rate for producing sheeting were \$1.00 in the United States and it took 0.0233 man-hour to produce one linear yard of sheeting. The labor

TABLE 2

Selected Cotton Fabrics: Man-hours per 100 Yards Produced, Labor Costs per Hour, United States and Japan, 1960

	Sheeting		*Print Cloth*		*Broadcloth*		*Gingham*	
	U.S.	Japan	U.S.	Japan	U.S.	Japan	U.S.	Japan
Man-hours per 100 yards produced	2.33	4.74	3.40	10.39	5.32	9.30	6.91	18.00
Labor costs per hour (U.S. \$)[a]	1.68	0.36	1.64	0.24	1.71	0.34	1.73	0.25

[a] Including "fringe" labor costs

cost for one yard of sheeting would be 2.33¢ per linear yard ($\$1.00 \times 0.0233 = 2.33¢$). On the assumption that all other costs remain unchanged or at least that there was only a small change in wage rate, we can figure out the total cost per linear yard of sheeting by summing (1) net cotton cost, 8.40¢ (Table 1); (2) labor cost as calculated above, 2.33¢; and (3) all other costs, 2.28¢ (Table 1). We have the total production cost per linear yard of sheeting in the United States equal to 13.01¢. Therefore, a wage rate–cost table can be constructed by assuming different wage rates as shown in Table 3 for sheeting. Column 1 consists of a schedule of our assumed hourly wage rate. Column 2 lists the labor cost per linear yard based on the assumed wage rate. Column 3 gives the sum of all costs, with the cost of cotton based on the U.S. domestic price. Column 4 is the same as column 3 except for the price of cotton, which is appropriately reduced to compensate the domestic producers for the export subsidy made to the purchasers of cotton outside the United States. A similar wage rate–cost table for sheeting can be constructed for Japanese producers by assuming various wage rates for sheeting workers and then calculating the total cost as shown in columns 5 and 6 of Table 3.

The essence of Table 3 can be grasped more easily if we present it in a diagram. Let the horizontal axis measure the hourly wage in dollars. Placing the origin of the diagram at the center, we measure the U.S.

wage rate from left to right and the Japanese wage rate from right to left. The vertical axis shows the price in cents per linear yard of sheeting. When we plot columns 1 and 3,[3] and 1 and 4[4] of Table 3, we have the two parallel straight lines on the right half of Figure 1. The upper line represents the U.S. cost–wage rate relationship when cotton price is at a higher domestic level, while the lower line shows the same relationship

TABLE 3

Costs of Producing Sheeting in United States and Japan, 1960
(U.S. cents per linear yard)

	United States			Japan	
		Total Cost			
Hourly wage rate, U.S. dollars (1)	Labor cost per linear yard (2)	Cotton price unreduced (3)	Cotton price reduced (4)	Labor cost per linear yard (5)	Total cost (6)
0.30				1.42	11.09
0.35				1.66	11.33
0.40				1.90	11.57
0.50				2.37	12.04
0.60				2.84	12.51
0.70				3.32	12.99
0.80				3.79	13.46
0.90	2.10	12.78	10.96	4.27	13.94
1.00	2.33	13.01	11.19	4.74	14.41
1.10	2.56	13.24	11.42	5.21	14.88
1.20	2.80	13.48	11.66	5.69	15.36
1.30	3.03	13.71	11.89		
1.40	3.26	13.94	12.12		
1.50	3.50	14.14	12.36		
1.60	3.73	14.37	12.59		
1.70	3.96	14.60	12.82		
1.80	4.19	14.83	13.45		

with cotton price reduced to the world level. The cost–wage rate relationship in Japan (columns 1 and 6[5] of Table 3) is represented by a straight line at the left-hand side of Figure 1. From the diagram, we see that the flatter the line, the more productive the workers are, because for a given increase of wage rate, a steeper line will result in a higher increase in cost per linear yard than will a flatter line. For this reason, we may call it the line of "labor productivity." Since American labor is more productive than the equivalent Japanese labor, the American productivity line is flatter than the Japanese counterpart.

The vertical distance between ATC_1 and ATC_2, i.e., *a b*, represents the impact of cotton price reduction on the price of sheeting per linear yard. However, if the same reduced cotton price is passed on to the workers in

[3] Equation 1.1 in Table 4.
[4] Equation 1.2 in Table 4.
[5] Equation 1.0 in Table 4.

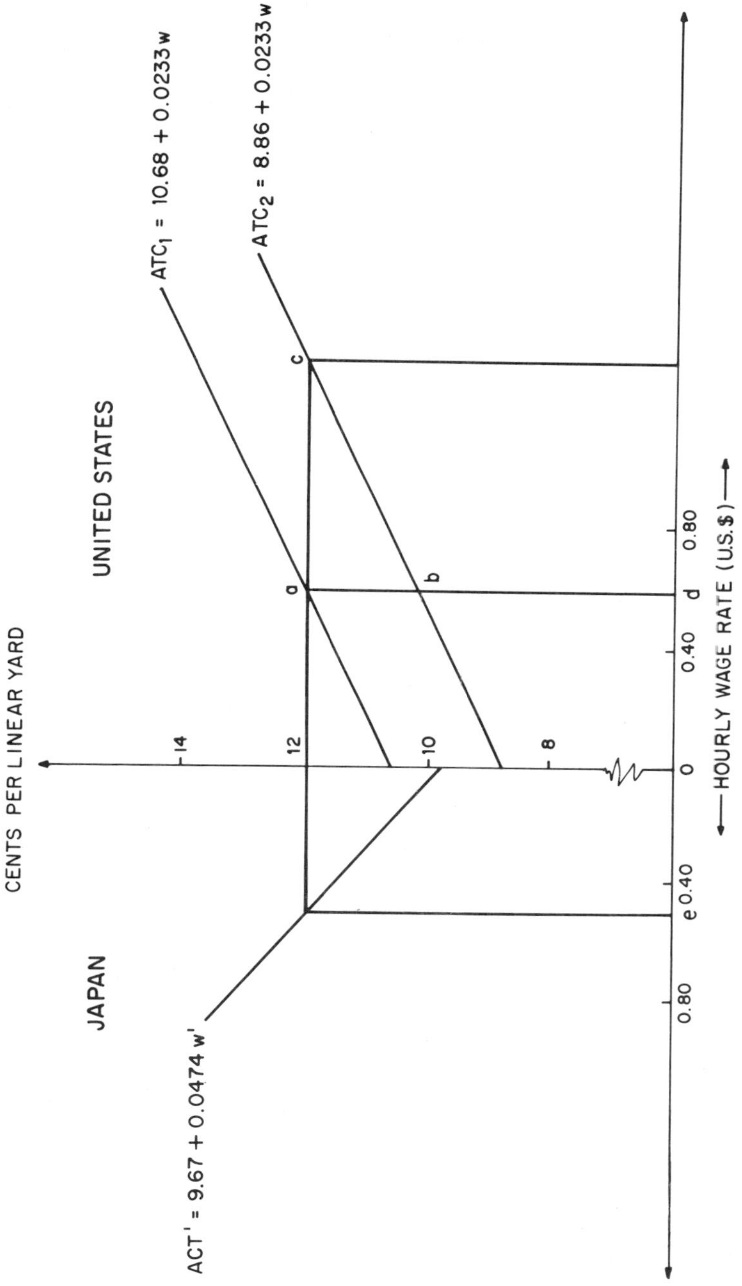
CENTS PER LINEAR YARD
UNITED STATES
JAPAN
ATC_1 = 10.68 + 0.0233 w
ATC_2 = 8.86 + 0.0233 w
ACT' = 9.67 + 0.0474 w'
14
12
10
8
0
0.40
0.80
a
b
c
d
e
HOURLY WAGE RATE (U.S.$)

FIGURE 1

the United States, with the price of sheeting per linear yard remaining the same, the workers in the United States will receive a wage rate increase represented by the horizontal distance *a c* in the diagram. Taking the productivity of U.S. workers as given, *od* and *oe* would be, respectively, the wage rates in the United States and Japan if the average total cost should be identical for both countries at, say, 12¢ per linear yard.

The relationship as shown by the figures in Table 3 and the lines in Figure 1 can be more compactly generalized in the form of equations. The basic assumptions underlying the relationships between the variables

TABLE 4

Linear Approximations to Total Cost Relations of Selected Fabrics in the United States and Japan, 1960

Fabrics	*United States*		*Japan*
	Cotton at domestic support price	Cotton at world price	Cotton at world price
Sheeting	(1.1) $ATC_1=10.68+0.0233w$	(1.2) $ATC_2=\ 8.86+0.0233w$	(1.0) $ATC'=\ 9.67+0.0474w'$
Print cloth	(2.1) $ATC_1=12.82+0.0340w$	(2.2) $ATC_2=10.94+0.0340w$	(2.0) $ATC'=12.09+0.1039w'$
Broadcloth	(3.1) $ATC_1=17.06+0.0532w$	(3.2) $ATC_2=14.51+0.0532w$	(3.0) $ATC'=15.17+0.0930w'$
Gingham	(4.1) $ATC_1=21.56+0.0691w$	(4.2) $ATC_2=19.33+0.0691w$	(4.0) $ATC'=17.56+0.1800w'$

w or w′ = hourly wage rate in U.S. dollars.

should be stated explicitly. We assume that the labor cost of prime concern to our analysis is considered a variable that depends on the wage rate. All other costs, including the cost of cotton, are considered fixed. Since in the short run, the labor productivity remains unchanged, the total cost per linear yard becomes a linear function of the wage rate. Table 4 lists twelve such equations for selected fabrics in the United States and Japan. For the United States we have two subsets, one on the assumption that the cotton price is at the domestic level (first column of equations in Table 4) and the other on the assumption that the cotton price is at the lower world level (second column of equations). Therefore the constant term in each of the equations represents the "fixed" costs as assumed above. For example, take equation (1.1). The average total fixed cost is 10.68¢ per linear yard. The slope coefficient of the linear equation represents the labor productivity or, to be precise, the man-hours required to produce one linear yard of a specific fabric. For equation (1.1), it is 0.0233 yard per man-hour. Thus by assuming the hourly wage rate to be the only unknown or independent variable in the equation, we can calculate the labor cost per linear yard. When it is added to the fixed cost, we have average total cost per linear yard.

Consequently, these equations give us a way to determine approximately the relative wage parity between the American and the Japanese textile workers if the price of textiles of the two countries should be set at a certain ratio to each other. Take the simple case of assuming an identical price for both American and Japanese textiles, say, 14.60 for each linear yard of sheeting. From equation (1.1), we know that for the price of 14.60¢ the wage rate in the United States will be \$1.68 per hour. To find the wage rate for the Japanese textile workers, we equate (1.1) (with $w = \$1.68$ inserted in the equation) with (1.0) and solve for w'. We get \$1.04 per hour for the Japanese textile workers.[6] If, instead, a wage rate of \$1.00 per hour is paid to the American workers, we can again derive the corresponding wage rate for the Japanese counterpart on the assumption that the prices of textiles remain the same for both countries in the export market. To do so, we equate (1.1) (with $w = \$1.00$ inserted) with (1.0) and solve for w', and get $w' = \$0.71$.[7]

SECTION III

On the basis of the 1960 survey, the production cost of Japanese textiles ranges from 78% of the U.S. costs, for sheeting, to 66% of the U.S. costs, for gingham.[8] What is the extent to which such cost differentials are reflected in the relative prices? Since relative price statistics for cotton cloth are more accessible in the form of a price index, we shall evaluate the differentials in terms of index points. Using the index number of wholesale prices of textile products as shown in Table 5, the 1960 price index reflects a 23% decline of the price of Japanese textiles relative to that of U.S. textiles from the base year 1953.

TABLE 5

Index Numbers of Wholesale Prices of Textile Products
(1953 = 100)

	1954	1955	1956	1957	1958	1959	1960
Japan (Tokyo)	92	87	88	82	75	77	77
United States	98	98	98	98	96	98	99
Japan (Tokyo) U.S. = 100	93.88	88.78	89.80	83.67	78.13	78.57	77.78

SOURCE: *U.N. Monthly Bulletin of Statistics*, March 1961.

[6] $10.68 + 0.0233 \times 168 = 9.67 + 0.0474\,w'$; or $w' = 4.93/0.0474 = \$1.04$.
[7] $10.68 + 0.0233 \times 100 = 9.67 + 0.0474\,w'$; or $w' = 3.34/0.0474 = 71¢$.
[8] Japanese costs were 73% of the American costs for print cloth, 70% for broadcloth.

Putting this in slightly different terms, we note that the competitive position of the Japanese textiles from the point of view of prices has improved since 1953 to the extent that in 1960 the Japanese price was 77% of the American price. The 77% figure, however, may not be a true reflection of the competitive advantage of the Japanese textiles in comparison with the cost data in 1960 because the price index measures from 1953 as the base period. It is reasonable to assume that Japanese textile producers had cost advantages over their counterparts in the United States as evidenced by a period of steady increase of the ratio of Japanese to American textile exports and accented by a marked change in 1953 (see Table 6). Of course a part of the cost differentials was ab-

TABLE 6

Export of Cotton Fabrics
(In millions of square yards)

Year	1954	55	56	57	58	59	60
U.S.	605.1	542.4	511.6	553.1	503.2	474.3	437.8
Japan	1278.1	1138.8	1262.1	1468.3	1245.3	1263.0	1424.0

SOURCE: *Commodity Yearbook*, Commodity Research Bureau, Inc., New York, 1961. *Statistical Digest of Japanese Textile Industry*, No. 10, Toyo Spinning Co., Ltd., Institute for Economic Research, Osaka, Japan, May, 1961.

sorbed by the American producers and consequently not fully reflected in prices. Price cutting is the major method used to answer competition in an industry such as the textile industry, the products of which are, in great measure, standardized. If the price reduction were inadequate relative to that of the Japanese counterpart, the American producers would be confronted with a partial or complete loss of the market to the Japanese. Whatever the American producers may decide—whether they reduce the price low enough in order to maintain their share of the market or give up a part of their share because of inadequate matching of price—a profit squeeze is inevitable. For example, between 1947 and 1957 in the United States, textiles profits declined from 8.2% to 1.9% when measured against sales and from 18.4% to 4.2% when measured against stockholders' equity.[9] Another way of telling the story is to look at the mill margin—the difference between the price of unfinished cloth made from a pound of raw cotton and the price of cotton. In 1947 the price of cotton cloth, based on 17 gray (unfinished) constructions, as compiled by the United States Department of Agriculture, was 88.99¢ per pound, and the price of cotton was 34.15¢ per pound. In 1957 the price of cloth based on the same 17 gray goods constructions had fallen to 61.37¢ per pound. At the same time the price of cotton increased to 34.42¢

[9] *Problems of Domestic Textile Industry*, Report of the Committee on Interstate and Foreign Commerce, U.S. Senate 85th Congress, Feb., 1959, Washington, D.C.

per pound. The higher raw cotton price and the lower cloth price resulted in a reduction of mill margins to 26.95¢.[10]

In spite of some cost absorption and profit reduction, exports of American textiles were declining absolutely and relatively to the main competitor, Japanese textiles, in many regions. However, the degree of competitiveness is by no means uniform in all parts of the world. Thus any consideration of relative American and Japanese exports in overall

TABLE 7

Correlation between Relative Exports and Relative Price Indices of the U.S. and Japanese Textiles, 1954–1960

Area	*Y*-intercept	Regression coefficient	*R*(corrected)	Relative price elasticity	von Neumann's ratio (ratio of mean square successive difference to variance)
.0 A Total exports	+8.0692 (1.3559)	−0.0710 (0.0160)	−0.8926	−2.88	2.1829*
B Total exports	+6.3200 (1.5071)	−0.0449 (0.0178)	−0.7480	−1.50	2.2137*
.1 Canada	+0.6965 (0.1978)	−0.0067 (0.0023)	−0.7878	−4.28	2.2000*
.2 S. America	+3.4026 (1.3802)	−0.0333 (0.0163)	−0.6741	−4.73	2.2484*
.3 W. Europe	−15.7384 (10.1503)	+0.2852 (0.1200)	+0.7283	(+2.89)	1.3151*
.4 Middle East	+46,2468 (18.7408)	−0.4798 (0.2216)	−0.6957	−7.02	3.1062*
.5 Asia	+19.6529 (3.1677)	−0.1848 (0.0374)	−0.9108	−3.84	3.4256*

NOTES: 1. The numbers in the parentheses are standard errors of estimates corresponding to the coefficients directly above them.
2. * Reject the hypothesis that the residuals are auto correlated.
3. Data: value of trade, U.N., *Commodity Trade Statistics;* price indices, U.N., *Monthly Bulletin of Statistics*. For sources of exports in square yards, see footnote, Table 5.
4. Regions are defined in terms of U.N. trade statistics classification.

terms tends to obscure significant details of the picture. To be more specific, we use the regression coefficients between relative exports and relative price indices of American and Japanese textiles, 1954–1960, as a measure of "elasticity of substitution."

In all the regression equations that follow (Tables 7 and 8), the independent variable is the relative price index of Japanese and American textiles. For the overall export equations, (1.0) and (2.0), we try two sets of dependent variables. We use first the relative value of exports deflated by price indices and then the physical quantity of textile exports in millions

[10] *Cotton Statistics*, U.S. Department of Agriculture, Washington, D.C. Since 1957 the method of computing monthly average mill margins has been revised. The cloth part of component now reflects wholesale prices for 20 combinations of cotton gray goods compared with 17 heretofore. The cotton prices component of the series was also expanded. For details, see *ibid.*, Sept., 1958, p. 11.

of square yards. For the regional equations, (1.1–1.5) and (2.1–2.5), because of unavailability of the figures in millions of square yards, we employ the deflated value of exports as an indicator of physical volume.

The difference in the "elasticities of substitution" between the linear and the logarithmic equations is, in general, minor as compared with that between the coefficients of the overall equations and the regional ones. The "elasticity of substitution" of equations (1.0B) and (2.0B) is only −1.5. Even if we take the larger figures (−2.88 to −2.98) of equations

TABLE 8

Correlation between Logarithms of Relative Exports and Logarithms of Relative Price Indices of the U.S. and Japanese Textiles (1954–1960)

Area	*Y*-intercept	Regression coefficient	*R*(corrected)	Relative price elasticity	von Neumann's ratio (ratio of mean square successive difference to variance)
2.0 A Total exports	+6.0414 (1.1438)	−2.9789 (0.5940)	−0.9133	−2.98	2.4117*
B Total exports	+3.2879 (1.0805)	−1.5006 (0.5612)	−0.7671	−1.50	2.3333*
2.1 Canada	+12.5893 (5.8608)	−7.0328 (3.0440)	−0.7186	−7.03	1.9051*
2.2 S. America	+9.8265 (4.0170)	−5.2499 (2.0864)	−0.7475	−5.25	2.2019*
2.3 W. Europe	−5.7106 (3.0207)	+3.4326 (1.5689)	+0.6994	(+3.43)	1.3389*
2.4 Middle East	+14.6113 (3.7948)	−7.2405 (1.9710)	−0.8542	−7.24	2.9193*
2.5 Asia	+8.6039 (2.0316)	−4.1652 (1.0552)	−0.8701	−4.17	3.3396*

(1.0A) and (2.0A), we find them exceeded by all the regional coefficients and in the case of the Middle East by more than 100%. The only exception is Western Europe where, because of severe import restriction of Japanese imports, particularly textiles, we arrive at an unmeaningful positive regression coefficient. Thus, inclusion of Western Europe in the calculation of total elasticity of substitution has definitely blurred the picture.

The range of the coefficients of substitution deserves a word of comment. Although we do not have very much confidence in the actual magnitude of substitution coefficients, some broad guidelines emerge from the analysis. On the basis of ranking of the coefficients, we can roughly group the regions into three categories: (a) Middle East, (b) Canada, South America, and Asia, and (c) Western Europe. If the American textiles should gradually be pushed out of the world market because of the competition of the Japanese textiles, then the Middle East, the "less accessible market" as Harrod puts it, will be given up first. The defense line of American exports of textiles against the approaching foreign competition may be penetrated in the order of regions listed above [2, p. 67].

The above discussion of competition in terms of comparative prices can be reformulated in terms of comparative costs if we accept as an approximation the wage rate and average cost relationship as shown in the equations in Table 4. Suppose relative costs are rough indications of relative prices, then we can address ourselves to such questions as the following. To what extent will the exports of American textiles be reduced if there is a 10% wage increase for the textile workers in the United States? Take sheeting for example. From equation (1.1), (Table 4) we know that wage–cost (price) elasticity is $W/P \times dP/dW = (168/14.6)(0.0233) = 0.27$. Or, a 10% increase of wage rate will result in a 2.7% increase of price of sheeting, other things being equal. This means there will be roughly an 8% reduction of exports in some regions where we assume that the "coefficient of substitution" is in the neighborhood of -3.

Suppose Japan at the same time raises the wage rate by 10%. According to equation (1.0), the price of sheeting per linear yard will go up by 1.5% $W/P \times dP/dW = (36/11.39)(0.0474) = 0.15$. Taking the difference between the 2.7% increase in the American costs and the 1.5% in the Japanese costs, we have a net increase of the relative price of approximately 1%. Accordingly, the export sales of the U.S. textiles will be reduced by 3%, notwithstanding the simultaneous increase in wage rates of 10% in each country. In this numerical example, it is not so much the magnitudes of various coefficients that are of significance as the principle we try to illustrate. In an industry with widely different cost structures in two countries, as the American and Japanese producers of textiles, the absolute as well as the relative wage rates are crucial in determining the competitive positions of the exports.

SECTION IV

Our study points out that cost plays an important role in determining the relative textile exports of the United States and Japan to the world market. The wage rate of the textile industry is but a part of the heirarchy of the general wage structure in each country. The higher U.S. textile wage rate relative to that of Japan reflects the higher U.S. productivity in other industries—those in which the United States has a comparative advantage. The productivity in the U.S. textile industry, however, has not been in line with the wage differentials between Japan and the United States, thus making textiles an industry in which Japan has a comparative advantage.

The effectiveness of cost (price) differentials varies from region to region depending upon the extent to which the exports can be brought into the country without direct impediment. In an extreme situation, the theory of comparative cost can be completely nullified as an explanation of the movement of a particular commodity, as for example, Japanese textiles, if quota restrictions against textiles originating in Japan were practiced by the countries in Western Europe. But this finding in no way impugns the theory in explaining the total textile exports of Japan relative to those of the United States. It only uncovers the areas that blunt the sensitivity of price in determining the relative exports of these two countries in the world market.

Where the labor cost is an important segment of the total cost picture as in the textile industry, the percentage change as well as the absolute level of wage rates is crucial in deciding the competitiveness of one country's output *vis a vis* that of other countries in the world market. In fact, given the labor productivity in the short run, relative wages must be such that the costs of textiles produced in Japan and in the United States can be maintained at a certain ratio to each other. Thus by making use of the regional trade statistics and analyzing the various cost components, our study has provided us with a method for examining in a more detailed fashion than before the role of comparative costs in determining the course of international trade.

REFERENCES

[1] Harris, S. E., *Report on the New England Textile Industry* (Cambridge, Mass. 1952), p. 29 in particular.

[2] Harrod, R. F., *International Economics* (Chicago, 1958).

[3] Kravis, I. B., "'Availability' and Other Influences on the Commodity Composition of Trade," *Journal of Political Economy*, Vol. 64 (April 1956), pp. 143–155.

[4] MacDougall, D., "British and American Exports: A Study Suggested by the Theory of Comparative Costs," Part I *Economic Journal*, Vol. 61 (December 1951), pp. 697–724, and Part II *Economic Journal*, Vol. 62 (September 1952), pp. 487–521.

[5] U.S. Department of Commerce, *Comparative Fabric Production Costs in the United States and Four Other Countries*, 1961.

CHAPTER 22

Moderating Economic Fluctuations in the Underdeveloped Areas

By EDWARD MARCUS*

Most of the underdeveloped economies are dependent primarily on agriculture as their major source of income. It is not surprising, therefore, to find exogenous forces as the dominant influence on domestic well-being. The two major factors in this category are the expected components of income—world prices of their major export commodities and the volume harvested, in turn a component of local weather conditions and domestic effort. This last comprises both the intensity of cultivation by the individual farmer and the extension of acreage devoted to such crops.

Sectionalism in the economic impact of the exogeneus factors is also typical. Many farmers limit their cash crops, usually to one, because of climatic and soil factors. For example, in Nigeria cocoa is overwhelmingly produced in the Western Region, whereas groundnuts and cotton are grown mainly in the Moslem North, with the East the primary source of palm products. This regional specialization thus gives rise to two different sets of fluctuations, which could be called general and particular.

A generalized fluctuation would be an upswing or downswing affecting commodity prices in general, such as occurred in the depressed thirties or the Korean Boom. Almost all commodity prices rise or fall, though perhaps unevenly in amplitude. As a result, all the export crops prosper or suffer simultaneously.

A particularized fluctuation affects only one crop. For example, cocoa prices might decline because of unusually good crops in a competing producer area, while at the same time an upswing in industrial demand in the Western World might increase palm oil prices. As a result, if we use Nigeria as an illustration, the Western Region would encounter curtailed incomes while in the East farmers would be relatively affluent. It is thus obvious that particularized fluctuations are less apt to lead to countrywide instability, since the chances of all major crops being subject to similar specific factors is less; a generalized fluctuation would usually result from the workings of a worldwide cycle and would be less likely to affect the various crops differently.

The period since 1945 has seen a new force added—development expenditures. Part of this is a reflection of extended aid, and is often

* Brooklyn College.

Although this article is based on the economies of West Africa, it is believed that the generalizations and suggestions apply to most underdeveloped areas.

unrelated to the state of the domestic economy.[1] But a large segment is financed by local revenues, in turn a function of the country's agricultural well-being. Often this revenue is derived from an export tax based on price (P) and/or volume (Q) exported; as a result, a shift in either P or Q will change government income in the same direction. This, in turn, will affect the volume of government spending so that fluctuations arising from the variation of the latter may reinforce those arising from the rural economy. On the other hand, to the extent that reserves are accumulated in the boom to be spent in the recession, development spending can moderate the changes that the farm community is spawning, though in practice such smoothing can be only minor in extent.

In any analysis of the agricultural economy, an essential element is the time factor. One obvious reason is the crop cycle itself. The farmer's cash needs tend to be continuous throughout the year, although they may peak at planting and harvesting. His income, however, will not be known until his crop is sold, usually after the harvest. Except to the extent that the result can be predicted successfully, a good or bad year—the sum of the price and quantity effects—will not be known until the end of the crop year, so a rise or fall in income will not be reflected in spending until after that period.

In practice, this time sequence is much more complicated, particularly for the cash crops. These are almost always for export. In West Africa, for example, the small peasant farmer predominates. As a result, the flow of the produce goes through several middlemen. Licensed buying agents usually are the first purchasers. Their operations tend to amalgamate the many small lots into larger quantities. As these aggregates swell, they pass into the exporters' hands, either large private firms or government marketing boards.

Over the crop year, before the harvest, financing for the grower is a necessity. Hence, at the start of the crop year the buying agent or a local cooperative will undertake to make the initial advances to the farmer. This sum, of course, is only a portion of the estimated final crop. Much of this financing, in turn, will probably be obtained from the local banks. In effect, the credit of the intermediary is the banks' reliance; it is rare for the indigenous farmer to have direct access to the original source of funds. Unlike the bank, the intermediary, through long years of dealing, has gotten to know his debtors; the magnitude of the task of assessing so many small risks is beyond the competence of the banks. Even with this division the bad debt record is still one that would make the orthodox Western banker shudder; it is not unusual for half the loans to be "slow,"

[1] At times, however, aid is extended to help alleviate depressed conditions.

and ultimate collection often rests upon various unsubtle coercive devices. (As a result, loans at 100± per annum are not unknown in rural areas.)[2]

The use of advances results in a two-peaked cash flow over the year. First, as the crop season begins, the initial movement of cash to the farmer starts as a loan from the buying agent. This cash is then gradually disbursed for the farmer's (cash) living expenses and operating costs. Then, usually three or four months after the loan, the start of the harvest begins. The first fruits are used to pay off the advances. It is not until the next quarter, when the bulk of the crop is marketed, that there occurs a second inflow of cash to the growing area. Once enough of the output has been sold to repay the loan, the balance can then be used to bring in money for the farmer's own wants. It is at this point that a good or bad year shows up; if prices are high and/or the quantity sold is above average, more is left over after liquidating obligations—conversely for poor years. In other words, it is in this final quarter that will determine the degree of rural well-being—how much of a surplus remains to be spent in the period until the next year's crop advances once more start the influx of cash.

One qualification that should be made arises from the possibility of a bad year for food crops. In many areas the small farmers grow their own basic foods in addition to the cash crop. But it is also possible that they are dependent to so great an extent on their marketable sector that they are also buyers of their necessities from the food-growers. If a good year for the cash crop coincides with a bad year for the food crops, then the result may be excessively high living costs, thus diverting much of the additional cash to the food-growing areas. This would alter the sectional impact of any increased spending from the export crop region to the food producers, and, to the extent that the two groups had different spending patterns, the "mix" of the demand component would also differ.

Unlike its role in variations in the industrial economy, the banking system plays a relatively passive role in the agricultural shifts of the underdeveloped country. Two factors that narrow its participation are the limited use of checks and the divorce of the original lender—the bank—from the final borrower—the grower. As a result, the banks function merely as cash depositories for the rural area. In the lending season cash flows out via the buying agents to the farmer; in the harvest period the tide is reversed. Consequently, the loan expansion does not build a credit pyramid, reinforcing the initial expansion (or contraction). Instead of the loan proceeds flowing into other banks, there to start another cycle as in the United States and Great Britain, cash flows out and remains outside the banking system until the end of the crop year.

To meet these cash movements, banks in the underdeveloped areas tend to look to their head office balances as the adjusting device. In most

[2] J. C. de Graft-Johnson, *African Experiment* (London: Watts, 1958), p. 120.

countries the major banks are branches or affiliates of American, British, or Continental European units; the adjustment is therefore in New York, London, Paris, Brussels, etc. In effect, as will be shown, the head office (abroad) undertakes the role that would otherwise fall to the (local) central bank. At the lending season, balances are drawn down, and the funds are converted into the local currency to meet the resulting drain. During the repayment period, when cash flows into the banks, the currency is then sold for deposit at the head office (or for the purchase of treasury bills in the parent country).

TABLE 1

Licensed Banks in Nigeria
(£ 000)

1960	Balances held abroad	Invest-ments abroad	Total foreign assets	Balances held for banks abroad	Capital	Due to foreigners	Net assets abroad
Jan.	25,564	329	25,893	10,151	2,695	12,846	13,047
Feb.	30,244	309	30,553	8,766	2,702	11,468	19,085
Mar.	29,274	282	29,556	7,371	2,752	10,123	19,433
Apr.	31,577	272	31,849	7,313	2,815	10,128	21,721
May	29,116	252	29,368	7,663	2,849	10,512	18,856
June	28,438	727	29,165	8,776	2,867	11,643	17,522
July	28,702	232	28,934	10,392	2,869	13,261	15,673
Aug.	29,851	179	30,030	15,193	2,871	18,064	11,966
Sept.	25,850	179	26,029	17,050	2,886	19,936	6,093
Oct.	24,043	179	24,222	20,857	2,886	23,743	479
Nov.	24,867	179	25,046	24,889	2,966	27,855	– 2,809[a]
Dec.	21,254	179	21,433	18,368	2,966	21,334	99

[a] (–) Net foreign investment in Nigerian banks.

Source: Federation of Nigeria, *Official Gazette* (Lagos, 1961), August 3, 1961, Vol. 48, No. 56, pp. 1043–1044.

It should be noted that one result of this use of head office balances is to reduce virtually to zero the contribution of the banks to the local capital supply. Quite often the net amount of balances held abroad exceeds the parent offices' capital invested in the local branches; in other words, the underdeveloped economy is, in effect, a net short-term lender to the financial parent. When to this is added the divorcement from the farmer—much of the banks' advances are to the large foreign trade houses, rather than to the indigenous community—the resulting hostility to the alien-owned (or controlled) banker can be understood. Rather, the foreign bank is at best only a seasonal supplier of capital; at the peak cash needs, balances abroad may be drawn down so low as to become negative, at which time the head office is a net lender to the underdeveloped economy. For example, if we net inter-country bank balances plus capital for the Nigerian licensed banks, the 1960 monthly figures show only one month in which Nigeria was a net recipient of foreign banking capital (Table 1).

It would seem that the cash cycle would be reflected in a lagged effect on foreign trade. That is, the initial increase in cash in the rural sector would first show up in an increase in orders from the local traders; a similar but increased reaction would also characterize the latter part of the harvest period. This step-up in sales would then stimulate new orders from domestic wholesalers, who, in turn, would then increase their purchases of imported goods. Since transportation and communication facilities are so poor in the underdeveloped areas, especially outside the urban centers, the interval between the initial cash flow and the effect on payments for imports could well be up to two years.[3] In other words, a good crop this year could mean an adverse balance of trade two years hence. (It is undoubtedly true that the deterioration in Nigeria's foreign position in the late fifties reflected the lag behind the good crops of the previous years.) A succession of good crops or high prices, of course, would continue to sustain export earnings, but once this item fell back, imports would still continue high under the impetus of earlier good years.

In practice, however, the activity of the major import houses on the domestic side tends to reduce this lag appreciably. Quite often they are also active as buying agents for the major domestic crops. As a result, they soon learn whether the signs are for a good or small crop, and these signs, combined with their price forecasts, can then be built into a projection of rural purchasing power. With this as a guide, imports can be ordered forward and their arrival timed to coincide with the cash movement at harvest time. The greater the forecasting accuracy, therefore, the closer the movement of exports and imports, thus reducing the gap in the foreign balance that the lagged effects would otherwise have introduced. Actually, it is probably easier to anticipate the crop size than the world price level at selling time; quantity changes in the availability of export crops are thus more likely to influence import shifts and advance ordering.

Within the underdeveloped economy the influence of these agricultural movements is not uniform. Even within the producing sector the impact may vary, especially in a period of low selling prices. This difference, essentially, is the reaction of the large plantation in contrast to that of the small peasant farmer.

The large plantation—usually run by expatriates—is often encumbered with a considerable overhead cost. In part this element represents amortization and depreciation of the original capital cost, in part the use of machinery and other expensive capital equipment, and in part a costly expatriate managerial class. Hence, when prices drop, there is a narrowing

[3] Cf. the 33-month cycle for a textile transaction intended for the Chad, reported in the United Africa Company, *Statistical and Economic Review*, No. 5, p. 31.

of gross profit margins and a tendency to step up output, the added quantity somewhat offsetting the lower per unit return.[4] Often there is sufficient casual labor in the vicinity to permit such a temporary expansion in the work force. In contrast, the relatively self-sufficient peasant meets much of his needs, such as food, from his own effort; the cash crop is his source of surplus income, and variations are thus only of marginal importance. Moreover, any attempt on his part to expand output would involve hiring labor; this would raise his direct costs, which, unlike those of the plantation, could not be offset by a resulting greater spread of overhead, since he uses practically no capital items. As a result, in periods of depressed prices he is less likely to expand production, and indeed this is usually the period when casual labor deserts the native economy for the plantation.[5] The supply of peasant-grown produce is thus relatively elastic with respect to price.[6]

Because of the overwhelming importance of agriculture, local industry must look to this sector for the bulk of its sales. Hence, the shifts in rural well-being are reflected in the newly industrializing urban centers. In view of their newness, it is unlikely that any are manufacturing for export; their market is the few emerging cities and the vast farm hinterland. As a result, there is no important alternative outlet to which sales could be shifted when agricultural purchasing power declines; only if government spending can be stepped up is there a balancing possibility.

As already indicated, trading activity would also reflect the shifts in the farm areas. Here, however, a distinction should be drawn between the large trading firm and the petty operator. The former, as discussed above, is able to anticipate to some extent the expected shifts in purchasing power, and to gear its purchases accordingly. The small merchant has no such ability, and is more dependent on the immediate shifts. When trade is in the doldrums his stock of goods turns over more slowly, and a brisk trade shows up in a higher frequency of sale. Even his pricing policy acts as an obstacle, being based on a crude first-in-first-out approach. If prices are rising, the corner trader does not raise his resale price until he goes out to buy replacements. Then, when he actually encounters the higher cost, he responds by raising his selling prices accordingly. As a result, since his replacement cost is rising, the physical volume of his stock may well decline in an upswing, as his limited capital fails to grow as

[4] Cf. J.-J. Poquin, *Les relations économiques extérieures des pays d'Afrique Noire de l'Union française, 1925–1955* (Centre d'Etudes Economiques et Mémoires 37) (Paris: Libraire Armand Colin, 1957), p. 120.

[5] Cf. International Labour Office, *African Labor Survey* (Geneva, 1958), p. 82, and the East Africa Royal Commission, *Report* (Cmd. 9475) (London, 1955), p. 148.

[6] Cf. the experience in Uganda with cotton and coffee, summarized in E. and M. Marcus, *Investment and Development Possibilities in Tropical Africa* (New York: Bookman Associates, 1960), pp. 264–265.

rapidly as the rising inventory costs. Turnover is more rapid, but the base stock is less. The reverse, of course, holds in the downswing.

However, the response of the banks is tied almost exclusively to the large trading firms; the small trader—and indeed most of the indigenous community—is virtually without credit standing. Trade loans are thus given mainly to the well-established houses oriented primarily towards foreign trade. The mass of domestic traders have to rely either on their own resources or the local extortionate money-lender. Some elasticity in this native sector, it should be noted, does arise from the agricultural shifts. In boom times some of the increased agricultural income is invested in trade, thus permitting it to expand at the same time, while in the recession funds flow back to mitigate the effects of the downturn. (This same relationship holds for capital available to small-scale indigenous industry: an expansion of availability of funds when agriculture is booming and a shrinkage or even reverse flow in a recession.)

Given the agricultural origin of its economic fluctuations, what can the underdeveloped country do to control and mitigate these forces? Or is it so helplessly dependent that its only salvation lies in relying on the stabilization policies of the industrial nations—its major customers—hoping that in this way demand and competing supplies may become less irregular, thus allowing it to experience less unevenness? Surprisingly, these underdeveloped economies can do much to moderate these various influences—although many have in practice yet to act—so as to utilize all the possible weapons in the armory. In particular, greater use of financial devices could help bring about some desired improvement in internal stability.

The most widely known instrument is the government-controlled marketing board. Using the services of licensed buying agents to do its field work, such a board buys the crop locally and sells abroad at world prices. As a stabilization measure it attempts to set a buying price that will—over the cycle—about equal the average of its selling prices (less costs of operation of the Board, shipping charges, etc.). In times of high world prices the Board builds up a reserve, to be drawn on when selling prices dip below its purchase price. In this way, one of the two variables affecting the farmer's income (P) is stabilized; he is still subject, of course, to income variations arising from good or bad harvests and other forces affecting his marketable volume. By having a stable price, he is also encouraged to expand operations, knowing that such a step will not end in disaster should its fruits be realized in a time of depressed prices.[7]

[7] Cf. F. J. Pedler, *Economic Geography of West Africa* (London: Longmans, Green & Co., 1955), Ch. 17, and P. T. Bauer, *West African Trade* (Cambridge: Cambridge University Press, 1954), Ch. 23.

(Sometimes the stabilized buying price is set below the expected average selling price, less handling costs, in order to build up a permanent reserve that can be diverted to other development needs; in effect the farmer is taxed to subsidize the other programs.)

A variant is the use of a progressive export duty. Above some normal or average price the government imposes a tax, which rises in proportion to the world price. This results in less net income to the domestic supplier, thus reducing the proportionate rise in income. As prices drop a progressive subsidy is paid, the funds coming from the reserve obtained when prices were high. As a result, prices to the domestic producer fluctuate in the same direction, but with less amplitude, as do world prices.

To illustrate: assume $1.00 is the accepted normal price. Up to $1.10 the excess is taxed 20%, from $1.10 to $1.20 the addition is taxed 40%, and from $1.20 to $1.30 it is taxed 60%. Similarly, from 90¢ to $1.00 a 20% subsidy is paid, rising to 40% when prices are 80¢–90¢, and to 60% in the 70¢–80¢ range. As a result, a rise in world prices from 70¢ to $1.30—almost double—would raise prices to the farmer from 82¢ to $1.18, or less than half. (For this simple example it is assumed that the quantities sold above $1 equal those sold below $1, so that the subsidy and tax are equal in amount.)

Essentially, the objective is to siphon off income in boom periods and supplement purchasing power when prices decline. But income variations can also arise from quantity shifts, and to control this element other forms of intervention are required.

One possibility is a road tax proportionate to the quantity marketed. Obviously, all the produce for export must go by rail, road, or water. When incomes are high because of good harvests such a tax would automatically rise, thus capturing a larger portion of the cash that would otherwise accrue to the farmer. At the same time it would not touch his subsistence crops—those items raised for his own consumption—since these do not use public thoroughfares. Where marketing boards exist, this objective can be achieved by widening the spread between the purchase price and the world price; the tax would be an additional deduction from the proceeds, analogous to insurance and handling costs, thus reducing the per unit take-home amount. A minimum exemption, such as the amount sold in the previous crop year, would remove from tax those not fortunate enough to enjoy increased yields, thus limiting the tax impact to those actually enjoying an increased volume of sales.

Another feasible tax would be education fees. In many underdeveloped countries there is a tremendous drive on the part of parents to send their children to school, for this is regarded as the road to material success and prestigious employment. In boom times, when incomes are up, more pupils would result in increased government revenue if a fee were imposed for such attendance. Universal free schooling, in contrast, would not affect the spending power in the producer's hands. However, there is the drawback that in a year of depressed incomes, with less money available for such uses, would mean an interruption in the child's education, possibly discouraging any return for further attempts at self-improvement.

A less direct approach would be to increase tariff duties in the boom period, especially on consumer goods. This would absorb some of the extra purchasing power and perhaps divert some consumption into savings. Even if consumers switched to domestic supplies in competition with these imports there would be a net gain, for this added demand for locally produced goods would stimulate industry, a development objective in all these countries. However, the increased costs may touch off a wage spiral in the urban areas, not only reducing the impact on local purchasing power, but also lessening the competitiveness of the nascent industries. While trade unions are generally weak, the government as the leading employer is particularly susceptible to such wage pressures, and once an increase is granted, all other employers are soon forced to follow.

Another control possibility, although of limited influence, is the use of pay-as-you-go taxation. Nigeria, for example, is experimenting with such a policy. In practice, it can be operated only by the manufacturing and trading establishments, since the self-sufficient farmer is not an amenable tax subject. While the taxable sectors form a very small minority of the total gainfully employed, average money incomes are usually well above those in the rural areas, thus giving urban workers a disproportionate share of the underdeveloped country's tax capacity. To the extent that the urban worker is a buyer of locally grown produce, his purchasing power can affect agricultural income. Since even the well-to-do expatriate element spends the bulk of his income on local goods and services,[8] this can amount to a significant though marginal factor. Variations in the rate of taxation can thus be used in a manner similar to those proposed for the more advanced economies, as long as it is recognized that by itself this tool is of only limited usefulness.

A particularly promising possibility lies in an active policy by both the government authorities and the central bank[9] to promote the creation and growth of an indigenous money market. In particular, the development of an active trade in government short- and long-term issues should be emphasized.

[8] Cf. G. B. Stapleton, *The Wealth of Nigeria* (London: Oxford University Press, 1958), p. 95. Cf. also Ghana's compulsory saving plan, which is to include the farmers in its coverage (*West Africa*, No. 2309, September 2, 1961, p. 979).

[9] If none exists then implied in this discussion is the immediate establishment of one.

At first glance such a suggestion seems far-fetched. Even so advanced an economy as that of Canada between the wars had only the rudiments of such a financial center. Since the use of checks in the underdeveloped countries is still so restricted, efforts to introduce the more sophisticated financial instruments would seem futile, indeed.

Yet closer examination of these economies reveals that there is a potentiality for monetary control if the required structure exists, and that this can be implemented despite the relative poverty of the people. Treasury bills, for example, are now in existence in West Africa, and their greater use could help the control objective.

Before examining some specific proposals, a brief summary of the current banking operations will indicate the path to be taken. To be more specific, the discussion will illustrate the current practice in West Africa. In general, the major banks in the area are associated with leading banks in the United States and Western Europe. In British Tropical Africa, for example, they amount to branches of London, and, to a lesser extent, Paris, and New York banks.[10] Most of their business is with the local expatriate community and the various government and quasi-government funds. Credit to the private sector is mainly short-term financing of local and foreign trade and of construction.

In British Tropical Africa the banks keep the bulk of their primary reserves with the overseas head office. In Nigeria they do maintain small balances with the central bank of that country, but these are mainly for check-clearing use. In practice the head office functions as each bank's "lender of last resort," the ebb and flow of local currency needs being reflected in the movement of head office balances.[11]

At the beginning of the crop year advances step up to finance the local crop purchases. Because currency is still so important for the rural elements, this amounts to an equal outflow of cash; very little remains as checkbook money, either with the lending or other banks. To meet this cash drain, balances at the head office (London mainly) are drawn on. These are sold for local currency, or presented to the currency-issuing authority. (Because of their dominant position, only two banks need be considered, Barclays Bank D.C.O. and the Bank of West Africa.)

To meet the local cash need, sterling balances in London are turned over to the central bank (for Ghana and Nigeria) or the West African Currency Board (for Sierra Leone and Gambia). Thus, West African sterling balances in London are unchanged, though ownership has shifted

[10] Three American banks—the Bank of America, Chase Manhattan, and First National City—are in West Africa.

[11] Another potential advantage in the minimization of the head office's stake in the country is to lessen the possible loss should the local government attempt nationalization with inadequate compensation.

from the commercial banks to the official currency authority. Currency is then paid locally to the drawing bank, increasing the amount of money outstanding. It should be noted that in this procedure there is no change in the fiduciary backing; the increase in currency liabilities is matched by the rise in sterling assets.

The currency so obtained is then used to meet the aforementioned seasonal drain. In practice, a good deal of this outflow is often in coin, since paper is still regarded with suspicion. (One British branch in Northern Nigeria literally keeps a roomful of coins!) As recently as 1955 the value of coins in circulation in Nigeria exceeded the currency outstanding, although the ratio of coin to currency has since declined to 2:5 (as of mid-1961). The drain, it should be noted, arises from both the rise in advances and the drawing down of deposits—balances that the crop buyers had built up in prior seasonally slack periods. The extent of this variation can be illustrated from Nigerian data: in 1960 demand deposits declined from £49.2 million in April to £40.7 million in November, while short-term advances rose from £31.8 million in May to £54.6 million in December. As a result, balances abroad declined from £30.5 million in April to £20.2 million in December while balances due to the head offices rose from £7.2 million in April to £24.7 million in November. Roughly, ignoring the one-month discrepancies, the net drain required a shift of about £25 million (cf. Table 1) in foreign balances. Money in circulation (outside the banks) rose some £20 million.

With the harvest, of course, the reverse flow occurs. Advances are paid off and deposit balances are built up, cash flowing back to the banks. The excess is then turned into the currency authorities and retired, while sterling balances in London shift back to the commercial banks.

An additional influence is the interest rate in the head office money market (here, London). A high rate induces the local (overseas) branch to send surplus funds there to get the extra income, or to curtail its borrowing from the head office to cut costs, since this latter charge is related to the going rate of interest; a low London rate acts in the reverse way. To offset this factor, rates in West Africa must follow suit, rising when money is tight in London, thus countering the "pull" of the head office, and declining when money is easy. (For example, the 2% rise in the Bank of England rate in 1961 was followed by an equivalent rise in prime rates charged by the Nigerian banks.) In effect, therefore, the monetary position in the overseas money market determines the level of interest rates in Africa. Local money conditions are dependent functions. As a consequence, should tight money in London coincide with a depressed economy in Africa, the latter would experience rising interest rates and still more pressure. Similarly, monetary ease in London and a local boom in Africa would be met with a local cheap money policy, fanning the expansion still further.

Working in the same direction is the reserve position of the head office. If money is tight in London, it is likely that the reserves there are at a minimum. Expansion of the African branches would mean a drain of reserves to the African currency authorities. To avoid compounding an already difficult head office position, the "export" of a high interest rate to the African branches would reduce lending there, curtailing their need for reserves, and thus lessen the pressure on London. The opposite chain of events would ensue when London reserves were ample. Parenthetically it might be added that it is this dominating influence of the money market abroad that has supplied much of the pressure for the establishment of central banks in the newer countries; this economic dependence prevents them from having a monetary policy different from that dictated by head office conditions.

The reader will note that the traditional functions of the central bank are being exercised by the head office. Its lending rate corresponds in influence with the discount rate, while it is looked to by the overseas branches as a source of needed reserves or as a depositary for excess reserves. In a sense, even the open-market operation can be brought within this analogy, for the head office could increase or decrease the capital allotted to its various branches. In practice this last step is accomplished by shifting funds from surplus to deficit areas; Nigerian funds, for example, may well have financed Rhodesian and South African business in the past (via the London office), since lending opportunities in the latter countries have usually been much more attractive. It can also be readily understood how these ties hinder development of local money markets.

In all fairness, one significant fact must be added to offset the criticism implied in this description: in the past there were no local financial instruments that the branches could use as a means of investment. Treasury bills were unknown and stock exchanges did not exist. The amount of good commercial paper available was limited; the credit standing of the indigenous trader was too questionable to permit reliance on such paper. As a result, the only outlet for excess reserves was the head office, and these reserves could always be drawn on when local requirements rose. Hence, by default, the head office became the center through which the overseas branches balanced their various reserve positions.

As can be inferred, the suggestions for increasing the possibility of monetary measures to moderate economic fluctuations involve various steps to lessen the ties with overseas money markets and the establishment, insofar as possible, of a local money market. Admittedly this is no easy task; poverty and a lack of financial sophistication are two seriously inhibiting forces. Yet limited steps leading to an increase in the degree of local control can still be attempted.

Foremost among such efforts would be the requirement that the commercial banks keep their reserves locally. Included in this suggestion is that the central bank should have the power to vary the required ratios. While a portion should be kept with the central bank, too great an amount in this form may reduce the commercial banks' income unduly. Hence, it might be simpler to keep a portion of such required reserves in government bills or longer-term obligations. As an incidental aid, this latter alternative would also broaden the market for these securities.

The changeover would involve the surrender to the central bank of head office (e.g., sterling) reserves in equal amount to the reserves to be maintained with the central authority. The central bank, of course, would have an equal increase in deposit liabilities and foreign exchange assets.

The reserve portion to be kept in government securities would involve a somewhat different procedure. The simplest arrangement would be if the central bank already had a portfolio of such assets; these could be sold to the commercial banks in exchange for a corresponding amount of external balances. For the central bank this would mean the substitution of non-earning assets, unless these new reserves were invested in overseas securities, e.g., British Treasury bills.[12]

If, however, the central bank did not have sufficient securities to sell, two alternatives would be necessary: the commercial banks could purchase such securities from other holders, either within the country or abroad, paying for them by drawing on their head office balance, or the government could issue additional obligations, to be taken up by the banks.[13]

Commercial bank purchases of securities from internal holders could be paid for either by drawing on head office balances or by domestic check, swelling demand deposit liabilities. Many secondary reactions could then ensue, and only a few will be mentioned as illustrations.

If the sellers are expatriate corporations, they may prefer to be paid abroad; in effect, as the commercial banks are bringing in capital to purchase securities, the sellers would be withdrawing capital. Or, alternatively, the funds so received might be ploughed back for expansion in

[12] There would still be some loss of income if the local government securities that had been sold carried a higher yield than the overseas bills acquired.

[13] Or the government could increase its capital investment in the central bank, paying for this purchase by the deposit of appropriate new securities. The central bank, in turn, could then sell these newly acquired obligations to the commercial banks in exchange for overseas reserves. Since the central banks in the newly independent countries are government-owned, this approach could be effected easily. Note that the end result is to increase the central bank's foreign exchange reserves and domestic liabilities equally. The government has added to its interest-bearing debt, offset by whatever income accrues to it from its added investment in the central bank.

other directions, either in the host country or in other areas wherein the company operates.

If the sellers are residents, it is probable that they would use the proceeds within the country. Here the net effect of the banks' action is to raise their deposit liabilities and domestic investments, leaving the external reserves unchanged. It is also possible that the sellers will draw currency from the banks; this would necessitate the banks turning their external reserves into the central bank in order to replenish their cash. In either event the money supply (M and/orM') will expand, subsequent multiplier effects would raise imports and thus react on the external balances.

If the government issued additional securities[14] it would acquire foreign balances in exchange. These could be deposited with the central bank, either to reduce the latter's holdings of government securities or to raise the government's working balances.

At the same time as this suggested concentration of reserves takes place, similar steps should be taken with reserves owned by quasi-governmental organizations, such as the marketing boards. (In Nigeria, for example, an end-of-1960 census listed official and semi-official sterling balances of £83.3 million, compared with £77.6 million owned by the central bank.) While it is true that most of these units do need funds overseas, it would still seem wiser to concentrate external assets with the central bank, with freedom to utilize them as at present, thus strengthening the position and information of the monetary authorities. And, to the extent that some of the units' operations offset each other, on the insurance principle, it is possible that a smaller overall foreign balance would be required.

Assuming the transition has been effected smoothly, an additional measure will still be needed to weaken the pull of the head office: discriminatory taxation. This could be in the form of either a special tax on income from overseas investments—specifically on income from "paper" holdings—or an added foreign exchange fee charged each time the bank transfers a large amount to or from its head office. (Exemption of smaller transactions would eliminate an effect on the normal flow of business payments.) Whatever the form, the objective would be to reduce the relative attractiveness of the foreign money market as a temporary resting place for short-term surplus funds.

Since an integral part of the program for increased central bank control requires the development of an active local money market, the bringing home and concentration of bank reserves should be coupled with a vigorous use of treasury bills, especially to smooth seasonal influences. It will be recalled that the outstanding monetary characteristic of the underdeveloped economies is the cash shift over the crop year; by appropriate

[14] Other than as in note 13.

use of treasury bills these swings could be utilized to extend the effectiveness of the nascent money market.

Essentially, the suggestion involves the appropriate timing of treasury bill operations. At present the commercial banks experience a cash drain at the start of the crop year and a return flow at harvest time. In response, overseas reserves are drawn on to meet the cash drain and then are rebuilt as the crops are sold. Since one suggested goal is to break this tie with the overseas centers, treasury bills should be employed in place of foreign balances. (It should be recalled that in concentrating bank reserves with the central bank an additional portion was to be held in government securities.) At the beginning of the crop year the commercial banks would reduce their holdings of treasury bills and in cash surplus periods they would be net buyers.

To effect this timing, two approaches appear feasible. If the government's fiscal operations show a corresponding seasonality—excess revenues at the start of the crop year, excess expenditures at harvest time—then the run-off of bank-held bills could be through the use of the revenue excess to redeem the maturing issues. At harvest time the government seasonal deficit could be met by a step-up in new bill issues, at the same time absorbing the excess cash with the banks.

However, it is more likely that the fiscal seasonality will not dovetail so neatly; harvest time means rising exports, with an accompanying boost in export tax receipts. It is thus even possible that the government would experience excess revenues simultaneously with the return flow of cash to the banks, and a need for funds at the start of the crop year, when the banks were under pressure. If such were the case, a second approach—intervention by the central bank—would be required. At the start of the crop season it would have to buy enough treasury bills to bridge the fiscal gap and supply the required cash to the commercial banks, and at harvest time it would have to allow its bills to run off, in part by using excess government revenues to reduce the outstanding volume, in part by selling to the commercial banks to absorb the seasonal excess of cash. As an incidental by-product, the transfers between the central bank and the commercial banks, by increasing the frequency of quotation, would add another dimension to the trade of the local financial center in such securities.

To supplement its use of the treasury bill, the central bank could encourage the use of its advances by the commercial banks. Since the cash drain coincides with the rise in their advances to the crop buyers, the advances could be pledged with the central bank for additional reserves. The repayment of advances at harvest time would then allow a reduction in central bank borrowing. This, of course, is the approach envisioned in more settled agricultural economies, such as that of the

United States at the time the Federal Reserve System was being established. (In Nigeria, for example, the six-month swing could easily involve a tripling of bank advances; for groundnuts in 1957 and 1958 this item jumped more than tenfold, or from 3% of all advances to 20%.)[15]

The broadening of the money market need not involve only the banks. Even in relatively underdeveloped economies non-banking funds are growing—social security, pension, insurance, etc. These too could become a source of activity in the government securities sector, though again some government prodding is necessary, mainly through discriminatory taxation. For example, income from funds invested outside the country could be assessed a penalty rate. Alternatively, funds invested locally—or, more narrowly, in government securities—could be tax-exempt. Thus, not only would there be an incentive to repatriate overseas balances, but there would also be an attraction to enter the local securities market. The expansionary effect on the latter would grow as these various funds increased in size, particularly the more sophisticated ones such as those of insurance companies operating locally and the pension funds of large companies.

While the preceding discussion has been aimed at a broadening of the monetary sphere, an additional gain would accrue from the concentration of foreign balances in the central bank. As indicated, these would be matched by an equivalent increase in central bank liabilities; this one-for-one rise would thus add appreciably to the "cover." Since only a portion of these liabilities need foreign exchange (or gold) backing, part of the newly acquired assets could be utilized for development projects. The foreign exchange portion could be siphoned off through government issues to the central bank, raising the fiduciary portion accordingly. As a consequence, low-yielding assets abroad would be replaced by domestic (real) capital formation likely to yield a far greater income. In general it can be said that the newly formed central banks—certainly in Africa—have been excessively cautious in utilizing the fiduciary portion; in Nigeria the coverage often approaches 100%.

Assuming the suggested measures are adopted, how could the central authorities employ them to mitigate the economy's swings, and how effective could this use be?

One definite gain has already been noted: the seasonal swings within the crop year would no longer show up in the overseas head offices, but rather at the central bank. That alone would be an advantage, especially because of the politically explosive implications of the present relationships. With nationalism so rampant in the areas under discussion, head

[15] Cf. the Federation of Nigeria, *Annual Abstract of Statistics 1960* (Lagos: Federal Office of Statistics, 1960), Table 12.

office dominance is too easily looked on as a new form of continued "economic colonialism." Hence, resort to a domestically located central bank as the source of financing and depositary of excess funds should help ward off attacks on this score.

Furthermore, the central bank would have added powers to influence the lending ability of the banks. Obviously, its control over reserve ratios would enable it to increase or ease pressure on the commercial banks, exerting an influence similar to that of the Federal Reserve Board. In fact, since there are so few banks serving any one underdeveloped country, it is probable that this weapon would operate more efficiently. And, where the system is so concentrated, moral suasion could be used as an effective supplement.

In addition, a coordinated treasury bill policy could also implement a desired credit policy. To tighten money at the start of the crop year, for example, it would only be necessary to reduce the seasonal purchases of bills from the banks. Similarly, at harvest time money could be made easier by reducing the volume of treasury bills normally offered for sale by the central bank. At other times a more active use of a market yields would be necessary; a tight money policy at harvest time would call for bill prices low enough (yields high enough) to persuade the banks to buy, running down cash. An easy money policy at the start of the crop year would call for a greater-than-usual purchase of bills; in order to induce the commercial banks to respond, prices would have to be bid up sufficiently to accomplish the switch. Of the four possibilities touched on, the last one would be the most difficult, for even high yields might not tempt the banks to sell if no sufficiently attractive income alternative were available. However, this is a possibility facing all central banks—in depressed times an easy money policy has not the chance of success that a tight money policy has in boom times.

It must be remembered, however, that economic disturbances in the underdeveloped area start primarily in the agricultural sector, either through price shifts or variations in the size of the harvest. As a result, the first impact is on rural incomes. Hence, as already explained above, the marketing board has been devised to even out the price fluctuations, while various taxes on quantity help smooth out the income effect of these variations. Monetary policy can be an influence in the next stage, as these variations affect the urban centers and the balance of payments.

If, for example, high world prices for primary products have touched off a rural boom, consumption of manufactured goods would increase, and, given the relative lack of self-sufficiency in these areas, imports would rise sharply. To counteract this in order to conserve foreign exchange for recession periods, or for the purchase of capital goods of higher priority, a tight money policy could be instituted. The resulting curtailment of loans would then hit the trading community, since usually almost half the loans are to this group.[16] The restriction of credit would compel the import houses to reduce their purchases abroad, lessening the volume of foreign supplies (except to the extent that these houses could call on funds abroad, such as from the head office). Since the volume of rural purchasing power will not have been affected, the three resulting possibilities are: (1) a reduction in rural consumption (and increase in savings); (2) a shift in rural spending to the unrestricted domestic supplies; and (3) a rise in the domestic price of imports and accompanying increase in middleman profits. Possibility (1) is obviously desirable in the boom; (2) would act as a stimulus to the newly industrializing sector, again a desirable developmental objective; and (3) should result in increased government revenues, since the import houses are generally the large expatriate firms that can be reached by the country's income tax machine; it is the petty traders whose operations are difficult to audit. In addition, if the check on imports is not sufficiently effective, an additional measure could be a tariff increase, which would work in the same direction.

The other major sector that would probably feel the credit squeeze would be the construction industry, since it is a heavy borrower. (About 10% of the Nigerian banks' advances are to this category.) The resulting curtailment of activity would cut urban incomes; in effect, the boom–tight money sequence would redistribute the overall income from the urban to the rural population. However, this would not be as painful as in the more advanced economies; in the underdeveloped country many urban residents are migrants from the farms, and the shift in relative prosperity back to agriculture would probably be accompanied by a migration of the unemployed back to their original homes; they would thus share the increased incomes that accrue there. In Africa, it should be noted, the typical urban worker always has "one foot in the reserves," as insurance against unemployment or old age. As a consequence, these population shifts are easier and much more prominent than in Western Europe or the United States.

It is also possible that the tight money policy could be used to increase rural savings in another way. If the restriction is continued into the new crop year, there would be a decline in bank advances to the buying agents. This, in turn, would reduce the initial flow of cash to the producers. If noticing of this tightening were communicated sufficiently in advance of the actual implementation—by rumor, word-of-mouth, etc.,—the farmers might reduce their harvest-time spending in order to conserve cash for the leaner quarter coming. If, at the same time, the supply of imports were restricted, there would be an added inducement to save.

[16] *Ibid., loc.cit.*

The answer really depends on the providence of the farmer; local opinion differs on this point, and little can be said definitively. Certainly, the low income level of the peasant would seem to preclude any margin for savings. But in Nigeria there has been a steady rise in both savings deposits and money in circulation; much of the latter may represent small-saver hoarding. Whether the farmer could be influenced to look ahead is thus problematical, although the rural grapevine is still a notoriously efficient instrument, and in Africa even the subsistence farmer is growing in financial sophistication.

It might be asked if symmetry implies that an easy money policy would work in exactly the opposite fashion. If the rural elements were depressed, could effective monetary action offset the lag? Certainly it should help to stimulate construction loans, thus tending to shift incomes—and workers—back to the city. There are now enough urban dwellers whose income comes from government and service employment to make their earnings independent of the rural economy; any easing of lending conditions should stimulate borrowing to build.[17]

How the import houses would react is more difficult to answer, since so much of their purchasing volume is geared to rural spending. In all probability an easier money policy would allow them to increase inventories. Hence, the reaction on the balance of payments would be somewhat noticeable, although the resulting increase in imports would be less marked than the decrease associated with tight money in the boom.

For greater effectiveness the development program's spending should also be used to offset fluctuations in the other sectors. This is admittedly a heroic recommendation, for the urge to accelerate growth is likely to predominate over any cyclical considerations. Yet the dimensions are such that this possibility should not be ignored; in Nigeria, as one example, government investment probably accounts for more than half the gross capital formation. Since probably more than 50% of these outlays are for construction,[18] there should be some scope for introducing sufficiently large programs that can be postponed or accelerated to help smooth out the economy's movements.

One stricture should be added as a concluding remark. As long as the underdeveloped economies are dependent primarily on the export of primary products, it is doubtful that they can hope to mitigate the major part of the fluctuations to which they are subject. World prices and demand are too important as influences and are beyond their effective control.

[17] In Nigeria probably half the paid employees—excluding peasant farmers—are in this category. Cf. the *1960 Abstract*, Table 4.

[18] Cf. the Federation of Nigeria, *Digest of Statistics*, vol. 10, No. 2, April 1961, Table 55.

Furthermore, the pressures on the marketing boards are in practice too strong to permit them to act as buffers; in good times the farmers clamor for higher prices, while governments hungry for development funds are constantly tapping the accumulated reserves. In Ghana and Nigeria increased self-rule has seen a sharp decline in the boards' foreign balances, so much so that their future effectiveness is questionable.

The recommendations in this paper are therefore more of a marginal nature. Greater local financial control and a vigorous use of its powers could enable the central bank to dampen the swings somewhat. But as long as more than 90% of the people depend on agriculture, the country must resign itself to being subject to fluctuations of economic conditions overseas, especially to those in its main markets. It can only be effective in ameliorating the effects of these alien forces. While this would be an improvement over the present state of dependency, it should not be looked to as leading to stability comparable with that achieved in the advanced industrial countries.

PART VII

Econometrics

CHAPTER 23

The Cost of Living Index

By S. N. AFRIAT*

1. INTRODUCTION

The value of an amount of money, since this arises from what can be got with it, depends on prices, and how this dependence can be decided practically is the question to be considered here. This requires the comparison of one amount of money at one set of prices with another amount at another set of prices. But the question as it stands is indefinite in meaning, even though in practical economic life it is unavoidable. Everything turns on the manner in which the question is to be made intelligible; and when this manner is decided in certain ways, the question is going to appear as capable of definite answer.

What can be obtained by purchase with a certain amount of money at certain prices is a commodity *composition* of goods whose cost at those prices does not exceed that amount of money. The restriction on composition will be called an expenditure *balance*,[1] and it depends jointly on the prices and the available money expenditure, but no more than to the extent of the ratio of prices to expenditure. A composition will be said to be within, on, or over a balance, determined by prices and an expenditure, according to whether its cost at those prices is at most equal to, or greater than, that expenditure. A balance and composition, such as could be found from market data for an occasion, together define an expenditure figure; a collection of such figures, as could be associated with a variety of occasions, defines an expenditure configuration. Such a configuration is the form of the data for the question. The question is to be made intelligible in terms of a hypothetical preference system, which determines a best composition on, or what is equivalent, within any balance. Such a preference system that obtains a choice for every balance will be called a proper preference system. It is admissible in respect to any observed expenditure figure, composed of a balance with a composition on it, if

* Purdue University.

I wish to express my indebtedness and thanks to Professor Oskar Morgenstern, who sponsored the investigation which is reported here and to whom this paper is dedicated. Thanks are also due to the Office of Naval Research and the National Science Foundation for grants in support of this work.

[1] The term is fitting with regard to the duality in the roles of balance and composition, and the equilibrium into which they enter together. Both are represented by vectors of *n* elements; and to every proposition involving balance and composition there is a dual proposition in which the terms are interchanged. While a balance can take on the role of a budget restriction, in many discussions it is not tied to this, and needs to be named differently.

that composition is, according to the system, the best one on, or within the balance. Thus, when preference is measured by a differentiable preference function, and preference direction is defined by the direction of its gradient, this criterion is Gossen's law: price and preference directions coincide in equilibrium. However, if any expenditure configuration admits a proper preference system as a hypothesis, by which condition it will be said to be consistent, it must admit an infinity of such hypotheses. Analysis of the data will be an analysis of such a class of preference hypotheses, which is either empty or infinite. The question may now be put in a more explicit form, by asking which is better, the best composition of goods that can be obtained with one expenditure at one set of prices, or the best that can be obtained with another expenditure at another set of prices, the decision of the better and the best being according to some preference hypothesis admissible on given data. On any occasion, expenditure is made at certain prices to obtain goods for the process of living, supposedly in such a manner as to achieve the highest standard attainable with the expenditure at the prices. If prices change, then expenditure must change compensatingly, if the same standard is to be maintained. To specify this compensating change, the cost-of-living index, with one occasion as base and another as current occasion, is defined as the multiplier of expenditure that will compensate it for the price-change from the base to the current occasion, according to some hypothetical preference system. But if this preference system is free to range in the admissible class for some data, this index will be free to describe a certain range. The problem presented, and to which a solution will be given, is to determine the range of a cost-of-living measurement thus prescribed by given data. Samuelson[2] has pointed out the indeterminacy of the cost-of-living measurement; but this leaves open the question of how to determine the extent of the indeterminacy. Some formulae that will be shown achieve this determination, but they are difficult to evaluate and no calculations have been made with them. However, in the familiar case of making the cost-of-living measurement between two periods on the basis of expenditure data just for those periods, the answer is simply that the Laspeyres index gives an upper limit, which can be approached arbitrarily closely. That is, no better upper limit can be obtained without more data, or without more restrictive assumptions about the underlying preference structure beyond that it is of the proper type. Similarly, there is a lower limit, though this turns out not to be the often assumed Paache index.[3] An important fact about index-number

[2] P. A. Samuelson, *Foundations of Economic Analysis*, (Cambridge, Mass.: Harvard University Press, 1947).

[3] S. N. Afriat, "The Method of Limits in the Theory of Index Numbers," presented at the Joint European Conference of the Institute for Mathematical Statistics and the

theory is that it is impossible to have a developed theory with the usual restriction to two-period data, and this is altogether confirmed in the history of the problem. However, the allowance of more data opens the way to further theoretical development, as is exemplified in a particular way by the method of Wald[4] and in a more general way by the method proposed here. The fittingness of such an enlargement of index-number theory is evident in that expenditure data for any period are as relevant to preference structure as those for the two periods between which the index is to be determined.

Wald has considered preference systems on a quadratic model, represented by a quadratic function of composition the magnitudes of which decide the preference relation between compositions. Now a quadratic function can represent a proper preference system in a region only if it is strictly convex and increasing in that region. This is because only a strictly convex quadratic can have strictly convex levels. Further, the expansion curves, which are the loci of composition as expenditure varies while prices are fixed, are straight lines concurrent at a point that defines the center of such a system. Thus, not any pair of linear expansion curves can belong to a quadratic system: firstly, they have to be concurrent, and second, their intersection has to be at a point over the balances associated with the region in which the system is valid. However, with these matters neglected, Wald has shown that if linear expansion curves corresponding to two sets of prices are known, then the cost-of-living measurement is determinate between occasions which have those prices, on the hypothesis of a quadratic model. He then shows that if these expansion curves pass through the origin, the determination reduces to that provided by Fisher's "ideal index."[5] But it should be noted that a proper quadratic preference hypothesis cannot have its center at the origin. Samuelson's revealed preference axiom[6] can be used to decide this. The justification of the Fisher index by its interpretation on the quadratic model, which could seem to arise here, and which was proposed by Buscheguennce[7] and Konüs,[8] is

Econometric Society, Copenhagen, 1963; to be published in *Econometrica.* "An Identity Concerning the Relation Between the Paasche and Laspeyres Indices" *Metroeconomica*, XV (1963), II–III, 136–140.

[4] A. Wald, "A New Formula for the Index of Cost of Living," *Econometrica*, Vol. 7, No. 4 (October, 1939), pp. 319–335.

[5] Irving Fisher, *The Making of Index Numbers* (New York: Houghton Mifflin Company, 1927).

[6] P. A. Samuelson, "Consumption Theory in Terms of Revealed Preference," *Economica* 15 (1948), pp. 243–253.

[7] S. S. Buscheguennce, "Sur une classe des hypersurfaces. A propos de 'l' index idéal' de M. Irv. Fisher," *Recueil Mathematique*, XXXII, 4 (1925), Moscow.

[8] A. A. Konüs, "The Problem of the True Index of the Cost of Living," *The Economic Bulletin of the Institute of Economic Conjuncture*, Moscow, No. 9–10 (36–37) (September–October, 1924), pp. 64–71.

altogether illegitimate. Far from there being such arguments in favor of the Fisher index, there are precise arguments on the same basis refuting its validity.

Consider now an expenditure configuration of four figures, corresponding to four sets of prices. If the sets of prices are parallel in pairs, and a quadratic model is assumed, then the two lines joining corresponding compositions must be expansion curves, and we have the case of Wald, with a pair of expansion curves hypothetically known. But again, for the quadratic hypothesis to be legitimate these lines must intersect at a point with a proper location.

Wald's case can be considered a special case of a general configuration of four figures. Investigation of this more general case shows that though the configuration may be consistent in the sense of consistency already described, it need not be quadratically consistent, in the sense of including in the class of admissible hypotheses a subclass on the quadratic model. However, if the quadratic class exists, it will be infinite; and correspondingly, a cost-of-living measurement is ranged in an interval, rather than having the point determination usually sought, and which is obtained in Wald's special case. Point determination is also obtained in the case intermediate between the general case and Wald's case, in which just one expansion curve is known, that being for the object occasion, and there are data for two other occasions, in which the prices need not be parallel. When these prices are parallel, there is a return again to Wald's case.

The formulae obtained for ranging the cost of living on the quadratic preference model are bound up with the general formulae for ranging the cost of living, that is, without any further assumption on preference structure beyond that it is of the proper type. The connection is such as to exhibit the quadratic hypotheses as accommodated by the data particularly well, by selecting them from within a wider subclass of admissible hypotheses which are distinguished in the entire class as in a well-defined sense, being accommodated by the data most perfectly. The cost-of-living interval determined on the quadratic normal model thus has a peculiar meaning, in relation to the larger interval determined just on the normal model, giving it a special importance.

A computer program has been prepared, by Mr. Harold Samuels, for the criteria for the admissibility of preference hypotheses on the quadratic model, and for the cost-of-living calculations based on that model when it is admissible. It has been applied to a number of numerical examples, including one in two dimensions which admits a graphical representation.

2. METHOD FOR THE INDEX

In the market there are supposed to be some n goods available for purchase in any amounts, at prices $\pi_1, \ldots, \pi_n$ given on any occasion and forming a vector $p = \{\pi_1, \ldots, \pi_n\}$. A collection of amounts $\xi_1, \ldots, \xi_n$ of the different goods defines a commodity *composition* in the goods, and is represented by a vector $x = \{\xi_1, \ldots, \xi_n\}$. The purchase (p, x), of a composition x at prices p, requires an *expenditure* $e = p'x = \pi_1\xi_1 + \ldots + \pi_n\xi_n$, which is the scalar product of the vector p with the vector x. When the expenditure e is taken as the unit of money, prices p become *relative prices* $u = p/e$, where $p/e = \{\pi_1/e, \ldots, \pi_n/e\}$. The vector u defines the expenditure *balance* determined by an expenditure e with prices p. A composition x is said to be *within, on,* or *over* a balance u according as $u'x \leq 1$, $u'x = 1$, or $u'x > 1$. A purchase (p, x) has associated with it the balance $u = p/e$, where $e = p'x$; so that $u'x = 1$. Thus the composition in a purchase lies on the associated balance. The compositions y attainable at the same prices for no greater expenditure are those such that $u'x \leq 1$. These compositions could be had as well as x without loss by further expenditure; but in the purchase, x is singled out from among them. The method of revealed preference[9] is to suppose that this is because x is preferred to all these other compositions, in some hypothetical preference system represented by an order relation S on the set of all compositions, which is admitted by the data.[10] Thus, from the purchase (p, x) is made the inference xSy for all $y \neq x$ such that $u'y \leq 1$, where $u = p/e$ and $e = p'x$. In this inference, prices p enter only to the extent of determining the balance u. Such an order S of compositions, which thus regulates market choice, defines a *preference hypothesis*. It selects a composition x on a balance u as preferred to all other compositions within u. In potentiality u can be any vector with positive elements; a *proper preference system* is defined as a minimal order that is sufficiently complete to select a composition from all those compositions within any balance. Thus if P is a proper preference system, it is an order with the property that to every balance u there corresponds a composition x, which may be denoted by $x = f_P(u)$, such that $u'x = 1$, and xPy for every $y \neq x$ such that $u'y \leq 1$. Any functional dependence $x = f(u)$ of composition x on balance u such that $u'x = 1$, that is, which determines a composition on every balance, defines an *expenditure system*. Generally, u and x can be considered confined to certain domains B and C, each being a region in the non-negative orthant of a Cartesian space of dimension n. An expenditure system of the form $f = f_P$, that is, which is derived from some proper preference system P, defines a *proper expenditure system*. If $f = f_P$, where P is a proper preference system, then it appears that

[9] P. A. Samuelson, *loc.cit.*

[10] It is recognized that to make choices is not necessarily to have preferences. Preferences can only be entertained as a hypothesis which is admitted by the data, and then they are *revealed* by Samuelson's principle. If the data reject the hypothesis, the principle has no application.

$P = P_f$, where $P_f = Q_f$ is the chain extension (transitive closure) of the relation Q_f defined by

$$xQ_fy \equiv \bigvee_u x = f(u) \wedge y \neq x \wedge u'y \leq 1.$$

It follows that f is proper if and only if P_f is an order; and since, in its construction, P_f is in any case transitive, all that need be asked is that it be irreflexive.[11] The proper preference systems are thus of the form $P = P_f$, where f is an expenditure system for which P_f is irreflexive.

Consider an expenditure system f, such that

$$\left| f\left(\frac{u}{\rho}\right) - f\left(\frac{u}{\sigma}\right) \right| < M \, |\rho - \sigma|$$

by which condition (introduced by Houthakker)[12] f might be called *regular* (following Uzawa).[13] Now let a *regular preference system* be one of the form $P = P_f$, where f is a regular expenditure system. Then, for such a preference system P, the following can be said: it is a proper preference system for which there exists a numerical function $\varphi = \varphi(x)$, which is strictly increasing, whose levels are strictly convex, and has the property $\varphi(x) > \varphi(y) \Leftrightarrow xPy$, by which it represents P. By this property, φ might be said to measure P. If φ measures P, then so does $\omega(\varphi)$, where $\omega(t)$ is any strictly increasing function. So the measures of P form an infinite class of functions. But any one determines a partition of compositions into a completely ordered set of classes. Such a particular kind of partial order, defined by a complete order of classes, is to be called a *scale*. A preference system such as is associated with a regular expenditure system has the advantage that it is a scale, representable by a numerical function that is strictly increasing and whose levels are strictly convex. Such a type function will be said to be a *normal preference function* and to correspond to a *normal preference scale*, and it is to be the basis for analysis. It is desirable to proceed only with the most essential suppositions. But while a proper preference system appears as the essential concept, it can be seen that some further conditions to be imposed, like normality make for no new departure from the essential when reference is to empirical, and therefore finite, data.[14]

[11] S. N. Afriat, "Preference Scales and Expenditure Systems." *Econometrica* 30 (1962), pp. 305–323.

[12] H. S. Houthakker, "Revealed Preference and the Utility Function," *Economica* 17 (1950), pp. 159–174.

[13] H. Uzawa, "On the Logical Relation Between Preference and Revealed Preference," *Mathematical Methods in the Social Sciences*, Stanford Mathematical Studies in the Social Sciences V (Stanford, 1959).

[14] S. N. Afriat, "The Algebra of Revealed Preference." Presented at the Winter Meeting of the Econometric Society, at Pittsburgh, Pennsylvania, December 1962. It appears that if the most general form of preference hypothesis is admissible on any

Suppose prices are fixed within certain periods, say years, and to vary between years, and that consumption is measured as the aggregate throughout a year. Consider two years, to be indicated by 0 and 1, and which are to have the roles of a *base* and a *current* year. Consumption in these years appears with composition x_0, x_1 and is obtained at prices p_0, p_1 and therefore for expenditures $e_0 = p_0'x_0$, $e_1 = p_1'x_1$; so the balances are $u_0 = p_0/e_0$, $u_1 = p_1/e_1$. It is asked what is the minimum cost $e_{1,0}$ at the current prices p_1 of maintaining a standard not inferior to that found in the base year. This minimum cost $e_{1,0}$ defines the cost of living for the current year relative to the base year. The multiplier ρ_{10}, such that $e_{1,0} = e_1\rho_{10}$ of current expenditure which will thus compensate for the price change between the base and current year, defines the cost-of-living index for the current year relative to the base year. It is defined by the condition that $f_P(u_0)$ and $f_P(u_1/\rho_{10})$ be on the same preference level.

The history of the cost-of-living problem has very largely been a search for an algebraic formula, involving expenditure data belonging just to the base and current occasion, which could with proper reason be considered as an index of the cost of living. An index of such a form and with such a justification has never been found. One may ask why should the index be given by an algebraic formula; also why should it involve just the expenditure data for the base and object occasions; and why in view of the necessary indeterminacy arising from incomplete knowledge of preferences should it be given a point determination at all; finally why should it always be properly definable seeing that it depends for its meaning on preferences, the existence of which requires a consistency of behavior that may not be borne out by the data? These questions all have a negative answer.

A vector pair u, x such that $u'x = 1$ represents a balance u together with a composition x which is on it, and it defines an expenditure figure $E = [u; x]$, which geometrically can be viewed as a hyperplane, the locus of compositions on the balance u, together with a particular point x on that hyperplane. Hence with any purchase (p, x), there is associated the expenditure figure $[u; x]$, where u is the associated balance. Now, given an expenditure system f, to every u there corresponds an expenditure figure $E = [u; x]$ where $x = f(u)$, which is said to *belong* to f. Any collection of expenditure figures defines an *expenditure configuration*. The collection $\mathscr{E}_f = \{E\}$ of expenditure figures E belonging to an expenditure system f defines the expenditure configuration associated with f. Every expenditure

finite data, then so it is equivalently in the special proper and normal forms. A further account "The Construction of Utility Functions from Expenditure Data" (Cowles Foundation Discussion Paper No. 177, October 9, 1964. Yale University), with revision to incorporate a development in the point of view, is to be published in the *International Economic Review*.

configuration $\mathscr{E}$ has a preference relation $P_{\mathscr{E}}$ defined by $P_{\mathscr{E}} = Q_{\mathscr{E}}$, where

$$xQ_{\mathscr{E}}y \equiv \bigvee_u [u; x] \in \mathscr{E} \wedge y \neq x \wedge u'y \leq 1.$$

Then, for an expenditure system f, with associated configuration $\mathscr{E}_f$, $P_f = P_{\mathscr{E}_f}$.

The data of observation of the kind to be admitted are expenditures on a variety of goods on a variety of similar occasions, to obtain goods in certain amounts at certain prices. Let there be some $k + 1$ occasions, for example years $r = 0, 1, \ldots, k$. Let $M_r = (p_r, x_r)$ be the market purchase in year r of a certain composition in some n goods at certain prices given by vectors x_r, p_r of order n. Then the data consists in a scheme $M = \{M_r\}_{r=0,1,\ldots,k}$. Associated with the purchase M_r, there is the expenditure figure $E_r = [u_u; x_r]$, where $u_r = p_r/e_r$, and $e_r = p_r'x_r$ is the total money expenditure in the purchase. These expenditure figures together form a finite expenditure configuration $\mathscr{E} = \{E_r\}_{r=0,1,\ldots,k}$. The expenditure configuration $\mathscr{E}$ is derived from the market data M. But the same configuration $\mathscr{E}$ could be derived from different possible data M. While M has a direct observational meaning, it is taken as relevant to preference structure only in reduction to $\mathscr{E}$; that is, the preference analysis of M is to be just an analysis of $\mathscr{E}$. All kinds of other data, answers to questionnaires, the weather, and so forth, could be considered relevant. But, in order to define our subject, it is taken that what is significant is not what an agent such as a consumer *says*, which may not be believed and is, in any case, generally of complex significance, but what that agent *does*. Assuming the usual picture of the market, whatever the difficulties, what the consumer does is something in an observable framework with coordinates that are, at least in principle, definite. As for the weather and such factors which are of undoubted relevance to preference, it is not yet known how to take them advantageously into account, and a necessary step is to ignore them. It is manifest in immediate experience that such factors are forces for change of preference, but any entity has to be known in some proper sense before it can advantageously be said to undergo change. Hence the first task before there can be any analysis of change of preference is to have the framework in which the preferences are to be taken as known.

There is to be considered a totality of normal preference hypotheses for the consumer admitted on the data. The totality requires not the entire data M but the configuration derived from it. It is to be supposed that observed acts which each single out a composition x_r on a certain balance u_r are elements in a system of behavior that in potentiality singles out a composition $x = f_S(u)$ on every possible balance u according to a proper expenditure system f_S, associated in some preference scale S. Thus there is considered the totality $\mathscr{S}_{\mathscr{E}}$ of normal scales S admitted by $\mathscr{E}$, being such that $x_r = f_S(u_r)(r = 0, 1, \ldots, k)$. If φ measures S with gradient $g = g(x)$, then, since $\varphi = \varphi(x)$ is to be a minimum under the constraint $u'x = 1$, there have to be the equilibrium conditions $g_r = u_r\lambda_r(r = 0, 1, \ldots, k)$, where $g_r = g(x_r)$, and $\lambda_r = x_r'g_r$ since $u_r'x_r = 1$. The function $\varphi = \varphi(x)$ determines the composition x on any balance u as the maximum of φ under the constraint $u'x = 1$. For this determination to always be possible, φ has to be strictly increasing, and have strictly convex levels, in the domain of x. Let x be confined to a compact composition domain, C. Then it can be shown that φ can be chosen to be a convex function in that domain.[15] Such a function can be smoothed into a continuously differentiable function, or even a function with any number of derivatives, which approximates it arbitrarily closely; so, empirically, there is no further limitation in considering only continuously differentiable preference functions, though the analysis, while then somewhat simplified, is not therefore more confined. A proper preference system P such that $P_{\mathscr{E}} \subset P$ is to be considered a preference hypothesis admissible on the data $\mathscr{E}$. It can be shown that such exists if and only if P is an order, by which condition the configuration $\mathscr{E}$ is said to be *consistent*. Instead of considering all proper preference systems that are admissible on $\mathscr{E}$, there is considered the class $\mathscr{S}_{\mathscr{E}}$ among these which are normal preference scales, these including at least all the admissible preference scales associated with expenditure systems that are, for instance, regular. For these preference systems, it is known that they are represented in some convex neighborhood of the points x_r by a numerical function φ which is strictly increasing and, beyond having strictly convex levels, is strictly convex. The same scale S corresponds to a variety of functions. But to each function there corresponds just one scale measured by it. The totality $\mathscr{S}_{\mathscr{E}}$ of hypotheses admissible on $\mathscr{E}$ can be approached by way of the totality $\Gamma_{\mathscr{E}}$ of such convex functions φ which represent them.

The analysis of data is to consist in analysis of the totality $\mathscr{S}_{\mathscr{E}}$ of preference hypotheses admissible on the configuration $\mathscr{E}$ and investigation of the range of any characteristic of a scale S as S ranges in $\mathscr{S}_{\mathscr{E}}$. Mathematically this depends on investigation of the class $\Gamma_{\mathscr{E}}$ of function φ that are strictly increasing and convex and under the constraints $u_r'x = 1$ $(r = 0, 1, \ldots, k)$ have maxima $x = x_r$.

3. THE GENERAL LIMITS OF THE INDEX

Given an expenditure configuration $\mathscr{E}$, which presents the data of observation, there is a class $\mathscr{S}_{\mathscr{E}}$ of preference systems that are admissible

[15] S. N. Afriat, "The Cost of Living Index," *Research Memorandum* No. 27 (April, 1961) Econometric Research Program, Princeton University.

hypotheses, and if there are any, each, without any further principle of limitation, is equally a candidate for construction as a hypothesis. Relative to any one system $S \in \mathscr{S}_{\mathscr{E}}$, there is a determination $\rho_{rs}(S)$ of the multiplier of expenditure in occasion r so as to maintain, at the prices of occasion r, a standard of living equivalent to that found in occasion s. The range of $\rho_{rs}(S)$ as S ranges in $\mathscr{S}_{\mathscr{E}}$ is an interval, whose lower and upper limits $\rho_{rs}{}^{i}(\mathscr{E})$, $\rho_{rs}{}^{0}(\mathscr{E})$ can be determined from $\mathscr{E}$ by certain formulae. These formulae will now be described. They have an elaborate derivation, which has been treated elsewhere[16] but which has still to be presented in a unified account.

The *cross-structure* of the expenditure configuration $\mathscr{E} = \{E_r\}$ with figures $E_r = [u_r; x_r]$ is defined by the array $D = \{D_{rs}\}$ of *cross-deviations* $D_{rs} = u_r'x_s - 1$ between its figures. $(r \neq s; r, s = 0, 1, \ldots, k)$. The following conditions in the cross structure are equivalent.

I. $$D_{rs} \leq 0, D_{st} \leq 0, \ldots, D_{qr} \leq 0$$

impossible for distinct $r, s, t, \ldots, q$ taken from $0, 1, \ldots, k$.

II. $$\lambda_r D_{rs} + \lambda_s D_{st} + \ldots + \lambda_q D_{qr} > 0$$

for some $\lambda_r > 0$ and for all distinct $r, s, t, \ldots, q$ taken from $0, 1, \ldots, k$.

III. $$\lambda_r > 0, \lambda_r D_{rs} > \varphi_s - \varphi_r (r \neq s; r, s = 0, 1, \ldots, k)$$

for some (λ_r, φ_r).

Let $\Lambda = \{\lambda_r\}$ and $\Phi = \{\varphi_r\}$. Then, moreover, Λ satisfies (II) if and only if (Λ, Φ) satisfies (III) for some Φ.

Condition I, which is the Houthakker condition[17] applied finitely, to the finite configuration $\mathscr{E}$, is the condition that $P_{\mathscr{E}}$, which is the minimal transitive relation with the property

$$D_{rs} \leq 0 \Rightarrow x_r P x_s,$$

be irreflexive, and therefore an order. It is a necessary condition that $\mathscr{E}$ admit a normal preference hypothesis, that is $\mathscr{S}_{\mathscr{E}} \neq 0$; and it can also be shown to be a sufficient condition.[18]

Thus all these three conditions are equivalent to the condition $\mathscr{S}_{\mathscr{E}} \neq 0$ for the existence of a normal preference scale S which is an admissible hypothesis for $\mathscr{E}$. In this and only this case, Λ which satisfies (II) and (Λ, Φ) which satisfy (III) exist.

[16] S. N. Afriat, "The System of Inequalities $a_{rs} > x_r - x_s$", *Proc. Cambridge Phil. Soc.*, 59, (1963), 125–133, and *Research Memoranda* Nos. 21 (February, 1961) and 27 (April, 1961), Econometric Research Program, Princeton University.

[17] H. S. Houthakker, "Revealed Preference and the Utility Function," *Economica*, 17 (1950) pp. 159–174.

[18] S. N. Afriat, "The Algebra of Revealed Preference," *loc.cit.*

Let (Λ, Φ) now denote any solution of (III), so that Λ denotes any solution of (II). Then Λ, Φ define *multiplier and level sets* for $\mathscr{E}$. Let $\alpha = \{\alpha_r\}$ be any set of numbers $\alpha_r \geq 0$ such that $\Sigma\, \alpha_r = 1$, and let

$$x_\alpha = \sum x_r \alpha_r,\ \varphi_\alpha = \sum \varphi_r \alpha_r.$$

Then

$$\rho_{rs}{}^{i}(\mathscr{E}) = \min_{\Lambda,\Phi} \min_{x} \{u_r' x; (x - x_t)' u_t \lambda_t \geq \varphi_s - \varphi_t (t = 0, 1, \ldots, k)\}$$

and

$$\rho_{rs}{}^{0}(\mathscr{E}) = \max_{\Lambda,\Phi} \min_{\alpha} \{u_r' x_\alpha; \varphi_\alpha \geq \varphi_s\}.$$

For any ρ_{rs} there exists a scale $S \in \mathscr{S}_{\mathscr{E}}$ such that $\rho_{rs}(S) = \rho_{rs}$ if and only if

$$\rho_{rs}{}^{i}(\mathscr{E}) < \rho_{rs} < \rho_{rs}{}^{0}(\mathscr{E}).$$

Let this range for the cost-of-living index ρ_{rs} be called the *absolute range*, on the data provided by the expenditure configuration.

Some of the preference systems in $\mathscr{S}_{\mathscr{E}}$ will be just barely admitted by $\mathscr{E}$. We may attempt to be more selective and consider systems which, in some sense, are amply admitted, and through this obtain a narrower range for ρ_{rs} corresponding to these more agreeable systems in $\mathscr{S}_{\mathscr{E}}$. Instead of asking of level and multiplier sets Λ, Φ that

$$\lambda_r D_{rs} > \varphi_s - \varphi_r > -\lambda_s D_{sr},$$

in other words that the differences $\varphi_s - \varphi_r$ lie in the intervals $[-\lambda_s D_{sr}, \lambda_r D_{rs}]$, one may ask more restrictively that they lie at the mid-points of these intervals. That is,

$$\varphi_s - \varphi_r = \tfrac{1}{2}(\lambda_r D_{rs} - \lambda_s D_{sr}).$$

Such a Λ, Φ will be called a *median solution*. Of course, a median solution may not exist. But if it does, with $\lambda_r > 0$, in which case *median consistency* may be said to hold for $\mathscr{E}$, the range of ρ_{rs} may be narrowed to correspond to it, defining a *median range*. A median solution always exists for $k \leq 3$, but need not have $\lambda_r > 0$. If $k = 3$, an essentially unique median solution always exists. So for $k = 3$, we only have to ask if $\lambda_r > 0$. The concept of a median solution may be given a general scope by suitably defining a median discrepancy for any level and multiplier sets, and seeking those with mimimum discrepancy.

4. THE QUADRATIC LIMITS OF THE INDEX

The consistency of $\mathscr{E}$ gives basis for the definition of an absolute range for ρ_{rs}, which is, however, hard to calculate. The more restrictive condition of median consistency, if it holds, provides the narrower median range, which is somewhat easier to calculate. The calculations are not made according to an explicit algebraic formula, but are of a combinatorial type, involving solutions of systems of linear inequalities. It would be of advantage if an algebraic formula locating a range within the absolute or even the median range could be found, and especially so if this corresponded to some peculiarly interesting class of hypotheses. Such a class is provided by preference scales that can be measured in a convex neighborhood of the points x_r by a quadratic function. If any such are admissible, then they will be said to satisfy the condition of *quadratic consistency*. Such a quadratic hypothesis can perhaps be taken as the model of the smoother and also the most centrally accommodated of the admissible hypotheses. The ranging of the cost of living with respect to this class is especially significant in this respect. Also the quadratic hypothesis is a constructive algebraic model which is the natural model for a statistical analysis, in which it has a role perhaps analogous to that of the linear hypothesis in regression analysis.

The algorithm for ranging a cost-of-living measurement on the quadratic hypothesis applies to a configuration of four figures ($k = 3$). This is connected with the fact that only in this case of four figures does a median solution generally exist and is essentially unique; and also it is bound up with the peculiar relation between median consistency and quadratic consistency. Given quadratic consistency, the *quadratic range* for ρ_{rs} will be determined as a subinterval of the median, and hence also of the absolute range. Thus there are three consistency conditions of increasing strength and, correspondingly, three intervals for ranging the cost of living, one lying inside another. But the stronger the condition required, the narrower the scope of the associated method, so the calculation on the quadratic method has the least scope of the three. Nevertheless, it has special importance, for the reasons indicated and since any configuration can be regarded as obtained by disturbance of a quadratically consistent configuration. This, however, is not the concern here, where discussion is to be confined to an exact rather than an approximate kind of analysis.

ALGORITHM.[19] *Let* $\mathscr{E} = \{E_r\}$ *be an expenditure configuration of four figures* $E_r = [u_r; x_r](r = 0, 1, 2, 3)$, *with cross deviations determined by* $D_{rs} = u_r'x_s - 1$. *Let* $\Lambda = \{\lambda_r\}$ *be four multipliers whose three independent ratios are determined from the cycle-reversibility equations*

$$C_{012} = 0,\ C_{023} = 0,\ C_{031} = 0$$

where

$$C_{0rs} = (\lambda_0 D_{0r} + \lambda_r D_{rs} + \lambda_s D_{s0}) - (\lambda_s D_{sr} + \lambda_r D_{r0} + \lambda_0 D_{0s}).$$

[19] S. N. Afriat, "The Cost of Living Index," *Research Memorandum* No. 24 (March, 1961) Econometric Research Program, Princeton University. This memorandum gives a first account of the derivation. For an investigation of the fundamental problem on which the derivation depends, reference is made to S. N. Afriat, "Gradient Configurations in Quadratic Functions," *Proc. Cambridge Phil. Soc.* 59 (1963) 287–305.

Then let $\Phi = \{\varphi_r\}$ *be four levels whose intervals are determined from the median equation*

$$\varphi_r - \varphi_0 = \tfrac{1}{2}(\lambda_0 D_{0r} - \lambda_r D_{r0}).$$

These multipliers and levels are uniquely determined by arbitrarily taking $\lambda_0 = 1, \varphi_0 = 0$. *Now let* $g_r = u_r\lambda_r$ *and form the matrices*

$$X_0 = \{x_r - x_0\},\ G_0 = \{g_r - g_0\}$$

of order $n \times 3$. *The* 3×3 *matrix* $X_0'G_0$ *should be symmetric, by consequence of the conditions determining the numbers* λ_r.

The criterion for quadratic consistency is $\lambda_1, \lambda_2, \lambda_3 > 0$, *and that* $X_0'G_0$ *be negative definite.*

Calculate

$$\tilde{\delta}_{rs} = \frac{\varphi_s - \varphi_r}{\lambda_r}.$$

Now calculate

$$\hat{M} = \varphi_0 + \tfrac{1}{2}g_0'X_0(X_0'G_0)^{-1}X_0'g_0.$$

Quadratic consistency provided, it is necessary that $\varphi_r < \hat{M}$, *so it is possible to calculate*

$$\hat{X}_r = -\{-2(\varphi_r - \hat{M})\}^{\frac{1}{2}}, \qquad \hat{U}_r = -\frac{\hat{X}_r}{\lambda_r},$$

and then $\hat{\delta}_{rs} = U_r(X_s - X_r)$.

With quadratic consistency given, it should be found automatically that $\tilde{\delta}_{rs} < \hat{\delta}_{rs} < D_{rs}$.

Assuming $n > 4$, *quadratic consistency implies the existence of an infinity of admissible quadratic preference functions. Any one of these gives a determination of the fractional change* δ_{rs} *in the expenditure of occasion r which exactly compensates for the price-change from occasion s. The totality of these determinations describes the open interval* $\tilde{\delta}_{rs} < \delta_{rs} < \hat{\delta}_{rs}$.

5. COMPUTER PROGRAM

This program for carrying out the cost-of-living algebraical algorithm is written in the Bell II language for the IBM 650 computer. All standard console settings for Bell II should be used.

INPUT. The input consists of the program deck, a control-card, and the price and quantity data. The control-card stores +0 000 000 N and, +0 000 N 000 in addresses 299 and 300, respectively. If there are 36 commodities in our example, addresses 299 and 300 will read +0 000 000 036 and +0 000 036 000; if two commodities, they will read +0 000 000 002 and +0 000 002 000. The maximum number of commodities that this program can accommodate is 50. The price vectors are put in consecutive addresses beginning at 301, 351, 401, and 451, while the quantity vectors start at 501, 551, 601, and 651.

For example, when there are 30 commodities, the control-card would read:

in columns 7–9	299
in column 10	2
in columns 11–21	+0 000 000 030
in columns 22–32	+0 000 030 000

The price and quantity vectors will be located as follows:

P_0	in addresses 301–330
P_1	in addresses 351–380
P_2	in addresses 401–430
P_3	in addresses 451–480
x_0	in addresses 501–530
x_1	in addresses 551–580
x_2	in addresses 601–630
x_3	in addresses 651–680

The input should have the following order:

1. The Bell II deck
2. The problem number card
3. The program deck
4. A "zero card"
5. For each standard problem
   a. The control-card
   b. All price and quantity data

OUTPUT. The output will have the following form:

1. 701	
2. 705	4×4 D_{rs} row-wise matrix
3. 709	
4. 713	
5. 732	1×4 λ vector
6. 832	1×4 φ vector
7. 720	

8.	723	3×3 $G_0'X_0$ row-wise matrix
9.	726	
10.	719	value of $G_0'X_0$ determinant
	901	901 if $G_0'X_0$ is negative definite
11.	or	
	902	902 if $G_0'X_0$ is not negative definite
12.	836	value of $\hat{M}$
13.	751	
14.	755	4×4 $\hat{\delta}_{rs}$ row-wise matrix if all $\hat{M} > \varphi_r$
15.	759	
16.	763	
17.	776	
18.	780	4×4 $\tilde{\delta}_{rs}$ row-wise matrix
19.	784	
20.	788	

The first column, which is the number of the output card, has no significant meaning for our problem. The second column lists the addresses of the first number in the given row. When we are calculating more than one standard problem, this column will be repeated for each standard problem. (See page 350.)

EXAMPLE. In the following example we show the price and quantity data (Table 1), how it should be punched on the input cards (Table 2), the output from the computer (Table 3), and the translated output (Table 4).

TABLE 1

Price and Quantity Data

1A. *Price Data*

	p	Year 0	1	2	3
Commodity	1	5.0636	5.2906	5.6218	6.5089
	2	1.27	2.07	3.58	2.40
	3	3.03	4.78	10.66	7.00
	4	0.1233	0.1217	0.1242	0.1695

1B. *Quantity Data*

	x	Year 0	1	2	3
Commodity	1	6.90	6.75	7.83	7.69
	2	477.	502.	522.	583.
	3	46.3	54.5	56.7	55.0
	4	10,400.	12,640.	14,880	15,680.

Column Card No.	Loc. 7-9	N 10	Field 1 11-21	Field 2 22-32	Field 3 33-43	Field 4 44-54	Field 5 55-65	Field 6 66-76
01	001	6	+9800049000	+7000299300	+6006299006	+9800047000	+9100011000	+7000301300
02	007	6	+8150008004	-6006299006	+9800048000	+1900900832	+9800001000	+9100110000
03	013	6	+9200110000	+9300001000	+4301501801	+0200300000	+8101000011	+9800002000
04	019	6	+9100101000	+9200101000	+9300010000	+3301801301	+0200300000	+8101000018
05	025	6	+8250004026	+9800003000	+8301004009	+9800004000	+1900900899	+9800005000
06	031	6	+9100110000	+9200010000	+9300001000	+9400100000	+9500001000	+4301501701
07	037	6	+0200300000	+8101000030	+9300101000	+9500101000	-1701901701	+8250004043
08	043	6	+9800028000	+8301004028	+8450004046	+9800029000	+8504004028	+7300701704
09	049	6	+7300705708	+7300709712	+7300713716	-1707708720	-1712710721	-1714715722
10	055	6	+7202899723	-1709712725	-1715713726	-1703704727	-1708705728	+1900900729
11	061	6	-1713714730	-1704702731	+3720728732	+9800006000	+9100101000	+9200010000
12	067	6	+2728732728	+9100111000	+9200100000	-1720728728	+8101004064	+9200000010
13	073	6	+3725729736	+8204002064	+7201901732	+3731730735	-5726727000	-3000725734
14	079	6	-5709703738	-3738902834	+2722735000	+4721734722	-3722720733	-5705702737
15	085	6	-3737902833	+2713735739	-1739704739	-3739902835	+7300732735	+7300832835
16	091	6	+9800009000	+9100101000	+9200101000	+9300010000	+2351733351	+9100111000
17	097	6	+9200101000	-1351301351	+9100111000	+9200101000	-1551501551	+0200300000
18	103	6	+8101000091	+8250003105	+9800011000	+8301003091	+9800012000	+1900900736
19	109	6	+9800013000	+9100110000	+9200010000	+9300001000	+9400100000	+9500001000
20	115	6	+4351551720	+0200300000	+8101000109	+8250003119	+9800030000	+8301003107
21	121	6	+8450003122	+9800031000	+8503003107	+7300720722	+7300723725	+7300726728
22	127	6	+2720724000	-4721723717	+2721726000	-4720727716	+2721725000	-4722724715
23	133	6	+2716901714	+2720728000	-4722726713	+2722727000	-4721728712	+2715901711
24	139	6	+2712901710	+2724728000	-4725727709	-2717901718	+2722711000	+4725714000
25	145	6	+4728717719	+7300719719	+8700718152	+8700719152	+8700720152	+7300901901
26	151	6	+8000000154	+9800014000	+7300902902	+9800015000	+9100101000	+3709719709
27	157	6	+8101009154	+9800016000	+1900900000	+9800017000	+9100110000	+9300010000
28	163	6	+9400001000	+4301551720	+0200300000	+8101000160	+9800018000	+9100011000
29	169	6	+9400100000	+9500011000	+2720709709	+9200011000	+9400111000	+2720709709
30	175	6	+8101003176	+9800019000	+8203003167	+8350003179	+9800020000	+8401003181
31	181	6	+9800027000	+8503003158	+9800021000	+9100111000	+1709710710	+8101008183
32	187	6	-3717902836	+7300836836	+9800022000	+9100010000	-1717736000	+8700000217
33	193	6	-2901000000	+9100001000	+0300000740	+9100111000	+3740732744	+8101004189
34	199	6	+9800023000	+9100010000	+9200100000	-1740740000	+9100001000	+9200100000
35	205	6	+9300001000	+5744900776	+8101004199	+8201004209	+9800024000	+8304004199
36	211	6	+8600899227	+7216776751	+7300751754	+7300755758	+7300759762	+7300763766
37	217	6	+9800041000	+9600001000	+9800025000	+9100011000	+3901732744	+9100101000
38	223	6	+3736902740	+8101004219	+1899901899	+8000000199	+9800026000	+7300776779
39	229	4	+7300780783	+7300784787	+7300788791	+8000000001		

PROGRAM

TABLE 2

Price and Quantity Input Card Form

	Loc.	N	Field 1	Field 2	Field 3	Field 4
Columns	7-9	10	11-21	22-32	33-43	44-54
Card No.						
1	299	2	+0000000004	+0000004000		
2	301	4	+5063600050	+1270000050	+3030000050	+1233000049
3	351	4	+5290600050	+2070000050	+4780000050	+1217000049
4	401	4	+5621800050	+3580000050	+1066000051	+1242000049
5	451	4	+6508900050	+2400000050	+7000000050	+1695000049
6	501	4	+6900000050	+4770000052	+4630000051	+1040000054
7	551	4	+6750000050	+5020000052	+5450000051	+1264000054
8	601	4	+7830000052	+5520000052	+5670000051	+1488000054
9	651	4	+7690000050	+5830000052	+5500000051	+1568000054

TABLE 3

Output Card Form

	Loc.	N	Field 1	Field 2	Field 3	Field 4
Columns	7-9	10	11-21	22-32	33-43	44-54
	701			+1609182049	+3314315049	+3954783049
	705		-1262368049		+1365295049	+1896547049
	709		-2103895049	-1088228049		+4279990048
	713		-2705695049	-1601283049	-4389760048	
	732		+1000000050	+1149631250	+1295415350	+1164408250
	832			+1530219849	+3019866449	+3552658249
	720		-1579233448	-2934717248	-3231949948	
	723		-2934716848	-5888978048	-6749290048	
	726		-3231940948	-6749286848	-8042486248	
	719		-1092431043			
	901		+1000000050			
	836		+9772975049			
	751			+1595324649	+3298124249	+3952232849
	755		-1274422149		+1360278449	+1882811949
	759		-2116393749	-1092680049		+4197390048
	763		-2707884549	-1614843049	-4481645448	
	776			+1530219849	+3019866449	+3552658249
	780		-1331052849		+1295760449	+1759206349
	784		-2331195649	-1149937549		+4112903448
	788		-3051041949	-1736880949	-4575644548	

TABLE 4

The Translated Output
(rounded to 4 decimal places)

4A. *D-Matrix*

	D	Year 0	1	2	3
Year	0		0.1609	0.3314	0.3955
	1	−0.1262		0.1365	0.1897
	2	−0.2104	−0.1088		0.0428
	3	−0.2706	−0.1601	−0.0439	

4B. *λ and φ-Vectors*

	Year 0	1	2	3
λ	1.0	1.1496	1.2954	1.1644
φ	0.0	0.1530	0.3020	0.3553

4C. *$G_0'X_0$ Matrix*

	$G_0'X_0$	Year 1	2	3
Year	1	−0.0158	−0.0293	−0.0323
	2	−0.0293	−0.0589	−0.0675
	3	−0.0323	−0.0675	−0.0804

$|G_0X_0' = -1.092 \cdot 10^{-7}$

$G_0'X_0$ is negative definite

$\hat{M} = 0.9773$

4D. *$\hat{\delta}$-Matrix*

	$\hat{\delta}$	Year 0	1	2	3
Year	0		0.1595	0.3298	0.3952
	1	−0.1274		0.1360	0.1883
	2	−0.2116	−0.1093		0.0420
	3	−0.2708	−0.1615	−0.0448	

4E. *$\tilde{\delta}$-Matrix*

	$\tilde{\delta}$	Year 0	1	2	3
Year	0		0.1530	0.3020	0.3553
	1	−0.1331		0.1296	0.1759
	2	−0.2331	−0.1150		0.0411
	3	−0.3051	−0.1737	−0.0458	

6. ILLUSTRATIONS

A number of numerical examples will now be shown, to illustrate the algebraic method of ranging the cost of living. The calculations have been carried out on the IBM 650 computer, using the program which has just been given.

EXAMPLE 1. In the construction of this example, there is first taken, somewhat arbitrarily, a price and a quantity vector $p_0 = (2, 3, 4)$, $x_0 = (5, 6, 7)$, making a total expenditure of

$$p_0'x_0 = (2 \times 5) + (3 \times 6) + (4 \times 7) = 56$$

in an initial year, which is to be kept fixed as the prices subsequently change. Now each of the prices is raised one unit, in turn. When one price is raised a unit, the corresponding quantity is diminished a unit, and the other quantities raised a unit. The resulting quantities are then scaled proportionately to bring the expenditure to the fixed total.

All that is intended in this procedure is to construct artificially a scheme of data that conforms to the normal idea that when a commodity rises in price while other prices remain fixed, less of this commodity is bought, and instead there is an increase in the quantities of other commodities, whose prices have remained the same.

In this way, there is obtained the following example, involving three goods 1, 2, 3 and four years 0, 1, 2, 3.

p	0	1	2	3
1	2	3	2	2
2	3	3	4	3
3	4	4	4	5

x	0	1	2	3
1	5	3.446	5.250	5.333
2	6	6.031	4.375	6.222
3	7	6.892	7.000	5.333

The cross structure is given by the array of cross-deviation

$$D_{rs} = \frac{p_r'x_s}{p_r'x_r} - 1:$$

D	0	1	2	3
0	0.0	−0.0615	−0.0781	−0.0952
1	0.0893	0	0.0156	(0)
2	0.1071	0.0462	0	0.0159
3	0.1250	0.0615	0.0469	0

From this is constructed the graph of the preference relations.

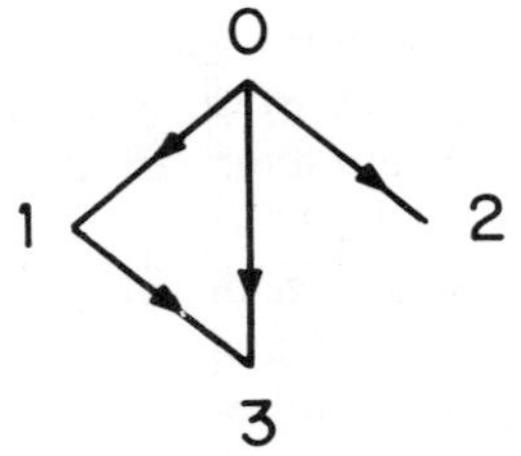

The preferences are thus consistent, and the relation is an order. It is a partial order, since year 2, determined in relation to year 0, is undetermined in relation to years 1 and 3.

The multipliers and levels are

	λ	φ
0	1	0
1	1.416	−0.0940
2	1.438	−0.1161
3	1.461	−0.1389

The multipliers are all positive.

The level-order of the years, following the φ-magnitudes, is thus 0, 1, 2, 3. This is a relation with graph

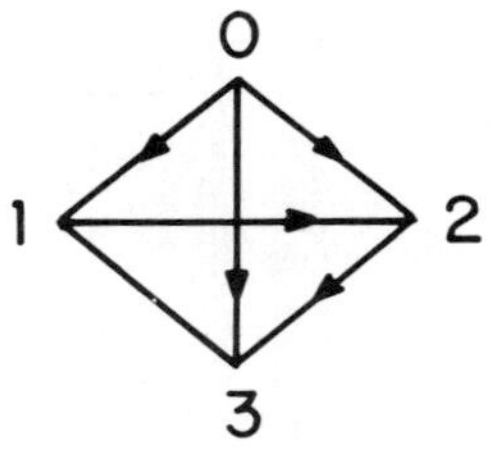

It is thus a complete order, which is a refinement of the partial order first obtained.

The matrix $G_0'X_0$ is

$G_0'X_0$	1	2	3
1	−0.0649	−0.0262	−0.0312
2	−0.0262	−0.0759	−0.0360
3	−0.0312	−0.0360	−0.0874

This is negative definite.

Since the multipliers λ are positive and the matrix $G_0'X_0$ is negative definite, the condition of quadratic consistency is satisfied.

The peculiar method of constructing the example was followed in the hope that it would provide quadratically consistent data, and here that hope is fulfilled. This does not mean the hope would always be fulfilled: that is a plausible conjecture which does not happen to be true, as can be shown by a counter-example. All the same, it would be interesting to have general rules for artificially constructing "good" examples.

Accordingly, to the question, "Do there exist normal preference systems which, on the data, are admissible hypotheses, and which, in a convex neighborhood of the points x_r ($r = 0, 1, 2, 3$), are represented by quadratic functions?" the answer now is "Yes."

It may be added further to this answer that there is an infinity of such quadratic functions; however, rather remarkably, when they are each normalized by addition and multiplication with suitable constants, they all take the same values φ_r at the points x_r ($r = 0, 1, 2, 3$). Thus the order determined for the years relative to each admissible quadratic preference hypothesis is invariant throughout the entire infinite class, and identical with the φ-order determined by the algorithm. This φ-order is a refinement of the preference order P determined directly from the cross-structure. While P is often a partial order, and can even be null, this φ-order is almost always a complete order.

It may now be asked what is the range of a cost-of-living measurement δ_{rs} corresponding to all admissible quadratic preference hypotheses. In view of quadratic consistency, admissible quadratic hypotheses do exist for this example, so this range is defined for all $r, s = 0, 1, 2, 3$. The answer is an interval whose lower and upper limits $\breve{\delta}_{rs}$, $\hat{\delta}_{rs}$ according to the algorithm, are:

$\hat{\delta}$	0	1	2	3
0	0.0	−0.0739	−0.0877	−0.0112
1	0.0806	0	−0.0151	−0.0298
2	0.1004	0.0159	0	−0.0150
3	0.1209	0.0327	0.0161	0

$\breve{\delta}$	0	1	2	3
0	0	−0.0940	−0.1161	−0.1389
1	0.0664	0	−0.0156	−0.0317
2	0.0807	0.0154	0	−0.0159
3	0.0951	0.0308	0.0156	0

It is verified in every case $\breve{\delta}_{rs} < \hat{\delta} < D_{rs}$.

Since $D_{rs} = (e_{rs}/e_{rr}) - 1$, where $e_{rs} = p_r'x_s$ is the cost of purchases in year s evaluated at the prices in year r, so that

$$e_{rs} = e_{rr} + e_{rr}D_{rs},$$

then D_{rs} is the fractional increment of the expenditure e_{rr} in year r needed to raise it to the expenditure e_{rs} needed to purchase in year r which exactly compensates for the change of prices from year s to year r. Hence $\epsilon_{rs} = e_{rr} + e_{rr}\delta_{rs}$ is a determination of the cost, at prices of year r, of maintaining a standard of living equivalent to that in year s, and is a determination of the cost of living with years s and r as base and current periods relative to a preference hypothesis that is admissible on the data. And it appears for any such δ_{rs}, that $\delta_{rs} < D_{rs}$. The numbers $\check{\delta}_{rs}$, $\hat{\delta}_{rs}$ calculated by the algorithm thus range the cost-of-living measurement δ_{rs} relative to a certain class of admissible hypotheses, and they bound it definitely less than the number D_{rs} which represents the conventional measurement.

EXAMPLE 2. In constructing examples, instead of producing them first and then making a trial to see if the admissible hypotheses exist, it is possible to start with a normal hypotheses on the quadratic model and produce examples directly from it, which will automatically have this desired consistency.

This will be done in two dimensions, for the advantage that graphical representation and direct geometrical measurement will then be possible.

For a particularly simple procedure (see Figure 1), let preference level be measured in some suitable region by the distance squared from some suitable point. Take four points in the region, and the directions to the

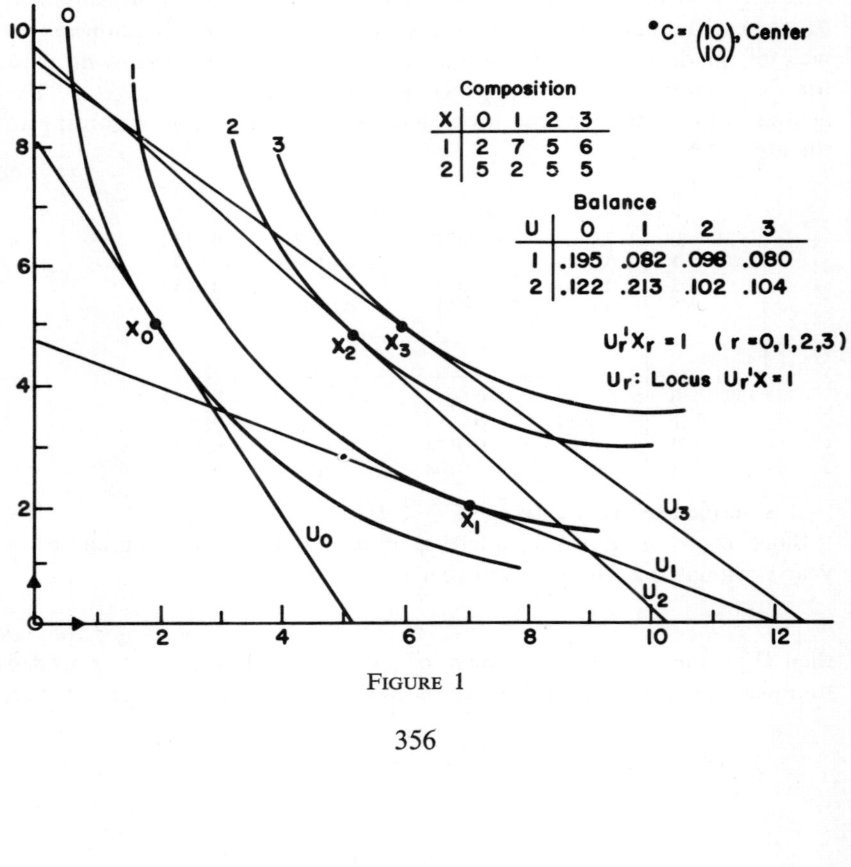

FIGURE 1

radii to the center from which level is measured. The point-coordinates provide quantities, and these directions determine ratios of prices since they are the directions of the gradient of level at those points.

Then, with $\hat{c} = (10, 10)$ as the center of measurement, and

$$\varphi = -\{(x_1 - 10)^2 + (x_2 - 10)^2\}$$

and the points

x	0	1	2	3
1	2	7	5	5
2	5	2	5	6

there can correspond any prices parallel to the gradients

g	0	1	2	3
1	8	3	5	4
2	5	8	5	5

The price vectors normalized to make unit total expenditure are the uniquely determined balance vectors

u	0	1	2	3
1	0.195	0.082	0.098	0.080
2	0.122	0.213	0.102	0.104

The directly calculated multipliers $\lambda = x'g$ and levels φ are

	λ	φ
0	41	−44.5
1	37	−36.5
2	50	−25
3	49	−20.5

which are such that $g_r = u_r \lambda_r$, and these, normalized to make $\lambda_0 = 1$, $\varphi_0 = 0$, are

	λ	φ
0	1	0
1	0.902	0.195
2	1.220	0.476
3	1.195	0.585

This normalization replaces φ by the function $(\varphi + 44.5)/41$ with maximum $\hat{M} = 44.5/41 = 1.085$ at the center $\hat{c} = (10, 10)$.

The algorithm, proceeding indirectly on just the cross-structure, will be found to reproduce these (λ, φ)-values exactly, the cross-structure being:

D	0	1	2	3
0	0	0.610	0.585	0.780
1	0.243	0	0.486	0.568
2	−0.300	−0.100	0	0.100
3	−0.327	−0.224	−0.082	0

The order determined just from cross-structure is the partial order P with graph

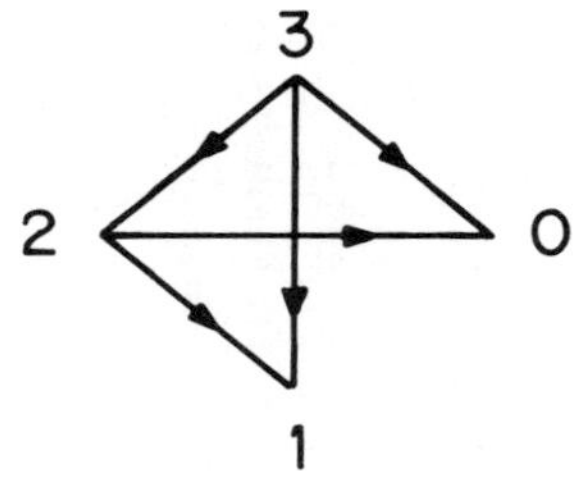

and the φ-order is the complete order 0, 1, 2, 3 with graph

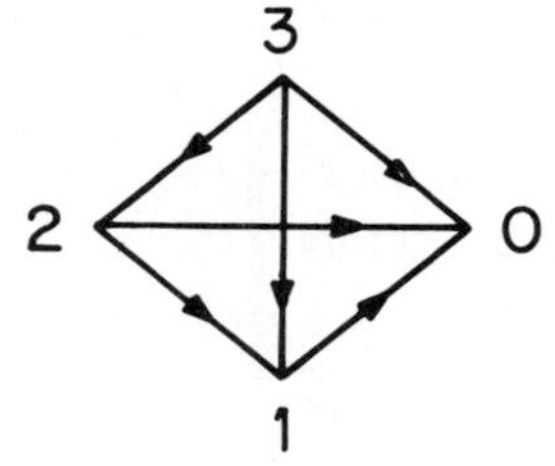

which is seen to be a refinement of P.

Either by arithmetical calculation or by graphical measurement, there is obtained the table

δ	0	1	2	3
0	0	0.205	0.543	0.697
1	−0.206	0	0.340	0.494
2	−0.334	−0.208	0	0.094
3	−0.396	−0.280	−0.087	0

for the compensation number δ_{rs} based on the preference system with level measure by distance to the point $\hat{c}$.

The algorithm cannot apply directly, since here $n = 2$, and it generally requires $n \geq 3$. With $n = 2$, the matrix $G_0{}'X_0$ must be singular. Here it is

$G_0{}'X_0$	1	2	3
1	−0.829	−0.366	−0.488
2	−0.366	−0.220	−0.293
3	−0.488	−0.293	−0.390

But for rounding error, it is found that $G_0{}'X_0$ is non-positive definite, and that its determinant vanishes. With the multipliers λ all positive, the criterion for quadratic consistency is satisfied, in the form it takes for two dimensions.

The following procedure will not be explained here, but it has its reason in the theory of calculation which underlies the algorithm. Delete the last row and column from $G_0'X_0$, to make

$$G_0'X_0 = \begin{pmatrix} -0.829 & -0.365 \\ -0.365 & -0.219 \end{pmatrix} \qquad (G_0'X_0)^{-1} = \begin{pmatrix} -4.555 & 7.593 \\ 7.593 & -17.210 \end{pmatrix}$$

Then

$$\begin{aligned} \hat{c} &= x_0 - X_0(G_0'X_0)^{-1}X_0'g_0 \\ &= \begin{pmatrix} 2 \\ 5 \end{pmatrix} - \begin{pmatrix} 5 & 3 \\ -3 & 0 \end{pmatrix}\begin{pmatrix} -4.555 & 7.593 \\ 7.593 & -17.210 \end{pmatrix}\begin{pmatrix} 0.609 \\ 0.585 \end{pmatrix} \\ &= \begin{pmatrix} 9.994 \\ 10.003 \end{pmatrix} \quad \left(\text{approximately } \begin{pmatrix} 10 \\ 10 \end{pmatrix}!\right) \end{aligned}$$

Also

$$\begin{aligned} \hat{M} &= -\tfrac{1}{2}(g_0'X_0)(G_0'X_0)^{-1}(x_0'g_0) \\ &\quad -2(0.609 \quad 0.585)\begin{pmatrix} -4.555 & 7.593 \\ 7.593 & -17.210 \end{pmatrix}\begin{pmatrix} 0.609 \\ 0.585 \end{pmatrix} \\ &= 1.084 \quad (\text{approximately } 1.085!) \end{aligned}$$

What has been done here is to reconstruct the center $\hat{c}$ and the maximum value $\hat{M}$.

So far the following has been done: We had a circular preference system and derived some data from it, associated with four points. We had a function φ, representing the preference system, and calculated the multipliers and levels λ and φ at the four points, the function being normalized to make $\lambda_0 = 1$, $\varphi_0 = 0$. This function had maximum value $\hat{M} = 1.085$ at the center $\hat{c} = \{10, 10\}$. Also, from the preference system, arithmetically or by geometrical measurement from the graph, the wanted numbers δ_{rs} are determined.

Now we take the data derived from the preference system (and, so to speak, forget how we got it). We verify quadratic consistency so we know it can be derived from some preference system on the quadratic model, but we do not know the system (since we have forgotten it). Since $n = 2$, there can generally be at most one such system. From the data, we reconstruct the λ and φ values we had before, and also the center $\hat{c}$ and maximum value $\hat{M}$ of some quadratic preference function, about which we know nothing further, beyond that it exists (not remembering it is the distance-squared from $\hat{c}$). The center $\hat{c}$ was reconstructed just out of curiosity to show how the calculation works. Our serious interest is in the numbers δ_{rs}, and we are not concerned with any other features of the underlying preference system, let alone the trouble of actually determining it. (We do not need to reconstruct the preference system we started with, which we pretend to have forgotten, and in practical situations never even know, to determine the numbers δ_{rs}, in which we are interested.)

With the formulae

$$\hat{X}_r = -\{2\hat{M} - 2\varphi_r\}^{\frac{1}{2}}, \qquad U_r = -\hat{X}_r/\lambda_r$$

and

$$\hat{\delta}_{rs} = \hat{U}_r(\hat{X}_s - \hat{X}_r),$$

and using the numbers λ_r, φ_r, and $\hat{M}$ we now have, we find $\hat{\delta}_{rs}$ reproduces the numbers δ_{rs} we determined directly from the preference map with which we started.

The last two examples have been artificially constructed to display the principles and practical working of the calculations. But these calculations have to apply to natural data. We shall use data from Irving Fisher's *The Making of Index Numbers.* He uses the data as a basis for the comparison of results obtained using several different formulae and the results here can also enter into this comparison. He gives prices and quantities for 36 commodities in the years 1913–1918.

EXAMPLE 3. We take Irving Fisher's data for the 36 commodities in the years 1913, 1915, 1916, and 1917. There is absolute consistency, and even median consistency, but not quadratic consistency, since the matrix $G_0'X_0$ has positive determinants. The results of the algorithm cannot therefore be given the strict interpretation we want; but we shall give them nevertheless. The years 1913, 1915, 1916, and 1917 are listed as 0, 1, 2, and 3.

D	0	1	2	3
0	0	0.089	0.187	0.194
1	−0.085	0	0.088	0.092
2	−0.160	−0.087	0	0.0036
3	−0.157	−0.085	0.0078	0

$\hat{\delta}$	0	1	2	3
0	0	0.090	0.188	0.186
1	−0.084	0	0.092	0.090
2	−0.158	−0.083	0	0.0019
3	−0.168	−0.087	0.00201	0

$\tilde{\delta}$	0	1	2	3
0	0	0.082	0.152	0.151
1	−0.093	0	0.081	0.080
2	−0.206	−0.096	0	−0.0019
3	−0.218	−0.100	0.0020	0

EXAMPLE 4. The algorithm is now applied just to food, that is to the ten commodities 1, 3, 4, 9, 13, 22, 24, 32, 35, and 36 in the years 1914, 1915, 1917, and 1918, which will be listed as years 0, 1, 2, and 3. Quadratic consistency is verified.

D	0	1	2	3
0	0	0.0643	0.0720	0.2223
1	−0.518	0	0.0143	0.1402
2	−0.0495	0.0066	0	0.1301
3	−0.1694	−0.1189	−0.1151	0

δ	0	1	2	3
0	0	0.0623	0.0666	0.2220
1	−0.0539	0	0.00374	0.1383
2	−0.0549	−0.00356	0	0.1280
3	−0.1698	−0.1222	−0.1189	0

$\tilde{\delta}$	0	1	2	3
0	0	0.0573	0.0609	0.1581
1	−0.0591	0	0.00371	0.1041
2	−0.0606	−0.00358	0	0.0968
3	−0.2852	−0.1819	−0.1754	0

EXAMPLE 5. Fuel, that is the four commodities 7, 8, 10, 23, for the years 1913, 1916, 1917, and 1918 (to be listed as 0, 1, 2, and 3). Quadratic consistency is verified.

D	0	1	2	3
0	0	0.161	0.331	0.395
1	−0.126	0	0.137	0.190
2	−0.210	−0.109	0	0.043
3	−0.271	−0.160	−0.044	0

δ	0	1	2	3
0	0	0.160	0.330	0.395
1	−0.127	0	0.136	0.188
2	−0.212	−0.109	0	0.042
3	−0.271	−0.161	−0.045	0

$\tilde{\delta}$	0	1	2	3
0	0	0.153	0.302	0.355
1	−0.133	0	0.130	0.176
2	−0.233	−0.115	0	0.041
3	−0.305	−0.174	−0.046	0

7. WALD'S INDEX-NUMBER METHOD

Let φ be a convex quadratic in several variables with gradient g. Then φ attains an absolute maximum, which can be assumed to be 0, at a point c, called its center, which is also the unique point at which its gradient vanishes. Thus $\varphi(x) \leq 0$; and the condition

$$\varphi(x) = 0, \qquad g(x) = 0, \qquad x = c$$

are equivalent. Also

$$\varphi(x) = \tfrac{1}{2}(x - c)'g(x)$$

and

$$(x - c)'g(y) = (y - c)'g(x).$$

Any quadratic which is increasing in any neighborhood having $g > 0$ (partial derivatives all positive), and has convex levels $\varphi = \varphi_0$, these being surfaces bounding convex regions $\varphi \geq \varphi_0$, has to be a convex function. These general facts about quadratics now will be taken for granted.

Wald [*op. cit.*] has considered the hypotheses that expenditures are regulated by a quadratic preference function. But a proper preference function must be increasing and have convex levels in the appropriate neighborhood, otherwise it could not have a unique maximum subject to any expenditure-price restriction; and therefore it must be convex if quadratic. If the prices are p and the expenditure is e, then the chosen composition of goods x such that $p'x = e$ is determined by Gossen's law that preference and price directions coincide in equilibrium, $g = p\mu$, when $e\mu = x'g$. As e varies with p fixed, the locus of x is a line through c, which is the expansion curve for prices p.

Wald has supposed linear expansion curves for prices p_0, p_1 on two occasions to be given, and has shown how cost-of-living measurements between the occasions can be made on the hypothesis of a quadratic preference function.

It has to be noticed that, for such a quadratic hypothesis to be admissible on the data, the given expansion curves have to intersect at some point c, which must moreover have a certain location, bounded away from the origin by the price-expenditure constraints; so certainly c cannot be at the origin. Wald's approach, thus qualified, can be greatly simplified; and a formula can be obtained which directly modifies Fisher's formula, in a way which should be considered a correction rather than a generalization. (See Figure 2.)

Suppose the data are that composition x_0, x_1 (vectors of quantities of goods) are purchased at prices p_0, p_1 (vectors of prices of goods) and that the expansion curves intersect in a point c, so they are the lines joining c to x_0, x_1. With hypothesis of a quadratic φ, Gossen's law gives

$$g(x_0) = p_0\mu_0, \qquad g(x_1) = p_1\mu_1,$$

for some positive μ_0, μ_1, which, in view of g being the gradient of a quadratic with center c, have their ratio determined by

$$(x_0 - c)'p_1\mu_1 = (x_1 - c)'p_0\mu_0;$$

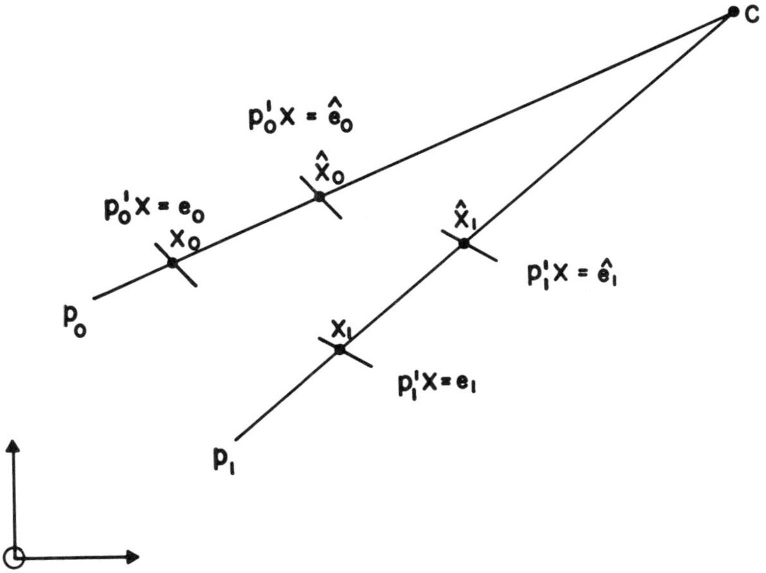

FIGURE 2

and they can be chosen otherwise arbitrarily, corresponding to the fact that the same expenditure system is obtained when the preference function is multiplied by an arbitrary constant.

Now

$$\varphi(x_0) = \tfrac{1}{2}(x_0 - c)'p_0\mu_0, \qquad \varphi(x_1) = \tfrac{1}{2}(x_1 - c)'p_1\mu_1$$

and hence the condition for

$$\varphi(x_0) = \varphi(x_1)$$

is

$$\frac{p_0'(x_0 - c)p_1'(x_0 - c)}{p_1'(x_1 - c)p_0'(x_1 - c)} = 1.$$

In the case $c = 0$, this becomes the Fisher criterion

$$\frac{p_0'x_0p_1'x_0}{p_1'x_1p_0'x_1} = 1,$$

establishing a relation, already well-known [Buschguennce, Konüs, *op. cit.*], between the Fisher index and a special but improper form of quadratic hypothesis, as has been remarked, this case must be excluded if a quadratic hypothesis is to be properly admissible.

It is now asked, what is the condition for an expenditure $\hat{e}_0$ at prices p_0 to have the same value, so far as what can be obtained with it, as an expenditure $\hat{e}_1$ at prices p_1. The compositions

$$\hat{x}_0 = x_0 t_0 + c(1 - t), \qquad \hat{x}_1 = x_1 t_1 + c(1 - t_1)$$

on the expansion curves for p_0, p_1 such that

$$p_0' \hat{x}_0 = \hat{e}_0, \qquad p_1' \hat{x}_1 = \hat{e}_1,$$

have to satisfy $\varphi(\hat{x}_0) = \varphi(\hat{x}_1)$, or the equivalent conditions just deduced. Apparently,

$$t_0 = \frac{\hat{e}_0 - p_0'c}{p_0'(x_0 - c)}, \qquad t_1 = \frac{\hat{e}_1 - p_1'c}{p_1'(x_1 - c)}$$

so that

$$\hat{x}_0 - c = (x_0 - c)\frac{\hat{e}_0 - p_0'c}{p_0'(x_0 - c)}, \qquad \hat{x}_1 - c = \frac{\hat{e}_1 - p_1'c}{p_1'(x_1 - c)}$$

and hence this condition is

$$\left(\frac{\hat{e}_0 - p_0'c}{\hat{e}_1 - p_1'c}\right)^2 = \frac{p_0'(x_0 - c)p_0'(x_1 - c)}{p_1'(x_1 - c)p_1'(x_0 - c)},$$

reducing to the condition in the form already found in the case $\hat{e}_0 = p_0'x_0$, $\hat{e}_1 = p_1'x_1$.

This formula determines the expenditure $\hat{e}_1$ at prices p_1 which is equivalent to any expenditure $\hat{e}_0$ at prices p_0; or, equivalently, it determines the multiplier $\rho_{01} = \hat{e}_1/\hat{e}_0 = \rho_{01}(\hat{e}_0)$ of the expenditure $\hat{e}_0$ which will compensate it for the price change from p_0 to p_1. This multiplier defines the cost-of-living index for base and object occasions with prices p_0 and p_1, corresponding to any level of expenditure $\hat{e}_0$. It thus determines a cost-of-living index corresponding to every standard of living, on a preference hypothesis which is both of the proper form and admissible on the data. The preference hypothesis has been further restricted to be of the special form which is represented by a quadratic preference function in a certain neighborhood containing x_0, x_1. In fact, if any, an infinite class of such hypotheses has to be admitted, if there are more than two goods, but each of these hypotheses leads to the same result for the wanted cost-of-living measurement, and therefore the choice between them is immaterial, as far as this measurement is concerned.

Naturally, the rather complicated quadratic interval algorithm already given for expenditure data of four occasions applies to Wald's case, since this is the special case in which the four price-vectors are parallel in two pairs. But peculiarly simple formulae go with this special case, so the more general formulae are not needed.

Three linear Engel curves, through points x_0, x_1, x_2, with corresponding prices p_0, p_1, p_2, intersecting in pairs and which are not coplanar must be concurrent in a single point, c; but still the quadratic hypothesis will not be admissible unless

$$\frac{p_0'(x_1 - c)}{p_1'(x_0 - c)} \frac{p_1'(x_2 - c)}{p_2'(x_1 - c)} \frac{p_2'(x_0 - c)}{p_0'(x_2 - c)} = 1.$$

Given any number of such curves, this condition applied to every three is necessary and sufficient for the admissibility of the quadratic hypothesis on the data they provide; so it is generally not admissible for more than two taken together. Nevertheless, the quadratic model can still be appropriate for analysis of data which are not exactly consistent on that model, though this makes for a statistical kind of analysis, and gives rise to a more elaborate algebraic formalism.

The extreme simplicity of the formulae here derived for Wald's case, and the fact that an intersecting pair of linear expansion curves gives a convenient statistical model for expenditure data in two periods, is a recommendation for practical usefulness in filling the same role, in the same two-period framework, as the conventionally used by conceptually inadequate Laspeyres index. However, when analysis has to span several years, such as would be the case in insurance policies based on "real-value"—then the analysis should be still more broadly based, and give proper recognition to all available data, which, together with the data in the base and object years, are all equally relevant to preference structure.

Bowley has given a method of approximate index-number construction based on a quadratic preference hypothesis. The method of approximation has been criticized by Frisch,[20] who has proposed another construction, also based on quadratics, known as the "double-expenditure method."[21] This method also relies on an approximation, the validity of which has been criticized by Wald [*op. cit.*]. That criticism can be linked to the criticism made here of Fisher's formula. However, the method based on quadratics which Wald proposes, when taken together with the further qualifications made here, is exact. Without these qualifications, it can only be valid in some approximate sense, since then no proper quadratic hypotheses which are exactly admissible on the data will exist.

[20] Ragnar Frisch, "The Problem of Index Numbers," *Econometrica*, 4 (1936), pp. 1–28.
[21] *Ibid.* "The Double-Expenditure Method," *Econometrica*, 6 (1938), p. 85.

CHAPTER 24

A Spectrum Analysis of Seasonal Adjustment

By MICHAEL D. GODFREY and HERMAN F. KARREMAN*

1.0. INTRODUCTION

The problem of dealing with the apparent seasonal variation in time series has been treated by a large number of research workers [15, 20, 27]. One of the most interesting of these treatments has been that of Abraham Wald [27]. On the basis of criticisms of the then current methods put forward by Oskar Morgenstern, Wald first analyzed the statistical basis of the most important of these methods and then developed a new method. This method has received relatively little attention and almost no practical utilization [20, pp. 151–176]. Therefore, we analyze this method and present its derivation (Appendix A). In addition we treat the most widely used method, the Census Method, along with a newer method put forward by Hannan [10]. Finally, we treat a fourth method which is developed on the basis of the analysis of the three previously mentioned methods.

This paper is principally concerned with the analysis of the comparative performance of various methods of seasonal adjustment. The approach taken is to generate the components of an additive seasonal model:

$$Y_t = C_t + S_t + I_t$$

where Y_t—observed time series
C_t—"trend-cycle" components
S_t—"seasonal" components
I_t—"irregular" components.

The "trend-cycle" plus "irregular" ($C_t + I_t$) and the "seasonal" (S_t) components are generated separately in order that each seasonal adjustment

* Princeton University; Mathematics Research Center, U.S. Army, University of Wisconsin.

The research reported in this paper was performed in the Econometric Research Program and originally appeared as Research Memorandum No. 64 of that organization. This research was supported by National Science Foundation Grant NSF-GS30, and by the Office of Naval Research, Contract No. Nonr 1858(16). The computer facilities used are supported by National Science Foundation Grant NSF-GP579. Reproduction, translation, publication, use and disposal in whole or in part by or for the United States Government is permitted.

While a number of people have commented on earlier drafts of this paper, the authors would particularly like to thank Professor John Tukey and Dr. David Brillinger for their many helpful suggestions. With respect to the computational facilities used, special acknowledgement is due to Professor Peter Warter, the author of the general purpose routine for output for the on-line graphical display and recording facilities of the Princeton University IBM 7094 computer, who provided extensive information and assistance in the utilization of this equipment.

method may be tested in terms of its effects upon a process, the components of which are known separately.

The experimental technique of actually generating series and applying the seasonal adjustment methods to them was adopted partly because of the analytic difficulty of fully analyzing the properties of these methods, some of which are intended to adjust non-stationary series. However it was also felt that this approach would help to give some further insight into the analytic problems of seasonal adjustment. It may also contribute in some way to a more direct understanding of the possible effects of seasonal adjustment as commonly applied to actual data of unknown composition.

In order to discuss the significance of the results of this analysis, it is first necessary to discuss the rationale of seasonal adjustment.

1.1 THE OBJECTIVES OF SEASONAL ADJUSTMENT. A heuristic statement of the objective of seasonal adjustment might be given thus: The elimination of variations in a time series which are attributable to predictable seasonal events in order to display more clearly the more important underlying variations. This statement, while not a rigorous definition, has two interesting implications. The first is that the seasonal variations are predictable and not of interest to analysis of the underlying system. The second is that these seasonal variations are separable from the rest of the series. The acceptance of these implications is a matter of open debate. Their acceptance is normally related to the use to which the data are to be put. If the data are to be used for the estimation of econometric models, it is normally assumed that the seasonal variation is of interest and its contribution to the explanation of the variation of the dependent variables is provided for through the use of seasonal variables and the application of the model to unadjusted data. If on the other hand the data are to be used for single time series extrapolation, one would be indifferent, within the context of linear theory, as to whether the seasonal variables were removed and treated separately or left in the data.

Finally, if the data are to be used to present a time series which "best" represents a realization from some underlying process (which does not involve strictly periodic terms of period one year) for purposes of, say, government policy decisions, then seasonal adjustment may be justified.

It is from this last viewpoint that we approach the problem of determining criteria for measuring the performance of seasonal adjustment methods. In a verbal form our criterion for the methods may thus be stated loosely as follows: That method is judged "best" which, when it operates on a time series composed of "seasonal" and other variations, produces a series which most closely approximates the other variations in the original series. This criterion will be more accurately specified and elaborated upon in Section 5.0. However, a difficulty with this criterion should be introduced at the point. If we consider the spectrum of a seasonally adjusted series, it is intuitively clear that the subjective significance which may be attached to contributions to the spectrum of "errors" in the seasonal adjustment will change considerably depending on the frequency at which the error occurs. It is natural to suppose that "errors" introduced by seasonal adjustment which occur at low frequencies would be considered as more undesirable than such "errors" occurring at high frequencies. Underlying this problem is the idea that information at certain frequencies is more important than information at certain other frequencies.

1.2 SCOPE OF THE METHOD OF ANALYSIS. Since we base our analysis on the analysis of spectra, we are restricted to the analysis of the linear information in the time series and linear dependence between two series. This restriction prevents the full analysis of non-linear operators. However, the severity of this restriction is reduced by the fact that nonlinear operators may be, at least partially, analyzed in terms of their effects on linear information and by the fact that in some cases transformations of the processes involved may yield linear relationships.

Another restriction is that of stationarity. Strictly, the spectrum estimates used only have meaning for second-order stationary processes. However, spectrum methods have been shown to give useful estimates under a fairly wide range of non-stationary conditions [12]. Thus we assume that our estimates have meaning if the time-averaging property of the pseudo-spectrum as described in [12] is kept in mind.

2.0. THE GENERATING PROCESS FOR THE TIME SERIES USED

The underlying process for the time series used is defined by a second-order autoregressive scheme of the following form:

$$y_t = \alpha_1 y_{t-1} + \alpha_2 y_{t-2} + \epsilon_t \tag{2.1}$$

where α_1, α_2—parameters

ϵ_t—normally distributed random independent numbers.

This corresponds to the conventional trend-cycle and irregular components of the seasonal adjustment model. The coefficients of the autoregressive scheme may be chosen to yield time series which, in terms of spectrum analysis, appear similar to many economic time series. It has been found in the course of various analyses of economic data that the series may frequently be well represented by a low-order autoregressive. It is most commonly found that a first- or second-order autoregressive may be used to approximate the estimated spectra quite closely. Interesting discussions of this point occur in [8 and 21], where it is mentioned that a characteristic root of the estimated autoregressive is frequently very close to unity. This point has further implications both in terms of the implied stability of the generating process and in terms of the sampling variance of the estimates of the process.

The seasonal component was generated by one of the two following processes:

The first process is defined by the following equations:

$$S_{i,k} = S_{1,k}(1 + v_{12i+k}) \qquad k = 1, 12; i = 1, l \tag{2.2}$$

$$v_j = v_{j-1} + \epsilon_j \tag{2.3}$$

where $S_{i,k}$—seasonal factor of kth month of ith year
v_j—disturbance term
ϵ_j—normally distributed random numbers.

This process produces a seasonal component with a constant pattern (given by $S_{1,k}$) but with a changing amplitude. The amplitude is determined by the first-order autoregressive scheme given by equation (2.3). The variance of the random term ϵ_i determines the amplitude of the variations in the seasonal amplitude.

The second process is defined by the equations:

$$S_{i,k} = \beta S_{i-1,k} + \rho_{i,k} \qquad k = 1, \ldots, 12; i = 1, \ldots, l \tag{2.4}$$

$$\rho_{i,k} = \rho_{i-1,k} + \sum_{j=1}^{12} C_j \, \epsilon_{12i+j+k} \tag{2.5}$$

where $S_{i,k}$—seasonal factor of kth month of ith year
$\rho_{i,k}$—disturbance term
ϵ_n—normally distributed random numbers
C_j; β—parameters.

Seasonal series generated from this process vary both in amplitude and pattern. Clearly the values for the ϵ_j coefficients determine the amount of correlation between a change in one month's seasonal and a change in the seasonal of neighboring months. At one extreme one may have simply twelve independent first-order autoregressive processes for the twelve monthly factors; or, on the other hand, one may have a high degree of correlation between months so that the seasonal pattern changes very little.

It is felt that these processes fairly describe a large class of possible seasonal components. In particular the first process is in close agreement with Wald's assumptions, while the second seems to fit the assumptions of the Census Method. The assumptions of Wald's method are discussed below (Section 4.2 and Appendix A). The Census Method is fully described in [20].

3.0. DIGITAL FILTERING AND EXTRAPOLATION FOR SEASONAL ADJUSTMENT

3.1 DIGITAL FILTERING. It has recently been recognized in the field of seasonal adjustment [10] that the application of moving averages is an example of the application of a time-invariant linear operator which may be characterized by its transfer function. From this observation stems Hannan's important observation that, since the effects of the operator are completely described by its transfer function, it is possible to correct for any unwanted effects when the operator is applied to a time series. Hannan specifically points out that the Spencer 15-point formula affects the amplitude of not only very low frequencies but also the seasonal and higher frequencies. Thus the simple estimate of the seasonal amplitude is biased by the attenuation introduced by the Spencer operator. Hannan goes on to show explicitly the calculations required to compensate for this attenuation.

Rather than use simple unweighted moving averages or particular weighted values (such as Spencer's formula) we have, using the spectrum approach, employed a somewhat more general filtering technique to eliminate low frequencies. These techniques have been used in conjunction with the Hannan, Wald, and rational-function methods, which were all programmed by the authors. The Census Method was received in completely programmed form from the Bureau of the Census and no functional modifications were performed on it. (Since the Census Method has been so extensively used for routine adjustment of actual data is may be used as a standard of comparison.)

Inspection of the transfer functions derived from the simple 12-month moving average and the 15-point Spencer formula (Figure 3.1) shows clearly that the latter is to be preferred on the basis of elimination of frequencies below the fundamental seasonal. The fact that it eliminates some of the variance at 1 cycle per year is unimportant because this is easily corrected for in the final estimate. However, the Spencer formula does have the disadvantage that more observations are lost at the end of the series. Thus the problem of producing estimates for the last months of the series when the seasonal is allowed to vary is aggravated. In addition one feels that it would be desirable to have a filter whose properties were, in some sense, optimal for the problem at hand. For the sake of simplicity in handling phase information it was decided to restrict the class of filters that would be considered, for the initial high-pass filtering operation, to symmetric moving averages. This class of filters is, of course, characterized by transfer functions with phase identically equal to zero at all frequencies. On the basis of previous analysis it seemed natural to adopt the minimum mean square error criterion put forward by Parzen [22].

Frequency response functions

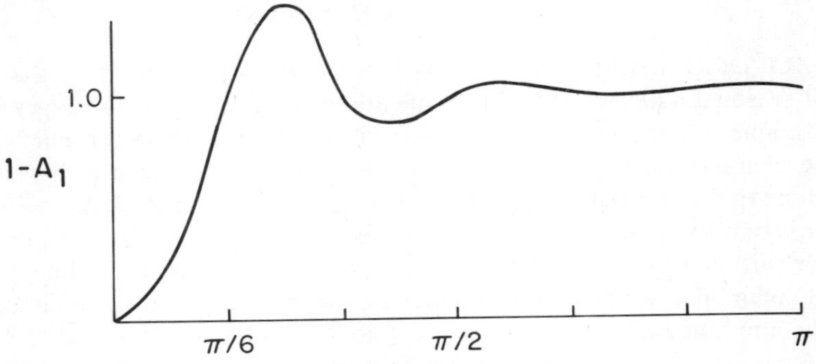

A_1 – simple 12 month moving average

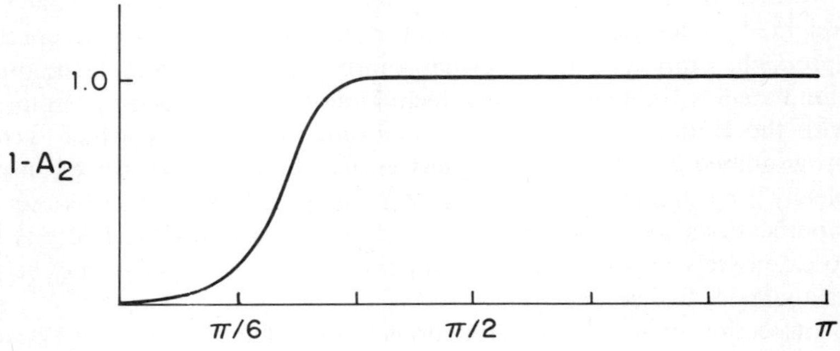

A_2 – 15 point Spencer formula

FIGURE 3.1

Thus we were led to use a filter of the form:

$$\begin{aligned} \lambda(v) &= 1 - 6\left(\frac{v}{m}\right)^2 + 6\left(\left|\frac{v}{m}\right|\right)^3 \qquad 0 \le v \le \frac{m}{2} \\ &= 2\left(1 - \left|\frac{v}{m}\right|\right)^3 \qquad \frac{m}{2} < v \le m \end{aligned} \tag{3.1.1}$$

which satisfies this criterion for a specified class of functions. This function is discussed by Parzen [22] in connection with spectrum windows. It was arbitrarily decided to restrict the filter to a 12-month average. However, it was still possible to vary the bandwidth of the filter to arrive at a "best" value. After some experimentation it was found that the properties of the estimates were not extremely sensitive to small changes in filter bandwidth, and the following transfer function was settled upon:

$$A(\omega) = I - \sum_{v=1}^{m} \frac{\sin kv}{v} \lambda(v) \cos \omega v \qquad (3.1.2)$$

where

$$k = \frac{\pi}{12}.$$

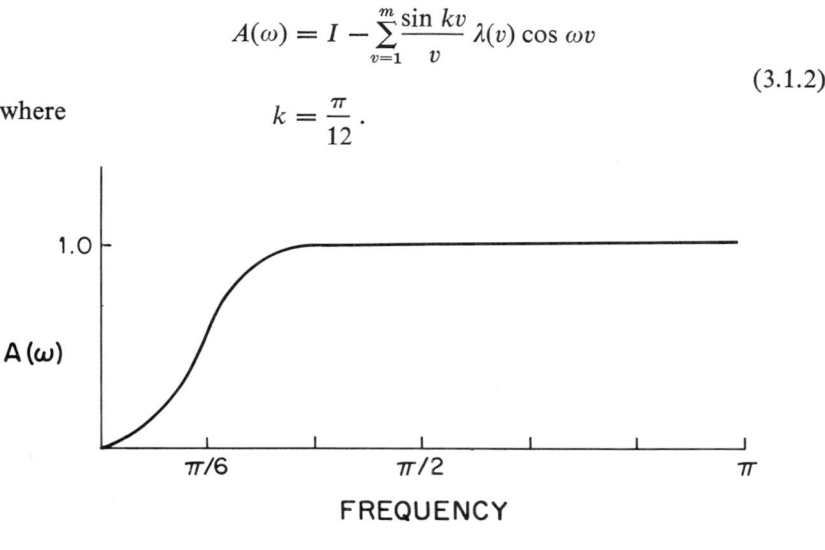

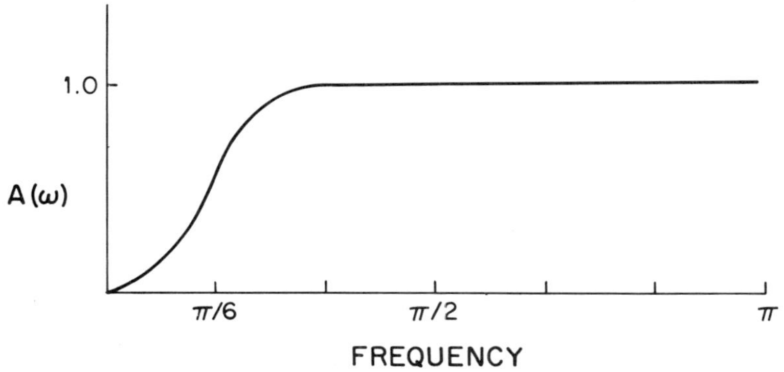

FIGURE 3.2

The moving average coefficients corresponding to $A(\omega)$ are obtained by taking the inverse Fourier transform:

$$D_j = \frac{2}{\pi} \sum_{\omega=0}^{\pi} A(\omega) \cos j\omega \qquad j = 0, 1, \ldots, 6 \qquad (3.1.3)$$

The gain of this filter is shown in Figure 3.2. The gain is about 0.57 at 1 cycle per year. Very little variation at frequencies below 1 cycle per year will pass through the filtering operation.

3.2 EXTRAPOLATION. After the filtering operation the resultant series is lacking points which correspond to the first and last six observations of the original series. In the context of seasonal adjustment this does not present a problem if only fixed seasonal factors are being estimated. However, if a moving seasonal is being estimated as in the Wald and rational-function methods, then the filtered series must be extended in some way to the last observation of the original series. Wald [26] analyzed this problem in connection with his seasonal adjustment method and incorporated extrapolation of the moving average into the method as it was used in Austria. The extrapolation method is fully described in [26]. However, like the monograph on seasonal adjustment [27], this paper has not been translated and no longer seems to be referred to in the literature. Therefore we will briefly outline the technique.

The extrapolation method is based upon three assumptions.

Define the 12-month moving average, $\psi(t)$, of a set of observations $\Phi(t)$ by:

$$\psi_{i,k} = \frac{\sum_{j=k-6}^{k+5} \Phi_{i,j} + \frac{1}{2}(\Phi_{i,k-6} + \Phi_{i,k+6})}{12}. \tag{3.2.1}$$

Defining $s(t)$ as the seasonal component and $z(t)$ as the random component of $\Phi(t)$ we may define a function $f(t)$ by:

$$f(t) = \Phi(t) - s(t) - z(t).$$

The first assumption is then that:

$$\frac{\sum_{j=k-l}^{k+l} f_{i,}}{2l+1} \sim \psi_{i,k} \qquad l = 2, 3, 4, 5. \tag{3.2.2}$$

The second assumption defines the seasonal component as it has been defined in Wald's seasonal adjustment method (see Section 4.2). Therefore the seasonal is given by:

$$s(t) = \lambda(t)p(t) \tag{3.2.3}$$

where $p(t)$ is a strictly periodic function and $\lambda(t)$ varies only slowly over time.

The third assumption restricts the random term by the following expression:

$$\frac{\sum_{j=k+l}^{k+m} z_{i,j}}{m} \sim 0. \tag{3.2.4}$$

The values of m for which assumption three may hold will vary with the kind of data being used.

From these assumptions the following expression for the extrapolated values, $\psi^*(t)$, of $\psi(t)$ is derived:

$$\psi^*_{i,k+6-l} = \alpha_{i,l} - \frac{\sum_{j=k-5}^{k+6} |\Phi_{i,j} - \psi_{i,k}|}{\sum_{j=k-5}^{k+6} |\Phi_{i-1,j} - \psi_{i-1,k}|}(\alpha_{i-1,l} - \psi_{i-1,k+6-l}) \tag{3.2.5}$$

where

$$\alpha_{i,l} = \frac{\sum_{j=k+6-2l}^{k+6} \Phi_{i,j}}{2l+1}.$$

This expression is valid under the three assumptions (equations 3.2.2, 3.2.3, 3.2.4) for $l = 3, 4, 5$, if (3.2.4) is assumed valid for $m \geq 7$; and is valid for $l = 2, 3, 4, 5$ if (3.2.4) holds for $m \geq 5$. For $l = 0, 1$ and possibly 2 this equation cannot be used. To arrive at estimates for these values of l Wald applies simple linear extrapolation of the last values of the series $\psi(t)$ and the extrapolated values given by equation (3.2.5). In the case of (3.2.4) being taken to be valid for $m = 7$ this leads to extrapolation using the five values

$$\psi_{i,k-1}, \quad \psi_{i,k}, \quad \psi^*_{i,k+1}, \quad \psi^*_{i,k+2}, \quad \psi^*_{i,k+3}.$$

Thus, defining

$$\Delta = \frac{2(\psi^*_{i,k+3} - \psi_{i,k-1}) + (\psi^*_{i,k+2} - \psi_{i,k})}{10} \qquad (3.2.6)$$

and μ = the arithmetic mean of the five values, the last three values of $\psi^*(t)$ are determined by:

$$\begin{aligned} \psi^*_{i,k+4} &= \mu + 3\,\Delta \\ \psi^*_{i,k+5} &= \mu + 4\,\Delta \\ \psi^*_{i,k+6} &= \mu + 5\,\Delta. \end{aligned} \qquad (3.2.7)$$

4.0. DESCRIPTION OF THE SEASONAL ADJUSTMENT METHODS

4.1 HANNAN'S METHOD. The first method that has been examined is the one developed by Hannan [10]. Here, only its main features will be discussed. The basic model is given by the additive relation:

$$y(t) = p(t) + s(t) + x(t)$$

where $p(t)$—the trend-cycle component
$s(t)$—the seasonal component
$x(t)$—the residual.

$x(t)$ is assumed to be stationary, though this is not essential for the method.

The seasonal component is assumed to be unchanging and of the form:

$$s(t) = \sum_{k=1}^{6} [\alpha(k) \cos \omega_k t + \beta(k) \sin \omega_k t], \qquad \omega_k = \frac{2\pi k}{12} \qquad (4.1.1)$$

To remove the low-frequency components (trend-cycle) from the series, Hannan, using the spectrum approach, considers several well-known operators. Using the notation $I - A$ (where A is a moving average operator) for an operator which removes low frequencies, the trend-cycle removed series, $y'(t)$, is given by:

$$y'(t) = [I - A]y(t). \qquad (4.1.2)$$

The preliminary estimates of the seasonal $\mu'(j)$ for $j = 1, 2, \ldots, 12$ are derived by calculating the mean for each calendar month:

$$\mu'(j) = \frac{1}{m} \sum_{t=1}^{m} y'_{12t+j} \quad \text{for} \quad j = 1, 2, \ldots, 12 \tag{4.1.3}$$

where m equals the number of (full) years for which the series $y'(t)$ is computed.

The 12 $\mu'(j)$'s are then adjusted to add to zero by subtracting their mean $\bar{\mu}$; the new estimates are called $\mu(j)$.

The final seasonal adjustments $\hat{a}(j)$ for $j = 1, 2, \ldots, 12$ are then estimated by taking a moving average, with weights $v(k)$, of the $\mu(k)$:

$$\hat{a}(j) = \sum_{k=1}^{12} \mu(k) v(k - j) \qquad j = 1, 2, \ldots, 12 \tag{4.1.4}$$

where $v(k) = v(k + 12)$ for $k \leq 0$. $v(k)$ is given by:

$$v(k) = \frac{1}{12} \sum_{s=1}^{11} \frac{1}{1 - h(\omega_s)} e^{-is\omega_k} \tag{4.1.5}$$

where $h(\omega)$ = transfer function of the operator A for frequency ω defined from 0 to π.

The function $v(k)$ is the transformed inverse of the transfer function of the operator $I - A$. Thus the application of $v(k)$ to the $\mu'(k)$ corrects the seasonal weights for any change in the amplitude of variation of the original series at the seasonal frequencies which the operator $I - A$ may have introduced.

4.2 WALD'S METHOD. The second seasonal adjustment method that has been analyzed is the one developed by Abraham Wald in 1936. Wald was at that time associated with the Austrian Institute for Business Cycle Research, under the direction of Oskar Morgenstern. The method was published as Contribution No. 9 of that Institute, under the title *Berechnung und Ausschaltung von Saisonschwankungen* [27]. There does not seem to be an English translation of this monograph,[1] which might help to explain why this method is not well known in English-speaking countries. The method is also known as the moving-amplitude method and has been characterized as being able to produce better results than other methods when the seasonal amplitude is thought to change relatively rapidly from year to year. Rapid amplitude changes have been noted in a certain number of agricultural crop series [20, p. 64].

The assumptions of Wald's method require that the seasonal pattern (i.e., the proportionality relationship between the seasonal at each month

[1] A very brief summary and application of the method can be found in G. Tintner, *Econometrics*, John Wiley & Sons, 1952, pp. 227–233.

and the seasonal at each other month) remain constant over time. This constant pattern is used to estimate changes in the amplitude of the seasonal. This approach permits the estimation of more rapid changes in the seasonal amplitude than would be possible if no assumption about the stability of the pattern were made.

Because Wald's method does not appear to be well known, we will give an outline of it here; a full description of the derivation is given in Appendix A. The technique which Wald developed after the publication of [27] for extrapolation of the moving-average series has been discussed separately in Section 3.2. This extrapolation method was incorporated in the adjustment method as it was employed in Vienna in order to improve the accuracy of the current estimates of the seasonal factors, and has also been used in our computations.

The basic model for Wald's method is that the functions of time which represent seasonal variations, the trend and business cycles, and the random terms, are additive. In Wald's notation

$$\varphi(t) = f(t) + s(t) + z(t) \qquad \text{for} \qquad t = 1, 2, \ldots, n \qquad (4.2.1)$$

where $\varphi(t)$—observed monthly series
$f(t)$—"trend cycle" component
$s(t)$—seasonal fluctuations
$z(t)$—residual.

The first step is the removal of the "trend-cycle," $f(t)$. This is accomplished by subtracting the 12-month moving average of $\varphi(t)$ from the original series $\varphi(t)$. This yields the series $\psi(t)$:

$$\psi(t) = \varphi(t) - \varphi^*(t) \qquad \text{for} \qquad t = 7, 8, \ldots, (n-6) \qquad (4.2.2)$$

where:

$$\varphi^*(t) = \frac{\varphi(t-6) + 2[\varphi(t-5) + \ldots + \varphi(t) + \ldots + \varphi(t+5)] + \varphi(t+6)}{24}$$

$$\text{for} \quad t = 7, 8, \ldots, (n-6). \qquad (4.2.3)$$

Replacing $\varphi(t)$ and $\varphi^*(t)$ in this expression by their components (where φ^* indicates the operation of taking the 12-month moving average), we find:

$$\psi(t) = f(t) - f^*(t) + s(t) - s^*(t) + z(t) - z^*(t). \qquad (4.2.4)$$

Since $f(t)$ represents the "trend-cycle", we may assume that $f(t)$ can be well approximated by a straight line over periods of 12 months. Therefore $f(t) \sim f^*(t)$.

With respect to the seasonal fluctuation $s(t)$, Wald rejects the assumption that it is merely a 12-month periodic function. The hypothesis that it is a periodic function, which is multiplicative with the original observations $\varphi(t)$ or the "trend-cycle" $f(t)$ is also rejected. Thus the models:

$$s(t) = p(t) \cdot \varphi(t); \qquad s(t) = p(t) \cdot f(t)$$

or

$$s(t) = p(t) \cdot \varphi(t) + q(t)$$

where $p(t)$ and $q(t)$ are 12-month periodic functions, are all considered to be unsatisfactory. Wald instead assumes that $s(t) = \lambda(t)p(t)$. $\lambda(t)$ is an arbitrary function, the value of which will slowly change over time, and $p(t)$ is a 12-month periodic function with mean $= 0$. In other words the intensity (*amplitude*) of the seasonal fluctuations, indicated by the function $\lambda(t)$, changes slowly with time, but is not systematically related to other variations in the series. The *pattern* of the seasonal fluctuations, indicated by the function $p(t)$, is however assumed to remain constant over time. The allowance for change in the intensity of the seasonal fluctuations is based on the observation that this intensity is influenced by the trend and the business cycle. However, since there is no a priori reason to expect that this influence follows a well-defined scheme, e.g., that the intensity of the seasonal fluctuations is proportional to the trend, the function $\lambda(t)$ is left arbitrary.

On the basis of the model $s(t) = \lambda(t)p(t)$ it is observed that $|s^*(t)|$, the absolute value of $s^*(t)$, will be smaller, the smaller the fluctuation of $\lambda(t)$ within the period of 12 months, and that $s^*(t) = 0$ if $\lambda(t)$ is constant. This leads to the assumption that $s^*(t) \sim 0$. Equation (4.2.4) is now reduced to:

$$\psi(t) \sim s(t) + y(t) \qquad \text{for} \quad t = 7, 8, \ldots, (n - 6) \tag{4.2.5}$$

where $y(t) = z(t) - z^*(t)$.

It is convenient at this point to introduce the matrix $\psi(i, k)$, the (i, k)th element of which refers to the kth month of the ith year. Let the corresponding values of $s(t)$ and $y(t)$ be similarly arranged in two matrices, the elements of which will be designated $s(i, k)$ and $y(i, k)$. Computing now the arithmetic mean of the values of the k-month of $\psi(i, k)$ as well as of $s(i, k)$ and $y(i, k)$ one obtains, after substituting $\lambda(i, k)p(i, k)$ for $s(i, k)$ in equation (4.2.4):

$$\frac{\sum_{i=1}^{m} \psi(i, k)}{m} \sim \frac{\sum_{i=1}^{m} \lambda(i, k)p(i, k)}{m} + \frac{\sum_{i=1}^{m} y(i, k)}{m} \tag{4.2.6}$$

$$\text{for} \quad k = 1, 2, \ldots, 12$$

where

$$m = \frac{n}{12} - 1.$$

From these assumptions Wald arrives at the following expression for the estimated seasonal:

$$s(i, k) = a(k)\frac{\sum_{j=k-6}^{k+5} \psi(i, j)a(j)}{\sum_{i=1}^{12} [a(l)]^2}$$

where $a(k) = \frac{1}{m}\sum_{i=1}^{m} \psi(i, k)$ for the m by 12 matrix of seasonal coefficients.

4.3 THE CENSUS METHOD. The third method that has been examined is Census Method II, which will be only briefly described here.[2] In this method the trend-cycle, seasonal and irregular components are assumed to be combined *multiplicatively*. It is sometimes indicated that this is the commonest form of seasonal relationship for the broad mass of economic time series [20, p. 58]. However, this model may be transformed into the additive model by taking logarithms. This transformation may introduce certain problems, particularly with respect to its effect on the distributions of the variables in the model. However, on the basis of computed comparisons of the performance of the Census Method with and without taking logarithms, there was no evidence that, for the kind of series dealt with here, any difficulties would be caused by the logarithmic transformation.

An important characteristic of the so-called "moving seasonality" which is incorporated in the Census Method is that no restriction is placed on the nature of any relationships between the changes in *amplitudes* in successive months. The method can therefore take care of changes in the *pattern* of seasonal variation over successive periods of twelve months as well as changes in amplitude [20, p. 259, footnote]. However, the changes in both amplitude and pattern of the seasonal ratios are assumed to be gradual and smooth. The method in its original version was not successful when applied to series with drastic changes in S-I (Seasonal-Irregular) ratios[3] as, for instance, total unemployment; nor could it satisfactorily adjust series with constant seasonal patterns but sharply varying amplitudes as, for example, agricultural stocks and farm employment series. However, later versions of the method contain devices which can take better care of series with extreme S-I ratios than could the original.[4] For this study version X-10 has been used as this was the version which was, according to our information, the most highly developed in early 1963.

[2] For a more elaborate description, the reader is referred to Julius Shiskin's paper, "Tests and Revisions of Bureau of the Census Methods of Seasonal Adjustments," *Bureau of the Census Technical Paper* No. 5, November 1960. This paper was incorporated (pp. 79–150) in [20].

[3] Seasonal-irregular ratios are the ratios of the original observations to the 15-term Spencer trend-cycle curve.

[4] Suggestions for modification of the method were given in [20], pp. 257–311.

4.4 A RATIONAL TRANSFER-FUNCTION METHOD. The purpose of this method has been to develop a simple method, based on spectrum concepts, for comparison with the other methods. The method involves the extension of Wald's method to treat a changing seasonal pattern, and the inclusion of the basic ideas of Hannan's method.

The first step of the method is the conventional one of applying a linear operator to remove low-frequency variations. As with the Hannan and Wald methods the operators and notation used are discussed in Section 3.0.

Next it is assumed that the seasonal pattern may be represented in the form of a set of twelve mixed moving average autoregressive processes with identical coefficients. This is a natural assumption if, for reasons of simplicity, the processes which generate the seasonal coefficients are taken to be linear. Thus the seasonal pattern coefficients are given by:

$$S_{i,k} = \sum_{j=1}^{n} A_j S_{i-j,k} + \sum_{j=0}^{m} B_j \gamma_{i-j,k} \tag{4.4.1}$$

where $S_{i,k}$—seasonal pattern coefficient of month k, year i

$\gamma_{i,k}$—random disturbance term

A_j, B_j—coefficients.

Thus it is natural to attempt to estimate $S_{i,k}$ from the filtered series by:

$$\hat{S}_{i,k} = \sum_{j=1}^{n} A_j \hat{S}_{i-j,k} + \sum_{j=1}^{m} B_j \psi_{i-j,k} \tag{4.4.2}$$

where $\hat{S}_{i,k}$—estimated seasonal

$\psi_{i,k}$—filtered series

A_j, B_j—coefficients.

If the Z-transform operator (defined by $Z^{-n} = x_{t-n}$) is applied to the above equation and the terms rearranged we have:

$$\hat{S}_{i,k} = \frac{\sum_{j=0}^{m} B_j Z^{-j}}{1 - \sum_{j=2}^{n} A_j Z^{-j}} \psi_{i,k}. \tag{4.4.3}$$

From equation (4.4.3) it is clear that we are simply applying a time invariant, linear, rational function operator to the 12 series $\psi_{i,k}$ ($i = 1, \ldots, l$; $k = 1, \ldots, 12$).

The coefficients A_j and B_j might be estimated for each series on the basis of a minimum mean square error criterion. However, in the interest of simplicity and generality the coefficients were in fact determined on the basis of more general, and in part heuristic, criteria. The transfer function defined by equation (4.4.3) should have a gain characteristic which in some sense minimizes the error of the estimate $\hat{S}_{i,k}$. One may assume

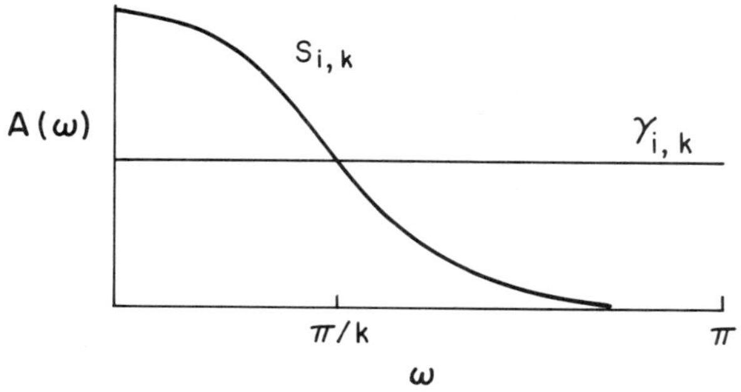

FIGURE 4.4.1

that $S_{i,k}$ has relatively high spectral density at low frequencies, while the spectrum of $\gamma_{i,k}$ is relatively flat. Then the two spectral densities will be of the form given in Figure 4.4.1.

The transfer function given in equation (4.4.3) should then take the form:[5]

$$A(\omega) = \frac{f_s(\omega)}{f_s(\omega) + f_\gamma(\omega)}$$

where $A(\omega)$—transfer function of filter

$f_s(\omega)$—spectrum of seasonal coefficients $S_{i,k}$ $i = 1, \ldots, l$ for all k

$f_\gamma(\omega)$—spectrum of random term $\gamma_{i,k}$ $i = 1, \ldots, l$ for all k.

Given the above assumptions $A(\omega)$ will take the form indicated in Figure 4.4.2.

It seems reasonable for many economic series to assume that π/k falls in the range of about 0.2 to 0.3 cycles per year. This gain characteristic could be approximated by a symmetric moving average filter in a way similar to that applied in the Census Method. However the rational

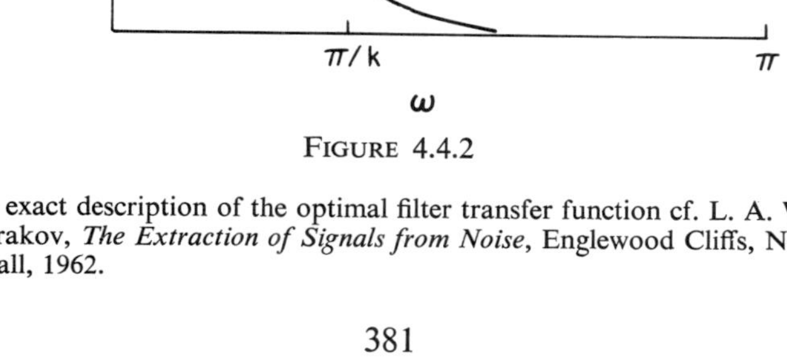

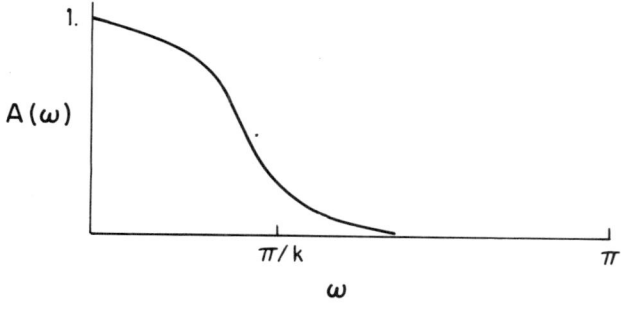

FIGURE 4.4.2

[5] For an exact description of the optimal filter transfer function cf. L. A. Wainstein, V. D. Zubrakov, *The Extraction of Signals from Noise*, Englewood Cliffs, New Jersey, Prentice-Hall, 1962.

function filter seems more natural, given the assumptions made about the way in which the seasonal variation is generated.

In deciding on the transfer function given in equation (4.4.3) it would be desirable to apply general analytic criteria in terms of both the gain and phase characteristics. However, appropriate general methods based on some practical error-of-estimate concept are not yet available. Therefore the coefficients were simply chosen on the basis of inspection of the transfer function and experimentation.

To this point the method has been analogous to the Census Method in that the seasonal coefficients have been estimated from the twelve annual series. However, we now make the assumption that the seasonal coefficients are intercorrelated. We will, in fact, assume that for relatively high-frequency variations the seasonal factors vary proportionally. This is analogous to Wald's assumption of a constant seasonal pattern. However, as previously assumed, we allow the seasonal pattern (i.e., the proportionality factors) to vary relatively slowly. Thus we attempt to combine the assumption of the Census Method that each seasonal coefficient may change very slowly but independently of the others with the assumption of Wald's method that the amplitude of the seasonal pattern may vary relatively rapidly.

In order to derive the estimate for the seasonal amplitude we have simply paralleled Wald's derivation. This results in the following expression:

$$\tilde{S}_{i,k} = \frac{\hat{S}_{i,k} \sum_{j=k-5}^{k+6} \hat{S}_{i,j} \psi_{i,j}}{\sum_{j=k-5}^{k+6} (\hat{S}_{i,j})^2}. \tag{4.4.4}$$

To complete the adjustment procedure the estimated seasonal, $\tilde{S}_{i,k}$ is corrected for any bias introduced by the original filtering operation and is then subtracted from the original data.

The actual equations computed are as follows:

First the low frequencies are removed by:

$$\psi_t = \Phi_t - \sum_{j=-6}^{6} D_j \Phi_{t+j}. \tag{4.4.5}$$

The last six values of ψ_t are extrapolated, and the D_j's are determined, as described in Section 3.0.

Then, starting values for the rational function filter are computed by averaging the first four available years of the series ψ_t

$$\hat{S}_{1,k} = \frac{1}{4} \sum_{i=2}^{5} \psi_{i,k} \qquad k = 1, \ldots, 12$$
$$\hat{S}_{2,k} = S_{1,k} \tag{4.4.6}$$

Next the rational function filter is applied according to

$$\hat{S}_{i,k} = A_1\hat{S}_{i-1,k} + A_2\hat{S}_{i-2,k} + B_1\psi_{i,k} + B_2\psi_{i-1,k} + B_3\psi_{i-2,k} \qquad i = 3, \ldots, l; k = 1, \ldots 12 \tag{4.4.7}$$

These estimates of the evolving seasonal pattern are then used to estimate the seasonal coefficients according to:

$$\tilde{S}_{i,k} = \frac{\hat{S}_{i,k} \sum_{j=k-5}^{k+6} \hat{S}_{i,j}\psi_{i,j}}{\sum_{j=k-5}^{k+6} (\hat{S}_{i,j})^2}. \tag{4.4.8}$$

Finally, the $S_{i,k}$ are corrected by:

$$\tilde{S}'_{i,k} = \sum_{j=1}^{12} \alpha_j \tilde{S}_{i,k-j} \tag{4.4.9}$$

where the α_j's are determined by the inverse of the transfer function of the D_j's given in equation (4.4.5).

The $\tilde{S}_{i,k}$'s are extrapolated to the ends of the series, and the seasonally adjusted series is formed by:

$$\Phi_t^s = \Phi_t - \tilde{S}_t'. \tag{4.4.10}$$

5.0. OUTLINE OF COMPUTATIONS

As shown in the flow chart given in Figure 5.1 the typical computation consisted of the following steps:

5.1 The computation of the underlying and seasonal series for a set of specified parameter values.

5.2 The seasonal adjustment of the sum of the underlying series and the seasonal.

5.3 The computation of the spectrum and cross-spectrum properties of the following pairs of series.:

5.3.1 The adjusted series and the underlying process.

5.3.2 The adjusted series and the sum of the underlying process and the seasonal.

5.3.3 The adjusted series and the seasonal.

5.3.4 The seasonal series and the difference between the sum of the underlying process and the seasonal on the one hand and the adjusted series on the other. This computation is, then, a comparison of the actual seasonal and the seasonal estimated by the seasonal adjustment method.

On the basis of the general statement that the quality of the seasonal adjustment method may be measured by its ability to accurately recover the underlying process from the sum of the seasonal and this process, the following ideal results may be identified in terms of these calculations. For calculation 5.3.1 the perfect adjustment method would produce spectra with identical shapes and a cross-spectrum with unit gain and zero phase at all frequencies. The coherence should be one at all frequencies. For calculation 5.3.2 the cross-spectrum will depend on the

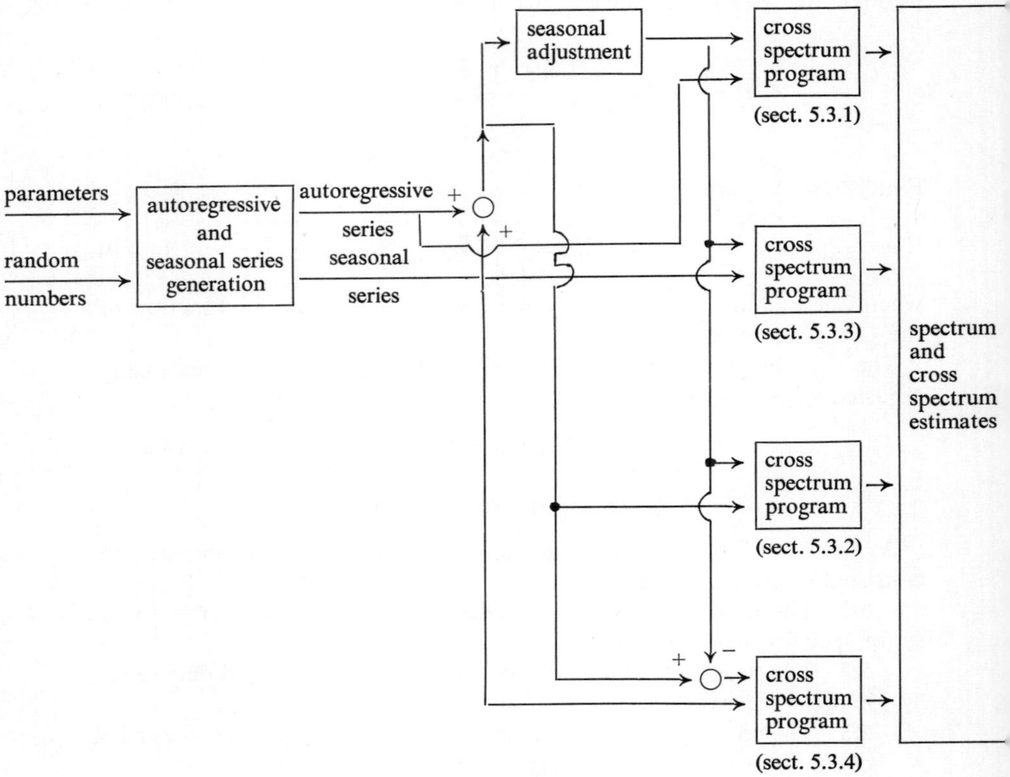

FIGURE 5.1

seasonal component. The reason for including this calculation is not to provide a direct test of the seasonal adjustment method, but to present the cross-spectrum for comparison with the cross-spectrum which may be computed from actual series with unknown seasonals. The perfect adjustment method would produce for calculation 5.3.3 a coherence consistent with the hypothesis of independence of the series. Therefore, the cross-spectrum should not be significant. Finally, in calculation 5.3.4 the perfect method should produce a unit correlation, as in calculation 5.3.2.

Imperfections in the various methods may show up in a considerable variety of ways as will be discussed in the next section.

6.0. EXPERIMENTAL RESULTS

In determining the experimental procedure for these computations it was necessary to decide on the values of two sets of parameters. One set involved the values of the autoregressive scheme, while the other was the parameter vector for the seasonal pattern. Through the computation of several pilot runs it was found that the seasonal adjustment methods were not very sensitive to changes in the autoregressive parameter values over a considerable range (i.e., those yielding characteristic periods from about 2 years to ∞) Therefore the calculations were mainly performed with only one set of values. The values of the parameters used in equation (2.1) are:

$$\alpha_1 = 1.4$$
$$\alpha_2 = -0.40.$$

It is interesting to note that the roots of the characteristic equation of equation (2.1), for these parameter values, are both real with one equal to 1.0 and the other equal to 0.4. Thus, the solution of the equation is not oscillatory. However, the time series generated by this process appear to be better approximations for a wide class of economic time series than series generated using parameter values which give an oscillatory solution.

For the seasonal processes it was, however, found that the methods were sometimes differentially sensitive to the process and parameter values used. Therefore the computations were performed with several sets of values.

The first process, equations (2.2), (2.3), gives a seasonal with a constant pattern. The only variable parameter is the variance of the error term. Several values for the variance were used. This process will be referred to as Type 1.

The second process, equations (2.4), (2.5), allows the seasonal pattern to vary slowly but may preserve some correlation between months. Two sets of values for the C_j were used. The first set is:

$$C(1) = 0.6$$
$$C(2) = 0.1$$
$$C(3) = 0.2$$
$$C(4) = 0.05$$
$$C(5) = 0.05$$
$$C(i) = 0 \qquad i = 6, \ldots, 12.$$

This process will be referred to as Type 2. The second set of values allows each month to vary independently of the other months. These values are:

$$C(1) = 1.0$$
$$C(i) = 0.0 \qquad i = 2, \ldots, 12.$$

This process will be referred to as Type 3. For each of the sets of values for the vector $C(i)$ several values of the variance of the term ϵ_j were used in order to give several different levels of amplitude change in the seasonal. In addition, values of β other than one were used in order to introduce trends in the seasonal amplitude.

The results of these computations show that all of the methods performed quite well when the generating process conformed to the assumptions on which the particular method was based. In the simple case of the constant seasonal (Type 1 with no error term) Hannan's method was definitely superior. Wald's method performed well for a constant seasonal pattern but changing amplitude. The Census Method performed well when both aplitude and pattern were changing if the changes in amplitude did not become large or relatively rapid. The rational-function method tended to combine the ability of the Census Method to adjust for very slow changes in pattern with the ability of Wald's method to estimate changes in amplitude. As the characteristics of each method are considerably different we will discuss each method separately in greater detail.

First, however, it is necessary to explain the general form of the graphical results. Each page of graphs (containing either two or four graphs) displays the results of the analysis of two time series. Viewing the page of graphs with the figure label at the bottom, the top graph displays the two time series used in the computation. The graph is divided in half horizontally. The upper half of the graph shows the series resulting from the seasonal adjustment process while the lower half shows the original generated series. The lower row of graphs gives the spectra of the two series and the cross-spectrum between them. The first graph in this row gives the spectra of the two series. The spectrum of the series from the seasonal adjustment process is labelled Y while the other series is labelled X. In some cases, as in Figure 6.01.1, these two spectra lie on top of each other and are not distinguishable. If, in addition, the two series were not significantly different at any frequency in terms of the cross-spectrum, as for Figure 6.01.1, then the graphs of the cross-spectrum are not shown. The deletion of the two graphs following the graph of the spectra implies that the coherence between the two series was not significantly different from unity at any frequency. In the cases where the two series were significantly different the graph following the graph of the spectra gives the coherence between the two series. The last graph shows the transfer function of the two series where the original series is taken as the input and the series from the seasonal adjustment process as the output. The upper half of the graph gives the gain while the lower half displays the phase.

The following index lists each of the figures and gives the two series used in each case as well as the type of seasonal and the name of the seasonal adjustment method.

INDEX TO FIGURES

6.1 HANNAN'S METHOD. As was mentioned above, Hannan's method provides the "best" seasonal adjustment of a series which contains a constant seasonal. The seasonally adjusted series is nearly identical to the original autoregressive series; this is clearly indicated in Figure 6.10.

When the seasonal changes, Hannan's method simply takes an arithmetic average of each month's values and estimates this constant average seasonal. Thus Hannan's method continues to perform satisfactorily only as long as the variations in the seasonal can be adequately represented by their time averages.

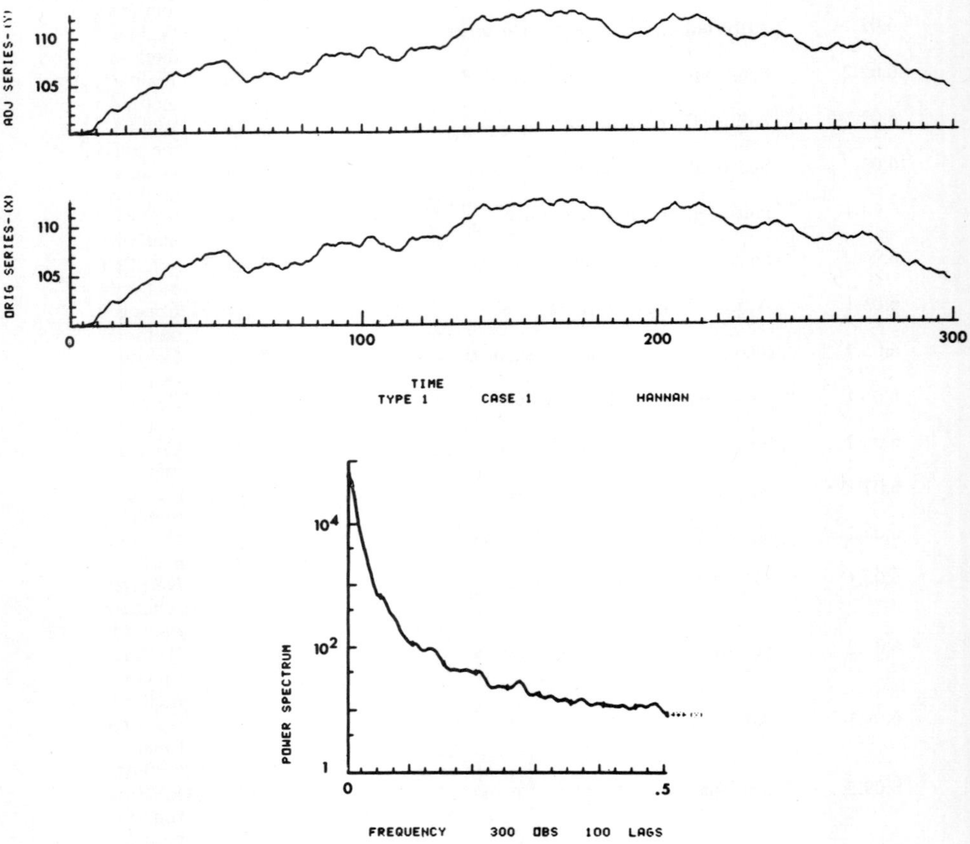

FIGURE 6.01.1

In this context it is important to mention the effect of the Hannan method on a non-stationary seasonal. To take the simplest case of a linear trend in amplitude, it is clear that the estimated seasonal has a constant amplitude equal to the average amplitude of the actual seasonal. Thus, for the case of a linear trend, the error of the estimate reaches a maximum at both ends of the series and a minimum at the mid-point. In this case the performance indicated by cross-spectrum analysis may

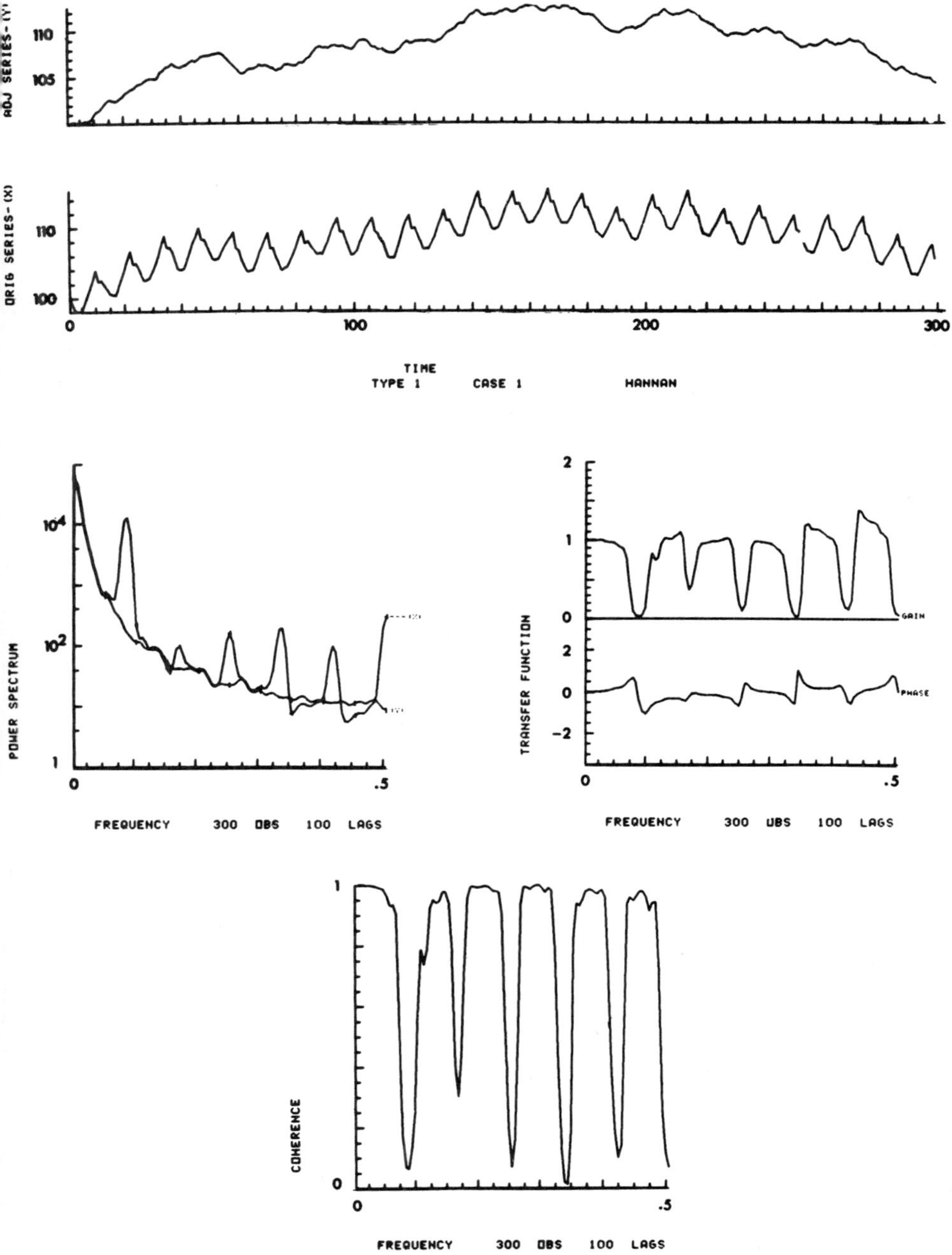
ADJ SERIES- (Y)
110
105
ORIG SERIES- (X)
110
100
0
100
200
300
TIME
TYPE 1
CASE 1
HANNAN
POWER SPECTRUM
10^4
10^2
1
0
.5
FREQUENCY
300 OBS
100 LAGS
TRANSFER FUNCTION
2
1
0
GAIN
2
0
−2
PHASE
0
.5
FREQUENCY
300 OBS
100 LAGS
COHERENCE
1
0
0
.5
FREQUENCY
300 OBS
100 LAGS

FIGURE 6.01.2

also be somewhat misleading due to the time-averaging property of the spectrum estimates. The cross-spectrum shows the average error of the method over the sample. If, however, one is not only interested in the average error but also in the expected error of the estimate for the last

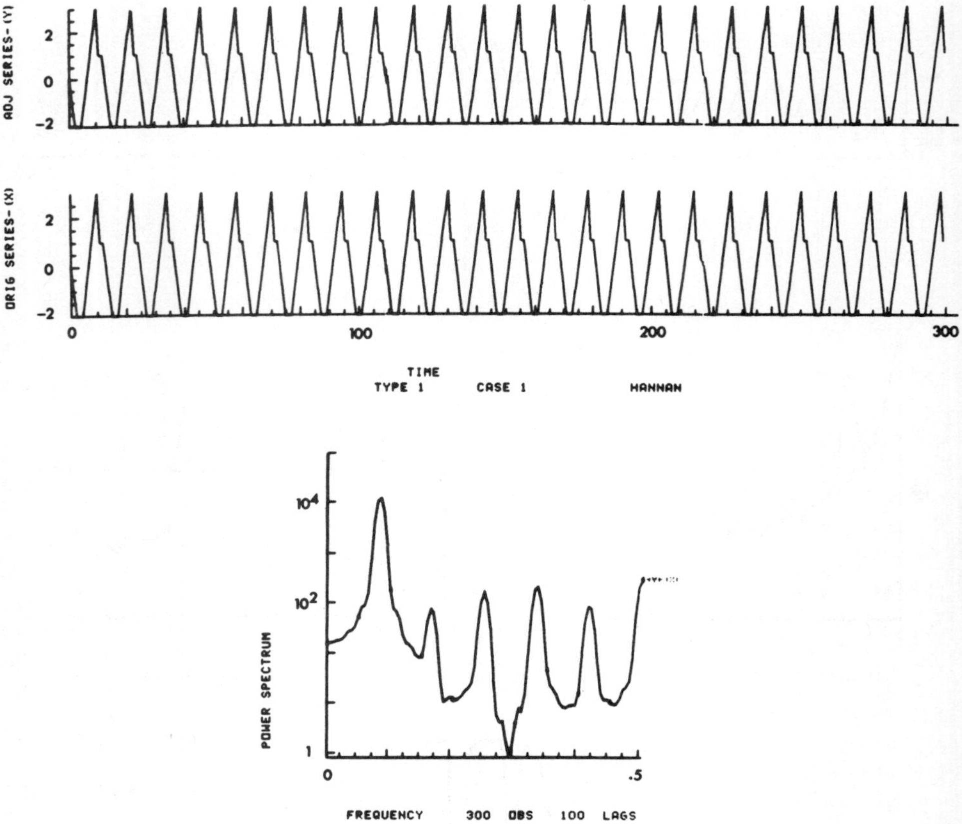

FIGURE 6.01.3

observation (or final set of 12 observations) then the cross-spectrum alone does not give a complete measure of performance. As was mentioned previously, this problem, and others relating to non-stationarity, are problems where the concept of the pseudo-spectrum may be applied. It has not been possible in the present paper to pursue this analysis in detail.

The practical limitations of Hannan's method are obvious. First, it is not often felt to be the case that the seasonal can adequately be represented by a set of constant monthly coefficients. Second, the fact that the use of stationarity leads in this case to the result that if the method is sequentially applied from year to year to a set of data to which each year 12 new observations are added, the seasonally adjusted series will in general have different values for corresponding months, not only in the current or recent years, but from the beginning of the series. As has been frequently observed this leads to the necessity of continuing revisions.

Clearly, the major contribution of Hannan's method is not to be given in terms of its potential application to actual data but rather in its explicit treatment of effects of linear operators in terms of frequency decomposition. Given a constant seasonal, Hannan's technique of adjusting the seasonal factors for any effects of the initial filtering operation leads to better estimates in terms of bias. This technique, obviously, is not restricted to Hannan's method. It is applicable to any method which employs a linear operator before estimation of the seasonal. In fact this adjustment technique, as mentioned previously, was experimentally applied to Wald's method and was used in the rational transfer function method. For a recent development by Hannan, which arrived too late for analysis in the present paper, see [11].

6.2 WALD'S METHOD. Wald's method, since it employs certain non-linear operators, affects the linear information in the series even when the actual seasonal is constant. However, these effects are typically quite small and are evident only at relatively high frequencies.

At the low-frequency end of the spectrum, Wald's method produces, for a constant seasonal, an estimate of the autoregressive series very nearly as good as Hannan's method. As is well known, the point at which Wald's method breaks down is when the pattern of the seasonal is allowed to change. Wald's method estimates a constant average pattern (as does Hannan's method) and uses this constant pattern to produce a "best" estimate of the change in amplitude of the seasonal for each year. This approach provides relatively very good estimates of the change in amplitude (particularly for rapid changes in amplitude) of the seasonal when the assumption of the constant pattern is met. In comparing Wald's method with a method which does not rely on the relationship of the seasonal in one month to the seasonal in each other month, it is clear that (again assuming the constant pattern) the use of the information contained in all twelve observations for a given year will lead to a better estimate than the use of only the series of annual observations for each month independently. Figure 6.02 shows the results of the application of Wald's method to a series containing a seasonal of constant pattern but varying amplitude. Figures 6.03 and 6.04 show the results of the use of Wald's methods when both the amplitude and pattern are allowed to change.

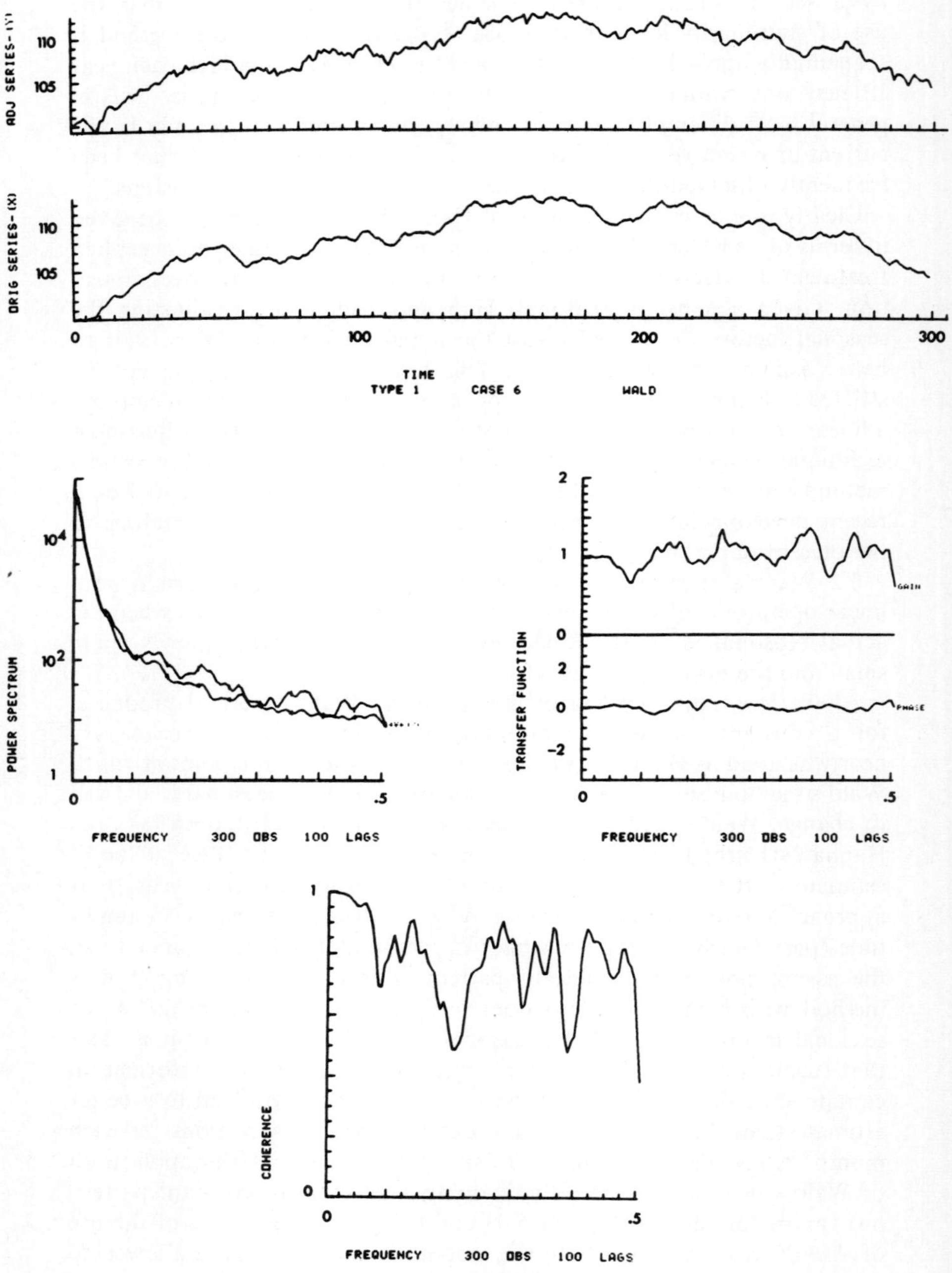
ADJ SERIES-(Y)
110
105
ORIG SERIES-(X)
110
105
0
100
200
300
TIME
TYPE 1
CASE 6
WALD
POWER SPECTRUM
10^4
10^2
1
0
.5
FREQUENCY
300 OBS
100 LAGS
TRANSFER FUNCTION
2
1
0
2
0
-2
GAIN
PHASE
0
.5
FREQUENCY
300 OBS
100 LAGS
COHERENCE
1
0
0
.5
FREQUENCY
300 OBS
100 LAGS

FIGURE 6.02.1

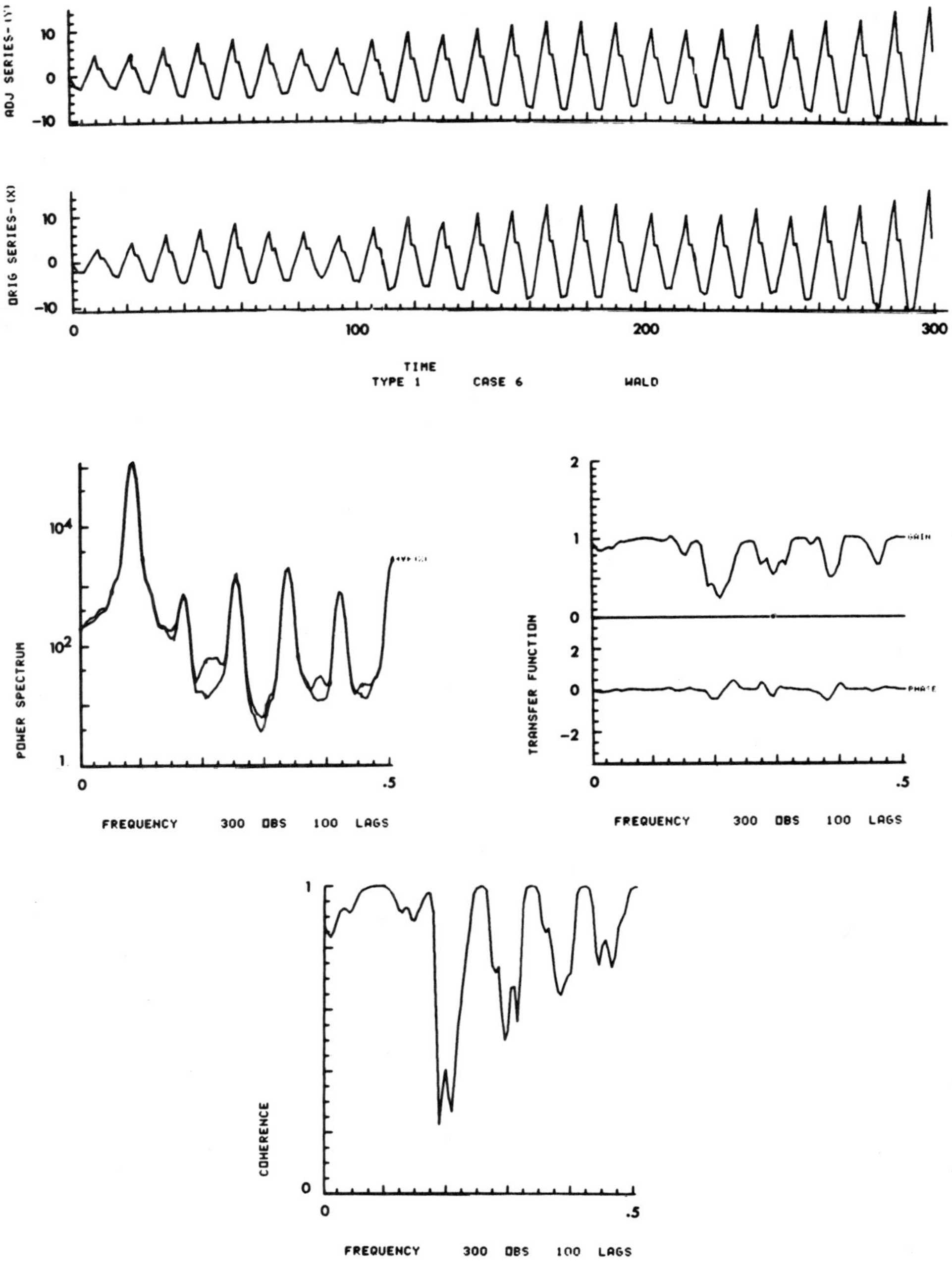
ADJ SERIES-(Y)
10
0
-10
ORIG SERIES-(X)
10
0
-10
0
100
200
300
TIME
TYPE 1
CASE 6
WALD
POWER SPECTRUM
10^4
10^2
1
0
.5
FREQUENCY
300 OBS
100 LAGS
TRANSFER FUNCTION
2
1
0
2
0
-2
GAIN
PHASE
0
.5
FREQUENCY
300 OBS
100 LAGS
COHERENCE
1
0
0
.5
FREQUENCY
300 OBS
100 LAGS

FIGURE 6.02.2

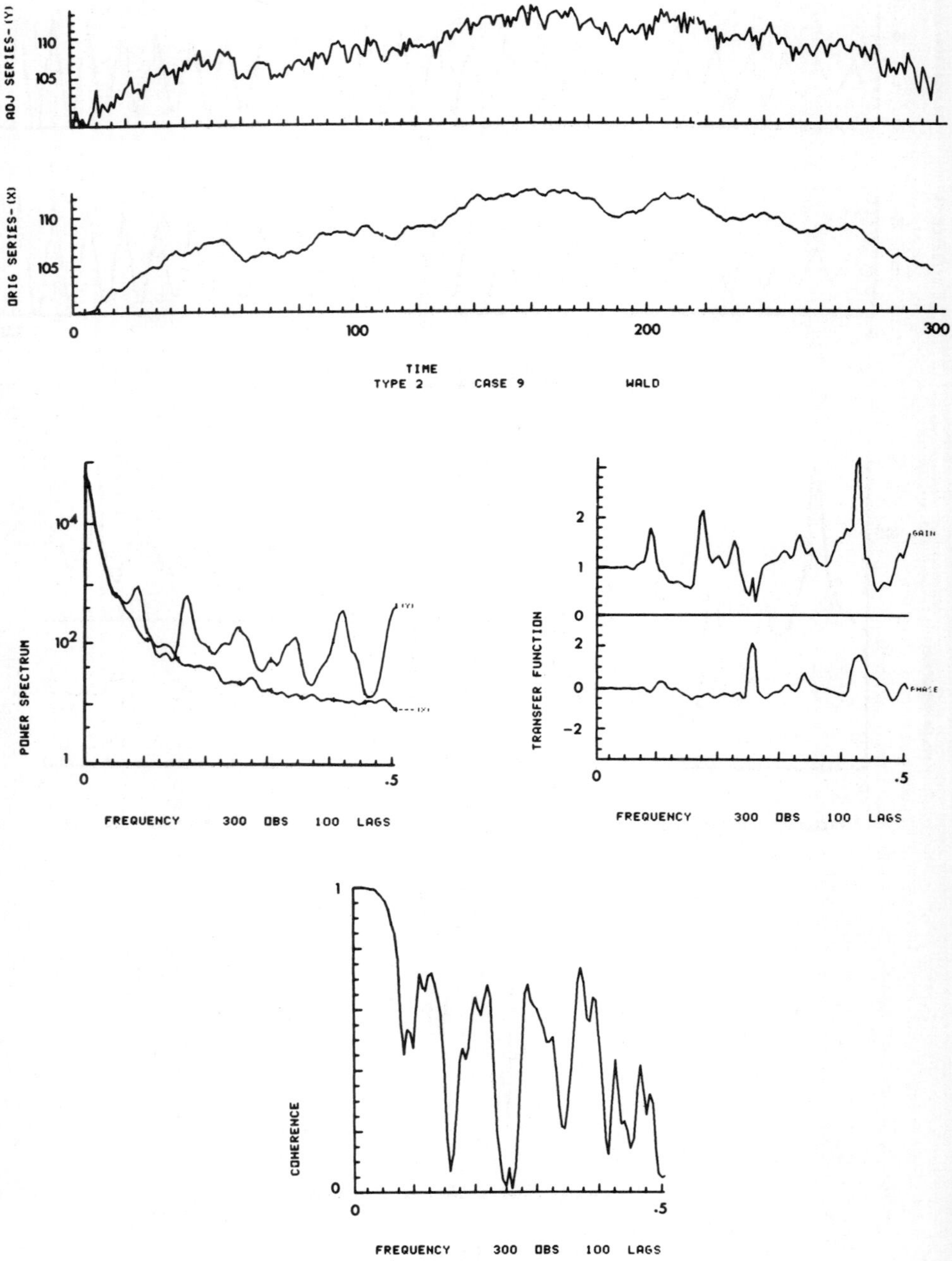
ADJ SERIES- (Y)
110
105
ORIG SERIES- (X)
110
105
0
100
200
300
TIME
TYPE 2
CASE 9
WALD
POWER SPECTRUM
10^4
10^2
1
0
.5
FREQUENCY
300 OBS
100 LAGS
TRANSFER FUNCTION
2
1
0
2
0
−2
GAIN
PHASE
0
.5
FREQUENCY
300 OBS
100 LAGS
COHERENCE
1
0
0
.5
FREQUENCY
300 OBS
100 LAGS

FIGURE 6.03.1

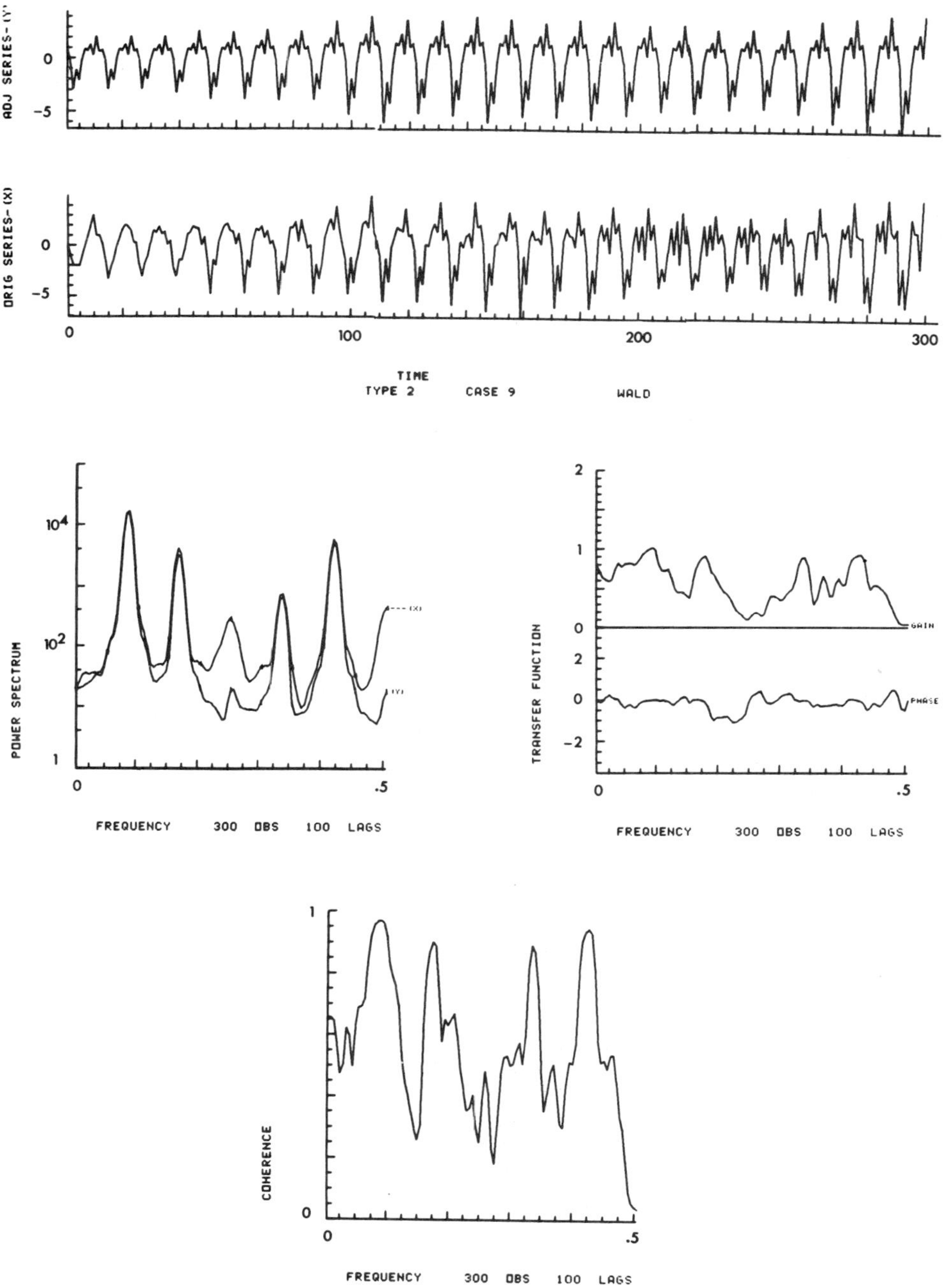
ADJ SERIES-(Y)
0
-5
ORIG SERIES-(X)
0
-5
0
100
200
300
TIME
TYPE 2
CASE 9
WALD
POWER SPECTRUM
10^4
10^2
1
(X)
(Y)
0
.5
FREQUENCY
300 OBS
100 LAGS
TRANSFER FUNCTION
2
1
0
GAIN
2
0
-2
PHASE
0
.5
FREQUENCY
300 OBS
100 LAGS
COHERENCE
1
0
0
.5
FREQUENCY
300 OBS
100 LAGS

FIGURE 6.03.2

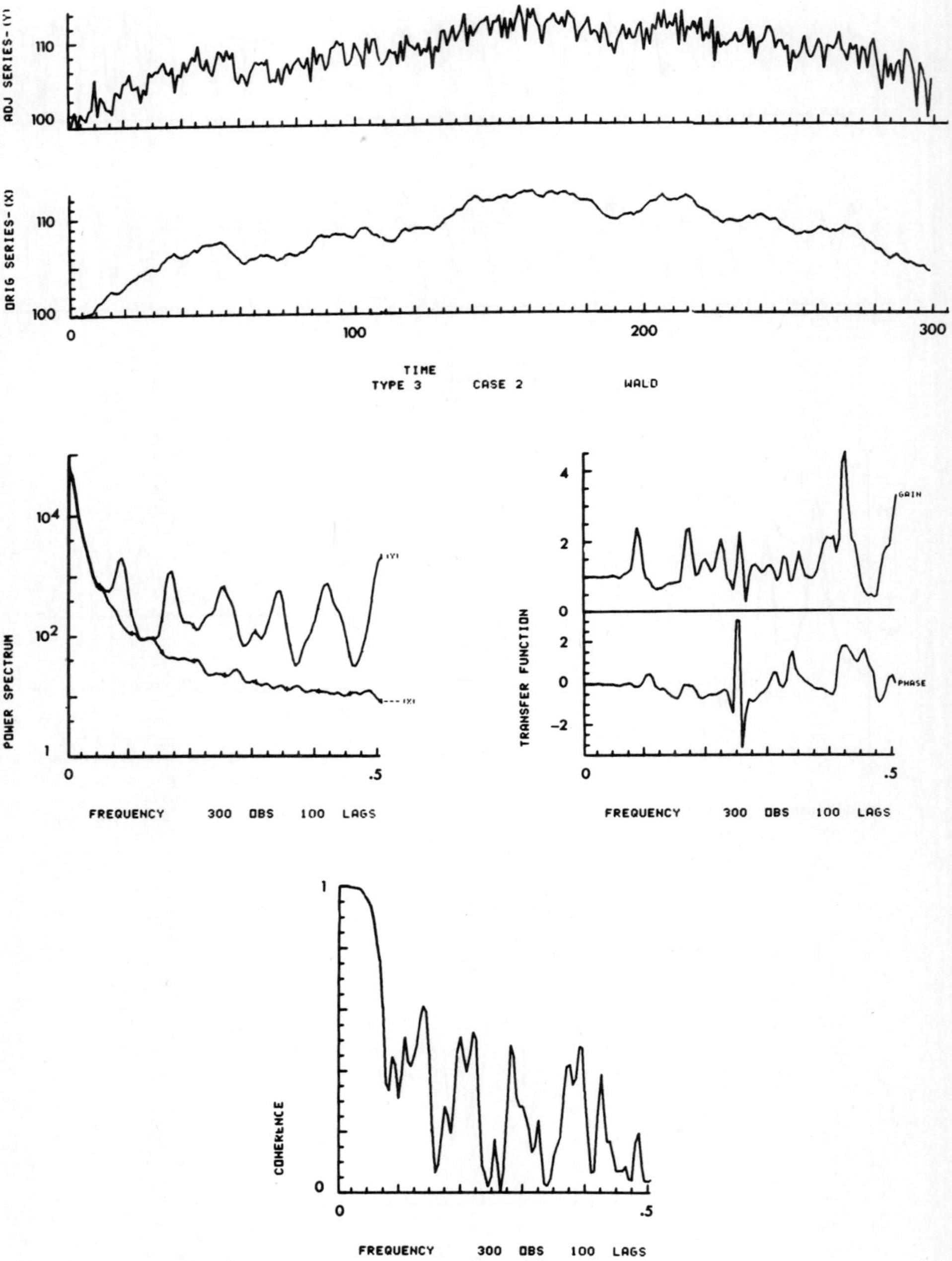
ADJ SERIES-(Y)
ORIG SERIES-(X)
110
100
0
100
200
300
TIME
TYPE 3 CASE 2 WALD
POWER SPECTRUM
10^4
10^2
1
(Y)
(X)
.5
FREQUENCY 300 OBS 100 LAGS
TRANSFER FUNCTION
4
2
-2
GAIN
PHASE
FREQUENCY 300 OBS 100 LAGS
COHERENCE
FREQUENCY 300 OBS 100 LAGS

FIGURE 6.04.1

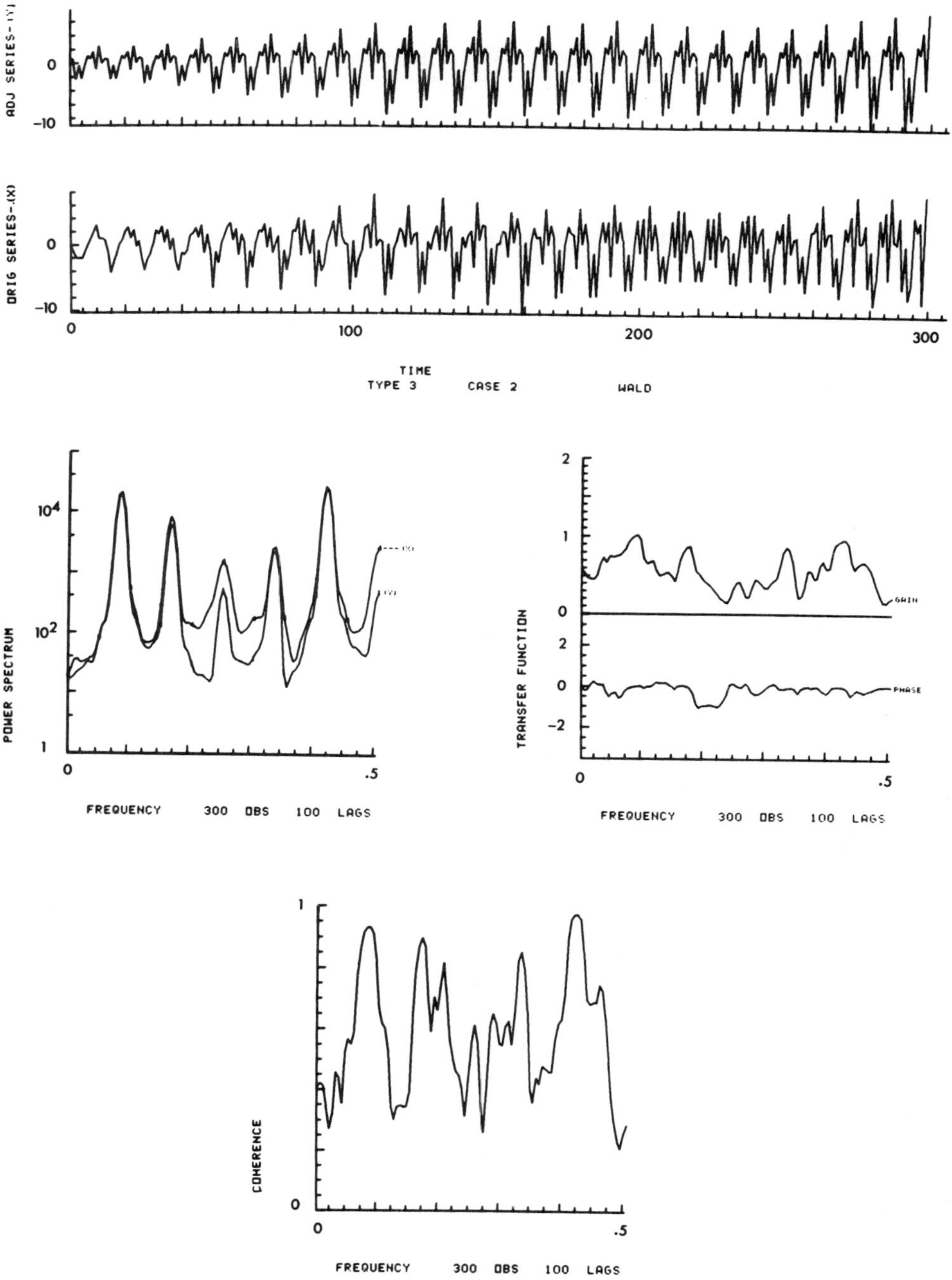
ADJ SERIES-(Y)
0
-10
ORIG SERIES-(X)
0
-10
0
100
200
300
TIME
TYPE 3
CASE 2
WALD
POWER SPECTRUM
10⁴
10²
1
0
.5
(X)
(Y)
FREQUENCY
300 OBS
100 LAGS
TRANSFER FUNCTION
2
1
0
2
0
-2
GAIN
PHASE
0
.5
FREQUENCY
300 OBS
100 LAGS
COHERENCE
1
0
0
.5
FREQUENCY
300 OBS
100 LAGS

FIGURE 6.04.2

6.3 THE CENSUS METHOD. The Census method is more general than either of the two previously discussed methods in that it does not make use of the assumption of either a constant seasonal amplitude or a constant seasonal pattern.

However, there are two respects in which the Census Method appears to be comparatively inferior to the two methods above. First no use is made of Hannan's technique of correcting for the bias introduced by the application of a linear operator to the original data. Since the Census Method uses the Spencer 15-point formula for this operation, the 12-month component is attenuated to 20% of its original value. This bias is not clearly evident in the results because Spencer's formula is applied only after most of the seasonal has been removed using a simple moving average. The use of Hannan's correction procedure eliminates the need for a two- (or more) stage process as used in the Census Method. The second weakness of the Census Method is its poor response, when compared to Wald's method, to relatively high-frequency variations of the seasonal amplitude. Part of the reason for this relatively poor frequency-response characteristic is the large number of observations for a single month which are required in order to form a stable estimate (in the sense that the estimate contains very little spectrum power at high frequencies) of each monthly factor.

The observation of overriding importance that must be made about the Census Method is, however, that it performs reasonably well under a wide range of conditions. Only when the seasonal amplitude changes very rapidly is the seasonal *not* completely removed, at least in the sense of the removal of the peaks in the estimated spectrum. However, the method does allow sharp changes in the seasonal to appear in the adjusted series. In addition, under certain conditions, the method alters the series at frequencies other than the seasonal. Thus the coherence between the adjusted and the autoregressive series is sometimes quite low at low (but non-seasonal) frequencies. The gain is also sometimes considerably different from 1 at low frequencies. However, the phase is uniformly close to zero even under extreme conditions. The fact that the phase of the autoregressive series is not significantly altered by the seasonal adjustment method is of greatest importance for the interpretation and use of the adjusted series. Since the series may be used by the government for stabilization policy measures, it may be important to the stability of the system that a phase lag not be introduced. Figure 6.05 shows the performance of the Census Method for a constant seasonal pattern, while Figures 6.06 and 6.07 show the result of the same method when both the amplitude and pattern of the seasonal change considerably.

6.4 THE RATIONAL-FUNCTION METHOD. This method produces results that are quite naturally similar to the results obtained by Wald's method.

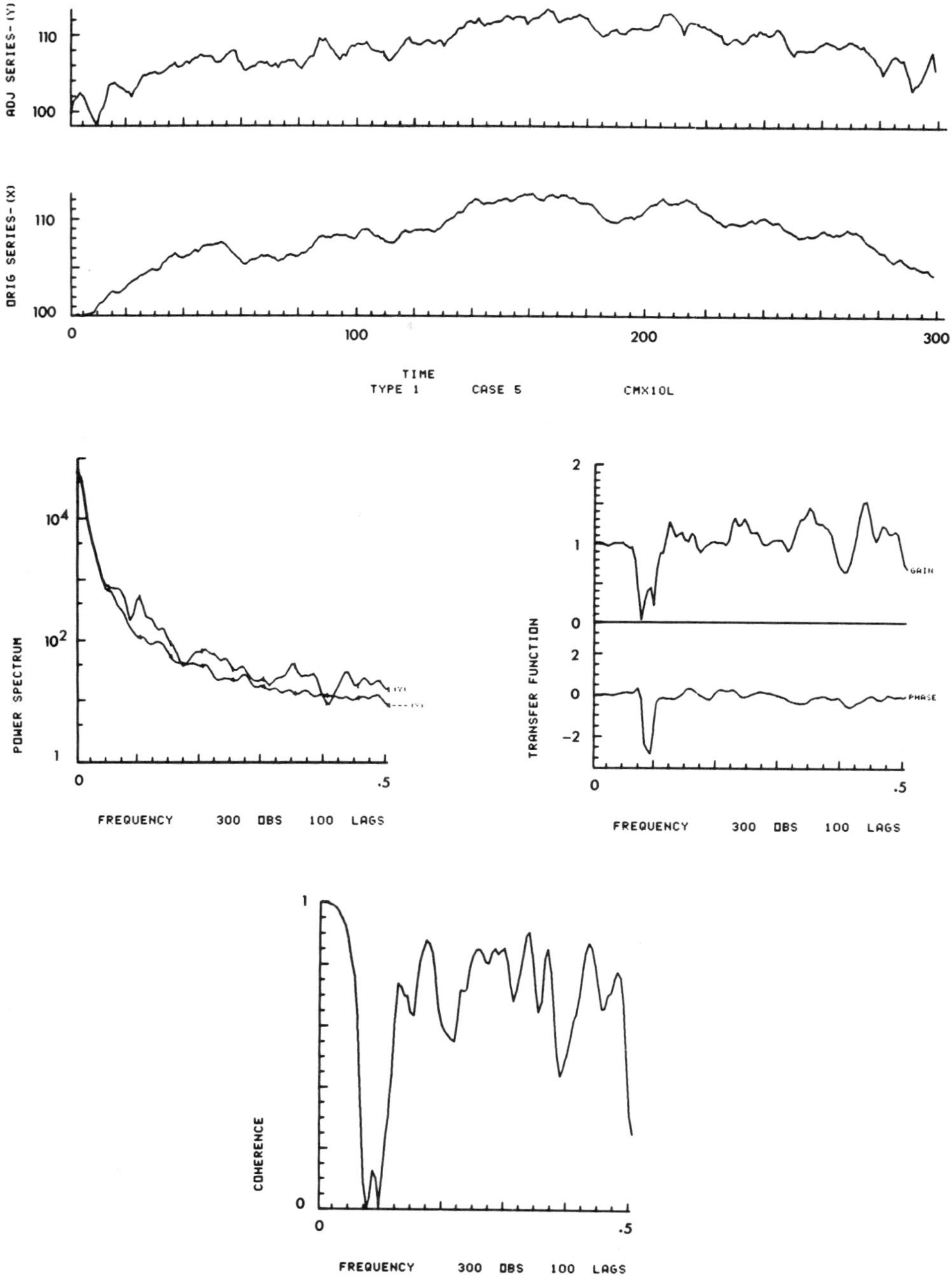
ADJ SERIES-(Y)
110
100
ORIG SERIES-(X)
110
100
0
100
200
300
TIME
TYPE 1
CASE 5
CMX10L
POWER SPECTRUM
10^4
10^2
1
0
.5
FREQUENCY
300 OBS
100 LAGS
TRANSFER FUNCTION
2
1
0
2
0
−2
GAIN
PHASE
0
.5
FREQUENCY
300 OBS
100 LAGS
COHERENCE
1
0
0
.5
FREQUENCY
300 OBS
100 LAGS

FIGURE 6.05.1

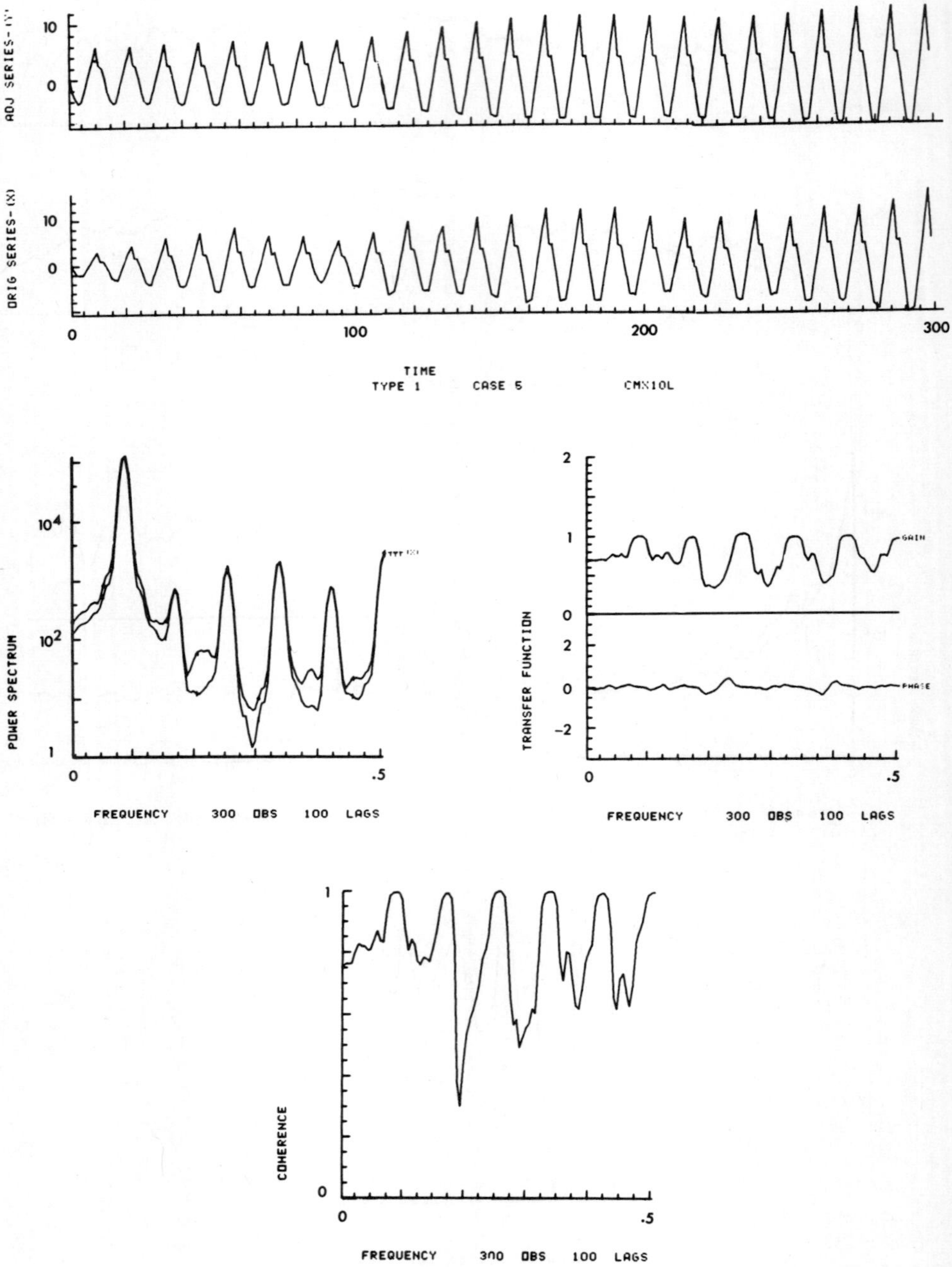
ADJ SERIES-(Y)
ORIG SERIES-(X)
10
0
100
200
300
TIME
TYPE 1 CASE 5 CMX10L
POWER SPECTRUM
10^4
10^2
1
0
.5
FREQUENCY 300 OBS 100 LAGS
TRANSFER FUNCTION
2
1
0
−2
GAIN
PHASE
COHERENCE

FIGURE 6.05.2

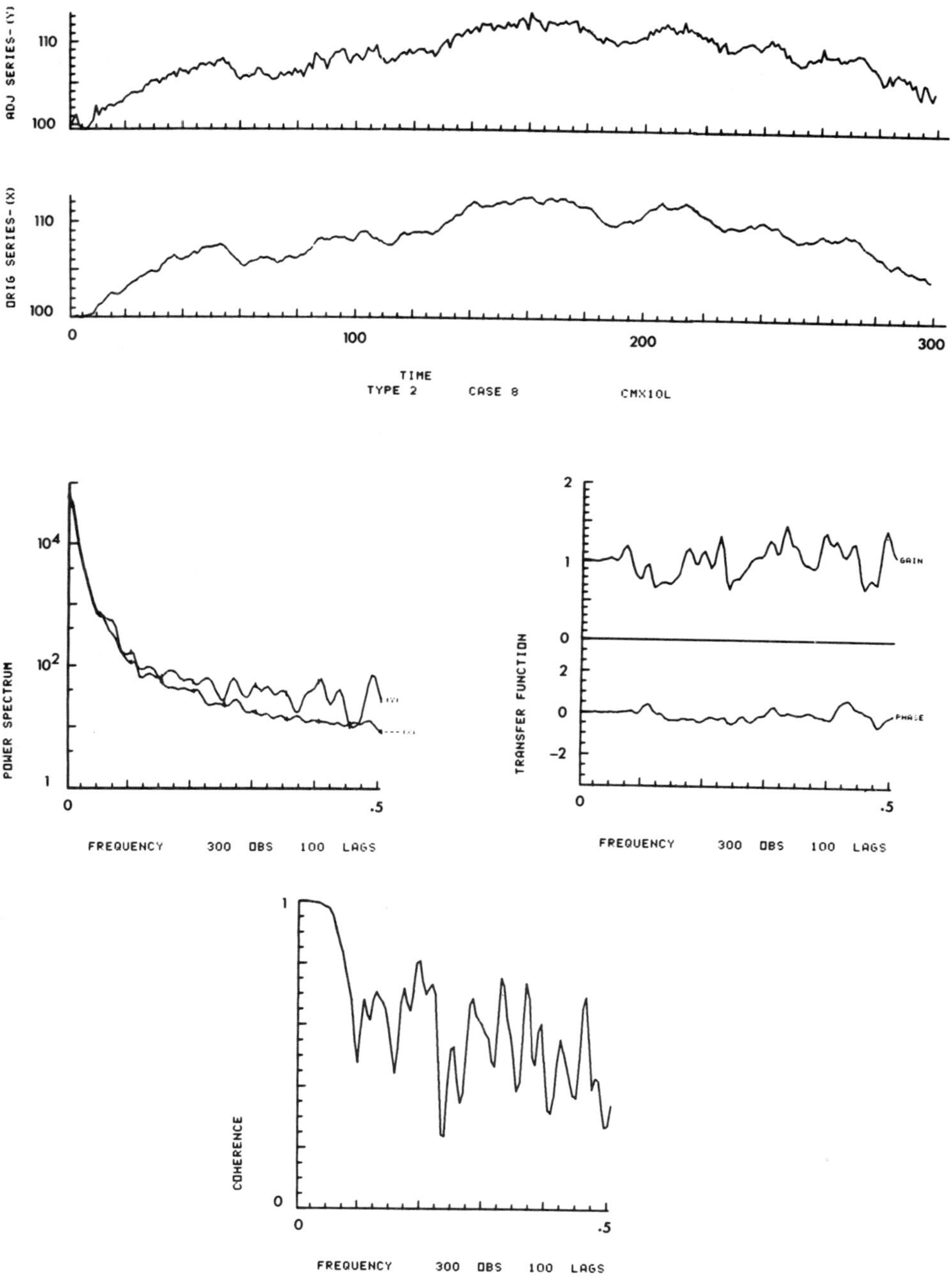
ADJ SERIES- (Y)
110
100
ORIG SERIES- (X)
110
100
0
100
200
300
TIME
TYPE 2
CASE 8
CMX10L
POWER SPECTRUM
10^4
10^2
1
0
.5
FREQUENCY 300 OBS 100 LAGS
TRANSFER FUNCTION
2
1
0
2
0
-2
GAIN
PHASE
0
.5
FREQUENCY 300 OBS 100 LAGS
COHERENCE
1
0
0
.5
FREQUENCY 300 OBS 100 LAGS

FIGURE 6.06.1

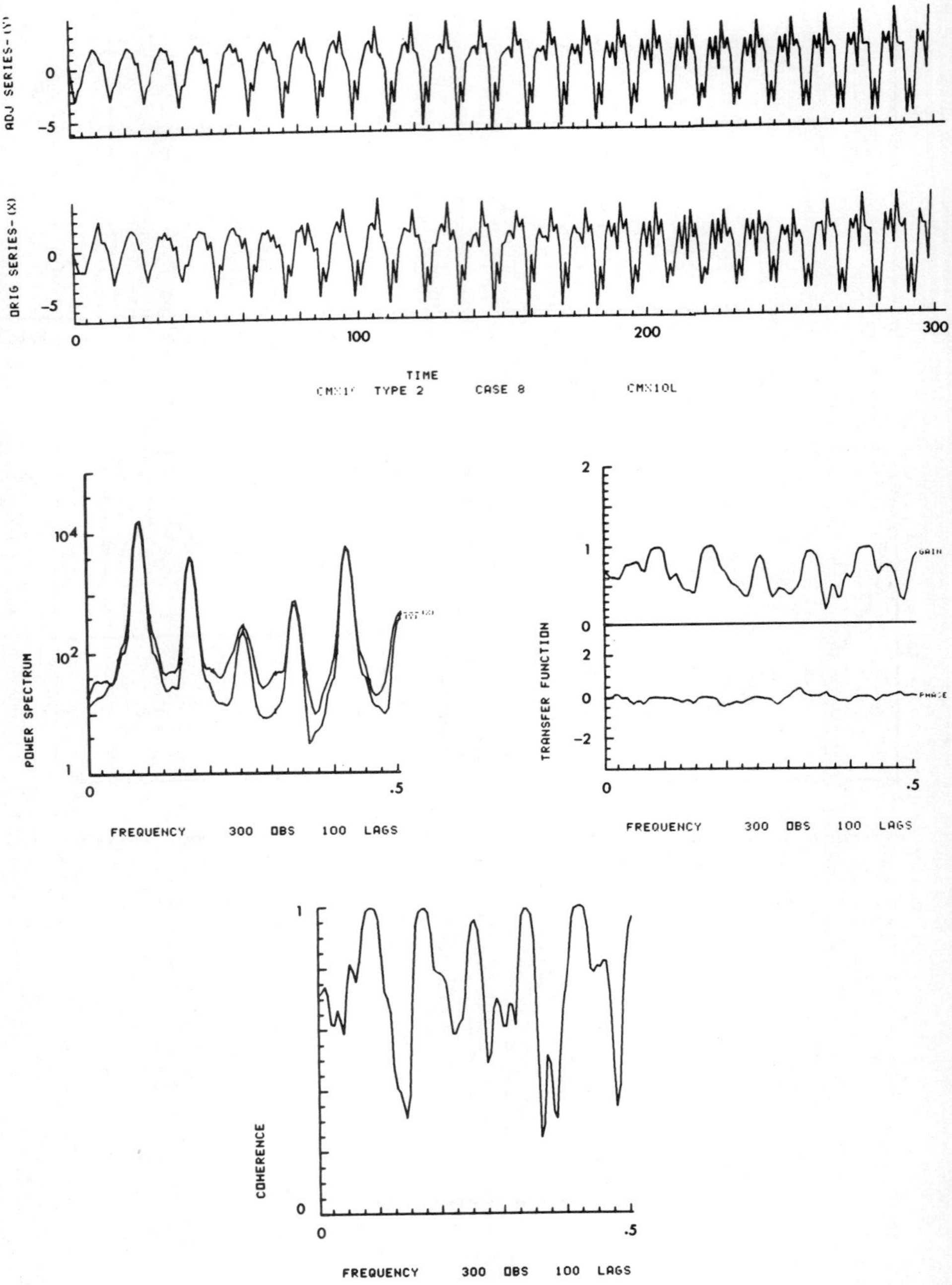
ADJ SERIES-(Y)
ORIG SERIES-(X)
0
-5
100
200
300
TIME
TYPE 2
CASE 8
POWER SPECTRUM
10^4
10^2
1
.5
FREQUENCY 300 OBS 100 LAGS
TRANSFER FUNCTION
2
-2
GAIN
COHERENCE

FIGURE 6.06.2

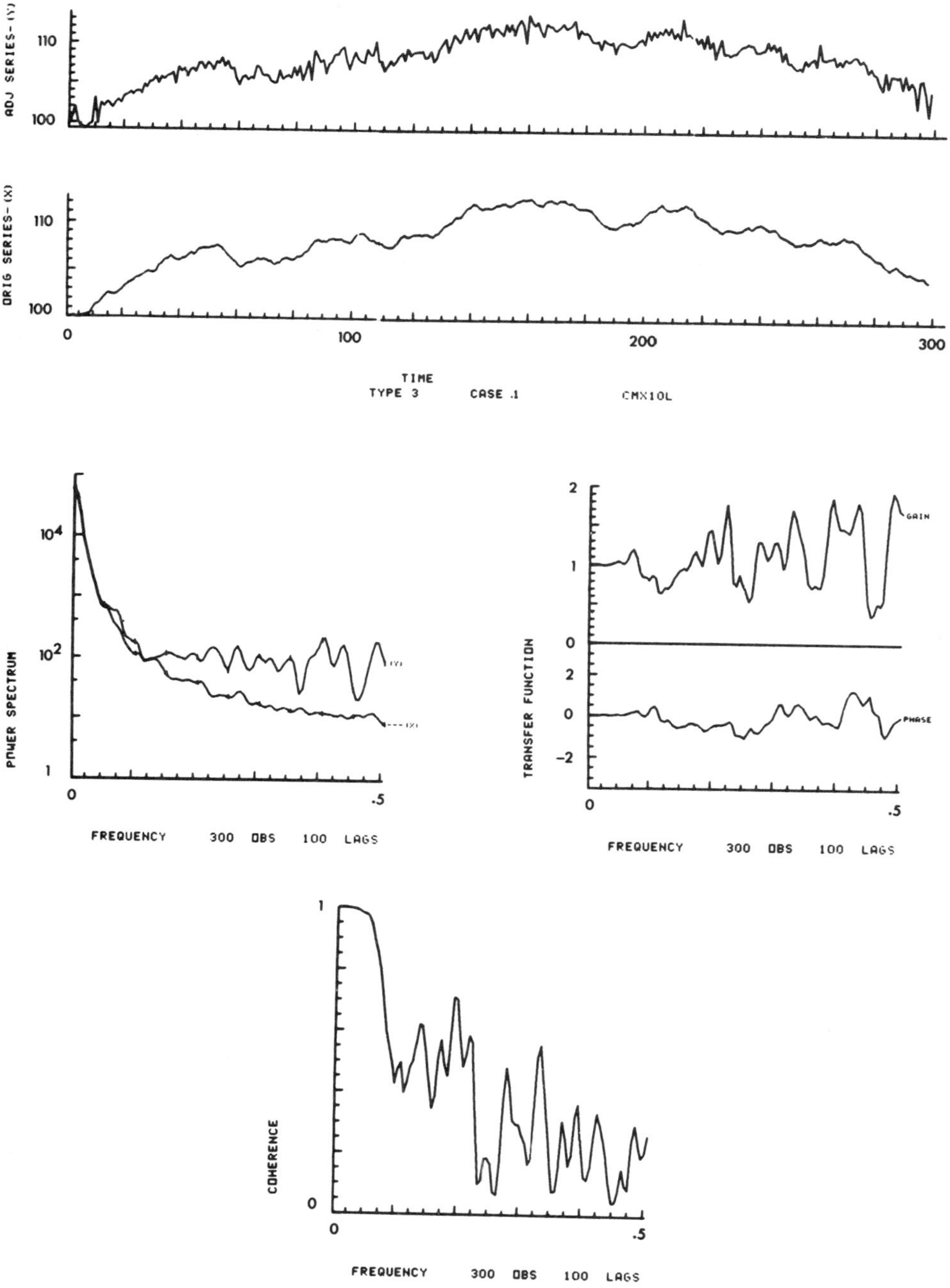
ADJ SERIES-(Y)
110
100
ORIG SERIES-(X)
110
100
0
100
200
300
TIME
TYPE 3
CASE .1
CMX10L
POWER SPECTRUM
10^4
10^2
1
0
.5
FREQUENCY 300 OBS 100 LAGS
TRANSFER FUNCTION
2
1
0
2
0
−2
GAIN
PHASE
0
.5
FREQUENCY 300 OBS 100 LAGS
COHERENCE
1
0
0
.5
FREQUENCY 300 OBS 100 LAGS

Figure 6.07.1

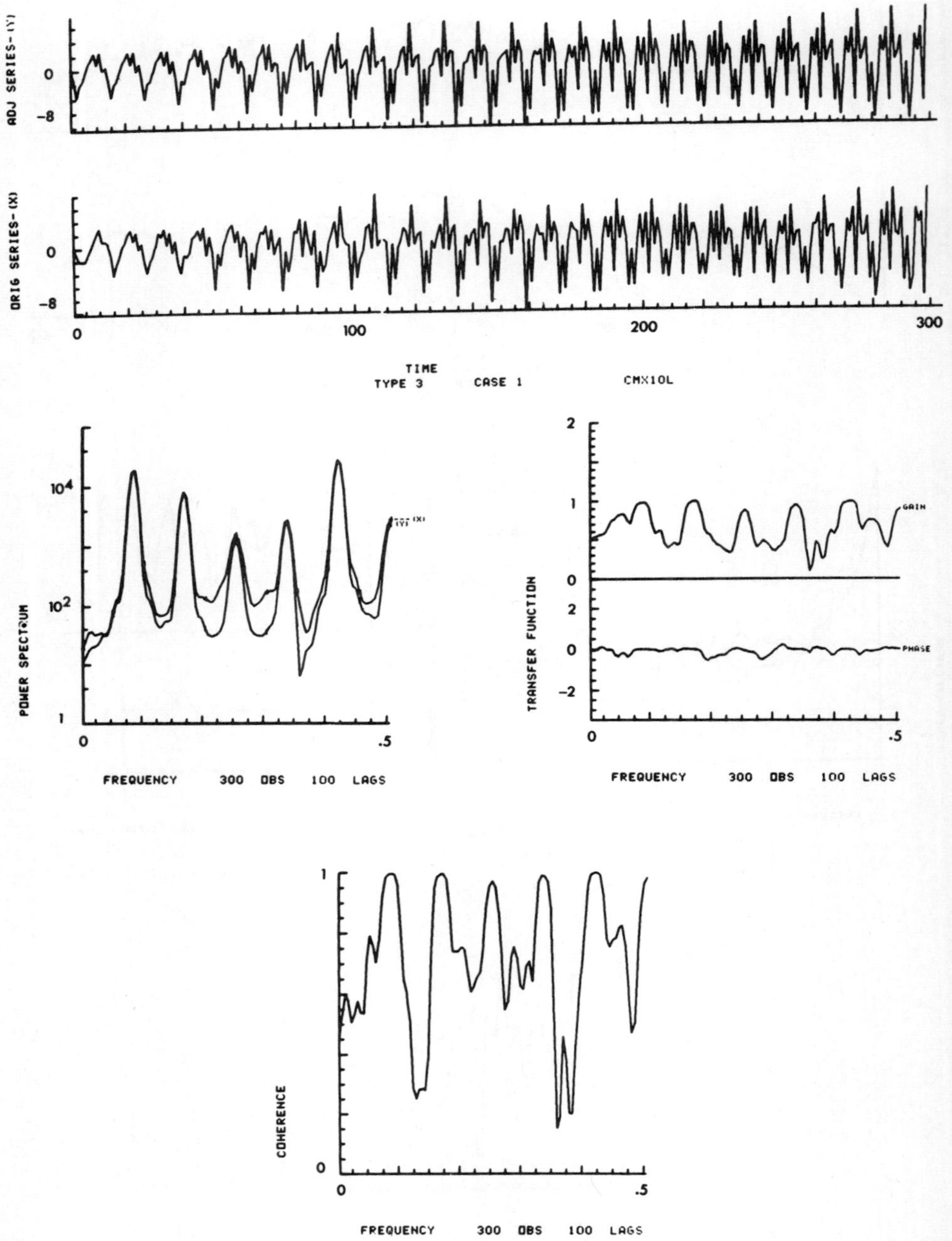
ADJ SERIES-(Y)
ORIG SERIES-(X)
0
-8
0
100
200
300
TIME
TYPE 3 CASE 1 CMX10L
POWER SPECTRUM
10^4
10^2
1
(X)
(Y)
0
.5
FREQUENCY 300 OBS 100 LAGS
TRANSFER FUNCTION
2
1
0
2
0
-2
GAIN
PHASE
0
.5
FREQUENCY 300 OBS 100 LAGS
COHERENCE
1
0
0
.5
FREQUENCY 300 OBS 100 LAGS

FIGURE 6.07.2

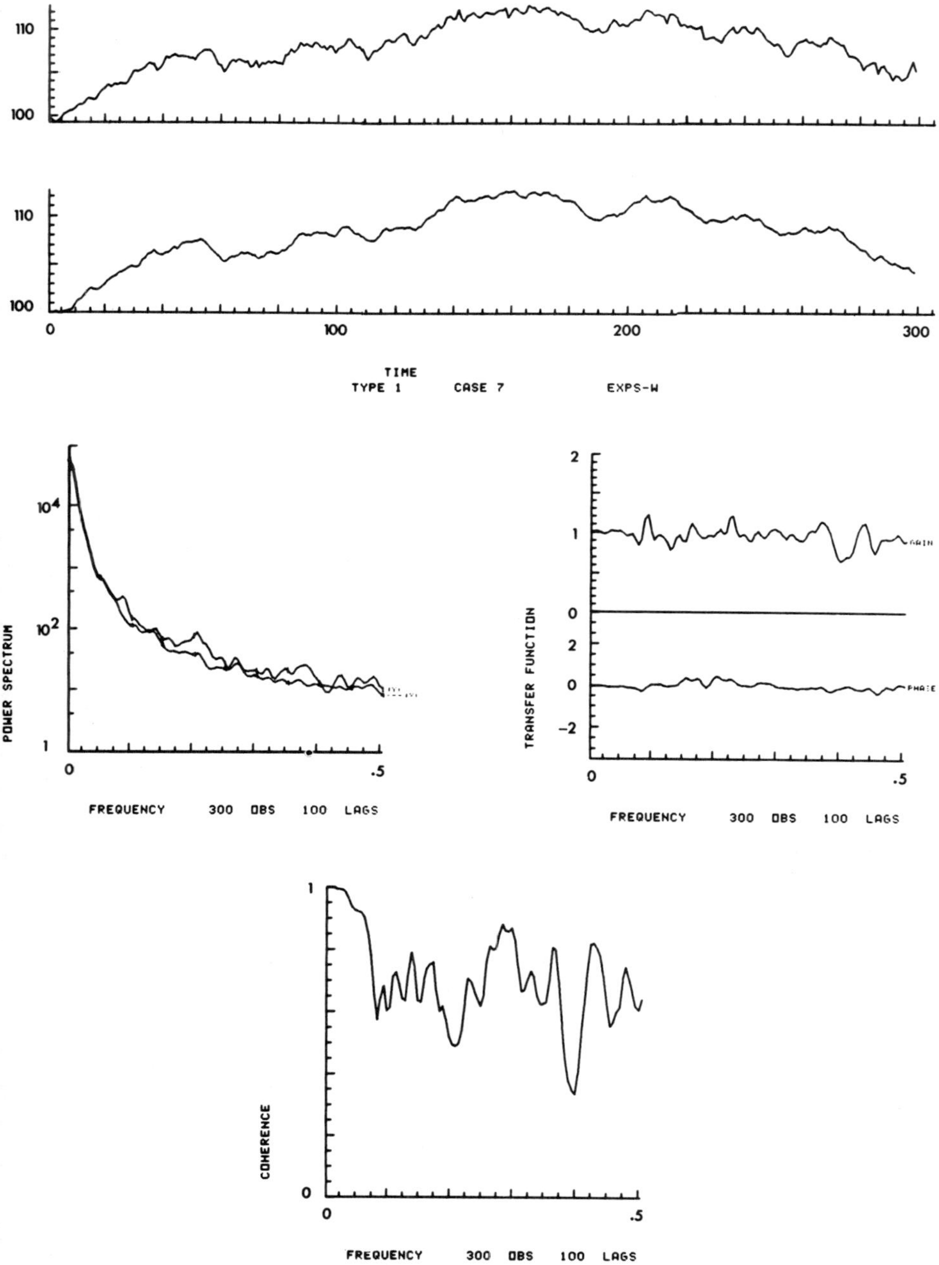
110
100
110
100
0
100
200
300
TIME
TYPE 1
CASE 7
EXPS-W
POWER SPECTRUM
10^4
10^2
1
0
.5
FREQUENCY
300 OBS
100 LAGS
TRANSFER FUNCTION
2
1
0
2
0
-2
0
.5
FREQUENCY
300 OBS
100 LAGS
COHERENCE
1
0
0
.5
FREQUENCY
300 OBS
100 LAGS

FIGURE 6.08.1

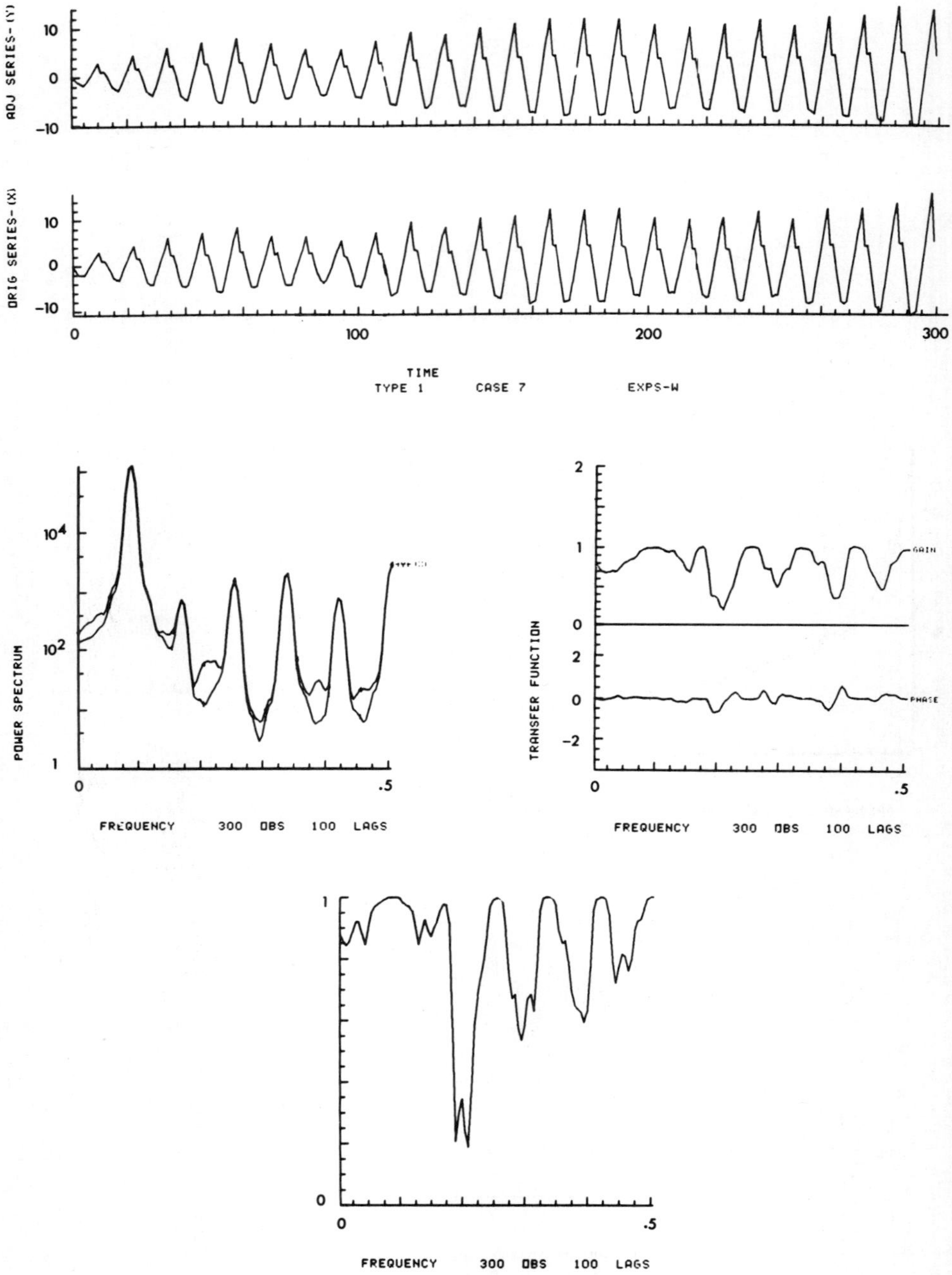
ADJ SERIES-(Y)
ORIG SERIES-(X)
10
0
-10
0
100
200
300
TIME
TYPE 1 CASE 7 EXPS-W
POWER SPECTRUM
10^4
10^2
1
0
.5
FREQUENCY 300 OBS 100 LAGS
TRANSFER FUNCTION
2
1
0
2
0
-2
GAIN
PHASE
0
.5
FREQUENCY 300 OBS 100 LAGS
1
0
0
.5
FREQUENCY 300 OBS 100 LAGS

FIGURE 6.08.2

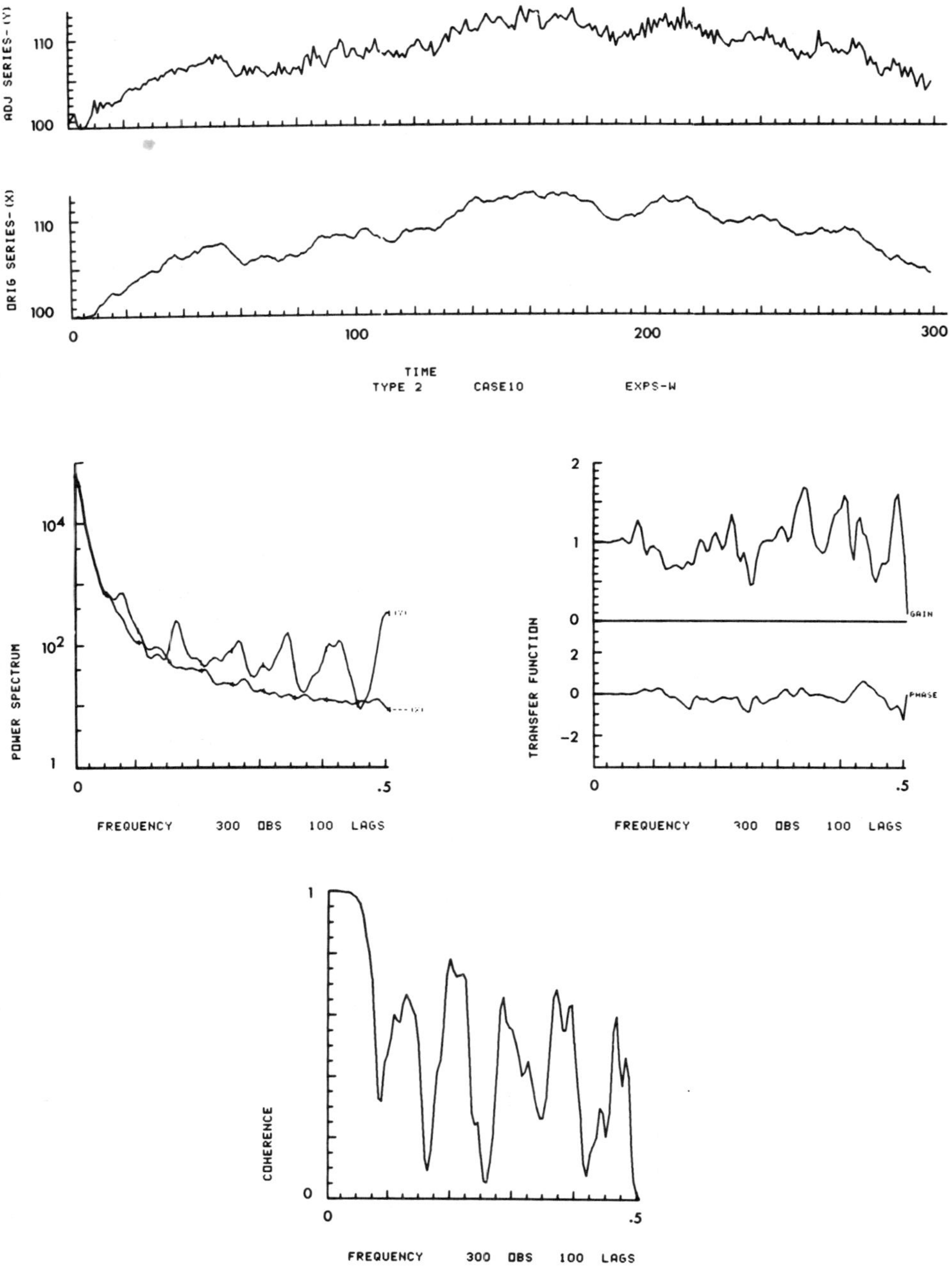
ADJ SERIES-(Y)
110
100
ORIG SERIES-(X)
110
100
0
100
200
300
TIME
TYPE 2
CASE10
EXPS-W
POWER SPECTRUM
10^4
10^2
1
0
.5
FREQUENCY
300 OBS
100 LAGS
TRANSFER FUNCTION
2
1
0
GAIN
2
0
-2
PHASE
0
.5
FREQUENCY
300 OBS
100 LAGS
COHERENCE
1
0
0
.5
FREQUENCY
300 OBS
100 LAGS

FIGURE 6.09.1

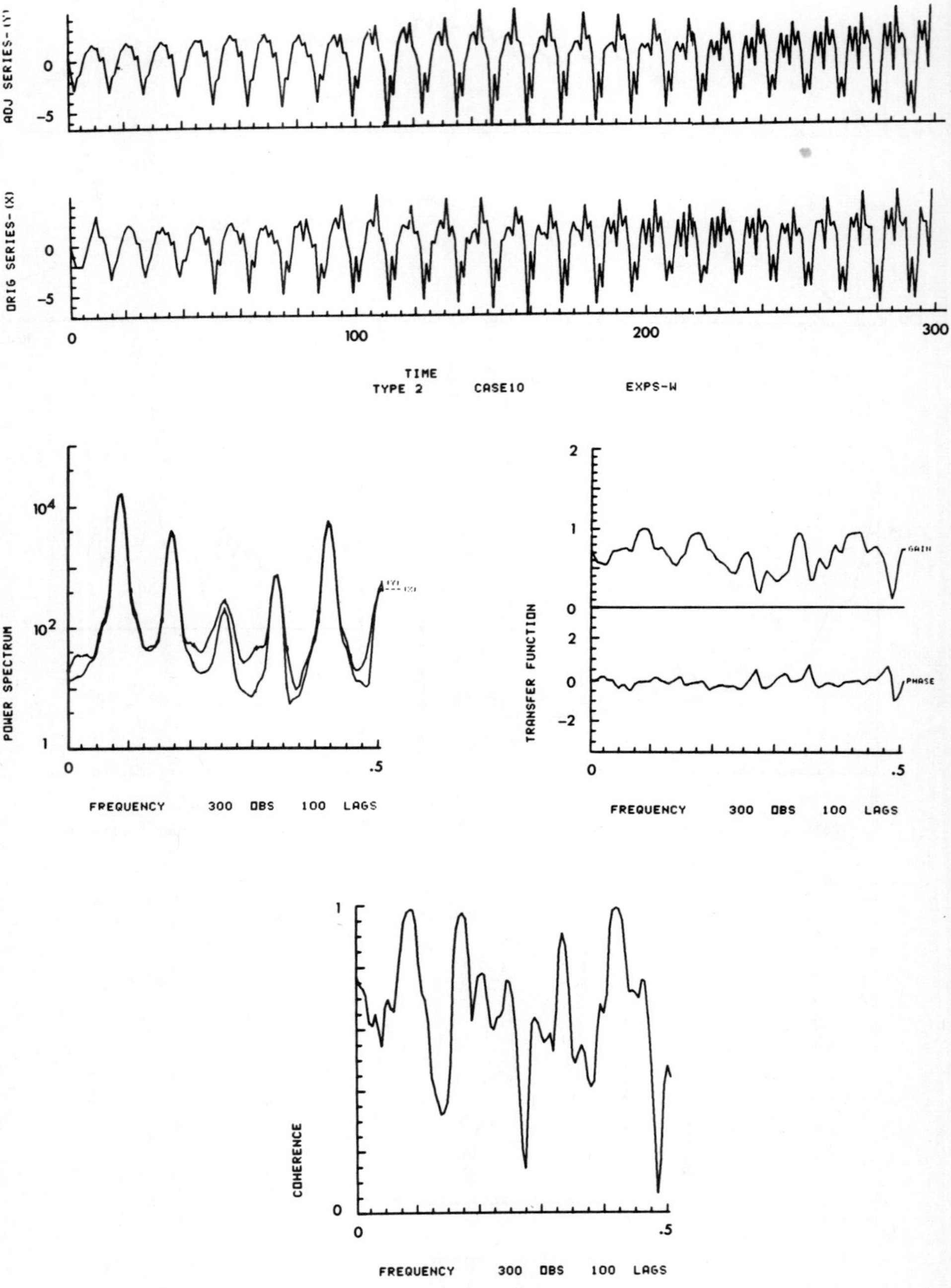
ADJ SERIES-(Y)
0
-5
ORIG SERIES-(X)
0
-5
0
100
200
300
TIME
TYPE 2
CASE10
EXPS-W
POWER SPECTRUM
10⁴
10²
1
0
.5
FREQUENCY
300 OBS
100 LAGS
TRANSFER FUNCTION
2
1
0
2
0
-2
GAIN
PHASE
0
.5
FREQUENCY
300 OBS
100 LAGS
COHERENCE
1
0
0
.5
FREQUENCY
300 OBS
100 LAGS

FIGURE 6.09.2

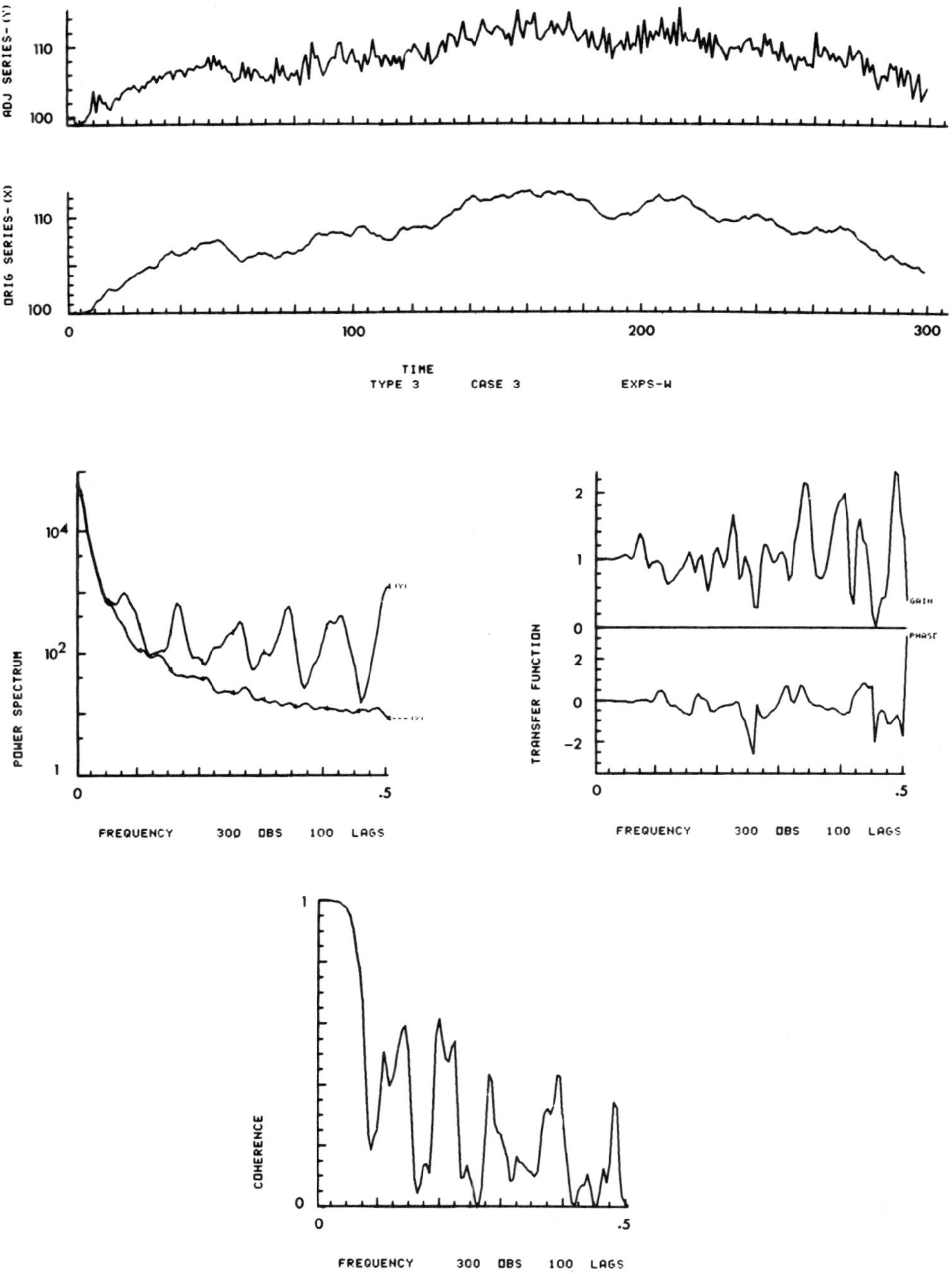
ADJ SERIES-(Y)
110
100
ORIG SERIES-(X)
110
100
0
100
200
300
TIME
TYPE 3
CASE 3
EXPS-W
POWER SPECTRUM
10^4
10^2
1
(Y)
(X)
0
.5
FREQUENCY
300 OBS
100 LAGS
TRANSFER FUNCTION
2
1
0
GAIN
PHASE
2
0
-2
0
.5
FREQUENCY
300 OBS
100 LAGS
COHERENCE
1
0
0
.5
FREQUENCY
300 OBS
100 LAGS

FIGURE 6.10.1

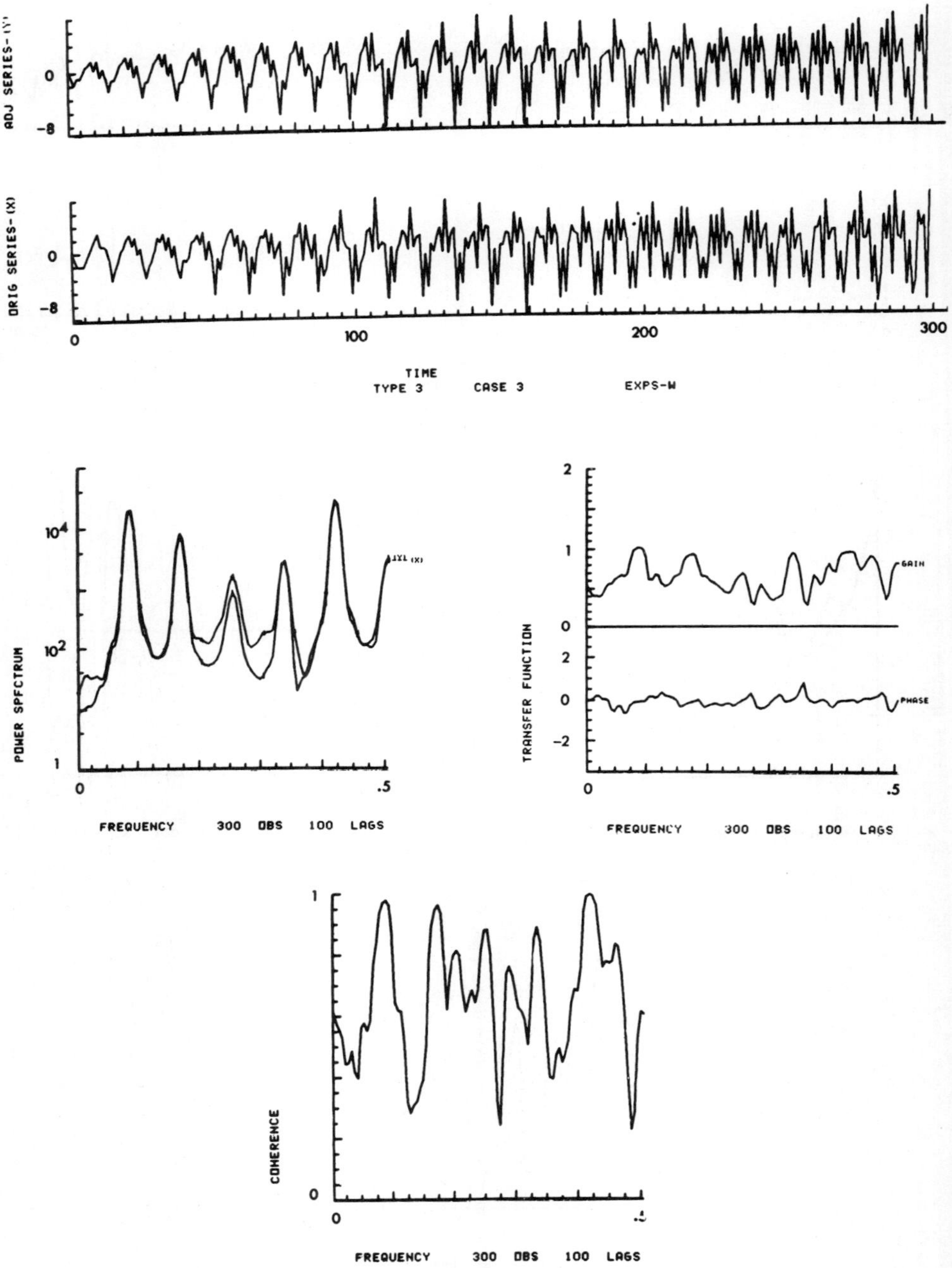
ADJ SERIES-(Y)
0
-8
ORIG SERIES-(X)
0
-8
0
100
200
300
TIME
TYPE 3
CASE 3
EXPS-W
POWER SPECTRUM
10^4
10^2
1
0
.5
FREQUENCY 300 OBS 100 LAGS
TRANSFER FUNCTION
2
1
0
2
0
-2
GAIN
PHASE
0
.5
FREQUENCY 300 OBS 100 LAGS
COHERENCE
1
0
0
FREQUENCY 300 OBS 100 LAGS

FIGURE 6.10.2

While the method is in many ways similar to Wald's method, it represents an innovation in the use of a rational-function filter for the estimation of the monthly seasonal factors. This removes the need for the use of a long moving average of the observations for each month and also reduces the requirement of extrapolation of the final values at this stage. The rational-function filter which has been used requires only observations that are coincident with or precede the estimate in time. Thus the estimates for the last twelve months in the series are produced exactly as the previous ones. Therefore no correction of the estimates is needed as more data become available, except the correction required by the initial filtering of the data. In addition only the last three years of data and the initial values for the filter estimates are required for the estimation of the adjusted series for the current year. After the adjustment method has been applied to a series, then the final values of the rational-function filter estimates may be saved so that as new data become available, it is only necessary to use these estimated values and the last three years of data for the new estimation. The only figures that will be revised in the entire computation are those for the last 12 months of each estimation. These are the estimates which are based upon extrapolations of the 12-month moving average which was applied to the original observations.

The use of Wald's least-squares technique for estimation of the seasonal amplitude makes the performance of the rational-function method comparable to Wald's method when the seasonal pattern is constant but the amplitude is undergoing rapid change. Figure 6.08 shows the performance of the method under this condition. Figures 6.09 and 6.10 show the performance of the method when both the seasonal pattern and amplitude are changing.

7. CONCLUSIONS

On the basis of the criteria put forward in the Introduction to this paper and the computations presented in Section 6.0, we have been able to compare the relative performances of various seasonal adjustment methods. The most important conclusions to be drawn from this analysis are as follows:

7.1 For the case of a constant, strictly periodic seasonal, it is difficult to see how it would be possible to improve upon Hannan's method. In addition, the use of transfer function analysis as applied in Hannan's method is of general applicability to seasonal adjustment methods which use moving averages.

7.2 Wald's method, which is unique in its use of the intercorrelation of the monthly seasonal factors, definitely displays the value of the use of this intercorrelation when the assumption of a constant seasonal pattern is met. When the amplitude (but not the pattern) of the seasonal varies relatively rapidly, Wald's method produces the best estimate of the underlying autoregressive series.

7.3 The value of the Census Method is that it fulfills the stated objectives reasonably well under all conditions. It continues to provide an adjusted series which closely approximates the autoregressive series even when both the amplitude and the pattern of the seasonal vary considerably. This result is not at all surprising given that the method uses a moving average estimate of the change in the seasonal and that it does not rely in any way upon intercorrelations between the monthly seasonal factors. However, there are two points at which the Census Method is relatively weak. The first is that no correction is made for the bias introduced by the use of the Spencer 15-point formula. As mentioned previously, this correction could easily be made in the Census Method without any change in the basic method. Therefore the presence of this bias in the current version of the method is not a fundamental criticism. The other weakness of the Census Method is that the frequency response of the moving average used to estimate the moving monthly seasonal factors is very low even at quite low frequencies. This characteristic produces a very stable estimate of the seasonal factors but a poor estimate of relatively rapid variations in the amplitude or pattern of the seasonal.

7.4 The rational-function method—basically an extension of Wald's and Hannan's methods—is an attempt, shown to be at least partially successful, to remove the two weaknesses of the Census Method mentioned above, while maintaining the generality of the assumptions on which the Census Method is based. The method is very simple both conceptually and computationally. In addition some of the problems of treatment of the end values of the series are avoided through the use of an asymmetric, single-sided, filter function. This method is not intended as a complete new method of seasonal adjustment, but is simply presented to show the potential value of the application of simple ideas of frequency decomposition to seasonal adjustment.

Finally, mention must be made of the cost involved in the use of a seasonal adjustment method which estimates relatively rapid changes in the seasonal. It is not possible to estimate rapid changes in the seasonal without, in some way, affecting the information in the series at frequencies near to the seasonal. Thus the more "flexible" the seasonal estimator, the greater the disturbance to the series at non-seasonal frequencies. In attempting to estimate a changing seasonal, it would generally seem to be desirable to use a method that does not completely remove the seasonal in order to reduce the amount of distortion at non-seasonal frequencies. The exact trade-off between removal of a varying seasonal and distortion of the series is something which needs ultimately to be determined in the context of the intended use of the information in the series.

APPENDIX A: DERIVATION OF WALD'S METHOD

This appendix presents the derivation of equation (4.2.7), which defines the estimated seasonal coefficients. We continue the notation established in Section 4.2 and presuppose the statement of the assumptions given in that section.

The values of $p(i, k)$ for the kth month are the same for all i since $p(t)$ is a periodic function. Replacing these values by a common value $p(k)$, we obtain from equation (4.2.6):

$$\frac{\sum_{i=1}^{m} \psi(i, k)}{m} \sim p(k) \frac{\sum_{i=1}^{m} \lambda(i, k)}{m} + \frac{\sum_{i=1}^{m} y(i, k)}{m} \quad \text{for } k = 1, 2, \ldots, 12. \tag{A.1}$$

Now let

$$\frac{\sum_{i=1}^{m} \lambda(i, k)}{m} = \lambda(k) \quad \text{for } k = 1, 2, \ldots, 12. \tag{A.2}$$

Also, let

$$\frac{\sum_{k=1}^{12} \lambda(k)}{12} = \lambda(0).$$

Then substituting in equation (A.2):

$$\frac{\sum_{k=1}^{12} \sum_{i=1}^{m} \lambda(i, k)}{12m} = \lambda(0). \tag{A.3}$$

Replacing the $\lambda(i, k)$ in (A.2) by $\frac{1}{12} \sum_{k=1}^{12} \lambda(i, k) + \delta(i, k)$, where $\delta(i, k)$ is the deviation of $\lambda(i, k)$ from its mean, results in:

$$\frac{\sum_{i=1}^{m} \left[\frac{\sum_{k=1}^{12} \lambda(i, k)}{12} + \delta(i, k) \right]}{m} = \lambda(k)$$

or

$$\frac{\sum_{i=1}^{m} \sum_{k=1}^{12} \lambda(i, k)}{12m} + \frac{\sum_{i=1}^{m} \delta(i, k)}{m} = \lambda(k).$$

Using (A.3) this becomes

$$\lambda(0) - \lambda(k) = -\frac{\sum_{i=1}^{m} \delta(i, k)}{m}. \tag{A.4}$$

Let us assume that the maximum deviation of $\lambda(t)$ within a year is $v\%$ of the mean of that year:

$$|\delta(i, k)| \leq \frac{v}{100} \frac{\sum_{k=1}^{12} \lambda(i, k)}{12}. \tag{A.5}$$

Taking the absolute values of the left and right sides of (A.4) one finds:

$$|\lambda(0) - \lambda(k)| = \left| \frac{\sum_{i=1}^{m} \delta(i, k)}{m} \right| \leq \frac{v}{100} \frac{\sum_{i=1}^{m} \sum_{k=1}^{12} \lambda(i, k)}{12m} = \frac{v}{100} \lambda(0). \tag{A.6}$$

In general, the $\delta(i, k)$ will have alternating signs and $\left| \frac{\sum_{i=1}^{m} \delta(i, k)}{m} \right|$ will therefore be considerably smaller than $\frac{v}{100} \lambda(0)$. This justifies the assumption that:

$$\lambda(k) \sim \lambda(0) \quad \text{for} \quad k = 1, 2, \ldots, 12. \tag{A.7}$$

Consequently, one can write for (A.1):

$$\frac{\sum_{i=1}^{m} \psi(i, k)}{m} \sim p(k)\lambda(0) + \frac{\sum_{i=1}^{m} y(i, k)}{m}. \tag{A.8}$$

With respect to the residuals $z(t)$ the following two assumptions are made:

$$\frac{\sum_{i=1}^{m} \sum_{k=1}^{12} z(i, k)}{12m} \sim 0$$

and further

$$\frac{\sum_{i=1}^{m} z(i, k)}{m} \sim 0. \tag{A.9}$$

By definition $y(i, k) = z(i, k) - z^*(i, k)$. Therefore:

$$\sum_{i=1}^{m} y(i, k) = \sum_{i=1}^{m} z(i, k) - \sum_{i=1}^{m} z^*(i, k). \tag{A.10}$$

Since

$$\sum_{i=1}^{m} z^*(i, k) = \frac{\sum_{k=1}^{12} \sum_{i=1}^{m} z(i, k)}{12},$$

equation (A.10), after dividing by m, becomes:

$$\frac{\sum_{i=1}^{m} y(i, k)}{m} = \frac{\sum_{i=1}^{m} z(i, k)}{m} - \frac{\sum_{i=1}^{m}\sum_{k=1}^{12} z(i, k)}{12m}.$$

From (A.9):

$$\frac{\sum_{i=1}^{m} y(i, k)}{m} \sim 0.$$

Hence, (A.8) can be replaced by:

$$\frac{\sum_{i=1}^{m} \psi(i, k)}{m} \sim \lambda(0)p(k) \quad \text{for } k = 1, 2, \ldots, 12. \tag{A.11}$$

The 12 sums, one for each month, on the left side of (A.11) can be obtained from the $\psi(i, k)$ series. What is left to be done is to split these sums into the two components $\lambda(0)$ and $p(k)$, where $\lambda(0)$ is a constant, being the arithmetic mean of all $\lambda(i, k)$. The $\lambda(i, k)$ and $p(k)$ are, however, still to be determined. From (4.2.5), we can write, after replacing $s(t)$ by $\lambda(t)p(t)$:

$$\psi(i, k) \sim \lambda(i, k)p(i, k) + y(i, k) \qquad \text{for} \quad \begin{aligned} i &= 1, 2, \ldots, m \\ k &= 1, 2, \ldots, 12, \end{aligned}$$

or, since the function $p(t)$ is periodic:

$$\psi(i, k) \sim \lambda(i, k)p(k) + y(i, k).$$

Then

$$\psi(i, k) \sim \frac{\lambda(i, k)}{\lambda(0)} \cdot \lambda(0)p(k) + y(i, k),$$

or, from (A.11)

$$\psi(i, k) \sim \frac{\lambda(i, k)}{\lambda(0)} \cdot \frac{\sum_{i=1}^{m} \psi(i, k)}{m} + y(i, k)$$

and

$$\psi(i, k) \sim \mu(i, k) \cdot a(k) + y(i, k) \tag{A.12}$$

where

$$\mu(i, k) \sim \frac{\lambda(i, k)}{\lambda(0)} \quad \text{and} \quad a(k) = \frac{\sum_{i=1}^{m} \psi(i, k)}{m}.$$

Now the $\mu(i, k)$ remain to be determined.

(A.12) may also be written:

$$y(i, k) \sim \psi(i, k) - \mu(i, k)a(k) \qquad \text{for} \quad \begin{aligned} i &= 1, 2, \ldots, m \\ k &= 1, 2, \ldots, 12. \end{aligned} \tag{A.13}$$

Since $y(t)$ may be assumed to be a normally distributed random variable, the $\mu(t, k)$ may be determined so as to minimize:

$$\sum_{i=1}^{m} \sum_{k=1}^{12} [\psi(i, k) - \mu(i, k)a(k)]^2.$$

The additional condition to be imposed on the function $\mu(t) = \dfrac{\lambda(0)}{\lambda(t)}$ is that its value will change only slowly over time. This leads to the assumption that the $\mu(i, k)$ may be approximated by minimizing:

$$\sum_{j=k-6}^{k+5} [\psi(i, j) - \mu(i, k)a(j)]^2 \qquad k = 1, 2, \ldots, 12 \tag{A.14}$$

subject to the condition that for each point t in time, the value of $\mu(t)$ will be nearly constant in each period $(t - 6, t + 5)$ for $t = 7, 8, \ldots, (n - 5)$.[1]

Rather than minimizing the expression $[\psi(i, k) - \mu(i, k)a(k)]^2$ with respect to $\mu(i, k)$ over the whole period $t = 7, 8, \ldots, (n - 5)$, it is minimized over consecutive 12-month periods. The first period includes the time points 7, 8, . . . , 18, the second one 8, 9, . . . , 19 etc., the last one $(n - 16), (n - 15), \ldots, (n - 5)$; there are, therefore, $n - 6$ periods all together. The condition imposed on the $\mu(t)$ assures that the 12 μ's related to the first period, $\mu(1), \mu(2), \ldots, \mu(12)$ will all have the same value, say $C(1)$. Similarly, the 12 μ's related to the jth period, $\mu(j)$ $\mu(j + 1), \ldots, \mu(j + 11)$, will have the same value $C(j)$, where, in general $C(1) \neq C(2) \neq \ldots \neq C(j)$. A particular μ, say $\mu(i)$, is determined when the minimization of equation (A.14) has been performed for the 12-month period i.

Replacing the 12 $\mu(i, k)$'s in (A.14) by a single $\mu(i, k)$ and differentiating that expression with respect to that $\mu(i, k)$ we have:

$$\mu(i, k) = \frac{\sum_{j=k-6}^{k+5} \psi(i, j)a(j)}{\sum_{j=k-6}^{k+5} [a(j)]^2}. \tag{A.15}$$

The seasonal fluctuations $s(i, k)$ can now directly be derived from these $\mu(i, k)$'s:

$$s(i, k) = a(k)\mu(i, k) = a(k) \frac{\sum_{j=k-6}^{k+5} \psi(i, j)a(j)}{\sum_{j=k-6}^{k+5} [a(j)]^2}. \tag{A.16}$$

[1] It should be noted that if in (A.14) the index j of $\psi(i,j)$ and $a(j)$ becomes ≤ 0, the former are replaced by $\psi(i - 1, j + 12)$ and $a(j + 12)$ and if $j \geq 12$, by $\psi(i + 1, j - 12)$ and $a(j - 12)$.

Since the same 12 $a(k)$'s appear in the denominator repeatedly, formula (A.16) can be still further simplified to:

$$s(i, k) = a(k) \frac{\sum_{j=k-6}^{k+5} \psi(i, j)a(j)}{\sum_{l=1}^{12} [a(l)]^2}. \tag{A.17}$$

Now, the $s(i, k)$ can be obtained from the original observations $\varphi(t)$ via the series of differences $\psi(t)$ and the $a(k)$ derived from the $\psi(t)$.

Wald also includes in his monograph the following scheme for carrying out the computational steps:

1. First $\varphi^*(t)$, the 12-month moving average of $\varphi(t)$, is computed.
2. Then the differences $\psi(t) = \varphi(t) - \varphi^*(t)$ are formed and put in the form of a matrix with 12 columns, one for each month. At this point, it should be remarked that if in actual situations the values of some of the $\psi(i, k)$ are too extreme, due to special circumstances (strikes, for example), then these $\psi(i, k)$ may be excluded from the rest of the computations.
3. For each month $k (k = 1, 2, \ldots, 12)$, the arithmetic mean $a(k)$ of the values of $\psi(i, k)$ which appear in the kth column of the matrix is then computed.
4. This is followed by an adjustment of the $a(k)$ according to the formula:

$$a'(k) = a(k) - |a(k)| \frac{a(1) + a(2) + \ldots + a(12)}{|a(1)| + |a(2)| + \ldots + |a(12)|} \tag{A.18}$$

so that

$$\sum_{k=1}^{12} a'(k) = 0.$$

5. Then the series $F(i, k) = b'(k)\psi(i, k)$ is formed where

$$b'(k) = \frac{a'(k)}{\sum_{l=1}^{12} [a'(l)]^2}.$$

6. Adding the first 12 values of the series $F(i, k)$ will give $\mu(1)$, the 1st term of the $\mu(t)$ series. Subtracting from $\mu(1)$ the 1st term of $F(i, k)$ and adding to it the 13th term of $F(i, k)$ will give $\mu(2)$, the 2nd term of the $\mu(t)$ series. This procedure is continued till the last term of $F(i, k)$ has been added and the (last $-$ 12)th term of $F(i, k)$ has been subtracted from $\mu(t - 1)$ to give $\mu(t)$. After the $\mu(t)$ series has been computed, it is arranged in the form of a matrix with 12 columns, one column for each month.

7. The seasonal fluctuations $s(i, k)$ are then computed according to the formula:

$$s(i, k) = a'(k)\mu(i, k). \tag{A.19}$$

It is not possible to obtain values for $\mu(t)$ and $s(t)$ for the last 11 months this way, which is a serious drawback in practical applications. Wald suggests in [27] that the simplest solution to this problem is to add 11 more terms to the $\mu(t)$ series, all equal to the last computed $\mu(t)$. This will then make it possible to compute the $s(t)$ for the last 11 months on the basis of (A.19). However, the extrapolation technique described in Section 3.0 is shown by Wald to be a better procedure.

8. Finally the seasonal series $s(t)$ is subtracted from the series $\varphi(t)$ to give the seasonally adjusted series.

APPENDIX B: DETAILS OF COMPUTATIONS

While analysis using the computation of spectra and cross-spectra is becoming more widespread in economics, the technique is neither completely accessible to all mathematical economists nor is it fully standardized. Therefore the following definitions and equations will be stated.

The serial cross-correlation coefficient for two series, x_t and y_t, (which may be identical) is defined by:

$$R_{xy}(s) = \frac{\sum_{t=1}^{N-s} (x_t - \bar{x})(y_{t+s} - \bar{y})}{\left[\sum_{t=1}^{N} (x_t - \bar{x})^2 \sum_{t=1}^{N} (y_t - \bar{y})^2\right]^{\frac{1}{2}}} \qquad s = 0, \ldots, m$$

where: N—number of observations.

$$\bar{x} = \frac{1}{N}\sum_{t=1}^{N} x_t$$

$$\bar{y} = \frac{1}{N}\sum_{t=1}^{N} y_t$$

$$R_{xy}(s) = R_{yx}(-s)$$

$$R_{xx}(s) = R_{xx}(-s).$$

The Parzen estimate of the normalized spectrum is then given by:

$$F_{xy}(\omega) = \sum_{s=-m}^{m} R_{xy}(s)\lambda(s)e^{i\omega s}$$

where:

$$\lambda(s) = 1 - 6\left(\frac{s}{m}\right)^2 + 6\left(\left|\frac{s}{m}\right|\right)^3 \qquad s \leq \frac{m}{2}$$

$$= 2\left(1 - \left|\frac{s}{m}\right|\right)^3 \qquad s > \frac{m}{2}.$$

Due to

$$R_{xx}(s) = R_{xx}(-s),$$

$$F_{xx}(\omega) = R_{xx}(0) + 2\sum_{s=1}^{m} R_{xx}(s)\lambda(s)\cos\omega s.$$

Writing $F_{xy}(\omega)$ in terms of its real and imaginary parts we have the real part, the co-spectrum:

$$C_{xy}(\omega) = R_{xy}(0) + \sum_{s=1}^{m} [R_{xy}(s) + R_{yx}(s)]\lambda(s)\cos\omega s$$

and the imaginary part, the quadrature spectrum:

$$Q_{xy}(\omega) = \sum_{s=1}^{m} [R_{xy}(s) - R_{yx}(s)]\lambda(s)\sin\omega s.$$

From these we define the following estimates:

1. Coherence

$$S_{xy}^2(\omega) = \frac{|F_{xy}(\omega)|^2}{F_{xx}(\omega)F_{yy}(\omega)}$$

2. Gain

$$G_{xy}(\omega) = \frac{|F_{xy}(\omega)|}{F_{xx}(\omega)}$$

3. Phase

$$\Phi_{xy}(\omega) = \tan^{-1}\frac{Q_{xy}(\omega)}{C_{xy}(\omega)}.$$

All of the series computed for this paper were made up of 300 observations. One hundred lags were used. Thus $N = 300$, $m = 100$.

All of the computations were Fortran programmed for the Princeton University IBM 7090. Extensive use was made of the on-line graphic display and recording facilities on the computer. In fact only summary statistics and identifying comments were output on tape for subsequent printing. All other results, of which the figures in the paper were examples, were recorded directly on 35 mm microfilm. Without this graphic output facility, the development of the various programs would have been much slower, the analysis of results vastly slower and more laborious, and the analysis and development of the rational-function method nearly impossible within the time available.

BIBLIOGRAPHY

[1] Anderson, O., *Zur Problematik der Empirischen Konjunkturforschung*, Veröffentlichungen der Frankfurter Gesellschaft für Konjunkturforschung, Heft 1, Bonn, 1929.

[2] Bowley, A. L., and Smith, K. C., *Seasonal Variations*, London and Cambridge Economic Service, Special Memorandum No. 7.

[3] Chaddock, R. E., *Principles and Methods of Statistics*, Boston, 1925.

[4] Crum, W. L., 'Progressive Variation in Seasonality," *Journal of the American Statistical Association*, vol. 20, 1925, pp. 48–64.

[5] Davies, G. R., *Introduction to Economic Statistics*, New York, 1922.

[6] Donner, O., "Die Saisonschwankung als Problem der Konjunkturforschung," *Vierteljahrshefte zur Konjunkturforschung*, Sonderheft 6, Berlin, 1928.

[7] Durbin, J., "Trend Elimination for the Purpose of Estimating Seasonal and Periodic Components of Time Series," Chapter 1, *Time Series Analysis*, ed. M. Rosenblatt, New York, John Wiley & Sons, 1963, pp. 3–16.

[8] Granger, C. W. J., "The Typical Spectral Shape of an Economic Variable," Technical Report No. 11, Department of Statistics, Stanford, California, Stanford University, 1964.

[9] Granger, C. W. J., in association with Hatanaka, M., *Spectral Analysis of Economic Time Series*, Princeton, Princeton University Press, 1964.

[10] Hannan, E., "The Estimation of the Seasonal Variation in Economic Time Series," *Journal of the American Statistical Association*, vol. 58, March 1963, pp. 31–44.

[11] Hannan, E., "The Estimation of a Changing Seasonal Pattern" (mineographed note) 1963.

[12] Hatanaka, M., and Suzuki, M., "A Theory of the Pseudospectrum and its Application to Non-stationary Dynamic Econometric Models," Research Memorandum 52, Econometric Research Program, Princeton University, Princeton, New Jersey.

[13] Kemmerer, E. W., *Seasonal Variation in the Relative Demand for Money and Capital in the U.S.A.*, Washington, 1910.

[14] King, W. J., "An Improved Method for Measuring the Seasonal Factor," *Journal of the American Statistical Association*, vol. 19, 1924, pp. 301–313.

[15] Kuznets, S., *Seasonal Variations in Industry and Trade*, New York, 1933.

[16] Meeting on the Measurement of Seasonal Variations, New York, May 1925, reported in *Journal of the American Statistical Association*, vol. 20, 1925, pp. 426–434.

[17] Morgenstern, O., *On the Accuracy of Economic Observations*, 2nd ed., Princeton, Princeton University Press, 1963.

[18] Nerlove, M., "Spectral Analysis of Seasonal Adjustment Procedures," 1962, to be published.

[19] Nettheim, N. F., "The Seasonal Adjustment of Economic Data by Spectral Methods," M.Sc. thesis, Australian National University, Camberra, 1963.

[20] O.E.E.C., *Seasonal Adjustment on Electronic Computers*, Proceedings of an International Conference on Seasonal Adjustment Methods Paris, November 1960.

[21] Parzen, E., "An Approach to Empirical Time Series Analysis," Technical Report No. 10, Department of Statistics, Stanford University, Stanford, California, 1964.

[22] Parzen, E., "Mathematical Considerations in the Estimation of Spectra," *Technometrices*, Vol. 3, No. 2, May 1961, pp. 167–190.

[23] Persons, W. M., "Correlation of Time Series," *Journal of the American Statistical Association*, vol. 18, 1923, pp. 713–740.

[24] Tintner, G., *Econometrics*, New York, John Wiley & Sons, 1952.

[25] Wainstien, L. A., and Zubrakov, V. D., *The Extraction of Signals from Noise*, Englewood Cliffs, New Jersey, Prentice-Hall, 1962.

[26] Wald. A., "Extrapolation des Gleitenden 12-Monatsdurchschnittes," Beilage Nr. 8 zur den *Monatsberichten des Österreichischen Institutes für Konjunkturforschung*, Heft 11, November 1937.

[27] Wald, A., *Berechnung und Ausschaltung von Saisonschwankungen*, Beiträge zur Konjunkturforschung No. 9, Verlag von Julius Springer, Vienna, 1936.

[28] Westergaard, H., "On Periods in Economic Life," *Metron*, vol. V, No. 1, 1925, pp. 3–26.

CHAPTER 25

New Techniques for Analyzing Economic Time Series and Their Place in Econometrics

By C. W. J. GRANGER*

INTRODUCTION

In September 1959 a research project under the direction of Professor Morgenstern at Princeton was instituted to consider the use of spectral methods for analyzing economic time series. Such methods had been used with considerable success in other fields and, although it was realized that economic data contained many difficulties not met with elsewhere, it was hoped that these new techniques would still extract new and useful information. The project was particularly fortunate in having available many unpublished results by J. Tukey concerning the measurement of relationships between economic series. The initial results and philosophy of the project are described in Granger [2] and Morgenstern [6], and a full account appears in Granger [3].[1] It is the object of this paper to present, in a precise mathematical form, the theoretical aspects of the new techniques and then to suggest where and how these techniques can best be used in the present framework of econometrics. The practical problems, such as estimation and how to deal with inaccurate or poor data, are not here discussed.

1. THE NEW TECHNIQUES—(a) SPECTRAL METHODS WHEN NO FEEDBACK IS PRESENT

A discrete stochastic process is defined as a sequence of random variables ordered with respect to some parameter, usually time, and will be denoted by $\{x_t, t \in T\}$. The set T can be infinite but is not necessarily so. When $-\infty \leq t \leq \infty$ the process will be denoted by $\{x_t\}$.

A process is said to be stationary to the second order if $E[x_t]$ and $E[x_t\bar{x}_{t-\tau}]$ are both independent of time. For such a process no generality is lost by assuming $E[x_t] = 0$ all t and we denote

$$\left.\begin{aligned} E[|x_t|^2] &= \sigma^2 \\ E[x_t\bar{x}_{t-\tau}] &= \sigma^2\rho_\tau \end{aligned}\right\} \qquad (1.1)$$

A process is said to be Gaussian if every finite subset $\{x_t, t \in k\}$ of the process is distributed in a multivariate normal distribution.

* The University of Nottingham.

[1] This reference will henceforth be G, thus G(9) will refer to the ninth chapter of this book.

Throughout the paper, we shall assume σ^2 to be finite but will only rarely require the process to be Gaussian.

Wold [9] has shown that every process stationary to the second order has the property

$$\sigma^2 \rho_k = \int_{-\pi}^{\pi} e^{ik\omega}\, dW(\omega) \tag{1.2}$$

where $W(\omega)$ is a monotone increasing function, and Cramér [1] has proved that all such processes can be represented by

$$x_t = \int_{-\pi}^{\pi} e^{it\omega}\, dz(\omega) \tag{1.3}$$

the equality being understood to hold as a limit in mean square, where $\{z(\omega)\}$ is also a stochastic process, being a process of noncorrelated increments, i.e.,

$$E[\{z(\omega_1) - z(\omega_2)\}\overline{\{z(\omega_3) - z(\omega_4)\}}] = 0 \quad \text{if} \quad \omega_1 \geq \omega_2 > \omega_3 \geq \omega_4 \tag{1.4}$$

Equation (1.2) and (1.3) are connected by

$$W(\omega_1) - W(\omega_2) = E[|z(\omega_1) - z(\omega_2)|^2] \tag{1.5}$$

Further, such processes can be decomposed into two noncorrelated components

$$x_t = x_{t,1} + x_{t,2} \tag{1.6}$$

where $x_{t,1}$ is a so-called non-deterministic component and $x_{t,2}$ is a deterministic component. The most important property of a deterministic series is that given information over the infinite past, one can predict its future with no error. If $\{x_t\}$ is a non-deterministic process it can be represented by

$$x_t = \sum_{j=0}^{\infty} \alpha_j \varepsilon_{t-j} \tag{1.7}$$

where $\{\varepsilon_t\}$ is a white-noise process, being stationary and, if $E[\varepsilon_t] = 0$, having $E[\varepsilon_t \varepsilon_{t+\tau}] = 0$ all $\tau \neq 0$. If $\{x_t\}$ is non-deterministic, it also has the property

$$\sigma^2 \rho_\tau = \int_{-\pi}^{\pi} e^{i\tau\omega} f(\omega)\, d\omega \tag{1.8}$$

where $f(\omega)$ exists everywhere, except possibly in a set of measure zero, and is called the power spectrum of the process.

A simple example of a deterministic process is one generated by

$$x_t = \sum_{j=0}^{m} k_j \cos(\omega_j t + \theta_j) \tag{1.9}$$

and important examples of non-deterministic processes are those generated by

$$x_t = \sum_{j=0}^{a} \alpha_j \varepsilon_{t-j} \tag{1.10}$$

called a moving-average process,

$$\sum_{j=0}^{b} \beta_j x_{t-j} = \varepsilon_t \tag{1.11}$$

called an autoregressive process, and

$$\sum_{j=0}^{b} \beta_j x_{t-j} = \sum_{j=0}^{a} \alpha_j \varepsilon_{t-j} \tag{1.12}$$

which may be called a linear regressive process.

A simple combination of the two main types is the linear cyclic process given by

$$x_t = \sum_{j=0}^{m} k_j \cos(\omega_j t + \theta_j) + \varepsilon_t, \tag{1.13}$$

ε_t being understood to be white noise in equations (1.10)–(1.13).

The power spectrum is easily interpreted from the fact that $f(\omega)\,d\omega$ represents that part of the total variance attributable to the components with frequencies in the range $(\omega, \omega + d\omega)$. It will be argued later that the concept of a power spectrum, although extremely useful in theoretical aspects of time series analysis, is generally of small use in practice when economic series are being considered.

If the white noise process $\{\varepsilon_t\}$ has Cramér representation

$$\varepsilon_t = \int_{-\pi}^{\pi} e^{it\omega}\, dz_\varepsilon(\omega)$$

then, from (1.10), we see that

$$x_t = \int_{-\pi}^{\pi} e^{it\omega} A(\omega)\, dz_\varepsilon(\omega)$$

where

$$A(\omega) = \sum_{j=0}^{a} \alpha_j e^{-ij\omega}$$

and thus

$$\begin{aligned} E[x_t \bar{x}_{t-\tau}] &= \int_{-\pi}^{\pi} \int_{-\pi}^{\pi} e^{it\omega} e^{-i(t-\tau)\lambda} A(\omega) \overline{A(\lambda)} E[dz_\varepsilon(\omega)\, \overline{dz_\varepsilon(\lambda)}] \\ &= \frac{1}{2\pi} \sigma_\varepsilon^2 \int_{-\pi}^{\pi} e^{i\tau\omega} A(\omega) \overline{A(\omega)}\, d\omega \end{aligned}$$

and it is then seen that the power spectrum of the moving-average process given by (1.10) is $\frac{1}{2\pi}\sigma_\varepsilon^2 A(\omega)\overline{A(\omega)}$. Similarly, one finds that the power spectrum of the linear regressive process given by equation (1.12) is

$$\frac{\sigma_\varepsilon^2}{2\pi}\frac{A(\omega)\overline{A(\omega)}}{B(\omega)\overline{B(\omega)}}$$

where

$$B(\omega) = \sum_{j=0}^{b} \beta_j e^{-ij\omega}$$

The power spectrum of the linear cyclic process (1.13) is a constant $\left(\frac{1}{2\pi}\sigma_\varepsilon^2\right)$ for all $\omega \neq \omega_j$, $0 \leq \omega \leq \pi$, and equals

$$\frac{\sigma_\varepsilon^2}{2\pi} + \tfrac{1}{2}k_j^2 \quad \text{at} \quad \omega = \omega_j, \qquad j = 0, 1, \ldots, m.$$

and is thus a mixed spectrum, containing both continuous and discrete parts.

When considering two stationary, non-deterministic processes $\{x_t\}$, $\{y_t\}$ the theory is generalized as follows: assuming both processes to have zero means and with power spectra $f_x(\omega)$, $f_y(\omega)$ respectively, if

$$E[x_t \bar{y}_{t-\tau}] = \mu_\tau \tag{1.14}$$

then

$$\mu_\tau = \int_{-\pi}^{\pi} e^{i\tau\omega} c_r(\omega)\, d\omega$$

where $c_r(\omega) = c(\omega) + iq(\omega)$ is called the power cross-spectrum. The real functions $c(\omega)$ and $q(\omega)$ are respectively the power co-spectrum and the power quadrature spectrum. The function

$$C(\omega) = \frac{c^2(\omega) + q^2(\omega)}{f_x(\omega) f_y(\omega)} \tag{1.15}$$

is known as the coherence and has the property $0 \leq C(\omega) \leq 1$. The plot of $C(\omega)$ against ω is called the coherence-diagram and a second important diagram is the plot against ω of

$$\psi(\omega) = \tan^{-1}\left[\frac{q(\omega)}{c(\omega)}\right] \tag{1.16}$$

called the phase-diagram.

The coherence $C(\omega)$ is interpreted as measuring how closely related are the components at frequency ω of the two processes $\{x_t\}$, $\{y_t\}$; it being strictly analogous to the square of the correlation coefficient. Thus a high value (near one) of $C(\omega)$ indicates that the components are closely related a low value (near zero) indicates that they are hardly connected at all.

The coherence diagram thus shows how this "correlation" varies with frequency.

The function $\psi(\omega)$ measures the phase-lag between the components of frequency ω of the two processes. This can be shown simply by considering a real process $\{x_t\}$ with Cramér representation

$$x_t = \int_{-\pi}^{\pi} e^{it\omega}\, dz(\omega)$$

and a second process

$$x_t^{\phi} = \int_{-\pi}^{\pi} e^{it\omega} e^{-i\Phi(\omega)}\, dz(\omega) \tag{1.17}$$

where

$$\begin{aligned} \Phi(\omega) &= \phi(\omega), & \omega > 0 \\ &= 0, & \omega = 0 \\ &= -\phi(-\omega), & \omega < 0. \end{aligned}$$

$\{x_t\}$ being a real process it is necessary that $dz(\omega) = \overline{dz(-\omega)}$ and, as $\Phi(\omega)$ is an odd function of ω (i.e., $\Phi(\omega) = -\Phi(-\omega)$), it follows that $\{x_t^{\phi}\}$ is also a real process.

We note that, if $E[x_t] = 0$, then $E[x_t^{\phi}] = 0$ and that

$$E[x_t \bar{x}_{t-\tau}] = E[x_t^{\phi} \bar{x}_{t-\tau}^{\phi}] = \int_{-\pi}^{\pi} e^{i\tau\omega} f(\omega)\, d\omega$$

and thus both processes have the same power spectrum and can differ only in phase. In fact, by its construction, $\{x_t^{\phi}\}$ is lagged to $\{x_t\}$ by an amount $\phi(\omega)$ at frequency ω.

Further,

$$\begin{aligned} E[x_t \bar{x}_{t-\tau}^{\phi}] &= \int_{-\pi}^{\pi} e^{i\Phi(\omega)} e^{i\tau\omega} f(\omega)\, d\omega \\ &= \int_{-\pi}^{\pi} e^{i\tau\omega} [\cos \Phi(\omega) + i \sin \Phi(\omega)] f(\omega)\, d\omega \end{aligned}$$

and so the power co-quadrature and quadrature spectral functions are given by

$$c(\omega) = f(\omega) \cos \phi(\omega), \qquad 0 < \omega < \pi$$

$$q(\omega) = f(\omega) \sin \phi(\omega), \qquad 0 < \omega < \pi$$

hence

$$C(\omega) = 1$$

and

$$\psi(\omega) = \phi(\omega), \qquad \text{as required.}$$

Two particularly simple cases of lagged processes are when $\phi(\omega) = b\omega$, which gives a time lag of b units, and when $\phi(\omega) = k$, k a constant, this being called a fixed-angle lag.

It is thus seen that the coherence and angle diagrams are both easily interpreted and provide important information. This is even more obvious when it is noticed that the short-run components of two economic variables need not be as highly related as the two long-run components, nor need the lags be the same for the two sets of components, as is assumed with a simple time lag.

The theory for more than two processes is a simple generalization of the above. Coherence- and phase-diagrams can be constructed for every pair of processes, and interpretation is as before. Further sophistication is possible by introducing the idea of a partial cross-spectrum, this being discussed in G(5).

If the spectral approach provides useful tools in the case of stationary series it is sensible to ask if these tools can also be used with non-stationary series. We introduce the concept of a smoothly changing, non-stationary series, having the representation

$$x_t = \int_{-\pi}^{\pi} e^{it\omega} a(t, \omega)\, dz(\omega), \tag{1.18}$$

$z(\omega)$ being a process with independent increments and assuming that the function $a(t, \omega)$ is continuous in ω (except possibly in a set of measure zero) and is only slowly changing with time. Such a process can be considered as the output of a time-changing black box, the change being a slow one. It is easily shown that most of the common models for non-stationary processes such as "explosive" autoregressive processes and autoregressive processes with parameters changing with time can be represented as in equation (1.18).

It is not suggested that all non-stationary series can be represented by models such as equation (1.18), but it is suggested that such non-stationary processes do form an important class and the assumption that a given economic series belongs to this class is not, without further information, an unreasonable one.

The "spectrum" of x_t at time t can be defined as $f(t, \omega) = |a(t, \omega)|^2$, and if one estimates the spectrum of.such a non-stationary process, it can be shown that one is approximately estimating $\frac{1}{n} \sum_{t=1}^{n} f(t, \omega)$ i.e., the "average spectrum." Further, the estimated coherence between two such processes is vaguely the "average coherence" and so both the spectral and coherence estimates still provide a certain amount of useful information. Of greater importance, however, is the fact that if two smoothly non-stationary processes can be represented by

$$x_t = \int_{-\pi}^{\pi} e^{it\omega} a(t, \omega)\, dz(\omega)$$

and

$$x_t^{\phi} = \int_{-\pi}^{\pi} e^{it\omega} e^{-i\Phi(\omega)} b(t, \omega)\, dz(\omega)$$

i.e., x_t^{ϕ} has an unchanging phase lag to x_t, then the phase-diagram still estimates $\phi(\omega)$. Proofs of these results have been indicated in G(9) and a series of experiments on constructed non-stationary series has provided further verification (same reference). Thus, cross-spectral methods will find the phase-lag relationship even between two non-stationary processes provided that the relationship itself is not changing with time.

A final problem which has to be considered is, given a process having a representation such as equation (1.18), how can one estimate the function $a(t, \omega)$? If it is assumed that $a(t, \omega)$ exists for all ω, it is clearly impossible, given only a finite amount of data, to estimate $a(t, \omega)$ exactly. However, it is possible to estimate the average of $a(t, \omega)$ over $m + 1$ small frequency bands, i.e.,

$$A(t, \omega_j) = \int_{\omega_j-\delta}^{\omega_j+\delta} a(t, \omega)\, d\omega$$

where

$$\omega_j = \frac{\pi j}{m}, \qquad j = 1, \ldots, m - 1$$

$$\delta = \frac{\pi}{2m}$$

$$A(t, \omega_0) = \int_0^{\delta} a(t, \omega)\, d\omega, \qquad A(t, \omega_m) = \int_{\pi-\delta}^{\pi} a(t, \omega)\, d\omega.$$

As in general $a(t, \omega)$ is a complex function, $A(t, \omega)$ will also be complex and we let

$$A(t, \omega) = c(t, \omega) + is(t, \omega).$$

Suppose that $F[\ \ ]$ is a filter having transfer function

$$g(\omega) = \frac{1}{2\delta}, \qquad |\omega| \leq \delta$$
$$= 0 \qquad \text{elsewhere}$$

i.e., if we have a stationary process $\{x_t\}$ with power spectrum $f(\omega)$ then the power spectrum of the new process generated by $y_t = F[x_t]$ is $g^2(\omega)f(\omega)$. Using the generalized Cramér representation (1.18), it is easily shown that, if $a(t, \omega)$ is changing only slowly with time and $F'[\ \ ]$ is a moving-average filter approximating $F[\ \ ]$, then $R(\omega_j, t) = 2[(z_t')^2 + (z_t'')^2]$ is an estimate of $[c^2(t, \omega_j) + s^2(t, \omega_j)]$ i.e., the "average" amplitude in the frequency band $(\omega_j \pm \delta)$ at time t and $\psi(t, \omega) = \tan^{-1} \dfrac{z_t'}{z_t''}$ is an estimate of $\tan^{-1}\left[\dfrac{s(t, \omega_j)}{c(t, \omega_j)}\right]$

i.e., the "average" phase in the frequency band $(\omega_j \pm \delta)$ where

$$z_t' = F'[x_t \sin \omega_j t]$$

$$z_t'' = F'[x_t \cos \omega_j t].$$

$R^2(t, \omega_j)$ may be called the "instantaneous spectrum," and the method is known as demodulation. By the use of this method we are able to estimate crudely how the complex function $a(t, \omega)$ changes with time and can thus estimate (very crudely) how the relationship (if any) between two non-stationary series is changing over time.

2. THE NEW TECHNIQUES—(b) CAUSALITY AND FEEDBACK[2]

Before considering the concepts of causality and feedback, some theory for multivariate stochastic processes is required. If the $q \times 1$ vector x_t, with

$$x_t' = \{x_{1t}, x_{2t}, \ldots, x_{qt}\}$$

is a multivariate stationary non-deterministic process, Zashain [10] has shown that it may be represented by

$$\mathbf{x}_t = \mathbf{B}(u)\boldsymbol{\epsilon}_t \tag{2.1}$$

where $\mathbf{B}(u)$ is the matrix

$$\mathbf{B}(u) = [B_{jk}(u)]$$

$$B_{jk}(u) = \sum_{m=0}^{\infty} b_{jkm} u^m,$$

u being the shift operator defined by

$$ux_t = x_{t-1}$$

Here, $\boldsymbol{\epsilon}_t$ is a $q \times 1$ multivariate white noise vector, i.e.,

$$E[\boldsymbol{\epsilon}_t\boldsymbol{\epsilon}_t'] = \mathbf{I}_q, \quad E[\boldsymbol{\epsilon}_t\boldsymbol{\epsilon}_{t+s}'] = \mathbf{O}_q, \qquad s \neq 0$$

where $\mathbf{I}_q$ is the $q \times q$ unit matrix and $\mathbf{O}_q$ the $q \times q$ zero matrix. (2.1 may be called the moving average representation of the process.)

If we assume $E[\mathbf{x}_t] = 0$ and define the theoretical autocovariances and power spectral functions by

$$\Gamma_{jk}(s) = E[x_{j,t+s}x_{kt}]$$

$$F_{jk}(\omega) = \sum_{s=-\infty}^{\infty} \Gamma_{jk}(s)e^{i\omega s}$$

$$\mathbf{F}(\omega) = [F_{jk}(\omega)],$$

it is easy to show that

$$\mathbf{F}(\omega) = \mathbf{B}(e^{i\omega})\mathbf{B}'(e^{-i\omega}).$$

[2] This section is a precis of G(7). The definition of causality used agrees with that introduced by Weiner [8].

If the equation in z, $|\mathbf{B}(z)| = 0$ has no roots on or within $|z| = 1$, we may invert (2.1) to get the "autoregressive representation"

$$\mathbf{A}(u)\mathbf{x}_t = \boldsymbol{\epsilon}_t \tag{2.2}$$

where

$$\mathbf{A}(u) = \mathbf{B}^{-1}(u)$$
$$= \sum_{j=0}^{\infty} \mathbf{A}_j u^j.$$

If we write (2.2) as

$$\mathbf{A}_0\mathbf{x}_t + (\text{past } \mathbf{x}_t) = \boldsymbol{\epsilon}_t \tag{2.3}$$

we have

$$\mathbf{x}_t = \mathbf{A}_0^{-1}\boldsymbol{\epsilon}_t + (\text{past } \mathbf{x}_t)$$
$$= \mathbf{B}_0\boldsymbol{\epsilon}_t + (\text{past } \mathbf{x}_t).$$

Thus, given all past x_t, we can predict $\mathbf{x}_t$ apart from the term $\mathbf{B}_0\boldsymbol{\epsilon}_t$. The determinant of the covariance matrix of these terms i.e.,

$$V = |E[(\mathbf{B}_0\boldsymbol{\epsilon}_t)(\mathbf{B}_0\boldsymbol{\epsilon}_t)']| = |\mathbf{B}_0\mathbf{B}_0'|$$
$$= |\mathbf{B}_0|^2 = |\mathbf{A}_0|^{-2} \tag{2.4}$$

is called the "total prediction variance."

We note that (2.3) is not a unique representation. If $\boldsymbol{\Lambda}$ is an orthogonal matrix so that $\boldsymbol{\Lambda}\boldsymbol{\Lambda}' = \mathbf{I}_q$, then as $\boldsymbol{\eta}_t = \boldsymbol{\Lambda}\boldsymbol{\epsilon}_t$ is also a white noise vector, we may write (2.3) in the similar representation

$$\boldsymbol{\Lambda}\mathbf{A}_0\mathbf{x}_t + (\text{past } \mathbf{x}_t) = \boldsymbol{\eta}_t.$$

It follows that, as any matrix can be written as the product of an orthogonal matrix and a triangular matrix, we lose no generality by assuming $\mathbf{A}_0$ to be triangular in representations such as (2.3). Clearly we are unable, in general to take $\mathbf{A}_0 = \mathbf{I}_q$ and still retain a representation such as (2.3) with $\boldsymbol{\epsilon}_t$ a white noise vector.

However, we shall for the time being consider multivariate stochastic processes having the representation

$$\mathbf{x}_t + (\text{past } \mathbf{x}_t) = \boldsymbol{\epsilon}_t \tag{2.5}$$

where

$$E[\boldsymbol{\epsilon}_t\boldsymbol{\epsilon}_t'] = \mathbf{V}, \quad \mathbf{E}[\boldsymbol{\epsilon}_t\boldsymbol{\epsilon}_{t+s}'] = \mathbf{O}_q, \qquad s \neq 0,$$

$\mathbf{V}$ being a diagonal matrix. Such processes can be considered as being "caused" entirely by past events apart from the white noise vector $\boldsymbol{\epsilon}_t$, about which it is impossible to obtain information prior to the time t. It should be noted that, if $\mathbf{V} = [V_j\, \delta_{ij}]$ where

$$\delta_{ij} = 0, \qquad i \neq j$$
$$= 1, \qquad i = j$$

then the total prediction variance is given by $V = |\mathbf{V}| = \prod_{j=1}^{q} V_j$ i.e., the total prediction variance is the product of the individual prediction variances for the q stochastic processes $\{x_{it}\}$, $i = 1, \ldots, q$.

Let us denote the set of q stochastic processes $\{x_{it}\}$, $i = 1, \ldots, q$ by Q, let $Q(j)$ be the set of processes Q excluding $\{x_{jt}\}$ and let $Q(j, k)$ be the set processes Q excluding $\{x_{jt}\}$ and $\{x_{kt}\}$. Define the optimum linear predictor of the process $\{x_{jt}\}$ involving some set S of the processes by

$$P_{jt}(s) = \sum_{s} \sum_{m=1}^{\infty} a_{sm} x_{s,t-m}, \qquad s \in S \tag{2.6}$$

where the coefficients a_{sm} are chosen such that the prediction error variance

$$V_j[s] = E[(x_{jt} - P_{jt}(s))^2] \tag{2.7}$$

is a minimum. Clearly, in the case being considered, $V_j[Q] = V_j$. We say that there is causality of the process $\{x_{jt}\}$ by the process $\{x_{kt}\}$ (within the set Q) if $V_j[\varphi(k)] - V_j[Q] > 0$ and that there is no causality if

$$V_j[Q(k)] = V_j[Q].$$

Thus, we are saying that there is causality from one variable $\{x_{kt}\}$ to another $\{x_{jt}\}$ if we are able to predict $\{x_{jt}\}$ better using the full set Q than when we use the set $Q(k)$, i.e., the set Q excluding the variable $\{x_{kt}\}$.

If both

$$V_j[Q(u)] - V_j[Q] > 0$$

and

$$V_k[Q(j)] - V_j[Q] > 0$$

we say that feedback is present.

We denote causality by $\{x_{kt}\} \Rightarrow \{x_{jt}\}$ and feedback by $\{x_{k,t}\} \Leftrightarrow \{x_{jt}\}$.

If $\{x_{kt}\} \Rightarrow \{x_{jt}\}$ and if the first non-vanishing coefficient in the sequence a_{km}, $m = 1, 2, \ldots$ occurring in the optimum linear predictor $P_{jt}(Q)$ is a_{km_0}, m_0 may be called the apparent causality lag. Here, we are saying that as we are unable to better predict $\{x_{jt}\}$ by knowing $\{x_{k,t-m}, m = 1, \ldots, m_0 - 1\}$ than by not knowing these figures, there is a lag of at least m_0 time units between the cause and the effect. (The actual causality lag will be $m_0 + a$ where $0 \leq a < 1$.) We may call $C(k_{ij}) = 1 - V_j[Q]/V_j[Q(u)]$ the "strength" of the causality $\{x_{kt}\} \Rightarrow \{x_{jt}\}$. The "apparent feedback lag" is the sum of the two apparent causality lags, and the "strength" of the feedback is the product of the two causality strengths.

It should be noted that all these concepts are defined with regard to the set Q. It is possible, for instance, for there to exist a further stochastic process $\{y_t\}$ outside of Q such that

$$\{x_{kt}\} \Rightarrow \{y_t\}$$
$$\{y_t\} \Rightarrow \{x_{jt}\}$$

and there to be no causality between $\{x_{kt}\}$ and $\{x_{jt}\}$, all these causalities being defined with regard to the set Q' consisting of Q plus $\{y_t\}$ whereas, with regard to Q only, we may have $\{x_{kt}\} \Rightarrow \{x_{jt}\}$.

It is clear that, in practice, we are unable to form an optimum linear predictor, as defined above, as we shall never have all past values of the processes, only a sample X_t, $t = 1, \ldots, n$. We are thus forced to use an approximate optimum linear predictor of the form

$$\tilde{P}_{jt}(s) = \sum_s \sum_{m=1}^{m_s} a_{sm} X_{s,t-m}, \qquad s \in S$$

where, the coefficients a_{jk} are chosen to minimize

$$\hat{V}_j[s] = \sum_{t=\bar{m}+1}^{n} [X_{jt} - \tilde{P}_{jt}(s)]^2$$

where

$$\bar{m} = \max\{m_s, s \in S\}.$$

A test for causality has been provided by Whittle [7]. If we assume all the processes involved are Gaussian and we form $\hat{V}_j[\varphi]$ and $\hat{V}_j[Q(k)]$ then under the null hypothesis of no causality, the statistic

$$\psi^2 = (n - q - M/q) \log_e (\hat{V}_j[Q(k)]/\hat{V}_j[Q]) \tag{2.8}$$

will be distributed as χ^2 with m_k degrees of freedom, where $M = \sum_{k=1}^{q} m_k$ i.e., the total number of parameters fitted when estimating $V_j[Q]$.

Suppose that, if causality is found, we now form an optimum linear predictor

$$\tilde{P}_{jt}(Q, a) = \sum_{s=1}^{q} \sum_{m=1}^{m_s} a_{sm} X_{s,t-m}$$

with $a_{km} = 0$, $m = 1, \ldots, a - 1$ minimizing

$$\hat{V}_j[\varphi, a] = \sum_{t=\bar{m}+1}^{n} [X_{jt} - \tilde{P}_{jt}(\varphi, a)]^2, \qquad \bar{m} = \max\{m_j\}$$

If we have the null hypothesis H_0 that $\tilde{P}_{jt}(Q)$ is no better predictor of x_{jt} than $\tilde{P}_{jt}(Q, a)$ then, under this hypothesis, the statistic

$$\psi^2 = (n - q - M/q) \log_e (\hat{V}_j[Q, a]/\hat{V}_j[a])$$

is distributed as χ^2 with $a - 1$ degrees of freedom. Thus, if we reject H_0 we know the apparent causality lag is less than or equal to a units.

We now return to the general representation

$$\mathbf{A}_0 \mathbf{x}_t = (\text{past } \mathbf{x}_t) + \boldsymbol{\epsilon}_t \tag{2.9}$$

If $\mathbf{A}_0$ is a diagonal matrix or if there exists an orthogonal matrix $\boldsymbol{\Lambda}$ such that $\boldsymbol{\Lambda}\boldsymbol{\Lambda}' = \mathbf{I}_q$ and $\boldsymbol{\Lambda}\mathbf{A}_0$ is a diagonal matrix, we then have the situation that has just been discussed. In general, however $\mathbf{A}_0$ will not have these properties, although we may take it to be triangular. A simple test can be constructed to find whether or not a given set of processes have a representation with $\mathbf{A}_0$ diagonal or not. Suppose that, given a sample $\mathbf{X}_t$, $t = 1, \ldots, n$ we fit an optimum linear filter

$$\tilde{P}_{jt}(Q) = \sum_{s=1}^{q} \sum_{m=1}^{m_s} a_{sm} X_{s,t-m}$$

to X_{jt} for all $j = 1, \ldots, q$. We now form the estimated error series

$$\hat{\varepsilon}_{jt} = X_{jt} - \tilde{P}_{jt}(Q)$$

As we are here using the representation

$$\mathbf{x}_t = (\text{linear function of some of past } \mathbf{x}_t) + \boldsymbol{\epsilon}_t$$

to approximate (2.9), we have two possible sources of error: (i) some or all of the m_s's are not large enough, i.e., the linear approximations involving past X_{st} do not involve enough terms for the approximation to be acceptable, and (ii) $\mathbf{A}_0$ is not a diagonal matrix.

The first of these errors is easily examined by testing if any of the series $\hat{\varepsilon}_{jt}$ contain serial correlation, and if no such serial correlations are found, the second type of error may be examined by forming the correlation coefficients between each of the series $\{\hat{\varepsilon}_{jt}\}$. If any of these coefficients are (significantly) non-zero, then $\mathbf{A}_0$ is not diagonal.

If the tests show that, in fact, $\mathbf{A}_0$ is not diagonal the whole problem of defining and testing for causality and feedback becomes not merely more difficult but perhaps impossible. This may be shown by considering a fairly simple example. Suppose we have two processes only, and that there is a causality $\{y_t\} \Rightarrow \{x_t\}$ but no feedback, and let the causality lag be exactly 1. If now, instead of sampling the processes at times $t = 1, 2, 3, \ldots$ we in fact only sample at times $t = 1, 3, 5, \ldots$ i.e., the basic time unit is twice the causality lag. Denoting the new processes by $\{x_T\}$, $\{y_T\}$ there is now a direct causality $\{y_{T-\frac{1}{2}}\} \Rightarrow \{x_T\}$. Thus, we would need a representation

$$x_T = \alpha_0 y_{T-\frac{1}{2}} + \sum_{j=1}^{\infty} \alpha_j y_{T-j} + a_0 x_{T-\frac{1}{2}} + \sum_{j=1}^{\infty} a_j x_{T-j} + \epsilon_T$$

$$y_T = b_0 y_{T-\frac{1}{2}} + \sum_{j=1}^{\infty} b_j y_{T-j} + \eta_T \tag{2.10}$$

Clearly, as we would not know $y_{T-\frac{1}{2}}$ the best linear "predictor" of would involve y_T, as y_T contains information about $y_{T-\frac{1}{2}}$ i.e., with forms such as

$$P = \sum_{j=0}^{\infty} \beta_j y_{T-j} + \sum_{j=1}^{\infty} \gamma_j x_{T-j}$$

and

$$P' = \sum_{j=1}^{\infty} \beta_j' y_{T-j} + \sum_{j=1}^{\infty} \gamma_j' x_{T-j}$$

we will be able to choose the coefficients β_j, β_j' etc. so that

$$\min E[(x_T - P)^2] < \min E[(x_T - P')^2]$$

Similarly, the best "predictor" of y_T will contain x_T, as x_T contains information about $y_{T-\frac{1}{2}}$. It is thus seen that the predictor approach to testing for causality will not work in this situation; we are able to show the processes to be interrelated but no direction of causality can be found. We may sum up this result by saying that when the causality lag is less than the time unit being used, the data is unsuitable for subtle causality and feedback analysis. We are, in fact, brought face to face with many of the identification and simultaneous-equation estimation problems of classical model-building. Thus, the problem is no longer one of analyzing the data but of using the data in model-fitting on the basis of a certain amount of prior knowledge or assumptions. It is true that much of the available data in economics is such that the causality lag is less than the time unit, and so the methods here presented are not of complete use, although the reason for this lies more in the unsuitability of the data than the theory. It is possible that the theory may be extended further to include such data but, in any case, the methods outlined ought to provide valuable information when model-building is contemplated.

The performance of the methods outlined in this section is uncertain when the assumption of stationarity is removed. As a definition of causality, the above is still appropriate once a generalized method of prediction with non-stationary series is available. It is hoped to inquire into these problems both theoretically and empirically in the near future.

3. ECONOMETRICS AND THE NEW TECHNIQUES

3.1 Introduction. The introduction of new techniques into a field of research is only justified if it can be shown that they either replace, complement, or, at the very least, satisfactorily compete with existing techniques. The two classical methods of analyzing time series are the detailed, historically concious but statistically unsophisticated methods exemplified by many of the National Bureau publications, and the model-building approach identified by the work of members of the Cowles Commission and the London School of Economics. It is the object of this part of the paper to show in which situations, in the author's opinion, the various techniques compliment each other and when they are alternatives.

As far as the analysis of data is concerned, economics can be considered to be a relatively undeveloped field. Research within such a field may be thought of as a game against some unknown opponent (who we may call "Nature") as it is impossible, at present, to use methods of complete generality. Thus, one must use a subset of the set of all possible realistic assumptions concerning the economy while having little information as to which of the assumptions are really the most realistic. Before more information is acquired, an investigator must choose a strategy, that is, a set of initial assumptions upon which the analysis is based. Just which strategy is chosen must depend to a considerable extent upon the opinions of the investigator, as he will doubtless wish to chose the set which, according to his personal experience, seems to be the most realistic while allowing the analysis to lie within the boundaries imposed by the data, methods, and computing machines that are available. It follows that any comparison of techniques ought to be based upon a given set of opinions and that these opinions may take the form "it is a good strategy to assume A." The ranking of the terms used below are: (i) "good strategy," (ii) "reasonable strategy," (iii) "poor strategy," and (iv) "very poor strategy." No attempt to assign probability measures to these terms will be made as such an attempt would imply considerably more experience than is available to the author.

3.2 SOME BASIC OPINIONS AND THEIR CONSEQUENCES.

Opinion 1. At nearly all moments of time, the economy is a noisy, linear black box.

A black box is a process whereby a set of inputs is transformed into a set of outputs. It is linear if every output can be expressed as the linear sum of past and present inputs, and it is noisy if every output needs to be expressed as a function of past and present inputs plus white noise. It seems likely that during major crises, the economy no longer behaves like a linear black box.

Opinion 2. The economy is a slowly changing, time-variant black box.

If a black box has a single input $e^{it\omega}$ which is transformed into the single output $A \exp [i(i\omega + \phi)]$ it is called time-invariant; i.e., the frequency of the output is the same as that of the input but with changed amplitude and phase and with these independent of time. When the black box produces an output with amplitude and phase which are slowly changing functions of time, it may be called slowly changing time-variant.

Such opinions lead one to consider how best to analyze a black box. The three most obvious ways seem to be:

(i) Take it apart and study each piece. Such a process is undoubtedly informative but as all the pieces will themselves be black boxes, the problem is not answered but is simplified. A main disadvantage of this procedure is the difficulty in constructing the properties of the whole box from knowledge of the properties of its components.

(ii) Try to build another black box having properties similar to the original box. Results obtained from such an analogous box, if it can be constructed, will be useful in prediction and control problems, but it is logically unallowable to imply that the original box and the new box have identical interiors. (Two television sets may produce almost identical results yet have very different circuits.) Attempts at model-building fall into this category.

(iii) Analyze directly the relationships between the inputs and outputs. This may be attempted either for the box as a whole or for any group of its components. The National Bureau and spectral methods both attempt such an analysis.

Which method is the most relevant must depend to a large extent upon the reason for the analysis and is discussed further in Section 3.4 below.

The "slowly changing" part of Opinion 2 has important implications concerning the possibility of prediction. Prediction with a time-invariant black box assumes the laws currently operating to be the same as those of the past. If a black box is time-variant but slowly changing, there is always the possibility that although the current laws are different from those of the past they may be "extrapolated" from them.

Opinion 3. It is very poor strategy to assume that the economy is not extremely complicated.

From a statistical point of view the economy is a particularly unpleasant kind of black box for reasons such as:

(i) It is time-variant.

(ii) Data from it may be contaminated by rare, highly disturbing occurrences such as strikes. During these occurrences the underlying economic laws may become nonlinear.

(iii) The box does not provide a continual, simple flow of output data, some series being recorded daily, others monthly, quarterly, annually, or even every decade (population figures).

(iv) Some of the input data are either not available or are hidden in constructed aggregative series.

(v) The data are frequently inaccurate or recorded in a time-variant manner or are based on definitions that change with time (see Morgenstern's report, *On the Accuracy of Economic Observations* [5]).

Further, the relationships between economic variables are almost certainly highly complicated even when only a small component of the black box is considered. (Whether components *can* be considered separately is itself debatable.) Because of this, whether it is worthwhile building "simple" models or attempting an analysis when only a small amount of data is available, is arguable.

One implication of the facts (i) to (v) above is that, in general, it is not possible to focus sharply on any relationship. An analogous situation is that of an astronomer observing a star through a slightly dusty, slowly swirling atmosphere, although the "atmosphere" obscuring the view of an econometrican is more likely to be highly-mobile multi-colored fog. Such thoughts imply that a very sophisticated method of extracting information from the available data will be required to discover anything except the most obvious facts.

A subsidiary opinion also follows:

Opinion 3a. It is poor strategy to attempt to explain the economy in terms only of a few variables.

It is conceivable that certain crude movements of the economy can be explained using only a few highly aggregative variables, but no subtle economic laws can be found by such a method.

Opinion 4. It is very poor strategy to assume anything other than that economic variables are non-deterministic stochastic processes.

It is surely obvious that no economic variable can be predicted perfectly, nor can it be represented as mean plus white noise, as prediction can be improved by using past values. Thus economic variables are stochastic processes and non-deterministic. It follows that it is incorrect to present economic laws as though the variables involved were mathematical variables. It is possible that the mean values of a set of economic variables do obey a mathematical law, but it must not be implied that the variables obey such laws. To do so would be to ignore the majority of the information that is contained in the data and which can only be extracted by considering the variance and possibly also the higher moments.

Opinion 5. It is poor strategy to assume an economic variable to be stationary.

This opinion follows directly from Opinions 2 and 4. If it is agreed that the economy is changing slowly with time, data obtained by frequently sampling on output, such as daily stock price data, can be treated as though they are stationary, but the majority of the important laws of economics involve middle- and long-run periods in which an assumption of stationarity will, almost always, be incorrect. It is for this reason that the effect on spectral techniques of removing the stationarity assumption was studied, as reported in Section 1 above.

Opinion 6. It is very poor strategy to assume that any mathematical transformation exists by which the data can be made stationary.

A trend in mean can be easily removed (mostly) by one of a number of efficient methods, such as filtering, but once this has been done the resulting data are still likely to have a spectrum changing with time. It has been suggested that the logarithm of the data followed by filtering, may remove the non-stationarity. This would only work if the scale (i.e., variance) of the process were changing with time but the structure of the economy, reflected in the auto-covariances and cross-variances, were invariant. Clearly this is an unrealistic assumption, particularly for aggregative data. It is agreed that such a transformation may make the non-stationarity less important, but whether or not the non-stationarity becomes negligible is uncertain. (A possible danger of the logarithmic transformation is that the resulting process could theoretically have an infinite variance.)

It should be pointed out that the large majority of the currently available techniques in econometrics assume either that the data are stationary or can be made stationary.

Opinion 7. It is reasonable strategy to assume that the only strictly periodic components ("cycles") present in an economic variable are those associated with annual variation.

Opinion 8. It is reasonable strategy to assume that all frequencies will be present to some extent.

Opinion 9. It is reasonable strategy to assume that the low frequencies will be considerably more important than the high frequencies.

These three opinions together indicate that the estimated spectra of (long- and middle-run) economics series will all have similar shapes—a continual and rapid decrease from the low frequencies to the high frequencies apart, possibly, from peaks at the annual component and its harmonics. This suggests that an estimated spectrum will not, in general, provide useful information. The cross-spectral diagrams have no such general shape.

Opinion 10. Non-normality is less important than non-stationarity.

Although no opinion is proposed as to whether or not economic data can be made Gaussian, it is nevertheless felt that non-stationarity is a more important complication than non-normality as the former makes various *techniques* inappropriate, whereas the latter makes various *tests* inappropriate, although the tests may still be asymptotically usable.

3.3 OBJECTS OF ANALYSIS. The method by which data are analyzed must depend upon the reason for the analysis and the objective which it is hoped will be reached. Thus, before comparing the available techniques for analysis, the various objectives should be discussed. There would seem to be four main objectives when analyzing economic data, although clearly these four are not independent.

(1) *Prediction.* Objective: to predict, as well as possible, future values of certain economic variables.

For stationary series, methods of prediction are well founded theoretically, but when the stationarity assumption is removed, the available techniques are less satisfactory. It might also be noted that often not all frequencies of the variables are equally important in a prediction; it may be desired, for instance, to predict only the long-run component.

(2) *Control.* Objective: to determine methods of controlling future values of certain economic variables.

There will almost certainly be various possible methods of control some, of which will prove to be more efficient or quicker acting than others. It is likely that control is required only of certain frequency components such as the "business cycle" component required by the government, or the high-frequency component of a production series required by a manufacturer.

(3) *Testing hypotheses.* Objective: to test assumptions, hypotheses, and theories used and produced by economists.

Scientific laws are achieved by the evolution of theories. An essential part of any such evolutions is the provision of information as to which of the currently available theories best fit the observed data.

(4) *Providing facts.* Objective: to provide observed "facts" about the economy that will have to be incorporated in and explained by future economic theories.

A second important component in the evolution of a theory is the feedback between the current stage of the theory and the observed facts of the economy. The current theory ought to take into account at least the majority of the available observed "facts" and as any theory will have a number of consequences, the investigation and testing of the data concerning these consequences are likely to throw up new facts, thus requiring the theory to be altered and improved.

Koopmans [4] once attacked the National Bureau method of analysis as being "measurement without theory," but theory without measurement is equally dangerous. Statistical investigations of economic data and economic theory should be continuously and intimately connected. In practice, if the data from some part of the economic system are being analyzed, an economist will be required to determine both the limits of the black box being studied and the particular aspects of the problem upon which the analysis should be focused.

How theories evolve in science is too complex to be mentioned in any detail here, but it might be pointed out that, although a pleasing analogy might be drawn between the development of the laws of astronomy and those of economics (see, for instance, Koopmans [4]), the analogy is a dangerous one, as most variables in astronomy are mathematical, and those of economics are statistical (i.e. stochastic processes).

3.4 COMPARISON OF THE AVAILABLE TECHNIQUES. For some of the objectives the previously available techniques and the new techniques are rival methods; for others they merely complement each other. The extent of the rivalry must depend upon one's personal opinions as to the "state of the game." The author's personal opinions have been given above, and the conclusions of this section are based upon these opinions.

(1) *Prediction.* Model-building is clearly theoretically the most suitable technique for prediction. When the available data may be considered stationary the only remaining problems are those of estimation, but these may be considerable due to the complexity of the economy. The causality and feedback techniques should prove extremely useful in determining which variables to include in the prediction model, and cross-spectral methods will provide information as to whether or not certain exogenous variables (which have no feedback properties) are worth including.

For the non-stationary case, when prediction is required only for certain frequency bands, the demodulation technique becomes appropriate, and coherence diagrams will still provide useful information for the construction of time-changing models.

For this objective, the new techniques would seem to complement the existing ones.

(2) *Control.* When all causality lags are greater than the time unit employed, the concepts of causality and feedback are basic in the problem of control methods. When causality only is present, the cross-spectral methods will indicate possible controls, their efficiencies (coherence), and their speed of operation (from phase-diagram). In particular, control at any required frequency can be considered.

In this situation, the new techniques and that of model-building may be considered direct rivals. It is suggested, however, that any "simple" model will be stochastically naïve whereas a complicated model is less easily interpreted in terms of the economic concepts of a long run and a short run.

When some causality lags are less than the time units, the two techniques complement each other once more.

(3) *Testing hypotheses.* If a theory is presented in terms of a model and is based upon certain assumptions and hypotheses, it would appear to be logically dangerous to test this theory by merely using model-building techniques. As, however, economic theories frequently are no (or should not be) expressed as simple mathematical laws, the model-building approach does not seem to be an appropriate method compared to the spectral methods, which are based on fewer assumptions. This is particularly true when non-stationary data are being used. The new techniques are well suited for testing theories and assumptions, particularly as they are easily interpreted in economic terms.

Doubtless many econometricians would consider the various techniques to be rivals with this objective, but it is the author's opinion that the new methods have overriding advantages.

(4) *Providing facts.* This objective has been much neglected by statisticians, model-building being inappropriate. The only available technique is the National Bureau method, which, although it has provided

much useful information, is likely to be too unsophisticated to achieve a great deal more, due to the economy being so highly complicated. It is likely that it is with this objective in mind that the new techniques will prove most useful, although it should be emphasized that a careful study of the data (charting, investigation of historical aspects) before the new techniques are used is necessary. It is as pointless to estimate spectra before plotting the data as it is to build a model before examining the estimated cross-spectral diagrams.

3.5 CONCLUSIONS. The above arguments are doubtlessly biased—one does not work in a field without believing in its usefulness—but an attempt has been made to base the arguments on a particular set of opinions, which, I hope, are considered to be reasonable. At the very least, I feel, it can be concluded that the new techniques have a real place among the available methods of econometrics; just how important a place only the future can tell.

REFERENCES

[1] Cramér, H., "On Harmonic Analysis in Certain Functional spaces," *Ark. Mat. Astr. Fys.* 28B (1942), 7 pp.

[2] Granger, C. W. J., "First Report of the Princeton Time Series Project," *L'Industria.* (1961), pp. 3–15.

[3] Granger, C. W. J., in association with M. Hatanka, *Spectral Analysis of Economic Time Series*, Princeton, Princeton University Press, (1964).

[4] Koopmans, T. C., "Measurement without Theory," *Rev. Econ. Stats.* 29 (1947), pp. 161–172.

[5] Morgenstern, O., *On the Accuracy of Economic Observations*, Princeton, Princeton University Press, (1950) revised edition (1963).

[6] Morgenstern, O., "A New Look at Economic Time Series," in *Money, Growth and Methodology*, In honour of J. Akerman, H. Hegeland (ed.) Lund (1961), pp. 261–272.

[7] Whittle, P., "The Analysis of Multiple Time Series," *J. Roy. Stat. Soc.* (*B*) 15 (1953), pp. 123–139.

[8] Wiener, N., "The Theory of Prediction," Chapter 8 of *Modern Mathematics for Engineers* (Series 1), E. F. Beckenback (ed.), (1956).

[9] Wold, H., "*A Study in the Analysis of Stationary Time Series*," Stockholm (1938) (2nd ed. 1953).

[10] Zashuin, *Comptes Rendus* (*Doklady*) *de l'Acad. Sc. de L'URSS* V. 23 (1941) pp. 435–437.

CHAPTER 26

A Theory of the Pseudospectrum and Its Application to Nonstationary Dynamic Econometric Models

By MICHIO HATANAKA and MITSUO SUZUKI*

1. THE PURPOSE AND CONCLUSIONS OF THIS PAPER

The spectrum of a stochastic process indicates its variance decomposition by frequencies. In mathematical analysis the spectrum has been defined for the stochastic process which has an infinitely long time domain and satisfies the stationarity conditions in the wide sense:

$E\{x_t\} =$ a constant
$E\{[x_t - E(x_t)]^2\} =$ a constant
$E\{[x_t - E(x_t)][x_s - E(x_s)]\}$ depends only on the time distance $|t - s|$.

The stationarity conditions mean, among other things, that the first two moments of the stochastic process do not depend upon the origin of time; i.e., the process is essentially "historyless" in the sense in which history is used in the social sciences.

On the other hand, in the practical applications of spectral analysis, samples of finite length have been used for the study of variance decomposition of the stochastic process which may not satisfy the stationarity conditions. As a first step toward filling the gap between the mathematical analysis and the practical applications, Blackman and Tukey[1] developed a method for making estimations of the spectrum of the stationary stochastic process by using samples of finite length. Thus, the practical applications of spectral analysis can be justified if the stochastic process

* The University of Rochester, Tokyo Institute of Technology.
This is a revised version of Research Memorandum No. 52 distributed by the Econometric Research Program, Princeton University, in 1963.

The authors are grateful to C. W. J. Granger for his many helpful comments on an earlier draft of this paper. However, any remaining errors are the authors'. The work presented in the present paper was initiated by J. W. Tukey's suggestion (in J. W. Tukey, "Discussion, Emphasizing the Connection Between Analysis of Variance and Spectral Analysis," *Techometrics*, Vol. 3, No. 2, 1961, p. 203) that the best way to understand the "spectrum" of the nonstationary process is to look upon it as some sort of average of changing spectra.

The authors wish to acknowledge the support given by the National Science Foundation and the Rockefeller Foundation. Reproduction in whole or in part is permitted for any purpose of the United States Government.

[1] R. B. Blackman and J. W. Tukey, *The Measurement of Power Spectra*. E. Parzen has also made an important contribution: E. Parzen, "Mathematical Considerations in the Estimation of Spectra," *Technometrics*, May, 1961.

is indeed stationary. In practice, however, spectral analysis has been used for the study of the stochastic process which is very unlikely to be stationary.

Particularly in economic applications we cannot assume that the stationarity conditions hold. The institutional, behavioral, and technological backgrounds of economic time series are always changing, and, therefore, there is no reason to believe that the first two moments of the economic stochastic process do not depend upon the origin of time.[2]

Under these conditions the finite length of the data has an entirely different implication than under the stationarity conditions. If the stationarity conditions hold, the estimate of the spectrum from samples of finite length can be considered as a kind of representation of the spectrum of the process having an infinitely long time domain. If the stationarity conditions do not hold, assuming that the "spectrum" of such a process would be meaningful, the estimate from samples of finite length cannot be considered a representation of the "spectrum," as long as the "spectrum" is a characteristic of the process over the entire time domain from $-\infty$ to $+\infty$.

There exist two ways to modify the concept of spectrum in order to make it applicable to nonstationary economic processes. One is to use the knowledge of economic theories about the structure of the nonstationarity. The other is to make the definition of spectrum completely free of any knowledge or assumptions about the structure of the nonstationarity and then to study special implications of this definition under special conditions of nonstationarity. The latter approach has been taken in this paper.

More specifically we have defined the pseudospectrum and the cross-pseudospectrum for those processes that are not necessarily stationary and that are defined over a finite time domain. We shall show first that the basic characteristics of spectrum as defined in the usual mathematical analyses are maintained for our definition of pseudospectrum and cross-pseudospectrum. By this we mean the following: The pseudospectrum is the frequency decomposition of the mean of the time-changing variance, just as the spectrum is the frequency decomposition of the (constant) variance. The mean of the time-changing autocovariance (or cross-covariance) and the pseudospectrum (or cross-pseudospectrum) is a Fourier transform pair, just as the (constant) autocovariance (or cross-covariance) and the spectrum (or cross-spectrum) is a Fourier transform pair. The coherence inequality holds for the cross-pseudospectrum and pseudospectra just as it does for the cross-spectrum and spectra. Further,

[2] Although we can apply transformations to a given time series with the purpose of eliminating nonstationarity, it is difficult to ascertain whether or not the series obtained by any transformation comes from a stationary stochastic process.

if the stochastic process is defined over an infinitely long time domain and if the stationarity conditions do hold, then the pseudospectrum is identical to the spectrum.

It is important to note that pseudospectrum and cross-pseudospectrum are, apart from the complications due to the spectral window, the mathematical expectations of the estimates of spectra and cross-spectra which we would obtain if we had made this estimation without considering the problems of nonstationarity. (The electronic computers "read out" some outputs even if they do not know whether the data come from a stationary stochastic process.)

Although the exact specification of the types of nonstationarity is not possible at the present stage of economics, some observations and fairly reasonable hypotheses have been presented to characterize vaguely the nonstationarity of the economic stochastic process. In the present paper we study the nature of the pseudospectrum assuming that these vague observations and hypotheses are acceptable. The nature of the pseudospectrum then obtained can be used for the interpretation of the estimates of spectra and cross-spectra that are derived by a standard method without considering the problems of nonstationarity. (The nature of the pseudospectrum derived from a too exact assumption as to the nonstationarity would not be useful because economics has not reached the stage in which an exact statement can be made as to the nonstationarity.)

First, we study the stochastic process in which the variance changes over time. The apparent variances of many economic time series for the United States show a secular change, usually with a clear discontinuity about World War II. If d_t is a deterministic function of time and x_t is a stationary stochastic process so that $d_t x_t$ is a nonstationary stochastic process whose variance changes with time in proportion to d_t^2, then the pseudospectrum of $d_t x_t$ is the convolution of the pseudospectrum of d_t and the spectrum of x_t. (A similar formula also holds for the cross-pseudospectrum.) Thus, if d_t is either a smooth function of time or involves one or two jumps in addition to a smooth trend so that the pseudospectrum of d_t is concentrated in the very low frequencies, the pseudospectrum of $d_t x_t$ is roughly equal to the mean of the time-changing, instantaneous spectra, $d_t^2 P_x(\omega)$, where $P_x(\omega)$ is the spectrum of x_t.

Second, we proceed to the study of the stochastic process in which the amplitudes and the phases of different frequencies change with time. We know, for example, that the amplitudes of the seasonal variations of many economic time series have a downward trend and also that the phases of the seasonal variations change from year to year. Further, the phases of the cyclical components are very likely to be affected by external events such as wars. The convolution theorem that is similar to the one mentioned above holds in this case too. If the phase changes are uniform over *all* different frequencies, and if the phase changes are either smooth or involve one or two jumps in addition to a smooth trend, then the phase of the cross-pseudospectrum between two such processes $y_t^{(1)}$ and $y_t^{(2)}$ shows the average of the time-changing differences between the phases of the two processes. If the phase changes are uniform over only a narrow frequency interval, the phase of the cross-pseudospectrum is more complicated, but its meaning is straightforward. Let $\varphi_\omega(t, v)$ represent the difference between the phase of $y_t^{(1)}$ at time t for frequency ω and the phase of $y_t^{(2)}$ at time $t + v$ for frequency ω. Then the phase of the cross-pseudospectrum at frequency ω is the double average of $\varphi_\omega(t, v)$ over t and $|v|$, where the averaging over v is done with weights that are roughly in inverse proportion to v.

Third, we apply the concept of the pseudospectrum to the time series generated by the nonstationary dynamic econometric models. As for the pseudospectral matrix of the endogenous variable in an explosive dynamic econometric model with constant parameters we can show that it is related to the spectral matrix of the random disturbance and the pseudospectral matrix of the exogenous variable, in the same way as the spectral matrix of the endogenous variable in a *stable* dynamic econometric model is related to the spectral matrix of the random disturbance.

Finally, we study the stochastic process of the deviation from the equilibrium solution of the model; this deviation is used to represent business fluctuations. For the case in which the parameters of the model change over time, we can show that the pseudospectral matrix of the deviation is a convolution which involves the spectral matrix of the random disturbance and a transfer function of the parameters. However, even in the case in which the parameters are smooth functions of time and the spectrum of the random disturbance is smooth, the pseudospectral matrix has a more complex form than an average of the time-changing, instantaneous spectra. For a given time point t, let us consider the impacts of the random disturbances in the periods prior to t (and including t) upon the values of the deviation at time t. Let $C_{t,j}$ be the impact of the random disturbance in period $t - j$, and, let $\tilde{G}(t, \omega)$ be the transfer function of $C_{t,0}, C_{t,1}, \ldots, C_{t,t-1}$ at frequency ω. If the parameters of the model are constant, $C_{t,0} = B^0, C_{t,1} = B^1, \ldots C_{t,j} = B^j, \ldots$. The transfer function is independent of t and can be written as $G(\omega)$. The spectral matrix of the deviation is $G(\omega)P_U(\omega)G(\omega)^*$, where $P_U(\omega)$ is the spectral matrix of the random disturbance and * means the Hermitian conjugate. If the parameters of the model are *not* constant, the pseudospectral matrix of the deviation under the same smoothness conditions is

$$\frac{1}{\pi N}\sum_t\left[\sum_v \tilde{G}(t, \omega)P_U(\omega)\tilde{G}(t + v, \omega)^* \frac{\sin \epsilon v}{v}\right],$$

a double average of

$$G(\omega)P_U(\omega)G(\dot{\omega})^*$$

over t and v.

So far we have mentioned the cases in which the pseudospectrum is directly amenable to a reasonable interpretation without the use of some specific a priori knowledge (e.g., the exact time function representing the changes in the parameters) about nonstationarity. There are many possible cases in which such an interpretation is not available. For example, as for the case of $d_t x_t$ mentioned above, if d_t is dominated by some irregular cycles, the pseudospectrum of $d_t x_t$ can be expressed as the result of smoothing the spectrum of x_t and then shifting it along the axis of ω. This follows from the convolution theorem which holds for any movements of d_t and from the fact that the pseudospectrum of d_t is not concentrated in the very low frequencies around zero but rather spreads around a certain non-zero frequency. For the other models of nonstationarity treated above, basically the same convolution theorem holds for any types of change of the parameters, although the relevant parameters vary from one model to another as described above. If the pseudospectra of the time series of the (changing) parameters are not concentrated in the very low frequencies, the interpretation of the spectral matrix is very difficult.

Thus the contribution of the present paper to economic applications of spectral analysis lies in the convolution theorem which holds for a broad class of nonstationary economic stochastic processes and which enables us to discern the cases in which a reasonable interpretation of the spectral matrix is available without the use of some specific a priori knowledge about nonstationarity from the cases in which such an interpretation is not possible.

II. PSEUDOSPECTRUM AND CROSS-PSEUDOSPECTRUM

One definition of the power spectrum (density) for a stationary stochastic process with zero mean is

$$f_x(\omega) = \lim_{N\to\infty} \frac{1}{2\pi(2N+1)} E\left\{\left| \sum_{t=-N}^{N} x_t e^{-i\omega t} \right|^2\right\}, \qquad -\pi \leq \omega \leq \pi \tag{1}$$

where ω is the (angular) frequency, and x_t is the discrete time series data produced by the stochastic process. The power spectrum is the decomposition of the variance of the process in terms of frequency ω. Equation (1) is equivalent to the more commonly used definition of the spectrum,

$$f_x(\omega) = \frac{1}{2\pi} \sum_{v=-\infty}^{\infty} e^{-iv\omega} r(v) \tag{1'}$$

where $r(v) = E\{x_t\overline{x_{t+v}}\}$. This is because

$$\begin{aligned}
&\lim_{N\to\infty} \frac{1}{2\pi(2N+1)} E\left\{\sum_{-N\le t,s\le N} x_t\bar{x}_s e^{-i\omega(t-s)}\right\} \\
&= \lim_{N\to\infty} \frac{1}{2\pi} \sum_{v=-2N}^{2N} \left(1 - \frac{|v|}{2N+1}\right) r(v) e^{-i\omega v} \\
&= \lim_{N\to\infty} \frac{1}{2\pi} \int_{-\pi}^{\pi} F(\omega-\omega'; N) \sum_{v=-2N}^{2N} r(v) e^{-i\omega' v}\, d\omega' \\
&\left(\begin{array}{l}\text{where } F(\omega-\omega'; N) = \displaystyle\sum_{v=-2N}^{2N}\left(1-\frac{|v|}{2N+1}\right)e^{-i(\omega-\omega')v} \\ \text{and } \displaystyle\lim_{N\to\infty} F(\omega-\omega';N) \text{ is a Delta function of } \omega-\omega',\end{array}\right) \\
&= \lim_{N\to\infty} \frac{1}{2\pi} \sum_{v=-2N}^{2N} e^{-iv\omega} r(v).
\end{aligned}$$

Let us consider a general (stationary or nonstationary) stochastic process $\{x_t\}$ and suppose that the data are available for a finite time period, $t = 1, 2, \ldots, N$. The pseudospectrum $p_x(\omega)$ of x_t is defined as

$$p_x(\omega) = \frac{1}{2\pi N} E\left\{\left|\sum_{t=1}^{N} x_t e^{-i\omega t}\right|^2\right\}, \qquad -\pi \le \omega \le \pi \tag{2}$$

where x_t is considered, in general, as a complex time series. For the purpose of our mathematical treatment, we find it convenient to work with the pseudospectrum for the variance about zero rather than the mean. This is why the mean is not subtracted in (2).

(a) The integral of $p_x(\omega)$ over the frequencies from $-\pi$ to π is the mean of the variance (about zero) over the given period.

$$\begin{aligned}
\int_{-\pi}^{\pi} p_x(\omega)\, d\omega &= \frac{1}{2\pi N}\int_{-\pi}^{\pi} E\left\{\left|\sum_{t=1}^{N} x_t e^{-i\omega t}\right|^2\right\} d\omega \\
&= \frac{1}{2\pi N}\int_{-\pi}^{\pi} \sum_{t=1}^{N}\sum_{s=1}^{N} E\{x_t\bar{x}_s\} e^{-i\omega(t-s)}\, d\omega \qquad (3)\\
&= \frac{1}{N}\sum_{t=1}^{N} E\{|x_t|^2\}
\end{aligned}$$

(b) Let us define the pseudoautocovariance $r(v)$ for lag v of a general process x_t as

$$r(v) = \frac{1}{N-v}\sum_{t=1}^{N-v} E\{x_t\overline{x_{t+v}}\} \qquad \text{for } v \geqq 0$$

$$r(v) = \frac{1}{N-|v|}\sum_{t=|v|+1}^{N} E\{x_t\overline{x_{t+v}}\} \quad \text{for } v < 0$$

Obviously $r(v) = \overline{r(-v)}$. Then the pseudospectrum $p_x(\omega)$ and the pseudo-autocovariance $r(v)$ are a Fourier transform pair when $r(v)$ is weighted[3] by $1 - \frac{|v|}{N}$, i.e.,

$$p_x(\omega) = \frac{1}{2\pi} \sum_{v=-(N-1)}^{N-1} \left(1 - \frac{|v|}{N}\right) r(v) e^{-i\omega v} \tag{4}$$

and

$$\left(1 - \frac{|v|}{N}\right) r(v) = \int_{-\pi}^{\pi} p_x(\omega) e^{i\omega v}\, d\omega \quad \text{for } v = 0, \pm 1, \ldots, \pm(N-1) \tag{5}$$

PROOF.

$$p_x(\omega) = \frac{1}{2\pi N} E\left\{\sum_t \sum_s x_t \bar{x}_s e^{-i\omega(t-s)}\right\}$$

$$= \frac{1}{2\pi N}\left[\sum_{v=0}^{N-1} \sum_{t=1}^{N-v} E\{x_t \overline{x_{t+v}}\} e^{-i\omega v} + \sum_{v=-(N-1)}^{-1} \sum_{t=|v|+1}^{N} E\{x_t \overline{x_{t+v}}\} e^{-i\omega v}\right]$$

$$= \frac{1}{2\pi} \sum_{v=-(N-1)}^{N-1} \left(1 - \frac{|v|}{N}\right) r(v) e^{-i\omega v}$$

$$\int_{-\pi}^{\pi} p_x(\omega) e^{i\omega v}\, d\omega = \frac{1}{2\pi} \int_{-\pi}^{\pi} \sum_{v'=-(N-1)}^{N-1} \left[1 - \frac{|v'|}{N}\right] r(v') e^{i\omega(v'-v)}\, d\omega$$

$$= \left(1 - \frac{|v|}{N}\right) r(v).$$

(c) The pseudospectrum $p_x(\omega)$ is the mathematical expectation of sample estimates of spectra[4] obtained by using $\left(1 - \frac{|v|}{N}\right)$ as the lag window for the estimation. When we define the sample estimates of $r(v)$ and $p_x(\omega)$

[3] Alternatively $r(v) = \frac{1}{N} \sum_{t=1}^{N-v} E\,\{x_t \overline{x_{t+v}}\}$ for $v \geq 0$

$$r(v) = \frac{1}{N} \sum_{t=|v|+1}^{N} E\,\{x_t \overline{x_{t+v}}\} \text{ for } v < 0$$

may be used. Then $p_x(\omega)$ and $r(v)$ are a Fourier transform pair without using the weight, $1 - \frac{|v|}{N}$.

[4] As Bartlett pointed out, the variance of $\hat{p}_x(\omega)$ is not reduced to zero when $N \to \infty$ even if x_t is stationary. This point, however, is totally irrelevant to the pseudospectrum of nonstationary process. The pseudospectrum is defined for and dependent upon a given finite length of time period covering a specific portion of our economic history; thus the consistency of the estimate $\hat{p}_x(\omega)$ is not a relevant problem.

respectively as

$$\begin{cases} \hat{r}(v) = \dfrac{1}{N-v}\sum_{t=1}^{N-v} x_t \bar{x}_{t+v} & v \geq 0 \\ \hat{r}(v) = \dfrac{1}{N-|v|}\sum_{t=|v|+1}^{N} x_t \bar{x}_{t+v} & v < 0 \end{cases}$$

and

$$\hat{p}_x(\omega) = \sum_{v=-(N-1)}^{N-1} \left(1 - \frac{|v|}{N}\right) \hat{r}(v) e^{-i\omega v},$$

the mathematical expectation of $\hat{p}_x(\omega)$ is

$$E(\hat{p}_x(\omega)) = \sum_{v=-(N-1)}^{N-1} \left(1 - \frac{|v|}{N}\right) E(\hat{r}(v)) e^{-i\omega v} = p_x(\omega).$$

(d) If x_t is stationary with zero mean and the time period is infinite in length, the pseudospectrum converges to the spectrum, as can be seen from (1) and (2).

We can now define the pseudospectrum $p_d(\omega)$ of a deterministic process d_t in the given time period $t = 1, \ldots, N$ as

$$p_d(\omega) = \frac{1}{2\pi N}\left|\sum_{t=1}^{N} d_t e^{-i\omega t}\right|^2$$

where d_t is considered, in general, a complex function of time. If a non-stationary process $y_t = d_t + x_t$, where d_t is a deterministic process and x_t is a stationary stochastic process with zero mean, the pseudospectrum of y_t is obviously the sum of the pseudospectra of d_t and x_t. Especially when d_t is a trend, it is well known that the pseudospectrum of d_t is significant only at the low frequencies. This will be elaborated on later in Section VI.

For the multi-variate stochastic process we can define the cross-pseudospectrum and pseudocovariance. The cross-pseudospectrum between two stochastic processes, $x_t^{(1)}$ and $x_t^{(2)}$, is

$$p_{12}(\omega) = \frac{1}{2\pi N} E\left\{\sum_{t=1}^{N}\sum_{s=1}^{N} x_t^{(1)} \bar{x}_s^{(2)} e^{-i\omega(t-s)}\right\}$$

and the cross-pseudocovariance is

$$r_{12}(v) = \frac{1}{N-v}\sum_{t=1}^{N-v} E\{x_t^{(1)} \bar{x}_{t+v}^{(2)}\} \qquad v \geq 0$$

$$r_{12}(v) = \frac{1}{N-|v|}\sum_{t=|v|+1}^{N} E\{x_t^{(1)} \bar{x}_{t+v}^{(2)}\} \qquad v < 0.$$

(e) The cross-pseudospectrum $p_{12}(\omega)$ and the cross-pseudocovariance $r_{12}(v)$ are a Fourier transform pair when $r_{12}(v)$ is weighted by $1 - \dfrac{|v|}{N}$.

The cross-pseudospectrum between two deterministic functions of time, λ_t and μ_t, is defined as

$$\frac{1}{2\pi N}\sum_t\sum_s \lambda_t\mu_s e^{-i\omega(t-s)}.$$

In general, let $x_t = (x_t^{(1)}, x_t^{(2)}, \ldots, x_t^{(m)})$ be a m-variate stochastic vector process. Then

$$R(v) = [r_{jk}(v)]$$

is the pseudocovariance matrix, and $R(-v) = \overline{R(v)}$.

The matrix of pseudospectral and cross-pseudospectral functions, i.e., the pseudospectral matrix, is defined as

$$P(\omega) = [p_{jk}(\omega)]$$

where

$$p_{jk}(\omega) = \frac{1}{2\pi}\sum_{v=-(N-1)}^{N-1}\left(1 - \frac{|v|}{N}\right) r_{jk}(v)e^{-i\omega v}$$

$$= \frac{1}{2\pi N}E\left\{\sum_{t=1}^{N}\sum_{s=1}^{N} x_t^{(j)}\bar{x}_s^{(k)}e^{-i\omega(t-s)}\right\}.$$

Another representation of the matrix $P(\omega)$ is

$$P(\omega) = \frac{1}{2\pi N}E\left\{\sum_i\sum_s X_t X_s{}^* e^{-i\omega(t-s)}\right\}$$

where X_t and $X_s{}^*$ are the vectors of m components, and where $X_s{}^*$ is defined as the complex conjugate of the transpose of X_s, i.e., the Hermitian conjugate of X_s. This representation will be used extensively in Sections IV and V.

(f) The coherence inequality holds for the pseudospectra and cross-pseudospectrum.

Let $f_j(\omega)$ and $f_k(\omega)$ be the Fourier transform of $x_t^{(j)}$ and $x_t^{(k)}$, i.e.

$$f_j(\omega) = \sum_{t=1}^{N} x_t^{(j)}e^{-i\omega t}$$

$$f_k(\omega) = \sum_{t=1}^{N} x_t^{(k)}e^{-i\omega t}$$

Then from the Schwarz's inequality, the inequality

$$E\{|f_j(\omega)f_k(\omega)|^2\} \leqq E\{|f_j(\omega)|^2\}E\{|f_k(\omega)|^2\}$$

holds. Then

$$E\left\{\left|\frac{1}{2\pi N}f_j(\omega)f_k(\omega)\right|^2\right\} \leqq E\left\{\frac{1}{2\pi N}|f_j(\omega)|^2\right\}E\left\{\frac{1}{2\pi N}|f_k(\omega)|^2\right\},$$

i.e.,

$$|p_{jk}(\omega)|^2 \leqq p_{jj}(\omega)\,.\,p_{kk}(\omega).$$

This is the coherence inequality.

III. THE PSEUDOSPECTRAL MATRIX OF THE STOCHASTIC PROCESS WITH CHANGING AMPLITUDES AND CHANGING PHASES

The present section deals with the stochastic process in which either the amplitudes or the phases of different frequencies change over time. It also serves as the mathematical background for Sections IV and V.

(a) The pseudospectral matrix of the stochastic process in which the amplitudes change uniformly over different frequencies but the phases do not change.

The nonstationary stochastic process in which the amplitudes change uniformly over different frequencies but the phases do not change can be represented by the product of some function of time and some stationary stochastic process.

THEOREM 1. *Let* $d_t^{(j)}$ $(j = 1, \ldots M)$ *be deterministic processes and* $x_t^{(j)}$ $(j = 1, \ldots M)$ *stationary stochastic processes with* $E(x_t^{(j)}) = 0$. *Then a column vector of the stochastic process*

$$y_t = \begin{bmatrix} y_t^{(1)} \\ \cdot \\ \cdot \\ \cdot \\ y_t^{(M)} \end{bmatrix} = \begin{bmatrix} d_t^{(1)} x_t^{(1)} \\ \cdot \\ \cdot \\ \cdot \\ d_t^{(M)} x_t^{(M)} \end{bmatrix}$$

$t = 1, 2, \ldots N$, *has the pseudospectral matrix* $P_y(\omega) = [p_{y_j y_k}(\omega)]$ *where*

$$p_{y_j y_k}(\omega) = \int_{\omega-\pi}^{\omega+\pi} p_{x_j x_k}(\omega - \omega') p_{d_j d_k}(\omega')\, d\omega'. \tag{6}$$

$p_{x_j x_k}(\omega)$ *and* $p_{d_j d_k}(\omega)$ *are respectively the cross-spectrum between* $x_t^{(j)}$ *and* $x_t^{(k)}$, *and the cross-pseudospectrum between* $d_t^{(j)}$ *and* $d_t^{(k)}$.

PROOF.

$$\begin{aligned} p_{y_j y_k}(\omega) &= \frac{1}{2\pi N} E\left\{\sum_t \sum_s d_t^{(j)} x_t^{(j)} \bar{d}_s^{(k)} \bar{x}_s^{(k)} e^{-i\omega(t-s)}\right\} \\ &= \frac{1}{2\pi N} \sum_t \sum_s d_t^{(j)} \bar{d}_s^{(k)} E\{x_t^{(j)} \bar{x}_s^{(k)}\} e^{-i\omega(t-s)} \\ &= \frac{1}{2\pi N} \int_{-\pi}^{\pi} \sum_t \sum_s d_t^{(j)} \bar{d}_s^{(k)} p_{x_j x_k}(\omega') e^{i\omega'(t-s)} e^{-i\omega(t-s)}\, d\omega' \\ &= \int_{-\pi}^{\pi} p_{x_j x_k}(\omega') p_{d_j d_k}(\omega - \omega')\, d\omega' \\ &= \int_{\omega-\pi}^{\omega+\pi} p_{x_j x_k}(\omega - \omega') p_{d_j d_k}(\omega')\, d\omega' \end{aligned}$$

q.e.d.

The pseudospectrum of $y_t^{(j)} = d_t^{(j)}x_t^{(j)}$ is the convolution of the pseudospectrum of $d_t^{(j)}$ and the spectrum of $x_t^{(j)}$, i.e., a sort of weighted moving average of the spectrum of $x_t^{(j)}$ by using the pseudospectrum of $d_t^{(j)}$ as the weights.

(i) If $x_t^{(j)}$ is a white noise, ϵ_t, with its variance σ^2, i.e., $y_t^{(j)} = d_t^{(j)}\epsilon_t$, then the pseudospectrum of $y_t^{(j)}$ is

$$p_{y_j}(\omega) = \frac{\sigma^2}{2\pi} \frac{\sum_t |d_t^{(j)}|^2}{N} .$$

This means that $p_{y_j}(\omega)$ is the mean of the changing, instantaneous spectra of $d_t^{(j)}\epsilon_t$, i.e., $\frac{\sigma^2}{2\pi} |d_t^{(j)}|^2$ over time.

(ii) Suppose that the pseudospectrum (for the variance about zero) of $d_t^{(j)}$ is significant only in the frequency band $[-\epsilon, \epsilon]$, where $\epsilon > 0$ is a certain small number. (This is possible when $d_t^{(j)}$ is either a very smooth function of time or $d_t^{(j)}$ involves one or two jumps in addition to the trend. The fact that $p_{d_j}(\omega)$ is a pseudospectrum for the variance about zero is important in judging the plausibility of the concentration of this spectrum in the low frequencies in economic studies.) Further assume that $p_{x_j}(\omega)$ is smooth so that

$$p_{x_j}(\omega) = p_{x_j}(\omega_0) \qquad \text{for all } \omega\text{'s in } (\omega_0 - \epsilon, \omega_0 + \epsilon).$$

Then

$$\begin{aligned} p_{y_j}(\omega_0) &\approx p_{x_j}(\omega_0) \int_{-\varepsilon}^{+\varepsilon} p_{d_j}(\omega')\, d\omega' \\ &\approx p_{x_j}(\omega_0) \int_{-\pi}^{\pi} p_{d_j}(\omega')\, d\omega' \\ &\approx p_{x_j}(\omega_0) \frac{\sum_t |d_t^{(j)}|^2}{N} \end{aligned}$$

This is again the mean of the changing spectra.

(iii) If the movements of $d_t^{(j)}$ are dominated by irregular cycles, the pseudospectrum of $y_t^{(j)} = d_t^{(j)}x_t^{(j)}$ can be expressed as the result of smoothing the spectrum of x_t and then shifting it along the axis of ω. When only one sample of $y_t^{(j)}$ is available and when no a priori information about d_t is given, there would be no easy way to interpret the pseudospectrum of $y_t^{(j)}$.

(b) The pseudospectral matrix of the stochastic process where the phases change uniformly over different frequencies but the amplitudes do not change.

Any real, stationary stochastic process with continuous spectrum can be represented as

$$\begin{aligned} x_t^{(j)} &= \int_{-\pi}^{\pi} e^{i\omega t}\, dz_j(\omega) \\ &= 2\int_0^{\pi} (\cos \omega t)\, du_j(\omega)\, d\omega + 2\int_0^{\pi} (\sin \omega t)\, dv_j(\omega), \quad j = 1, \ldots, M \end{aligned} \tag{7}$$

where

$$z_j(\omega) = u_j(\omega) - iv_j(\omega), \qquad u_j(\omega) \text{ and } v_j(\omega) \text{ being real functions}$$

$$u_j(\omega) = u_j(-\omega),$$

$$v_j(\omega) = -v_j(-\omega),$$

$$E\{(du_j(\omega)\, du_j(\omega')\} = E\{(dv_j(\omega)\, dv_j(\omega')\} = \begin{cases} 0 & \text{if } \omega \neq \omega' \\ \tfrac{1}{2}p_{x_j}(\omega)\, d\omega & \text{if } \omega = \omega' \end{cases}$$

$$E\{(du_j(\omega)\, dv_j(\omega')\} = 0, \qquad \text{for all } \omega, \omega'$$

$$E\{(du_j(\omega)\, du_k(\omega')\} = E\{(dv_j(\omega)\, dv_k(\omega')\} = \begin{cases} 0 & \text{if } \omega \neq \omega' \\ \tfrac{1}{2}R_e[p_{x_jx_k}(\omega)\, d\omega] & \text{if } \omega = \omega' \end{cases}$$

$$E\{(du_j(\omega)\, dv_k(\omega')\} = -E\{(dv_j(\omega)\, du_k(\omega')\} = \begin{cases} 0 & \text{if } \omega \neq \omega' \\ \tfrac{1}{2}I_m[p_{x_jx_k}(\omega)\, d\omega] & \text{if } \omega = \omega' \end{cases}$$

($R_e[\ \ ]$ and $I_m[\ \ ]$ mean respectively the real part and the imaginary part of [ ].) The phase of (7) is defined as[5]

$$E\left[\tan^{-1} \frac{dv_j(\omega)}{du_j(\omega)}\right].$$

We are concerned with a real, stationary process of which the spectrum is identical to $p_{x_j}(\omega)$ and of which the phase of frequency ω is greater than that of $x_t^{(j)}$ by $\varphi_j(\omega)$. $\varphi_j(\omega)$ is assumed to be a deterministic function of ω. If $\varphi_j(\omega) = -\varphi_j(-\omega)$, then such a process is represented by

$$\begin{aligned} x_t^{(j)}\{\varphi_j(\omega)\} &= \int_{-\pi}^{\pi} e^{i\omega t} e^{-i\varphi_j(\omega)}\, dz_j(\omega) \\ &= 2\int_0^{\pi} \cos \omega t[\cos \varphi_j(\omega)\, du_j(\omega) - \sin \varphi_j(\omega)\, dv_j(\omega)] \end{aligned} \tag{8}$$

$$+ 2\int_0^{\pi} \sin \omega t[\sin \varphi_j(\omega)\, du_j(\omega) + \cos \varphi_j(\omega) dv_j(\omega)]. \tag{8'}$$

Actually (8) is a real process, because $\cos \varphi_j(\omega)\, du_j(\omega) - \sin \varphi_j(\omega)\, dv_j(\omega)$

[5] To be more precise, take the principal value of $\tan^{-1} dv/du$ if $du \geq 0$, $dv \geq 0$
$\pi -$ the principal value of $\tan^{-1} dv/du$ if $du \leq 0$, $dv \geq 0$
$\pi +$ the principal value of $\tan^{-1} dv/du$ if $du \leq 0$, $dv \leq 0$
$2\pi -$ the principal value of $\tan^{-1} dv/du$ if $du \geq 0$, $dv \leq 0$.

is an even function of ω, and $\sin \varphi_j(\omega)\, du_j(\omega) + \cos \varphi_j(\omega)\, dv_j(\omega)$ is an odd function of ω.

The spectrum of $x_t^{(j)}\{\varphi_j(\omega)\}$ is identical to the spectrum of $x_t^{(j)}$ because

$$E\{|e^{-i\varphi_j(\omega)}(du_j(\omega) - i\, dv_j(\omega))|^2\} = E\{du_j(\omega)^2\} + E\{dv_j(\omega)^2\}.$$

The phase of $x_t^{(j)}\{\varphi_j(\omega)\}$ is greater than the phase of $x_t^{(j)}$ by $\varphi_j(\omega)$ at the frequency ω because

$$E\left\{\tan^{-1}\frac{\sin \varphi_j(\omega)\, du_j(\omega) + \cos \varphi_j(\omega)\, dv_j(\omega)}{\cos \varphi_j(\omega)\, du_j(\omega) - \sin \varphi_j(\omega)\, dv_j(\omega)}\right\} = E\left\{\tan^{-1}\frac{dv_j(\omega)}{du_j(\omega)}\right\} + \varphi_j(\omega).$$

The nonstationary, real stochastic processes, where the phases change with time, but the amplitudes do not change, can be represented as

$$y_t^{(j)} = \int_{-\pi}^{\pi} e^{i\omega t} e^{-i\varphi_{jt}(\omega)}\, dz_j(\omega) \tag{9}$$

$$\begin{aligned} &= 2\int_0^{\pi} \cos \omega t[\cos \varphi_{jt}(\omega)\, du_j(\omega) - \sin \varphi_{jt}(\omega)\, dv_j(\omega)]\, d\omega \\ &\quad + 2\int_0^{\pi} \sin \omega t[\sin \varphi_{jt}(\omega)\, du_j(\omega) + \cos \varphi_{jt}(\omega)\, dv_j(\omega)]\, d\omega. \end{aligned} \tag{9'}$$

In the present section (b), we shall study the special case of $y_t^{(j)}$,

$$\begin{aligned} \varphi_{jt}(\omega) &= \varphi_{jt} && \text{for} \quad 0 < \omega \leq \pi \\ \varphi_{jt}(\omega) &= 0 && \qquad \omega = 0 \\ \varphi_{jt}(\omega) &= -\varphi_{jt} && \qquad -\pi \leq \omega < 0 \end{aligned} \tag{10}$$

i.e., the case where the phases change uniformly over different frequencies.

COROLLARY 1. *The nonstationary stochastic process of* $y_t^{(j)}$ *and* $y_t^{(k)}$ ($j = 1, \ldots, M, k = 1, \ldots, M$) *defined by* (9) *and* (10) *have the cross-pseudospectrum* $p_{y_j y_k}(\omega)$,

$$p_{y_j y_k}(\omega) = \int_{\omega}^{\omega+\pi} p_{x_j x_k}(\omega - \omega') p_{\bar{\varphi}_j \bar{\varphi}_k}(\omega')\, d\omega' + \int_{\omega-\pi}^{\omega} p_{x_j x_k}(\omega - \omega') p_{\varphi_j \varphi_k}(\omega')\, d\omega' \tag{11}$$

where $p_{\varphi_j\varphi_k}(\omega)$ *and* $p_{\bar{\varphi}_j\bar{\varphi}_k}(\omega)$ *are the cross-pseudospectra, respectively, between* $e^{i\varphi_{jt}}$ *and* $e^{i\varphi_{kt}}$ *and between* $e^{-i\varphi_{jt}}$ *and* $e^{-i\varphi_{kt}}$.

PROOF. Since

$$y_t^{(j)} = e^{-i\varphi_{jt}} \int_{-\pi}^{0} e^{i\omega' t}\, dz_j(\omega') + e^{i\varphi_{jt}} \int_0^{\pi} e^{i\omega' t}\, dz_j(\omega'),$$

$$E(y_t^{(j)} \bar{y}_s^{(k)}) = e^{-i(\varphi_{jt} - \varphi_{ks})} \int_{-\pi}^{0} e^{i\omega'(t-s)} p_{x_j x_k}(\omega')\, d\omega' + e^{i(\varphi_{jt} - \varphi_{ks})} \int_0^{\pi} e^{i\omega'(t-s)} p_{x_j x_k}(\omega')\, d\omega'.$$

Therefore,

$$p_{y_jy_k}(\omega) = \frac{1}{2\pi N}\sum_t\sum_s E(y_t^{(j)}\bar{y}_s^{(k)})e^{-i\omega(t-s)}$$
$$= \int_{-\pi}^{0} p_{x_jx_t}(\omega')p_{\bar{\varphi}_j\bar{\varphi}_k}(\omega-\omega')\,d\omega' + \int_0^{\pi} p_{x_jx_k}(\omega')p_{\varphi_j\varphi_k}(\omega-\omega')\,d\omega'$$
$$= \int_{\omega}^{\omega+\pi} p_{x_jx_k}(\omega-\omega')p_{\bar{\varphi}_j\bar{\varphi}_k}(\omega')\,d\omega'$$
$$+ \int_{\omega-\pi}^{\omega} p_{x_jx_k}(\omega-\omega')p_{\varphi_j\varphi_k}(\omega')\,d\omega'.$$

q.e.d.

Suppose that the norms of $p_{\bar{\varphi}_j\bar{\varphi}_k}(\omega)$ and $p_{\varphi_j\varphi_k}(\omega)$ are relatively significant only in the frequency band $[-\epsilon, \epsilon]$ where $\epsilon > 0$ is a certain small number. This is possible when the phase changes of $y_t^{(j)}$ and $y_t^{(k)}$ are smooth, or, otherwise involve one or two jumps. Further assume that $p_{x_jx_k}(\omega)$ is smooth so that

$$p_{x_jx_k}(\omega) = p_{x_jx_k}(\omega_0) \quad \text{for all } \omega\text{'s in } (\omega_0 - \epsilon, \omega_0 + \epsilon).$$

Then if $\pi \geq \omega_0 > \epsilon$, the first integral of (11) for $\omega = \omega_0$ is not significant, because the interval $[\omega_0, \omega_0 + \pi]$ does not include the frequency band $[-\epsilon, \epsilon]$. Therefore,

$$p_{y_jy_k}(\omega_0) \approx p_{x_jx_k}(\omega_0)\int_{-\varepsilon}^{\varepsilon} p_{\varphi_j\varphi_k}(\omega')\,d\omega'$$
$$\approx p_{x_jx_k}(\omega_0)\int_{-\pi}^{\pi} p_{\varphi_j\varphi_k}(\omega')\,d\omega' = p_{x_jx_k}(\omega_0)\cdot\frac{1}{N}\sum_t e^{i(\varphi_{jt}-\varphi_{kt})}. \quad (12)$$

If $-\pi \leq \omega_0 < -\epsilon$, then the second integral of (11) for $\omega = \omega_0$ is not significant, and we obtain

$$p_{y_jy_k}(\omega_0) \approx p_{x_jx_k}(\omega_0)\int_{-\varepsilon}^{\varepsilon} p_{\bar{\varphi}_j\bar{\varphi}_k}(\omega')\,d\omega'$$
$$\approx p_{x_jx_k}(\omega_0)\cdot\frac{1}{N}\sum_t e^{-i(\varphi_{jt}-\varphi_{kt})}. \quad (12')$$

Thus, the phase of the cross-pseudospectrum between $y_t^{(j)}$ and $y_t^{(k)}$ differs from the phase of the cross-spectrum between $x_t^{(j)}$ and $x_t^{(k)}$ by

$$\tan^{-1}\frac{-\sum_t \sin(\varphi_{jt}-\varphi_{kt})}{\sum_t \cos(\varphi_{jt}-\varphi_{kt})} \qquad (\omega_0 > \epsilon) \quad (13)$$

or

$$\tan^{-1}\frac{\sum_t \sin(\varphi_{jt}-\varphi_{kt})}{\sum_t \cos(\varphi_{jt}-\varphi_{kt})} \qquad (\omega_0 < -\epsilon). \quad (13')$$

In order to understand the meaning of the phase of the cross-pseudo-spectrum between $y_t^{(j)}$ and $y_t^{(k)}$, let us consider a special case where $x_t^{(j)} \equiv x_t^{(k)}$ for all t's. Then, the phase difference between $y_t^{(j)}$ and $y_t^{(k)}$ is solely due to the fact that $y_t^{(j)}$ in (9) involves $e^{-i\varphi_{jt}(\omega)}$ whereas $y_t^{(k)}$ involves $e^{-i\varphi_{kt}(\omega)}$. Indeed, $(\varphi_{jt} - \varphi_{kt})$ is the instantaneous phase difference at t between $y_t^{(j)}$ and $y_t^{(k)}$. When $x_t^{(j)} \equiv x_t^{(k)}$, $p_{x_jx_k}(\omega_0)$ in (12) or (12′) is real, and, the phase of $p_{y_jy_k}(\omega_0)$ is given by (13) and (13′). Therefore, the phase of the cross-pseudospectrum is a kind of average (over time) of the instantaneous phase difference. (13) or (13′) shows that the average of any two angles θ_1 and θ_2 must be defined as

$$\tan^{-1}\frac{\sin\theta_1 + \sin\theta_2}{\cos\theta_1 + \cos\theta_2}.$$

(c) The pseudospectral matrix of the stochastic process where both the amplitudes and the phases change differently over different frequencies.

The real nonstationary stochastic process, $y_t^{(j)}$, where both the amplitudes and phases change differently over different frequencies, can be generated from a real stationary stochastic process

$$x_t^{(j)} = \int_{-\pi}^{\pi} e^{i\omega t}\, dz_j(\omega)$$

by

$$y_t^{(j)} = \int_{-\pi}^{\pi} e^{i\omega t}\, d_t^{(j)}(\omega)\, dz_j(\omega) \tag{14}$$

where

$$d_t^{(j)}(\omega) = a_{jt}(\omega)e^{-i\varphi_{jt}(\omega)}$$
$$\varphi_{jt}(\omega) = -\varphi_{jt}(-\omega)$$
$$a_{jt}(\omega) = a_{jt}(-\omega).$$

$a_{jt}(\omega)$ represents the amplitude of the frequency ω at time t, and $\varphi_{jt}(\omega)$ the phase difference of the frequency ω between $x_t^{(j)}$ and $y_t^{(k)}$ at time t.

COROLLARY 2. *The cross-pseudospectrum between the real non-stationary stochastic processes $y_t^{(j)}$ and $y_t^{(k)}$ defined by* (14) *can be represented as*

$$p_{y_jy_k}(\omega) = \int_{-\pi}^{\pi} p_{x_jx_k}(\omega')p_{f_jf_k}(\omega, \omega')\, d\omega' \tag{15}$$

where

$$p_{f_jf_k}(\omega, \omega') = \frac{1}{2\pi N}\sum_t\sum_s f_t^{(j)}(\omega')\overline{f_s^{(k)}(\omega')}e^{-i\omega(t-s)}$$

and

$$f_t^{(j)}(\omega') = d_t^{(j)}(\omega')e^{i\omega' t}$$
$$f_s^{(k)}(\omega') = d_s^{(k)}(\omega')e^{i\omega' s}$$

(*The proof is omitted.*)

$p_{f_jf_k}(\omega, \omega')$ is the cross-pseudospectrum at frequency ω between the deterministic processes $d_t^{(j)}(\omega')e^{i\omega' t}$ and $d_s^{(k)}(\omega')e^{i\omega' s}$.

Suppose (i) that $d_t^{(j)}(\omega')$ and $d_t^{(k)}(\omega')$ are smooth functions of time (for any ω') so that $|p_{f_jf_k}(\omega, \omega')|$ is significant only for the values of ω that are near or equal to ω', which means

$$\int_{-\pi}^{\pi} p_{f_jf_k}(\omega_0, \omega')\, d\omega' \approx \int_{\omega_0-\varepsilon}^{\omega_0+\varepsilon} p_{f_jf_k}(\omega_0, \omega')\, d\omega'$$

and (ii) that $p_{x_jx_k}(\omega')$ is smooth so that

$$p_{x_jx_k}(\omega') \approx p_{x_jx_k}(\omega_0)$$

for any ω' such that

$$\omega_0 - \epsilon \leq \omega' \leq \omega_0 + \epsilon.$$

Then from (15) we obtain

$$p_{y_jy_k}(\omega_0) \approx p_{x_jx_k}(\omega_0)\int_{\omega_0-\varepsilon}^{\omega_0+\varepsilon} p_{f_jf_k}(\omega_0, \omega')\, d\omega'.$$

Further, if we can assume, in addition to (i) and (ii), that (iii)

$$\left.\begin{aligned} d_t^{(j)}(\omega') &= d_t^{(j)}(\omega_0) \\ d_s^{(k)}(\omega') &= d_s^{(k)}(\omega_0) \end{aligned}\right\} \quad \text{for any } \omega' \text{ in } (\omega_0 - \epsilon, \omega_0 + \epsilon)$$

then

$$p_{y_jy_k}(\omega_0) \approx p_{x_jx_k}(\omega_0) \cdot \frac{1}{\pi N} \sum_t \sum_s d_t^{(j)}(\omega_0)\, \overline{d_s^{(k)}(\omega_0)}\, \frac{\sin \epsilon(t-s)}{t-s}.$$

The phase of $p_{y_jy_k}(\omega_0)$ differs from the phase of the $p_{x_jx_k}(\omega_0)$ by

$$\tan^{-1} \frac{\sum_t \sum_s a_{jt}(\omega_0)a_{ks}(\omega_0) \sin \{\varphi_{jt}(\omega_0) - \varphi_{ks}(\omega_0)\} \dfrac{\sin \epsilon(t-s)}{t-s}}{\sum_t \sum_s a_{jt}(\omega_0)a_{ks}(\omega_0) \cos \{\varphi_{jt}(\omega_0) - \varphi_{ks}(\omega_0)\} \dfrac{\sin \epsilon(t-s)}{t-s}} \tag{16}$$

Expression (16) is a generalization of (13) and (13′). Notice that

$$\frac{\sin \epsilon(t-s)}{t-s}$$

is inversely related to $|t - s|$ where ϵ is small.

In the special case in which the amplitude does not change at any frequency (16) becomes

$$\tan^{-1} \frac{\sum_t \sum_s \sin \{\varphi_{jt}(\omega_0) - \varphi_{ks}(\omega_0)\} \dfrac{\sin \epsilon(t-s)}{t-s}}{\sum_t \sum_s \cos \{\varphi_{ft}(\omega_0) - \varphi_{ks}(\omega_0)\} \dfrac{\sin \epsilon(t-s)}{t-s}} \tag{16′}$$

Let $v = t - s$ and $\varphi_{jk,\omega_0}(t, v) = \varphi_{jt}(\omega_0) - \varphi_{ks}(\omega_0)$. Then (16′) is a double average of $\varphi_{jk,\omega_0}(t, v)$ over t and v, where the averaging over v is done with the weights $\frac{\sin \epsilon v}{v}$.

IV. THE PSEUDOSPECTRAL MATRIX OF THE EXPLOSIVE DYNAMIC ECONOMETRIC MODEL

Most of dynamic econometric models are sets of linear difference equations such as

$$B_0 y_t + B_1 y_{t-1} + \ldots + B_p y_{t-p} = \Gamma_0 z_t + \Gamma_1 z_{t-1} \ldots + \Gamma_q z_{t-q} + v_t \quad (17)$$

where

$$y_t = \begin{bmatrix} y_t^{(1)} \\ \cdot \\ \cdot \\ \cdot \\ y_t^{(k)} \end{bmatrix}, \qquad z_t = \begin{bmatrix} z_t^{(1)} \\ \cdot \\ \cdot \\ \cdot \\ z_t^{(k')} \end{bmatrix}, \qquad \text{and} \quad v_t = \begin{bmatrix} v_t^{(1)} \\ \cdot \\ \cdot \\ \cdot \\ v_t^{(k)} \end{bmatrix} \quad (17')$$

represent respectively the vectors of k endogenous variables, k' exogenous variables, and k random disturbances, $B_0, B_1, \ldots, B_p$ and $\Gamma_0, \Gamma_1, \ldots, \Gamma_q$ are respectively $k \times k$ and $k \times k'$ matrices of the parameters. $|B_0| \neq 0$ is assumed.

If the exogenous variables are removed from (17) and the stability condition holds, (17) represents a stationary stochastic process of the endogenous variable y_t. It is well known[6] that the spectral matrix of y_t is related to $P_v(\omega)$ by

$$P_y(\omega) = (B_0 + B_1 e^{-i\omega} + \ldots + B_p e^{-i\omega p})^{-1} \cdot P_v(\omega) \cdot (B_0 + B_1 e^{-i\omega} + \ldots + B_p e^{-i\omega p})^{*-1} \quad (18)$$

If the stability condition does not hold, y_t represents an explosive stochastic process. In this case, if we take a finite time period, we can prove that the pseudospectral matrix of y_t is related to the pseudospectral matrix of z_t and the spectral matrix of v_t in just the same way as $P_y(\omega)$ is related to $P_v(\omega)$ in (18).

THEOREM 2. *In the system* (17), *let us assume* (i) *that the end effects for p time lag of y_t and for q time lag of z_t are negligible for the period,* $t = 1, \ldots, N$, (ii) *that z_t is deterministic and v_t is stationary with* $E(v_t) = 0$, *and* (iii) *that*

$$|B_0 + Be^{-i\omega} + \ldots + B_p e^{-i\omega p}| \neq 0, \qquad -\pi \leq \omega \leq \pi.$$

[6] Whittle, P., "The analysis of multiple stationary time series," *Journal of the Royal Statistical Society* (B) 15, (1953) pp. 125–139.

Then

$$P_y(\omega) = (B_0 + B_1 e^{-i\omega} + \ldots + B_p e^{-i\omega p})^{-1}$$
$$\cdot [(\Gamma_0 + \Gamma_1 e^{-i\omega} + \ldots + \Gamma_q e^{-i\omega q})$$
$$\cdot P_z(\omega)(\Gamma_0 + \Gamma_1 e^{-i\omega} + \ldots + \Gamma_q e^{-i\omega q})^* + P_v(\omega)]$$
$$\cdot (B_0 + B_1 e^{-i\omega} + \ldots + B_p e^{-i\omega p})^{*-1}$$

where $P_y(\omega)$, $P_z(\omega)$ and $P_v(\omega)$ *are respectively the pseudospectral matrices of* y_t, z_t, and v_t.

PROOF. The Fourier transforms of the left and right hand sides of (17) are, respectively,

$$\sum_{t=1}^{N} (B_0 y_t + B_1 y_{t-1} + \ldots + B_p y_{t-p}) e^{-i\omega t}$$
$$\approx (B_0 + B_1 e^{-i\omega} + \ldots + B_p e^{-i\omega p}) \sum_{t=1}^{N} y_t e^{-i\omega t}$$

and

$$\sum_{t=1}^{N} (\Gamma_0 z_t + \Gamma_1 z_{t-1} + \ldots + \Gamma_q z_{t-q} + v_t) e^{-i\omega t}$$
$$\approx (\Gamma_0 + \Gamma_1 e^{-i\omega} + \ldots + \Gamma_q e^{-i\omega q}) \sum_{t=1}^{N} z_t e^{-i\omega t} + \sum_{t=1}^{N} v_t e^{-i\omega t}.$$

This is because, for example, if the end effect for k time lag is negligible,

$$\sum_{t=1}^{N} B_k y_{t-k} e^{-i\omega t} = \sum_{t=1}^{N} B_k e^{-i\omega k} y_{t-k} e^{-i\omega(t-k)}$$
$$\approx B_k e^{-i\omega k} \sum_{t=1}^{N} y_t e^{-i\omega t}.$$

The pseudospectral matrices of both sides of (17) are, respectively,

$$\frac{1}{2\pi N} E\left\{(B_0 + B_1 e^{-i\omega} + \ldots + B_p e^{-i\omega p}) \sum_{t=1}^{N} y_t e^{-i\omega t}\right.$$
$$\left.\times \sum_{s=1}^{N} y_s e^{i\omega s} (B_0 + B_1 e^{-i\omega} + \ldots + B_p e^{-i\omega p})^*\right\}$$
$$= (B_0 + B_1 e^{-i\omega} + \ldots + B_p e^{-i\omega p}) P_y(\omega) (B_0 + B_1 e^{-i\omega} + \ldots + B_p e^{-i\omega p})^*,$$

and

$$(\Gamma_0 + \Gamma_1 e^{-i\omega} + \ldots + \Gamma_q e^{-i\omega q}) P_z(\omega) (\Gamma_0 + \Gamma_1 e^{-i\omega} + \ldots + \Gamma_q e^{-i\omega q})^* + P_v(\omega).$$

Since

$$|B_0 + B_1 e^{-i\omega} + \ldots + B_p e^{-i\omega p}| \neq 0 \quad \text{for } \omega \text{ in } [-\pi, \pi],$$

we can get the theorem.

q.e.d.

V. PSEUDOSPECTRAL MATRIX OF THE DEVIATION FROM THE EQUILIBRIUM SOLUTION IN THE DYNAMIC MODEL WITH CHANGING PARAMETERS

In econometric studies of business' fluctuations the deviation of the solution of (17) from its equilibrium solution are frequently used to represent business fluctuations. We shall consider this deviation in the dynamic econometric model whose parameters change over time.

To treat the model with changing parameters, let us change the form of the system from (17) to a form that is solvable even when the parameters change over time. Any linear difference equation of order p can be replaced by a system of p first order difference equations in p variables,[7] if we define the new variables as follows:

$$Y_t \equiv \begin{bmatrix} y_t \\ \cdot \\ \cdot \\ \cdot \\ y_{t-p+1} \end{bmatrix} \qquad U_t \equiv \begin{bmatrix} B_0^{-1}v_t \\ 0 \\ \cdot \\ \cdot \\ 0 \end{bmatrix}$$

$$Z_t \equiv \begin{bmatrix} B_0^{-1}\Gamma_0 z_t + B_0^{-1}\Gamma_1 z_{t-1} + \ldots + B_0^{-1}\Gamma_q z_{t-q} \\ 0 \\ \cdot \\ \cdot \\ \cdot \\ 0 \end{bmatrix} \tag{18}$$

$$B \equiv \begin{bmatrix} -B_0^{-1}B_1 & -B_0^{-1}B_2 & \ldots & -B_0^{-1}B_{p-1} & -B_0^{-1}B_p \\ I_k & 0 & & 0 & 0 \\ 0 & I_k & & 0 & 0 \\ \cdot & \cdot & & \cdot & \cdot \\ \cdot & \cdot & & \cdot & \cdot \\ \cdot & \cdot & & \cdot & \cdot \\ 0 & 0 & & I_k & 0 \end{bmatrix}$$

where y_t, z_t, and v_t are the column vectors defined by (17′), I_k is a $k \times k$ identity matrix, Y_t, U_t, and Z_t are $kp \times 1$ matrices, and B is a $kp \times kp$ matrix.

The system (17) can now be represented by

$$Y_t - BY_{t-1} = Z_t + U_t. \tag{19}$$

Let us consider the model with constant parameters by using the representation (19) in order to clarify the concepts that we use in this section.

[7] Paul A. Samuelson: *Foundations of Economic Analysis*, Harvard University Press, (1947) Mathematical Appendix B.

The solution of (19) with Y_0 as the initial condition is

$$Y_t = B^t Y_0 + \sum_{m=0}^{t-1} B^m (Z_{t-m} + U_{t-m}). \tag{20}$$

Let X_t be the deviation from the equilibrium solution, i.e.,

$$X_t \equiv Y_t - \left(B^t Y_0 + \sum_{m=0}^{t-1} B^m Z_{t-m} \right) = \sum_{m=0}^{t-1} B^m U_{t-m}.$$

This is a moving average of random disturbances U_t. Then we can get the pseudospectral matrix of X_t,

$$P_X(\omega) = \frac{1}{2\pi N} E\left\{ \sum_{t=1}^{N} \sum_{s=1}^{N} \sum_{m=0}^{t-1} \sum_{n=0}^{s-1} B^m (U_{t-m} U_{s-n}^*) B^{n*} e^{-i\omega(t-s)} \right\}$$

$$= \frac{1}{2\pi N} \sum_t \sum_s \sum_m \sum_n B^m \left[\int_{-\pi}^{\pi} P_U(\omega') e^{i\omega'(t-m-s+n)} \, d\omega' \right] B^{n*} e^{-i\omega(t-s)}.$$

Put

$$G(t, \omega') \equiv \sum_{m=0}^{t-1} B^m e^{-i\omega' m} \tag{21}$$

$$H(\omega - \omega', \omega') \equiv \sum_{t=1}^{N} G(t, \omega') e^{-i(\omega-\omega')t}. \tag{22}$$

We shall call $H(\omega - \omega', \omega')$ the double transfer function of the parameters $B^0, B^1, \ldots, B^{t-1}$. Then we obtain

$$P_X(\omega) = \frac{1}{2\pi N} \int_{-\pi}^{\pi} H(\omega - \omega', \omega') P_U(\omega') H(\omega - \omega', \omega')^* \, d\omega' \tag{23}$$

or

$$P_X(\omega) = \frac{1}{2\pi N} \int_{\omega-\pi}^{\omega+\pi} H(\omega', \omega - \omega') P_U(\omega - \omega') H(\omega', \omega - \omega')^* \, d\omega'.$$

In order to understand the meaning of (21), (22), and (23), let us assume that $\{B^m\}$ converges to zero when $m \to \infty$ and that, since X_t is now stationary, the time domain extends from $t = -\infty$ to $t = +\infty$, and accordingly the initial time point $t = 0$ is carried back to $t = -\infty$. Then

1. $$G(t, \omega') = \sum_{m=0}^{\infty} B^m e^{-i\omega' m} = G(\omega').$$

This is the transfer function of the coefficients of U's.

2. $H(\omega - \omega', \omega')$ is a Delta function of $\omega - \omega'$ such that $H(\omega - \omega', \omega') = 0$ except $\omega - \omega' = 0$.

3. Further (23) becomes

$$P_X(\omega) = \left(\sum_{m=0}^{\infty} B^m e^{-i\omega m}\right) P_U(\omega) \left(\sum_{n=0}^{\infty} B^{n*} e^{i\omega n}\right)$$
$$= G(\omega)P_U(\omega)G(\omega)^*, \tag{24}$$

which is a well-known formula of the spectral matrix of a multi-variate moving average process.

Going back to the general case where the above assumptions need not hold, we can regard $G(t, \omega')$ as the instantaneous transfer function at t of the moving average with the weights $B^{t-1}, B^{t-2}, \ldots, B^0$. We can also regard $H(\omega - \omega', \omega')$ for a given ω' as the transfer function of $G(t, \omega')$ evaluated at $\omega - \omega'$. If $G(t, \omega')$ is a smooth function of time t,

$$|H(\omega - \omega', \omega')|^2$$

is significant only in the neighborhood around $\omega - \omega' = 0$.

Now let us consider the dynamic econometric model with the parameters changing over time. The system (19) should be replaced by

$$Y_t - B_{t-1}Y_{t-1} = Z_t + U_t \tag{25}$$

Let

$$\bar{Y}_t = B_{t-1} \cdot B_{t-2} \cdot \ldots \cdot B_0 Y_0.$$

Then the deviation from the equilibrium solution with Y_0 as the initial condition is

$$\tilde{X}_t \equiv Y_t - \left(\bar{Y}_t + \sum_{m=0}^{t-1} C_{t,m} Z_{t-m}\right) = \sum_{m=0}^{t-1} C_{t,m} U_{t-m} \tag{26}$$

where

$$C_{t,m} = \begin{cases} I, & \text{at } m = 0 \\ B_{t-1} \cdot B_{t-2} \cdot \ldots \cdot B_{t-m}, & \text{at } m = 1, \ldots, t-1. \end{cases}$$

We can then get the pseudospectral matrix of the deviation from the equilibrium solution as a convolution of the spectrum of the random disturbance U_t and the double transfer function of the parameters.

THEOREM 3. *The pseudospectral matrix $P_{\tilde{X}}(\omega)$ of $\tilde{X}_t$ in* (26) *is,*

$$P_{\tilde{X}}(\omega) = \frac{1}{2\pi N} \int_{-\pi}^{\pi} \tilde{H}(\omega - \omega', \omega') P_U(\omega - \omega') \tilde{H}(\omega - \omega', \omega')^* \, d\omega' \tag{27}$$

where

$$\tilde{H}(\omega - \omega', \omega') \equiv \sum_{t=1}^{N} \tilde{G}(t, \omega') e^{-i(\omega - \omega')t}$$

$$\tilde{G}(t, \omega') \equiv \sum_{m=0}^{t-1} C_{t,m} e^{-i\omega' m}.$$

PROOF.

$$P_{\tilde{X}}(\omega) = \frac{1}{2\pi N} E\left\{\sum_{t=1}^{N} \sum_{s=1}^{N} \sum_{m=0}^{t-1} \sum_{n=0}^{s-1} C_{t,m} U_{t-m} U_{s-n}^{*} C_{s,n}^{*} e^{-i\omega(t-s)}\right\}$$

$$= \frac{1}{2\pi N} \sum_{t}\sum_{s}\sum_{m}\sum_{n} C_{t,m} \cdot \int_{-\pi}^{\pi} P_U(\omega') e^{i\omega'(t-m-s+n)}\, d\omega' \cdot C_{s,n}^{*} e^{-i\omega(t-s)}$$

$$= \frac{1}{2\pi N} \int_{-\pi}^{\pi} \sum_{t} \tilde{G}(t, \omega') P_U(\omega') \sum_{s} \tilde{G}(s, \omega')^{*} e^{-i(\omega-\omega')(t-s)}\, d\omega'$$

$$= \frac{1}{2\pi N} \int_{-\pi}^{\pi} \tilde{H}(\omega - \omega', \omega') P_U(\omega') \tilde{H}(\omega - \omega', \omega')^{*}\, d\omega'$$

$$= \frac{1}{2\pi N} \int_{\omega-\pi}^{\omega+\pi} \tilde{H}(\omega - \omega', \omega') P_U(\omega - \omega') \tilde{H}(\omega - \omega', \omega')^{*}\, d\omega'$$

q.e.d.

The pseudospectral matrix of the deviation of y_t in (17) from its equilibrium solution can be obtained as a portion of $P_{\tilde{X}}(\omega)$.

The meaning of $\tilde{G}$, $\tilde{H}$, and (27) are now obvious from the explanation given for G, H, and (23). Furthermore, we can derive the pseudospectral matrices of the following case.

If (i) $P_U(\omega')$ is smooth so that for some small number $\epsilon > 0$

$$P_U(\omega') = P_U(\omega_0)$$

where $|\omega_0 - \omega'| < \epsilon$, (ii) $\tilde{G}(t, \omega)$ is a slowly changing function of time so that

$$\tilde{H}(\omega - \omega', \omega') \cdot \tilde{H}(\omega - \omega', \omega')^{*}$$

is negligible when $|\omega_0 - \omega'| > \epsilon$, and (iii) $\tilde{G}(t, \omega') = \tilde{G}(t, \omega_0)$ for any t and ω such that $|\omega_0 - \omega'| < \epsilon$, then we obtain

$$P_{\tilde{X}}(\omega) \approx \frac{1}{\pi N} \sum_{t}\sum_{s} \tilde{G}(t, \omega) P_U(\omega) \tilde{G}(s, \omega)^{*} \frac{\sin \epsilon(t - s)}{t - s}. \tag{28}$$

The above procedure can be summarized as follows. For a given time point t, let us consider the impacts of the random disturbances in the periods prior to t (including t) upon the values of $\tilde{X}$ at time t. These impacts can be represented by $C_{t,0}, C_{t,1}, C_{t,2}, \ldots, C_{t,t-1}$ ($C_{t,j}$ is the impact of the random disturbance in period $t - j$.) The transfer function of $C_{t,0}, C_{t,1}, \ldots, C_{t,t-1}$ at frequency ω is $\tilde{G}(t, \omega)$. If the parameters of the model are constant, $C_{t,0} = B^0$, $C_{t,1} = B^1$, $C_{t,2} = B^2, \ldots$. The transfer function is $G(\omega)$ and the spectral matrix is $G(\omega)P_U(\omega)G(\omega)^{*}$. If the parameters of the model are not constant, the pseudospectral matrix is

$$\frac{1}{\pi N} \sum_{t}\left[\sum_{v} \tilde{G}(t, \omega) P_U(\omega) \tilde{G}(t + v, \omega)^{*} \frac{\sin \epsilon v}{v}\right]$$

It should be noted that the pseudospectral matrix of the deviation cannot be considered the mean of the time changing instantaneous spectra, i.e.,

$$\frac{1}{N}\sum_t\left(\sum_{m=0}^{t-1} B_t^{\,m} e^{-i\omega m}\right) P_U(\omega)\left(\sum_{n=0}^{t-1} B_t^{\,n} e^{-i\omega n}\right)^*.$$

This is because $C_{t,m}$ can be very different from $B_t^{\,m}$, particularly when B_t has a trend, and also because (28) has a double summation over t and s.

If the cyclical changes rather than the constants or trends dominate the changes of the parameters of the dynamic econometric model, then the interpretation of the pseudospectrum is difficult, just as in the case of the cyclically changing variance discussed in Section III.

VI. PSEUDOSPECTRUM OF A CONSTANT AND A TREND

In the previous sections we have mentioned that the pseudospectra of a constant and a trend are significant only in the narrow frequency band around zero. In the present section we investigate how narrow the band really is. Since the pseudospectrum of a trend depends upon where the origin of time is, it is difficult to present results that have significant generality. Furthermore, the pseudospectrum of $a + bt$ is not generally equal to the sum of the pseudospectra of a and bt unless the origin of time is centered over the time span. This diminishes the significance of the following presentation to some extent. Thus we merely try to present a clue which might be useful for forming some idea as to the pseudospectra of constant and trend.

Let us define $p_{c,n}(\omega)$ and $p_{t,n}(\omega)$ respectively, as

$$p_{c,n}(\omega) = \frac{1}{2\pi N}\left|\sum_{t=0}^{N-1} e^{-i\omega t}\right|^2$$

$$p_{t,n}(\omega) = \frac{1}{2\pi N}\left|\sum_{t=0}^{N-1} t e^{-i\omega t}\right|^2.$$

Let us define ω_0 for each preassigned value of α in such a way that

$$\int_{-\omega_0}^{\omega_0} p_{c,n}(\omega)\,d\omega = \alpha\int_{-\pi}^{\pi} p_{c,n}(\omega)\,d\omega = \alpha$$

$$\int_{-\omega}^{\omega_0} p_{t,n}(\omega)\,d\omega = \alpha\int_{-\pi}^{\pi} p_{t,n}(\omega)\,d\omega = \alpha\,\frac{(N-1)(2N-1)}{6}$$

We have studied the range of α between 90% and 99% and the range N between 200 and 1,000. When α is fixed, ω_0 depends upon N. This relation between ω_0 and N for a given α is almost an inverse proportion within the regions of α and N which we studied. Therefore, we can use $2\pi/N$ as the unit of frequency. As for $p_{c,n}(\omega)$, the relation between ω_0 and N for any given α is so nearly an inverse proportion that we have presented fairly

TABLE 1

$p_{c,n}(\omega)$

α	90%	95%	99%
ω_0 / $(2\pi/N)$	0.9	2.1	10.2

TABLE 2

$p_{t,n}(\omega)$

α	90%	95%	99%
ω^0 / $(2\pi/N)$	1.6	approximately 3	12–15

exact figures in Table 1.[8] For $p_{t,n}(\omega)$, the relation is slightly more complicated, and Table 2 for $p_{t,n}(\omega)$ does not have the same degree of exactness as Table 1.

It might be interesting to note that $p_{t,m}(\omega)$ near zero frequency has a violent movement, and we had to estimate the integral

$$\int_{-\omega_0}^{\omega_0} p_{t,n}(\omega)\, d\omega$$

by subtracting

$$\int_{-\pi}^{-\omega_0} p_{t,n}(\omega)\, d\omega + \int_{\omega_0}^{\pi} p_{t,n}(\omega)\, d\omega$$

from

$$\int_{-\pi}^{\pi} p_{t\,n}(\omega)\, d\omega.$$

In view of equations (6), (11), (15), and (27) for the pseudospectra of nonstationary processes, and in view of the fact that the pseudospectrum is the average of all simple estimates of the spectrum (apart from the complications due to the spectral window) obtained with no regard to the problem of stationarity, an interesting question is how wide the unit of the frequency interval should be in order to make this average of sample estimates independent of the slow changes in the parameters of the stochastic process. Obviously the small numerical study summarized in the above two tables is not sufficient to give an answer to this question for the general class of slow movements of parameters. Let us suppose, however, that Table 2 represents the nature of the pseudospectra of the linear trend in general. Then we can answer the above question. When m represents the number of lags used in the estimation of the spectrum, the unit of frequency interval is $2\pi \times (1/2m)$. If d_t in Section III, and $G(t, \omega')$ in Section V are linear functions of time, then 95% of the variance of d_t or $\tilde{G}(t, \omega')$ is contained in the frequency interval having the width $6 \times (2\pi/N)$, where N is the number of data. If $2\pi \times (1/2m)$ is equal to $6 \times (2\pi/N)$, i.e., $N = 12m$, then the pseudospectra represented in the equations (6), (11), (15), and (27) become practically independent of the parameter changes.

[8] If we *were* treating continuous time, this relation should be *exactly* an inverse proportion. However, time element is discrete in the pseudospectrum.

CHAPTER 27

New Formulas for Making Price and Quantity Index Numbers

By KAZUO MIZUTANI*

According to J. M. Keynes, "index numbers of prices" are a series of numbers indicative of changes in the given price-level and this concept is fairly widely accepted.

In accordance with this concept, we may define P_{ij}, the index number of prices of the jth year relative to the ith year as

$$P_{ij} = \frac{\pi_j}{\pi_i} \tag{1}$$

where π_i and π_j denote the price-levels of the ith and the jth years respectively. It is obvious that these index numbers should satisfy the so-called circular test, namely,

$$P_{ij}P_{jk}P_{ki} = 1 \tag{2}$$

But it is to be regretted that there are no existing index numbers of prices and/or quantities published by any country in the world which would satisfy this circular test. Therefore, there are no correct index numbers.[1]

It is well known that J. M. Keynes was dissatisfied with the then existing index numbers of prices and relied upon wage units in his *General Theory*.

One of the greatest defects of index numbers which do not satisfy the circular test is that they should not and cannot be used to deflate national

* Catholic University of Nagoya.

[1] cf. "The Price Statistics of the Federal Government," *National Bureau of Economic Research*, No. 73, General Series: "If a poll were taken of professional economists and statisticians, in all probability they would designate (and by a wide majority) the failure of the price indexes to take full account of quality changes as the most important defect in these indexes. And by almost as large a majority, they would believe that this failure introduces a systematic upward bias in the price indexes—that quality changes have on average been quality improvements." (p. 35)

"Great as the difficulties are, however, we think it is possible to go beyond the recommendation that more research be done on the problem." (p. 35)

"One form of the quality problem is the appearance of new goods: television, blankets made of synthetic fibers, new drugs. . . . The procedures we recommend will not take full account of new products, but will serve to reduce greatly a lag that is now too large."

"But the main quality problem will remain: how should one deal with the steady advance of medical knowledge. . .? In general, there is no known method of coping with these problems on a current basis, and the current price indexes must ignore them. (p. 35)

TABLE 1

Year	I	II
1935	50	200
1939	50	600
1944	150	1,000
1947	5,000	28,000
1953	15,400	53,600

accounts. Suppose we are given two series of prices of two commodities as listed in Table 1.

On the basis of these data, we shall proceed to construct index numbers of prices relative to the base year "1935."

In the first place, we calculate the individual price index numbers, taking 1935 as the base year, and we have the second and third columns of Table 2. Then, in each year, we take either their arithmetic mean or

TABLE 2

Year	I	II	P^M	P^G
1935	100	100	100	100
1939	100	300	200	173
1944	300	500	400	387
1947	10,000	14,000	12,000	11,832
1953	30,800	26,800	28,800	28,730

I signifies "Index Numbers of Commodity I"
II signifies "Index Numbers of Commodity II"

their geometric mean across each row. Thus, we have the column headed P^M (M meaning the arithmetic mean) and the column headed P^G (G meaning the geometric mean).

Now, it is very often necessary, for a certain purpose of study, to change the base year from 1935 to some other year—say, to 1944. In order to calculate the index numbers of prices of the data given in Table 1, taking 1944 as the new base year, we must proceed in the way shown in Table 3.

Comparing Table 3 with Table 2, we notice very easily the absurdity of the index numbers derived from the arithmetic mean. Since the number 400 for the year 1944 in Table 2 was changed to 100 in Table 3, the index numbers for those years 1935, 1939, 1947, and 1953, namely 100, 200,

TABLE 3

Year	I	II	P^M	P^G
1935	$33.\dot{3}$	20.0	$26.\dot{6}$	25.8
1939	$33.\dot{3}$	60.0	$46.\dot{6}$	44.7
1944	100.0	100.0	100.0	100.0
1947	$3{,}333.\dot{3}$	2,800.0	$3{,}066.\dot{6}$	3,055.1
1953	$10{,}266.\dot{6}$	5,360.0	$7{,}813.\dot{3}$	7,418.2

12,000, and 28,000 should have been changed to a quarter of their values, i.e., 25, 50, 3,000, and 7,200 respectively, if the arithmetic mean index numbers were correct. But, they are actually $26.\dot{6}$, $46.\dot{6}$, and so on as shown in Table 3, and this shows clearly that the arithmetic mean index numbers are erroneous in the sense that they do not fulfill this circular test. Therefore they should not be used to deflate national accounts.

On the other hand, the geometric mean index numbers as shown in column headed "P^G" of Table 2, namely, 100, 178, 387, 11,832, and 28,730 were changed to 25.8, 44.7, 100.0, 3,055.1, and 7,418.2 in Table 3, which are all 100/387 times the corresponding numbers in Table 2. Thus, the geometric mean index numbers do fulfill the circular test. It is, however, the total defect of the geometric mean index numbers that they do not fulfill the factor reversal test.

Let us, then, examine these several points, which we must take into consideration in constructing price and quantity index numbers.

1. THE CIRCULAR TEST. It must be clear from what has been stated above that this test must be fulfilled by price and quantity index-formula.

2. THE FACTOR REVERSAL TEST (in a broader sense). In economics, we are familiar with the equation

$$P_i O_i = Y_i \tag{3}$$

where P_i stands for the price-level for the ith year, O_i stands for the physical volume of the national product in the same year, and Y_i stands for the national product in terms of money in the same year.

Then, what the factor reversal test in a broader sense requires is nothing but the logical consequence of (3), that is:

$$P_{ij} Q_{ij} = V_{ij} \tag{4}$$

where P_{ij} is the price index number of the jth year relative to the ith year as basis, Q_{ij} is the corresponding quantity index number, calculated according to a formula which may be different from the formula for P_{ij}—this is what the annexed adjective phrase "in a broader sense" means—so we cannot necessarily get Q_{ij} from P_{ij} by simply interchanging p and q in P_{ij} as Irving Fisher did.

3. THE CONTINUITY TEST. What the price index numbers indicate must be the changes (e.g., yearly changes) in one and the same price level. The same holds for the quantity index numbers. But in the course of a long period, quality changes may occur with respect to some of the commodities, and new products often come into vogue. The price as well as quantity index numbers should be such as to allow for these changes properly. This is what is required by what may be called the continuity test. We are of the opinion that only an index-formula with variable weights can fulfill this test.

In the latter part of the 19th century in Tokyo, oil lamps were much in use, whereas we had no fluorescent lamps at that time. At present, on the contrary, oil lamps are out of vogue and the use of fluorescent lamps has grown. In order to fulfill the continuity test, we must suppose that both oil and fluorescent lamps were always included among the objects of the consumers' price survey. We suppose that our social choice function has changed gradually and continuously during this long period of time, and, consequently, the importance of these goods among our family expenditures has changed, where these changes find their expression only in terms of weight functions.

4. THE FALLACY OF "THE BEAUTY CONTEST". In price or quantity index number making, we must guard against implicit but improper weighting. Suppose that five judges gave the marks shown in Table 4 to seven candidates for "Pearl Queen."

TABLE 4

	J_1	J_2	J_3	J_4	J_5	S	A
C_1	80	75	80	85	30	350	70
C_2	75	70	75	80	30	330	66
C_3	70	80	70	75	30	325	65
C_4	75	70	70	80	30	325	65
C_5	70	65	70	70	30	305	61
C_6	65	65	65	70	100	365	73
C_7	60	65	65	70	30	290	58
S	495	490	495	530	280	2290	458
A	70.7	70.0	70.7	75.7	40.0	341.4	65.4

C: Candidate J: Judge
S: Sum A: Average

As a result, candidate No. 6 won the first prize. It is clear, however, from this table that candidate No. 1 would have won the first prize if there had not been J_5 among the judges. This is because the dispersion of the marks given by J_5 is unusually large as compared with those given by the other judges. This had the same effect on the resulting marks as an implicitly large weight given to the marks of J_5. Therefore, the marks given by J_5 dominated, so to speak, the other marks.

In order to avoid this fallacy of "the beauty contest," we should standardize the marks given by every judge. Thus, having once made the dispersions of all marks equal to each other, we may proceed to give proper weights to each of them. If we standardize the marks of Table 4, we get the result shown in Table 5.[2]

[2] Of course, this is only one of the various methods of avoiding the fallacy of "the beauty contest." We may, for example, begin by taking the common logarithms of the marks given to candidates, and then standardize these logarithms.

TABLE 5

	J_1	J_2	J_3	J_4	J_5	S	A
C_1	149	94	188	165	−41	555	111
C_2	69	0	87	76	−41	191	38
C_3	−11	187	−14	−13	−41	108	22
C_4	69	0	−14	76	−41	90	18
C_5	−11	−94	−14	−102	−41	−262	−52
C_6	−92	−94	−115	−102	245	−158	−32
C_7	−172	−94	−115	−102	−41	−524	−105

Thus, if we proceed properly and at the outset standardize the marks given by every judge, candidate No. 1 will get the first prize and candidate No. 6 shifts to the 5th rank.

5. TEST OF GROUPING. What the factor reversal test requires, for example, must be fulfilled not only by the whole series, taken in its entirety, but also by any component of the series. Suppose that we divide the whole series into several groups according to the kind of industry to which the commodities belong. Then in each group the factor reversal test should be fulfilled.

6. PROPORTIONALITY TEST. It seems to the author that it is commonly believed that it is quite natural for any price-index formula to fulfill the proportionality test. But we now must allow for quality changes and new products. Therefore, we are confronted with the changing significance of commodities in consumer preference patterns. To put it in terms of an index-formula, we are faced with a new index-formula with variable weights, and with variable weights we cannot and must not require that the index-formula fulfill the proportionality test.

Now we shall proceed to explain how to make price and quantity index numbers according to our new index-formulae, which fulfill the circular test, the factor reversal test, the continuity test, and the test of grouping and avoid the fallacy of "the beauty contest."

Suppose we are given: (a) a price matrix $[p_t^s]$, (b) a quantity matrix $[q_t^s]$, and (c) a value vector v_t, where p_t^s stands for the price of the sth commodity in the tth year $[s = 1 \sim n, t = 1 \sim T]$, q_t^s stands for the corresponding quantity, v_t stands for the total value transacted in the tth year, namely:

$$v_t = \sum_s p_t^s q_t^s \qquad \left[\sum_s = \sum_{s=1}^{n}\right] \tag{5}$$

Following custom, I do not consider the problem of whether or not the price and quantity surveys are statistically accurate.

To begin with, I assume that p_t^s as well as q_t^s and therefore v_t are continuous functions of time t and that all of them are continuously differentiable with respect to t.

Now, I take "semi-elasticity" of both sides[3] of (5) with respect to "t," namely:

$$\frac{d \log v_t}{dt} = \sum_s \alpha_t^s \frac{d \log v_t^s}{dt} \tag{6}$$

where $\log v_t$ signifies the natural logarithm of v_t

$$v_t^s = p_t^s q_t^s$$
$$\alpha_t^s = \alpha(s, t) = \frac{v_t^s}{v_t} = \frac{p_t^s q_t^s}{\sum_s p_t^s q_t^s} \tag{7}$$

Integrating both sides of (6) with respect to t from $t = k$ to l, we have

$$I = \sum_s I^s = \sum_s \sum_t I_t^s \qquad \sum_t = \sum_{t=0}^{m} \tag{8}$$

where

$$I = \int_k^l \frac{d \log v_t}{dt} dt = \log v_l - \log v_k = \log V_{kl} \tag{9}$$

$$I^s = \int_k^l \alpha_t^s \frac{d \log v_t^s}{dt} dt = \sum_t I_t^s \qquad \sum_t = \sum_{t=0}^{m} \tag{10}$$

$$I_t^s = \int_t^{t+1} \alpha_t^s \frac{d \log v_t^s}{dt} dt \tag{11}$$

$$t_0 = k, \qquad t_{m+1} = l$$

In this procedure, $\frac{d \log v_t^s}{dt}$ is assumed to be integrable in the sense of the Riemann integral over the whole interval $[k, l]$ and we have divided the whole interval into $m + 1$ sub-intervals $[t, t + 1]$, $(t = 0 \sim m)$, such that $\alpha_t^s = \alpha(s, t)$ will be a monotonic function of t in each sub-interval. Therefore, by the generalized second mean value theorem for integrals, we get:

$$\begin{aligned} I_t^s = {} & \alpha(s, t) \int_t^{\xi} \frac{d \log v_t^s}{dt} dt + \alpha(s, t + 1) \int_{\xi}^{t+1} \frac{d \log v_t^s}{dt} dt \\ & = \alpha(s, t + 1) \log v_{t+1}^s - \alpha(s, t) \log v_t^s \\ & \quad + [\alpha(s, t) - \alpha(s, t + 1)] \log v_{\xi}^s \\ & \quad \xi \text{ being } \xi_t \quad \text{and} \quad t < \xi_t < t + 1. \end{aligned} \tag{12}$$

[3] Let $y > 0$ be a real-valued, continuously differentiable function of x defined over a positive interval $[x_1, x_2]$ in two-dimensional real space R^2. Then $\frac{d \log y}{dx}$ shall be called the semi-elasticity of y with respect to x in the interval (x_1, x_2).

Consequently, we have from (8), (9), (10), (11), and (12):

$$I = \sum_s I^s = \sum_s \sum_t I_t^s = \sum_s [\alpha(s,\ l) \log v_l^s - \alpha(s,\ k) \log v_k^s] + R \quad (13)$$

where:

$$R = \sum_s \sum_t [\{\alpha(s,\ t) - \alpha(s,\ t + 1)\} \log v_\xi^s] \qquad (\xi = \xi_t) \qquad (14)$$

But $\alpha(s, t)$ is a weight-function. Therefore, we have:

$$\sum_s \alpha(s,\ t) = \sum_s \frac{v_t^s}{v_t} = \frac{1}{v_t} \sum v_t^s = 1 \qquad (15)$$

Hence:

$$\sum_s [\alpha(s,\ t) - \alpha(s,\ t + 1)] = 0 \qquad (16)$$

It is, moreover, usually assumed that $\alpha(s, t)$ changes very slowly. Consequently the difference $\alpha(s, t) - \alpha(s, t + 1)$ is not only small in numerical value, but also, if it is positive for some s, there must necessarily be some other s, for which the above difference is negative. Accordingly, we take this R to be practically zero. In any case, it suffices to consider only two cases:

1. Where R is actually or at least practically zero and
2. Where we may not assume this to be the case.

For the first case, the expression (13) yields:

$$\begin{aligned} \log V_{kl} &= \sum_s \alpha(s,\ l) \log v_l^s - \sum \alpha(s,\ k) \log v_k^s \\ &= \sum_s \alpha(s,\ l) \log p_l^s - \sum_s \alpha(s,\ k) \log p_k^s \\ &\quad + \sum_s \alpha(s,\ l) \log q_l^s - \sum_s \alpha(s,\ k) \log q_k^s \end{aligned} \qquad (17)$$

Consequently, we have our Natural Index Formulae P_{kl} and Q_{kl} as follows:

$$P_{kl} = \frac{\pi_l}{\pi_k} \qquad \text{and} \qquad Q_{kl} = \frac{V_{kl}}{P_{kl}} \left[= \frac{r_l}{r_k} \right] \qquad (18)$$

where

$$\log \pi_t = \sum_s \alpha(s,\ t) \log p_t^s \qquad (19)$$

$$\log r_t = \sum_s \alpha(s,\ t) \log q_t^s \qquad (20)$$

We must note here that Q_{kl} is not defined as $Q_{kl} = r_l/r_k$ because the $q_t^1, q_t^2, \ldots, q_t^n$ are not always commensurable. Some of them may be measured in kilogram-tons, some others may be measured in kilometers. Therefore, we adopt for each commodity its proper unit of measure, namely, that quantity which we can buy for one dollar.

For example, if the price of an egg in the tth year is five cents, the egg's proper unit of measure is twenty eggs. In this way, we can take $v_t^s = p_t^s q_t^s$ to indicate the quantity of the sth commodity transacted in the tth year in its proper unit of measure.

But, v_t^s fluctuates depending upon the fluctuation of p_t^s. Therefore, we must deflate v_t^s by p_t^s. Using these deflated quantities $q_t^s = v_t^s / p_t^s$, we calculate the level of quantities p_t transacted in the tth year as their geometric mean with α_t^s's as their weights.

In the case where R is sufficiently close to zero, we have

$$\frac{r_l}{r_k} = \frac{V_{kl}}{P_{kl}} = Q_{kl} \tag{18}$$

where

$$\log r_t = \sum_s \alpha(s, t) \log q_t^s \tag{20}$$

For the second case, where R is not sufficiently close to zero, P_{kl} will not satisfy the factor reversal test in the broader sense. For the same reason, Q_{kl} will not fulfill the factor reversal test. But, in this case, it suffices to transform these "Natural Index Formulae P_{kl} and Q_{kl}" into our "New Index Formulae $\mathfrak{P}_{kl}$ and $\mathfrak{Q}_{kl}$" respectively, through the following formulae:

$$\begin{aligned} \log \mathfrak{P}_{kl} &= \tfrac{1}{2}[\log P_{kl} + \log V_{kl} - \log Q_{kl}] \\ \log \mathfrak{Q}_{kl} &= \tfrac{1}{2}[\log Q_{kl} + \log V_{kl} - \log P_{kl}] \end{aligned} \tag{21}$$

Now, we examine our Natural Index Formulae:

(I) $P_{kl} = \pi_l / \pi_k$ and $Q_{kl} = [V_{kl}/P_{kl}] = r_l/r_k$ are the ratios by which Keynes defined index numbers of prices and quantities. Therefore, it is clear that these fulfill the circular test, that is $P_{kl}P_{lm}P_{mk} = 1$.

(II) In the case where R is sufficiently close to zero $Q_{kl} = V_{kl}/P_{kl} = r_l/r_k$. Hence, it is also clear that these satisfy the factor reversal test, namely:

$$P_{kl}Q_{kl} = V_{kl} \tag{22}$$

(III) With regard to the continuity test, we shall take again the example of oil lamps and fluorescent lamps. We assume that both oil and fluorescent lamps were always included among the objects of the consumers' price survey. We suppose that our social choice function has changed gradually and continuously during this long period of time and, consequently, the importance of these goods among family expenditures has changed, where these changes find their expression only in terms of weight functions. Thus the weight of the oil-lamps $\alpha_t^s = p_t^s q_t^s$ has decreased during that period as the quantity sold has decreased considerably and has eventually become almost nothing and the weight of the fluorescent lamps, say $\alpha_t^m = p_t^m q_t^m$, on the contrary, has increased from zero to a large one.

This is the way of treating such new products as fluorescent lamps, and if we combine those two kinds of lamps under the head of lighting apparatus, we can take full account of the quality changes of the lighting apparatus through the method described above. Thus our "natural index formula" satisfies the continuity test.

(IV) In deducing our new index formulae of prices and quantities, we took the natural logarithm of both sides of the equation:

$$v_t = \sum_s p_t^s q_t^s \tag{5}$$

and differentiated the result with respect to time t. Namely, we took "semi-elasticity" of both sides of (5) with respect to t, and we obtained

$$\frac{d \log v_t}{dt} = \sum \alpha_t^s \frac{d \log v_t^s}{dt} \tag{6}$$

where

$$v_t^s = p_t^s q_t^s$$

Thus, we found the weight function of $v_t^s (= p_t^s q_t^s)$ to be $\alpha_t^s = v_t^s / v_t$ as a result of logical necessity. Hence the name "natural index formulae." Therefore, the weight functions to v_t^s's are absolutely right and proper. We can be fully assured that they properly avoided the fallacy of the "beauty contest."

(V) It is also not difficult to see clearly from the structure of the weight function that our natural index formulae fulfill the test of grouping.

Thus, our natural index formulae P_{ij} and Q_{ij} fulfill all five tests except the case where R is not sufficiently close to zero. In that case it suffices to use our new index formulae $\mathfrak{P}_{ij}$ and $\mathfrak{Q}_{ij}$ in their place.

Since the natural index formulae fulfill all the tests except the factor reversal test, it is easy to see that these new index formulae fulfill all these four tests which the natural index formulae always fulfill. It is also very clear from their structure that the new index formulae fulfill the factor reversal test.

Hence, in all cases the new index formulae $\mathfrak{P}_{ij}$ and $\mathfrak{Q}_{ij}$ fulfill all the tests we enumerated above.

But, the probability that R is sufficiently close to zero is very great as long as the number of commodities taken into the calculation of π_t and r_t is as large as is usually the case with index numbers of prices and quantities.

In order to alleviate the difficulty in calculating P_{kl} or Q_{kl}, we may apply the sequential-analytic device, adopting those goods whose weights are large and picking goods up one after another according to the order of the numerical value of their weights. We may expect, by this device, to decrease the number of goods adopted in the calculation of the index numbers.